MATHEMATICS FOR THE COLLEGE BOARDS

(Revised 1975)

PSAT • SAT

Preliminary Scholastic Aptitude Test

Scholastic Aptitude Test

By BARNETT RICH

Former Chairman, Department of Mathematics
Brooklyn Technical High School
New York City

Adjunct Assistant Professor
The City College
The City University of New York

Dedicated to serving

AMSCO

our nation's youth

When ordering this book, please specify:
either **R 38 W** *or*
MATHEMATICS FOR THE COLLEGE BOARDS

AMSCO SCHOOL PUBLICATIONS, INC.

315 Hudson Street New York, N. Y. 10013

ISBN 0-87720-201-X

Revised 1975

PRINTED IN THE UNITED STATES OF AMERICA

PREFACE

This book is intended to prepare students for the sections on mathematics of the following tests given by the College Entrance Examination Board: (1) Preliminary Scholastic Aptitude Test (PSAT) (2) Scholastic Aptitude Test (SAT).

The Scholastic Aptitude Test is a three-hour objective test designed to measure how well a student has developed his verbal and mathematical skills. The Preliminary Scholastic Aptitude Test is a slightly shorter version of the Scholastic Aptitude Test for use in high school guidance and counseling programs. Both the PSAT and the SAT are designed to test the ability to think rather than the ability merely to recall facts.

The SAT score in mathematics is expressed as a number on a scale extending from 200 to 800. This score indicates how well a student performed on the mathematics section of the test. The score is intended to help college admissions officers to judge a student's abilities in mathematics when compared with the performance of applicants from other high schools. The SAT score in mathematics is very helpful to the college counselor in the high school, since he needs to know all he can about a student's capacities before he can assist him in selecting a college to which he is suited. The score is also important in determining the college courses in mathematics and science for which the student is adequately prepared.

This is not to say that the SAT score alone is a precise indicator of mathematical ability. There is, in fact, no single method of precisely predicting a student's academic performance in mathematics. However, *the SAT score in mathematics, when used in conjunction with his mathematics record in high school, does furnish an accurate indication of how well a student is likely to do in college mathematical work.*

The mathematics questions in the SAT are designed to measure how well a student has mastered elementary mathematics. Elementary mathematics includes arithmetic, algebra, and geometry. The questions on the test remain within the scope of elementary mathematics and do not test advanced work. The SAT questions in mathematics require the application of symbolic, or algebraic, techniques; numerical, or arithmetic, techniques; spatial, or geometric, techniques; and also graphical techniques. The results of the test thus make it possible to determine how important a contribution the student can make not only in mathematics, but also in such widely varied fields as physics, chemistry, engineering, and astronomy.

A student preparing for the SAT test must keep in mind that the problems in the test present *new settings* as well as the familiar settings used in his textbooks. In fact, the SAT mathematics test focuses upon the student's ability to apply his knowledge in these unfamiliar settings and to *think logically* rather than to recall facts.

This book contains more than 300 types of problems in arithmetic, algebra, geometry, modern mathematics, and graphs. Each problem lies within the scope of the elementary mathematics tested by the Scholastic Aptitude Test. For each type of problem, there is at least one solved model problem, followed by at least one set of five related problems. These related problems offer further practice in applying the central principle of the solved problem. Many more practice problems are to be found at the end of each chapter. Finally, seven complete tests, each containing 30 questions, are included. The seven tests are similar in format and content to the mathematics section of the SAT test. In all, the book presents more than 2500 elementary mathematical problems.

Note. Answers to all the problems are supplied in the key at the back of the book. Additional help is provided by the cross-reference system in the practice problems at the end of each chapter. A student who has difficulty with a topic can refer to the bracketed number at the right of the practice problem to find the text dealing with that topic.

CONTENTS

PREVIEW

We shall begin with a preview of ten types of problems suggestive of the settings and the degrees of difficulty to be found on the SAT test:

1. Evaluating sums of products
2. Solving an equation by division
3. Finding sums of angles in figures of triangles and quadrilaterals
4. Finding the coordinates of the midpoint of a line segment
5. Finding the distance between two points on a graph
6. Determining relative positions
7. Evaluating changes in expressions
8. Finding ranges of values
9. Performing operations on inequalities
10. Solving data-sufficiency problems
11. Solving quantitative comparison problems

1. EVALUATING SUMS OF PRODUCTS

Model Problem

If $a = 17$ and $b = 23$, evaluate $35a + 41b + 65a + 59b$.

(A) 400 (B) 4000 (C) 5000 (D) 8000 (E) 40,000

Solution:

$$35a + 41b + 65a + 59b$$

Collect like terms. $35a + 65a + 41b + 59b$

Combine like terms. $100a + 100b$

Factor. $100(a + b)$

Substitute 17 for a and 23 for b. $100(17 + 23)$ $100(40) = 4000$ *Ans.* (B)

Matching Test

Evaluate each of the following as indicated. After completing this test, match your answers with the numbers in the box.

1. If $a = 23$ and $b = 37$, evaluate $25a + 32b + 75a + 68b$.

2. If $a = 17$ and $b = 43$, evaluate $21b + 29b + 15a + 35a$.

3. If $c = 27$ and $d = 13$, evaluate $130c + 90d + 110d + 70c$.

4. Evaluate $(65 \times 27) + (40 \times 35) + (60 \times 35) + (65 \times 73)$.

5. Evaluate $(15 \times 81) + (13 \times 69) + (35 \times 81) + (37 \times 69)$.

a. 7500
b. 8000
c. 6000
d. 10,000
e. 3000

2. SOLVING AN EQUATION BY DIVISION

Model Problem

Find x: $4(8)(16)x = (16)(16)(32)$

(A) $\frac{1}{16}$ (B) 4 (C) 16 (D) 32 (E) 480

Solution:

Divide each side by $4(8)(16)$, the coefficient of x.

$$x = \frac{(16)(16)(32)}{(4)(8)(16)}$$

Cancel common factors of the numerator and the denominator.

$$x = \frac{\overset{4}{(\cancel{16})}\,\overset{2}{(\cancel{16})}\,\overset{2}{(\cancel{32})}}{\underset{1}{(\cancel{4})}\,\underset{1}{(\cancel{8})}\,\underset{1}{(\cancel{16})}}$$

Multiply the remaining factors.

$$x = (4)(2)(2) = 16 \quad \textit{Ans.}\ \text{(C)}$$

Matching Test

Find the required value in each of the following. After completing this test, match your answers with the numbers in the box.

1. Find x: $4(8)x = 16(16)(8)$
2. Find y: $8(16)y = 16(16)(4)$
3. Find b: $16(8)(4) = 16(16)b$
4. Find (?): $16 \times 16 \times (?) = 8 \times 8 \times 4$
5. Find (?): $8 \times 4 \times 4 = 16 \times 16 \times (?)$

a. $\frac{1}{2}$
b. 64
c. 1
d. 8
e. 2

3. FINDING SUMS OF ANGLES IN FIGURES OF TRIANGLES AND QUADRILATERALS

Model Problem

In the straight-line figure at the right, what is the sum of the angles marked with an arc?

(A) 180° (B) 270° (C) 360° (D) 540° (E) 720°

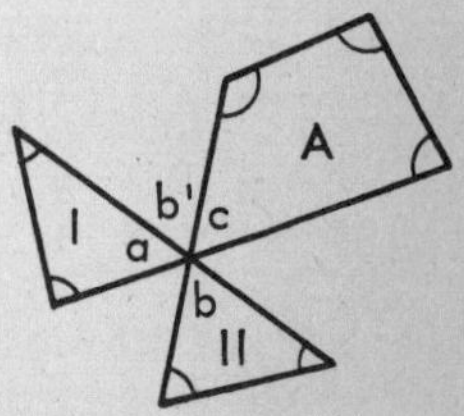

Solution:

Since the sum of the angles of a triangle equals 180° and the sum of the angles of a quadrilateral equals 360°, the sum of all the angles in triangles I and II and quadrilateral *A* is equal to 2(180°) + 360°, or 720°.

To find the sum of the marked angles of the figure, we subtract the sum of the unmarked angles, *a*, *b*, and *c*, from the total of 720°. Note that $\angle b$ and $\angle b'$ are equal vertical angles. Hence, the sum of unmarked angles *a*, *b*, and *c* = the sum of angles *a*, *b'*, and *c* = 180°. Subtracting, we find that the sum of the marked angles = 720° − 180° = 540°. *Ans.* (D)

Matching Test

In each of the straight-line figures, find the sum of the measures of the angles marked with an arc. After completing this test, match your answers with the numbers in the box.

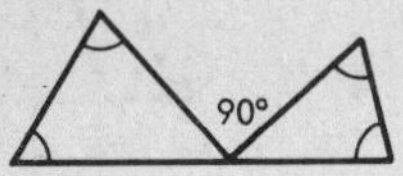

1.

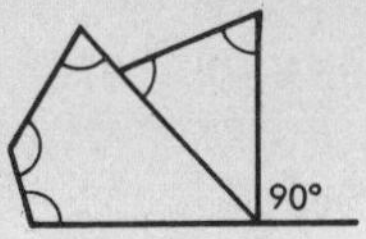

2.

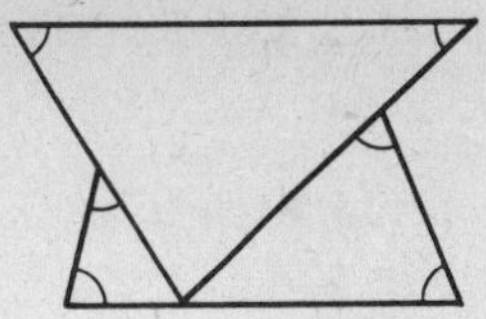

3.

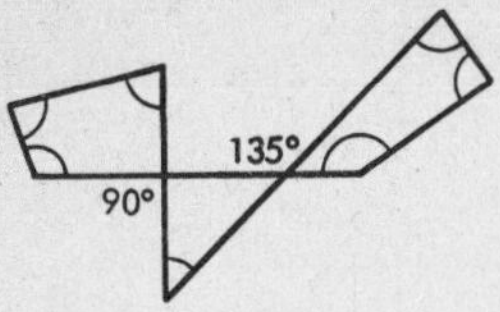

4.

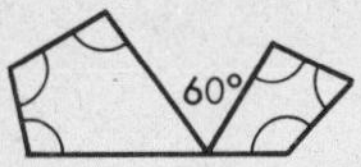

5.

a. 600	*b.* 360	*c.* 630	*d.* 270	*e.* 450

4. FINDING THE COORDINATES OF THE MIDPOINT OF A LINE SEGMENT

Model Problems

(1) Given points $A(-3, 9)$ and $B(11, -5)$. Find the coordinates of the midpoint of line segment AB.

(A) (2, 4) (B) (−4, −2) (C) (7, −7) (D) (−7, 7) (E) (4, 2)

Solution:

To find the coordinates of M, the midpoint of AB:

$B(11, -5) \rightarrow x_2 = 11,\ y_2 = -5$
$A(-3, 9) \rightarrow x_1 = -3,\ y_1 = 9$

midpoint formulas:

$$x_M = \frac{x_1 + x_2}{2} \qquad y_M = \frac{y_1 + y_2}{2}$$

$$x = \frac{(-3) + 11}{2} \qquad y = \frac{9 + (-5)}{2}$$

$$= \frac{8}{2} = 4 \qquad = \frac{4}{2} = 2$$

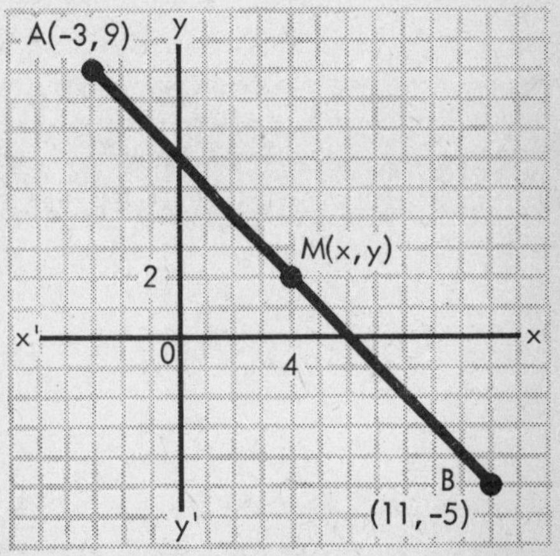

Ans. (E)

(2) AB is a diameter of the circle whose center is the point $(2, 0)$. If the coordinates of A are $(0, -2)$, find the coordinates of B.

(A) (4, 2) (B) (−2, −4) (C) (2, 4) (D) (−4, −2) (E) (2, −2)

Solution:

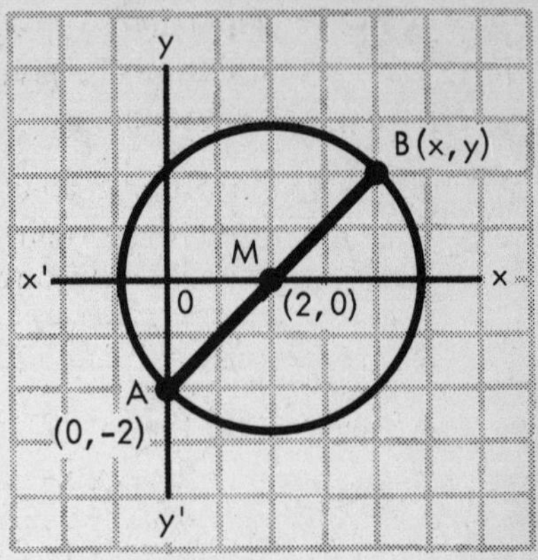

To find the coordinates of B, an endpoint of diameter AB:

midpoint, $M(2,0) \rightarrow x_M = 2,\ y_M = 0$
$A(0,-2) \rightarrow x_1 = 0,\ y_1 = -2$

midpoint formulas: $x_M = \dfrac{x_1 + x_2}{2}$ | $y_M = \dfrac{y_1 + y_2}{2}$

$2 = \dfrac{0 + x}{2}$ | $0 = \dfrac{(-2) + y}{2}$

Multiply each side by 2. | Multiply each side by 2.

$4 = x$ | $0 = -2 + y$
$2 = y$

Ans. (A)

Matching Test

In each of the following, find the required coordinates. After completing this test, match your answers with the pairs of coordinates in the box.

1. Find the coordinates of the midpoint of the line segment joining point (0, 0) to point (8, 6).

2. Find the coordinates of the midpoint of the line segment joining point (−4, 8) to point (10, −2).

3. Find the coordinates of point A of line segment AB if the midpoint of AB is point (5, 4) and the coordinates of point B are (6, 8).

4. Find the coordinates of the center of a circle if the endpoints of the diameter are points (−4, 6) and (8, 0).

5. Find the coordinates of point A of diameter AB if the center of the circle is point (3, 8) and the coordinates of B are (5, 10).

a. (4, 0)	*b.* (3, 3)	*c.* (1, 6)	*d.* (4, 3)	*e.* (2, 3)

5. FINDING THE DISTANCE BETWEEN TWO POINTS ON A GRAPH

Model Problems

(1) Which point lies farthest from the origin?

(A) (0, −9) (B) (−7, 6) (C) (8, 5) (D) (−2, 9) (E) (−9, 0)

Solution:

The lengths d_1, d_2, d_3, d_4, and d_5 determine which of the given points is farthest from the origin.

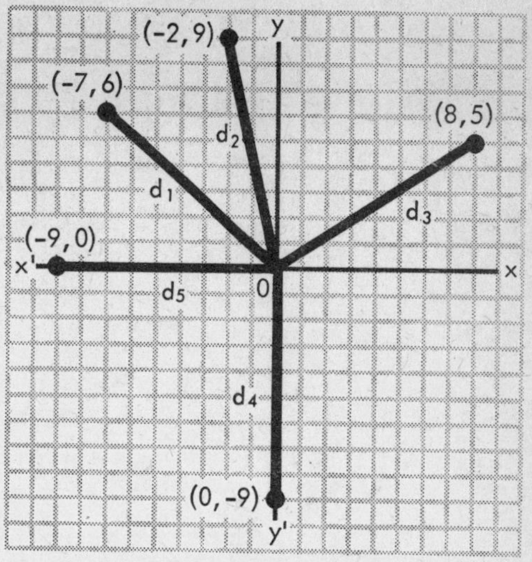

Find these lengths, which are measured from the origin (0, 0), by means of the simplified distance formula:

$$d = \sqrt{x^2 + y^2}$$

$$d_1 = \sqrt{(-7)^2 + 6^2} = \sqrt{49 + 36}$$
$$= \sqrt{85}$$

$$d_2 = \sqrt{(-2)^2 + 9^2} = \sqrt{4 + 81}$$
$$= \sqrt{85}$$

$$d_3 = \sqrt{8^2 + 5^2} = \sqrt{64 + 25}$$
$$= \sqrt{89}$$

$$d_4 = \sqrt{0^2 + (-9)^2} = \sqrt{0 + 81}$$
$$= \sqrt{81} = 9$$

$$d_5 = \sqrt{(-9)^2 + 0^2} = \sqrt{81 + 0}$$
$$= \sqrt{81} = 9$$

The longest line is d_3. Hence, (8, 5) is farthest from the origin. *Ans.* (C)

(2) Find the distance between the points whose coordinates are (−2, 4) and (2, 7).

(A) 3 (B) $3\sqrt{2}$ (C) 4 (D) 5 (E) 7

Solution:

Find d, the length of AB, as follows:

$$B(2, 7) \rightarrow x_2 = 2, \quad y_2 = 7$$
$$A(-2, 4) \rightarrow x_1 = -2, \; y_1 = 4$$

distance formula: $d = \sqrt{(x_2 - x_1)^2 + (y_2 - y_1)^2}$

$$= \sqrt{[2 - (-2)]^2 + (7 - 4)^2}$$
$$= \sqrt{4^2 + 3^2} = \sqrt{16 + 9}$$
$$= \sqrt{25} = 5 \quad \textit{Ans.}\ (D)$$

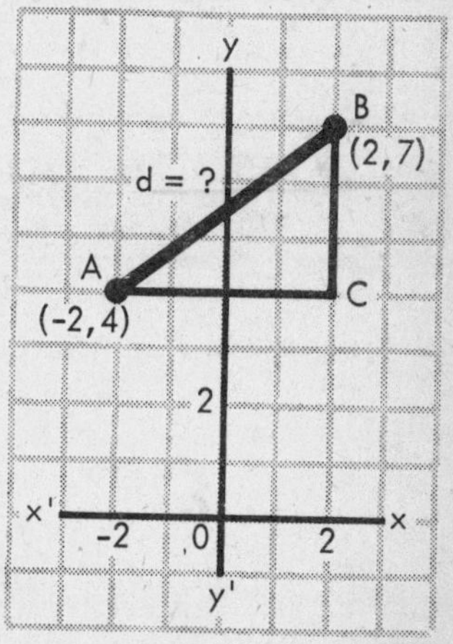

Note. $\triangle ABC$ is a 3-4-5 right triangle.

(3) Given the points $A(-2, 1)$ and $B(6, 7)$. Find the length of line segment AB.

(A) 14 (B) 12 (C) 4 (D) 100 (E) 10

Solution:

To find the length of AB:

$B(6,7) \rightarrow x_2 = 6, \quad y_2 = 7$
$A(-2,1) \rightarrow x_1 = -2, y_1 = 1$

distance formula:

$$d = \sqrt{(x_2 - x_1)^2 + (y_2 - y_1)^2}$$
$$= \sqrt{[6-(-2)]^2 + (7-1)^2}$$
$$= \sqrt{8^2 + 6^2} = \sqrt{64 + 36}$$
$$= \sqrt{100} = 10 \quad \textit{Ans.}\ \text{(E)}$$

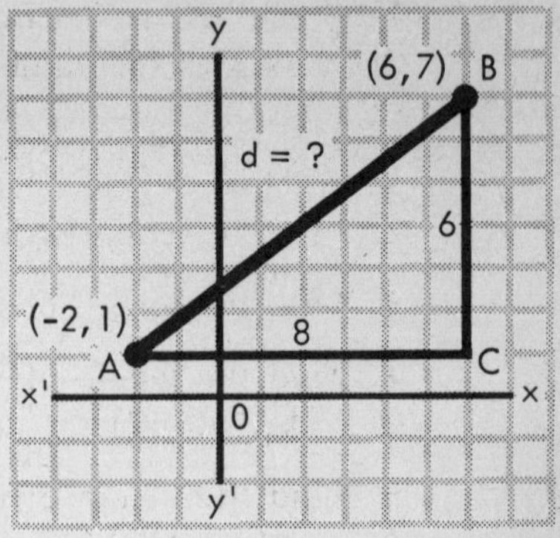

(4) The length of the line segment joining the points whose coordinates are $(1,-4)$ and $(4,2)$ is

(A) $3\sqrt{5}$ (B) $\sqrt{13}$ (C) 45 (D) 13 (E) $3\sqrt{2}$

Solution:

To find d, the length of P_1P_2:

$P_2(4,2) \rightarrow x_2 = 4, y_2 = 2$
$P_1(1,-4) \rightarrow x_1 = 1, y_1 = -4$

distance formula: $d = \sqrt{(x_2 - x_1)^2 + (y_2 - y_1)^2}$

$$= \sqrt{(4-1)^2 + [2-(-4)]^2}$$
$$= \sqrt{3^2 + 6^2} = \sqrt{9 + 36}$$
$$= \sqrt{45} = 3\sqrt{5} \quad \textit{Ans.}\ \text{(A)}$$

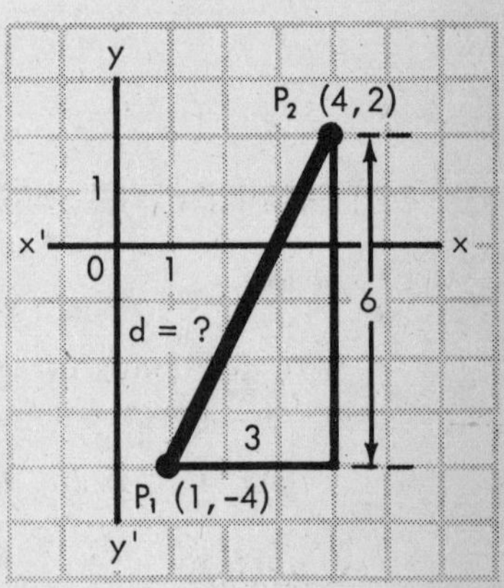

(5) Find the length of the longest side of the triangle whose vertices are $A(1,1)$, $B(9,7)$, and $C(9,16)$.

(A) 289 (B) 23 (C) $\sqrt{337}$ (D) 15 (E) 17

Solution:

AC, opposite the right angle, is the longest side of $\triangle ABC$. To find the length of AC:

$C(9,16) \rightarrow x_2 = 9, y_2 = 16$
$A(1,1) \rightarrow x_1 = 1, y_1 = 1$

distance formula: $d = \sqrt{(x_2 - x_1)^2 + (y_2 - y_1)^2}$

$$= \sqrt{(9-1)^2 + (16-1)^2}$$
$$= \sqrt{8^2 + 15^2} = \sqrt{64 + 225}$$
$$= \sqrt{289} = 17 \quad \textit{Ans.}\ \text{(E)}$$

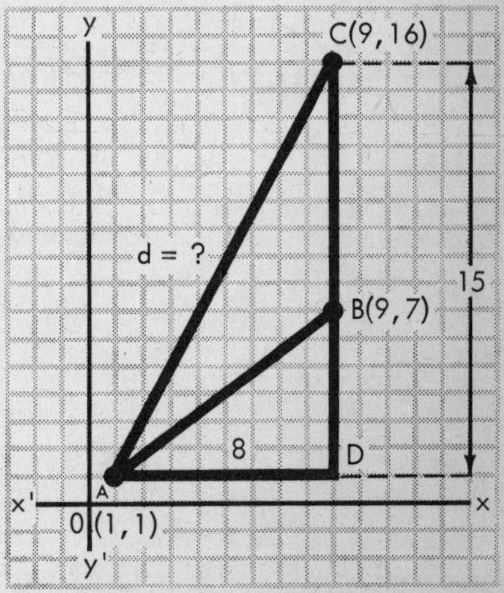

Matching Test

Find the required length in each of the following. After completing this test, match your answers with the numbers in the box.

1. Find the distance between the origin and point $(5, 12)$.

2. Find the length of line segment AB between $A(-1, 2)$ and $B(2, 6)$.

3. Find the length of the line segment joining point $(0, 1)$ and point $(6, -7)$.

4. Find the length of the longest side of triangle ABC if its vertices are $A(2, 2)$, $B(10, 8)$, and $C(10, 17)$.

5. Find the distance between the midpoint of line segment AB and point B, given $A(5, 7)$ and $B(8, 3)$.

a. 17	*b.* 5	*c.* $2\frac{1}{2}$	*d.* 10	*e.* 13

6. DETERMINING RELATIVE POSITIONS

Model Problem

A road running north and south passes through towns A, B, C, and D. Town D is north of B and C. Town B is south of A. Town D is between A and C. Which town is farthest north?

(A) A (B) B (C) C (D) D (E) It cannot be determined from the given information.

Solution:

Since the town farthest north is required, change the statement "Town B is south of A" to read "Town A is north of B."

Town D is north of C and between A and C. Hence, town A must be north of D. Therefore, town A must be the town farthest north. *Ans.* (A)

Note. The accompanying diagram shows the relative positions of the four towns. From the given data, we cannot tell whether B is north or south of C, but this information is not needed to solve the problem.

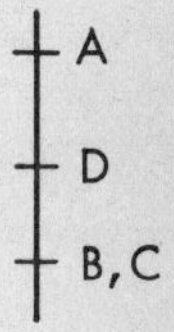

Matching Test

After completing this test, match your answers with the letters in the box.

1. A horizontal line contains the four points J, N, P, and Q. Q is to the right of both N and J. P is to the right of N and to the left of J. Which point is farthest left?
2. An east-west road passes through the towns B, E, H, and K. B is west of K, E, and H. E is between B and H and east of K. Which town is farthest east?
3. Four villages, P, Q, S, and T, lie along a north-south road. T is south of P. P is south of S. S is north of Q. T is between P and Q. Which town is farthest south?
4. A vertical line contains the four points K, B, C, and D. K is below C and above D. B is between C and D. Which is the lowest point?
5. Five camps, G, J, L, R, and W, are located along an east-west road. Which camp is farthest west if (1) G is east of J and R, (2) L is east of W and G, and (3) J is east of R and west of W?

a. R	*b.* N	*c.* Q	*d.* D	*e.* H

7. EVALUATING CHANGES IN EXPRESSIONS

Model Problem

Given the following expressions: I. $x - \frac{1}{x}$ II. $\frac{12}{x^2 - x}$ III. $2x^3 - x^2$

As x increases from 2 to 4, which of the three expressions increase?

(A) I only (B) I and II only (C) I and III only (D) II and III only (E) I, II, and III

Solution:

I. If $x = 2$, then $x - \frac{1}{x} = 2 - \frac{1}{2} = 1\frac{1}{2}$.

If $x = 4$, then $x - \frac{1}{x} = 4 - \frac{1}{4} = 3\frac{3}{4}$.

Since its value changes from $1\frac{1}{2}$ to $3\frac{3}{4}$, expression I increases by $2\frac{1}{4}$.

II. If $x = 2$, then $\frac{12}{x^2 - x} = \frac{12}{4 - 2} = 6$.

If $x = 4$, then $\frac{12}{x^2 - x} = \frac{12}{16 - 4} = 1$.

Since its value changes from 6 to 1, expression II decreases by 5.

III. If $x = 2$, then $2x^3 - x^2 = 16 - 4 = 12$.

If $x = 4$, then $2x^3 - x^2 = 128 - 16 = 112$.

Since its value changes from 12 to 112, expression III increases by 100.

Hence, only I and III increase. *Ans.* (C)

Matching Test

In each of the following, find the change in the value of each expression as x increases from 2 to 4. After completing this test, match your answers with the numbers in the box.

1. $x^2 - \frac{16}{x^2}$ **2.** $x^3 - 2x^2$ **3.** $\frac{24}{x^2 - x}$ **4.** $x^3 + \frac{12}{x}$ **5.** $100 - \frac{x^3}{2}$

a. 32	*b.* −10	*c.* −28	*d.* 15	*e.* 53

8. FINDING RANGES OF VALUES

Model Problem

If b has a value between 20 and 30 and a has a value between 2 and 10, what range of values does $\frac{b}{a}$ have?

(A) 2 to 30 (B) 3 to 10 (C) 2 to 15 (D) 18 to 20 (E) 10 to 28

Solution:

1. The value of a fraction decreases if the numerator is decreased or if the denominator is increased. Hence, to find the minimum value of $\frac{b}{a}$, select the smallest possible value for b and the largest possible value for a. Thus, the minimum value of $\frac{b}{a}$ is $\frac{20}{10}$, or 2.
2. The value of a fraction increases if the numerator is increased or if the denominator is decreased. Hence, to find the maximum value of $\frac{b}{a}$, select the largest possible value of b and the smallest possible value of a. Thus, the maximum value of $\frac{b}{a}$ is $\frac{30}{2}$, or 15.
3. Therefore, the value of $\frac{b}{a}$ ranges from a minimum of 2 to a maximum of 15. *Ans.* (C)

Matching Test

After completing this test, match your answers with the answers in the box.

1. If b has a value between 6 and 8 and a has a value between 1 and 2, what range of values does $\frac{b}{a}$ have?

2. If d has a value between 3 and 5 and c has a value between $\frac{1}{2}$ and 1, what range of values does $\frac{d}{c}$ have?

3. If y has a value between 14 and 16 and x has a value between 6 and 10, what range of values does $(y - x)$ have?

4. If y has a value between 1 and 3 and x has a value between 4 and 6, what range of values does $(y + x)$ have?

5. If y has a value between 4 and 6 and x has a value between 1 and 2, what is the range of values of $\frac{y + x}{y - x}$?

a. 4 to 10	*b.* 3 to 8	*c.* 5 to 9	*d.* 1 to 4	*e.* 3 to 10

9. PERFORMING OPERATIONS ON INEQUALITIES

Model Problem

Given the following statements in which w, x, y, and z are positive integers:

I. $x > z$ II. $x + w > y + w$ III. $w - y > w - z$

If $x > y$ and $y > z$, which of the three statements must be false?

(A) none of them (B) only I (C) only II (D) only III (E) all of them

Solution:

Statement I is true. Since x is greater than y and y is greater than z, x must be greater than z. (*Reason:* If the first of three numbers is greater than the second and the second is greater than the third, then the first must be greater than the third.)

Statement II is true. Begin with the given inequality $x > y$ and add w to each side to obtain $x + w > y + w$. (*Reason:* If equals are added to unequals, the sums are unequal in the same order.)

Statement III is false. Here the unequals y and z are being subtracted from the equals w and w. The differences must be unequal in the opposite, or reverse, order; that is, if $y > z$, then $w - y < w - z$. *Ans.* (D)

Matching Test

In each of the following, $x > y$ and $y > z$. Also, w, x, y, and z are positive integers. In each, determine which of the given three statements *must be false*. After completing this test, match your answers with the five choices in the box.

> *Note.* A statement that *may be true* is not to be considered as a statement that *must be false*. For example, in the problems below, some statements that may be true are:
>
> 2. (3) $y + z > x$ 3. (1) $y^2 > xz$ 4. (2) $z > x - y$

1. (1) $x + w > z + w$
(2) $w - x > w - z$
(3) $wx > wy$

2. (1) $2z > x + y$
(2) $x^2 > y^2$
(3) $y + z > x$

3. (1) $y^2 > xz$
(2) $x^2 > y^2$
(3) $\frac{w}{x} > \frac{w}{y}$

4. (1) $wx > wz$
(2) $z > x - y$
(3) $\frac{x}{w} > \frac{y}{w}$

5. (1) $\frac{w}{x} > \frac{w}{z}$
(2) $z^2 > xy$
(3) $x - w > y - w$

> *a.* none of them *b.* only (1) *c.* only (2) *d.* only (3) *e.* (1) and (2)

10. SOLVING DATA-SUFFICIENCY PROBLEMS

Directions: Each of the questions on the next page is followed by two statements, labeled (1) and (2), in which certain data are given. In these questions you do not actually have to compute an answer, but rather you have to decide whether the data given in the statements are *sufficient* for answering the question. Using the data given in the statement *plus* your knowledge of mathematics and everyday facts (such as the number of days in July), you are to select answer

(A) if statement (1) ALONE is sufficient, but statement (2) alone is not sufficient to answer the question asked;

(B) if statement (2) ALONE is sufficient, but statement (1) alone is not sufficient to answer the question asked;

(C) if BOTH statements (1) and (2) TOGETHER are sufficient to answer the question asked, but NEITHER statement ALONE is sufficient;

(D) if EACH statement is sufficient by itself to answer the question asked;

(E) if statements (1) and (2) TOGETHER are NOT sufficient to answer the question asked and additional data specific to the problem are needed.

Model Problem

In a four-volume work, how many pages does the fourth volume contain?

(1) The four volumes contain 3200 pages.
(2) The first three volumes contain 2500 pages.

Solution:

Since the first three volumes contain 2500 pages out of the total of 3200 pages, the fourth volume must contain the remainder, (3200 − 2500), or 700 pages. Hence, both statements (1) and (2) are sufficient to answer the question asked, but neither statement alone is sufficient. *Ans.* (C)

Matching Test

In each of the following data-sufficiency problems, select one of the preceding choices A, B, C, D, or E as the answer.

1. In a four-volume work, how many pages does the fourth volume contain?
(1) The four volumes contain 3200 pages.
(2) The first two volumes contain 1850 pages.

2. In a four-volume work, how many pages does the first volume contain?
(1) The four volumes contain 3200 pages.
(2) The last three volumes contain 1200 pages more than the first.

3. In a four-volume work, how many illustrations do the second and third volumes contain?
(1) Of the 500 illustrations in the four volumes, the ratio of the number of illustrations in each is 1:2:3:4.
(2) There are as many illustrations in the second and third volumes as in the first and fourth.

4. In a four-volume work, how much does the second volume weigh?
(1) The weight of the second volume is less than one-half the combined weight of the remaining volumes.
(2) The total weight of the four volumes is 10 pounds and that of the second volume is one-third of the total weight of the remaining volumes.

5. In a four-volume work, how many pounds does the entire set weigh?
(1) The first volume weighs $1\frac{1}{2}$ pounds, and each volume in succession weighs 2 ounces more than the previous one.
(2) The first two volumes are 4 ounces lighter than the remaining two volumes, while the ratio of the weights of the first two volumes to the last two is 39:41.

11. SOLVING QUANTITATIVE COMPARISON PROBLEMS

Directions: In this type of problem, a quantity in Column A is to be compared with a quantity in Column B. You are to select answer

(A) if the quantity in Column A is the greater;
(B) if the quantity in Column B is the greater;
(C) if the quantities are equal;
(D) if the relation cannot be determined from the information given.

Model Problem 1

Column A	*Column B*
$3 \times 352 \times 8$	$4 \times 352 \times 6$

Solution:

Each of the two quantities is a product of three factors. Since 352 is a common factor, it is not necessary to multiply all the factors. Rearrange the factors and simplify as follows:

$3 \times 8 \times 352$	$4 \times 6 \times 352$
24×352	24×352

Hence, the two quantities are equal. *Ans.* (C)

Model Problem 2

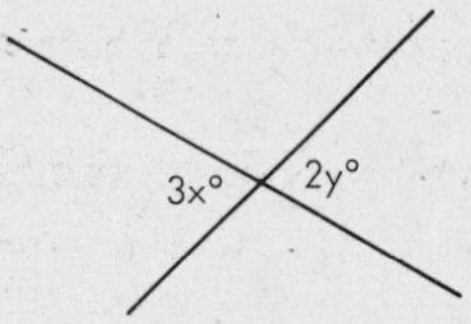

Column A	*Column B*
x	y

Solution:

The measures of the two angles in the figure are $3x$ and $2y$, and the quantities to be compared are x and y. Since vertical angles have equal measures, $3x = 2y$. Hence, $x = \frac{2}{3}y$. Therefore, x is less than y. *Ans.* (B)

Matching Test

In each problem, compare the quantities in Columns A and B and indicate the relation between them by choosing one of the answers—(A), (B), (C), or (D)—listed in the directions.

	Column A		*Column B*
1.	$2 \times 54 \times 67 \times 9$		$3 \times 67 \times 54 \times 6$
2.	$\frac{1}{2} \times \frac{1}{3} \times \frac{1}{4}$		$\frac{1}{3} \times \frac{1}{4} \times \frac{1}{5}$
3.	75% of $\frac{2}{5}$		$\frac{2}{3}$ of 75%
4.		$y = x^2$	
	y		x

I. REVIEWING ARITHMETIC

1. REVIEWING WHOLE NUMBERS

Whole numbers are the numbers used in counting (1, 2, 3, 4, etc.) and zero (0).

Each digit of a whole number represents one or more units, tens, hundreds, etc., depending on its place in the whole number. For example, in the whole number 345, the digit 3 represents 3 hundreds, 4 represents 4 tens, and 5 represents 5 units. Hence, $345 = 300 + 40 + 5$.

An approximate value of a whole number is obtained by rounding off the whole number to a multiple of 10, 100, etc., as indicated by the problem. For example, the approximate value of 312, to the nearest ten, is 310. To the nearest hundred, the approximate value of 312 is 300.

In estimating answers, use approximate values of the numbers involved. For example, to estimate the product of 59 and 72, multiply 60 by 70. Hence, $59 \times 72 \approx 4200$. The symbol $\approx$ means "is approximately equal to."

Terms Used in Fundamental Operations on Whole Numbers

The fundamental operations on whole numbers are addition, subtraction, multiplication, and division. The following terms are used in these operations:

ADDITION

Addends are the numbers being added. The ***sum*** is the answer in addition.

For example, if 45 and 12 are added, then 45 and 12 are addends and their sum is 57.

SUBTRACTION

The ***minuend*** is the number from which we subtract. The ***subtrahend*** is the number being subtracted. The ***difference*** is the answer in subtraction.

For example, if 14 is subtracted from 19, then 14 is the subtrahend, 19 is the minuend, and 5 is their difference.

MULTIPLICATION

The ***multiplicand*** is the number being multiplied. The ***multiplier*** is the number by which we multiply. The ***product*** is the answer in multiplication. The numbers that are multiplied are also ***factors*** of their product.

For example, if 12 is multiplied by 10, then 12 is the multiplicand, 10 is the multiplier, and 120 is their product. The numbers 12 and 10 are also factors of 120.

DIVISION

The ***dividend*** is the number being divided. The ***divisor*** is the number by which we divide. The ***quotient*** is the answer in division.

For example, if 15 is divided by 3, then 15 is the dividend, 3 is the divisor, and 5 is their quotient. However, if 17 is divided by 3, then 5 is the partial quotient and there is a ***remainder*** of 2. When 17 is divided by 3, the complete quotient is $5\frac{2}{3}$.

The ***multiples*** of a whole number are found by multiplying the whole number by 1 or by other whole numbers. For example, the multiples of 5 are 5, 10, 15, 20, 25, etc.

The ***common multiples*** of two or more whole numbers are numbers that are exactly divisible by each of the whole numbers. For example, some common multiples of 5, 10, and 15 are 30, 60, and 90. The ***least common multiple,*** or ***L.C.M.,*** is the smallest of the common multiples. Of the common multiples of 5, 10, and 15, the least common multiple is 30.

Any whole number except 1 is a ***prime number*** if it has no factors other than itself and 1. For example, 2, 3, 5, 7, and 11 are prime numbers. Prime numbers form the ***prime factors*** of a non-prime number. For example, the prime factors of 30 are 2, 3, and 5. (Note that 10 is a factor of 30, but is not a prime factor.)

1.1 ESTIMATING ANSWERS

Model Problem

Estimate the answer: $\dfrac{58 \times 810}{1989}$

(A) 2 (B) 2.4 (C) 24 (D) 200 (E) 240

Solution:

Round off each whole number to a multiple of 10, of 100, or of 1000. Thus, 58 becomes 60, 810 becomes 800, and 1989 becomes 2000.

$$\text{Hence, } \frac{58 \times 810}{1989} \approx \frac{60 \times 800}{2000} = 24 \quad \textit{Ans.} \text{ (C)}$$

Matching Test

Estimate each answer. After completing this test, match your answers with the numbers in the box.

1. $\dfrac{590 \times 2015}{29400}$ **2.** $\dfrac{587 \times 38}{303}$ **3.** $\dfrac{44950}{896}$ **4.** $\begin{array}{r}47985\\-36892\\\hline\end{array}$ **5.** $\dfrac{811 \times 4030}{196 \times 814}$

a. 11000	*b.* 40	*c.* 20	*d.* 80	*e.* 50

1.2 EQUIVALENT QUANTITIES

Model Problem

How many ten-gallon cans are needed to provide the same capacity as 20 six-gallon cans?

(A) 8 (B) 10 (C) 12 (D) 16 (E) 18

Solution:

The 20 six-gallon cans have a capacity of 20(6), or 120 gallons.

The number of ten-gallon cans needed $= \frac{120}{10}$, or 12. *Ans.* (C)

Matching Test

After completing this test, match your answers with the numbers in the box.

1. How many twelve-gallon cans are needed to provide the same capacity as 18 ten-gallon cans?
2. How many eight-pound weights are needed to balance 12 twelve-pound weights?
3. If 4 children can be seated in the space occupied by 3 adults, how many children can be seated in a bus that can hold 40 adults and already has 28 adults in it?
4. In a restaurant, a meat roast can be served to 9 adults or to 12 children. How many children can be served if there are 3 roasts and 12 adults are present?
5. A quart of milk is used to serve 6 cats or 8 kittens. If there are 4 quarts and 15 cats are on hand, how many kittens can be fed?

a. 12	*b.* 18	*c.* 16	*d.* 15	*e.* 20

1.3 LOWEST COMMON MULTIPLE

Model Problem

The denominators of three fractions are 12, 15, and 18. What is the lowest common multiple of these denominators?

(A) 12 (B) 18 (C) 15 (D) 180 (E) 720

Solution:

Obtain the prime factors of each denominator.

$$12 = 4 \cdot 3 = \underline{2 \cdot 2 \cdot 3} \qquad 15 = \underline{3 \cdot 5} \qquad 18 = 2 \cdot 9 = \underline{2 \cdot 3 \cdot 3}$$

In the three underscored groups, the prime factor 2 appears twice among the factors of 12, the prime factor 3 occurs twice among the factors of 18, and the prime factor 5 occurs only once. Hence, the lowest common multiple of the numbers 12, 15, and 18 is $2 \cdot 2 \cdot 3 \cdot 3 \cdot 5$, or 180. *Ans.* (D)

> *Note.* The lowest common multiple of a set of denominators is their lowest common denominator (L.C.D.). See Reviewing Fractions, pages 18–26.

Matching Test

After completing this test, match your answers with the numbers in the box. (*Hint.* Each answer in this test is a lowest common multiple.)

1. Two nearby lights are flashing, one every 7 seconds and the other every 5 seconds. How many seconds long are the intervals between simultaneous flashings?
2. A red light at one corner goes on every 6 minutes while another red light at the next corner goes on every 8 minutes. How many minutes long are the intervals between the times that the lights go on simultaneously?
3. One Ferris wheel makes a complete rotation every 20 minutes and another Ferris wheel completes a rotation every 30 minutes. If they start at the same time, how many minutes will it take before the wheels begin a new rotation together?
4. Two alarms are rung at regular intervals, one every 4 minutes, the other every 6 minutes. After they sound together, how many minutes will it take before they sound together again?
5. What is the least common multiple of the denominators of four fractions if these denominators are 6, 9, 14, and 21?

a. 24	*b.* 60	*c.* 12	*d.* 126	*e.* 35

1.4 POSTS AND SPACES

Model Problem

How many trees are there if the distance from the first tree to the last is 200 feet and the trees are 50 feet apart? (Assume the trees are in a straight line.)

(A) 5 (B) 4 (C) 3 (D) 6 (E) none of these

Solution:

The distance of 200 feet is divided by the 50-foot spacings into 4 spaces. If each space is paired with a tree, there must be one tree for which there is no space. Hence, there are 5 trees. *Ans.* (A)

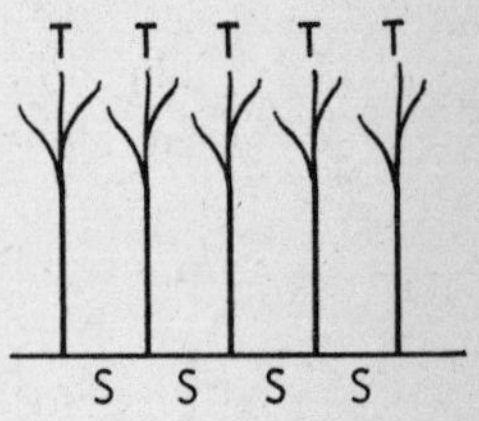

T = Tree
S = Space

Matching Test

After completing this test, match your answers with the numbers in the box.

1. How many trees are there if the distance from the first tree to the last is 200 feet and the trees are 25 feet apart?
2. How many posts are there in a fence if the posts are 8 feet apart and the distance from the first post to the last is 136 feet?
3. How many children are there in an aisle of an auditorium if the children are 3 feet apart, the aisle is 100 feet long, and no child is closer than 20 feet from each end of the aisle?
4. How many times does a bell ring from 9 A.M. to 3 P.M. if it rings every half-hour as well as at 9 A.M. and at 3 P.M.?
5. How many multiples of 5 are there between 15 and 95 if both 15 and 95 are included?

a. 13	*b.* 21	*c.* 18	*d.* 17	*e.* 9

1.5 SEQUENCES OF WHOLE NUMBERS

Model Problem

A ball rolling down an inclined plane rolls 10 feet during the first second, 30 feet during the second second, 50 feet during the third second, etc. How many feet will it roll during the eleventh second?

(A) 230 (B) 210 (C) 200 (D) 190 (E) 170

Solution:

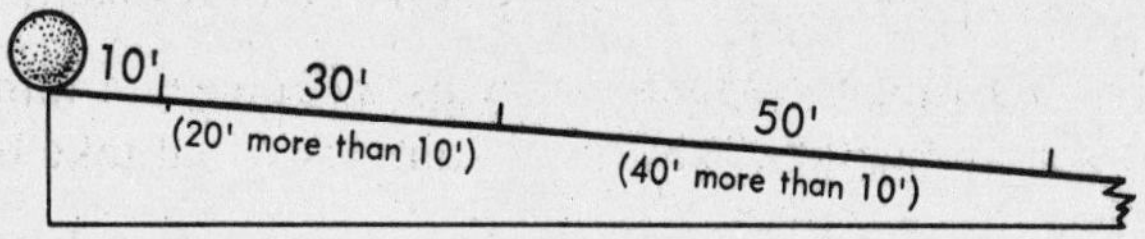

Note that each new distance is 20 feet more than the previous one. Compare each new distance with the first distance as follows:

The second distance is 20 feet more than the first.

The *third* distance is *2*(20), or 40 feet more than the first.

The *fourth* distance is *3*(20), or 60 feet more than the first.

The *eleventh* distance will be *10*(20), or 200 feet more than the first. Therefore, during the eleventh second, the ball rolls 200 + 10, or 210 feet. *Ans.* (B)

Matching Test

After completing this test, match your answers with the numbers in the box.

1. A ball rolling down an inclined plane rolls 5 feet during the first second, 15 during the second second, 25 feet during the third second, etc. How many feet will it roll during the fifth and sixth seconds?

2. In a meeting hall, the first row accommodates 20 persons, the second row 25, the third row 30, etc. How many seats are there in the last row, if the last row is the twentieth row?

3. How many inches are there in the next measurement of the following set if the increase between measurements remains the same: 2 feet, 1 yard 10 inches, 1 yard 2 feet 8 inches?

4. In a tennis tournament, there are 128 players to begin play. If one-half the players are eliminated during each round, how many rounds of play will it take to eliminate all but two of the contestants?

5. Find the missing term in the following set: 600, 600, 300, 100, 25, (?) (*Hint*. Each new divisor is 1 more than the previous one.)

a. 5	*b*. 115	*c*. 6	*d*. 100	*e*. 90

2. REVIEWING FRACTIONS

The ***terms*** of a fraction are the ***numerator*** and the ***denominator.*** For example, in the fraction $\frac{2}{3}$, the terms are 2, the numerator, and 3, the denominator. In arithmetic, the terms of a fraction are usually whole numbers.

A ***proper fraction*** is a fraction whose numerator is less than the denominator. Hence, the value of a proper fraction is less than 1. For example, the proper fraction $\frac{14}{15}$ is less than 1.

An ***improper fraction*** is a fraction whose numerator is equal to the denominator or more than the denominator. Hence, the value of an improper fraction is equal to 1 or is greater than 1. For example, the improper fraction $\frac{15}{14}$ is greater than 1.

Any whole number can be expressed as an improper fraction with a denominator of 1. For example, $6 = \frac{6}{1}$.

A ***mixed number*** is a number that is the sum of a whole number and a proper fraction. For example, the mixed number $5\frac{2}{3} = 5 + \frac{2}{3}$.

To change a mixed number into an improper fraction, add the numerator to the product of the whole number and the denominator. Then place the result over the denominator. For example, to change $5\frac{2}{3}$ into an improper fraction, add 2 to the product of 5 and 3. Then place the result, 17, over 3, giving $\frac{17}{3}$.

To change an improper fraction into a mixed number, divide the numerator by the denominator. Thus, to change $\frac{17}{3}$ into a mixed number, divide 17 by 3, giving $5\frac{2}{3}$.

The ***lowest common denominator,*** or ***L.C.D.,*** of two or more fractions is the least whole number exactly divisible by each of the denominators. For example, 12 is the lowest common denominator of $\frac{1}{2}$, $\frac{2}{3}$, and $\frac{5}{6}$. The denominators 2, 3, and 6 have many common denominators, such as 12, 24, 36, and 120. However, the least, or smallest, of these is 12.

The L.C.D. may be found as follows:

When the denominators have no common factor, the L.C.D. is the product of the denominators. For example, the L.C.D. of $\frac{1}{2}$, $\frac{2}{3}$, and $\frac{5}{7}$ is 2(3)(7), or 42.

When the denominators have a common factor, the L.C.D. is found by multiplying the common factor by all the other factors. For example, the denominators of $\frac{3}{5}$, $\frac{7}{10}$, and $\frac{4}{15}$ have the common factor of 5. To obtain the L.C.D., multiply the common factor, 5, by the other factors, 2 and 3. Hence, the L.C.D. is 5(2)(3), or 30.

Equivalent fractions are fractions having the same value. For example, $\frac{3}{5}$, $\frac{30}{50}$, and $\frac{300}{500}$ are equivalent fractions.

To change a fraction into an equivalent fraction, apply the following fundamental law of fractions:

Law of Fractions. **A fraction may be changed into an equivalent fraction by multiplying or dividing the numerator and the denominator by the same number.**

For example, if the terms of the fraction $\frac{3}{5}$ are multiplied by 10, then $\frac{3}{5}$ is changed to $\frac{30}{50}$, an equivalent fraction. Also, if the terms of the fraction $\frac{300}{500}$ are reduced by dividing by 100, then $\frac{300}{500}$ is changed to $\frac{3}{5}$, an equivalent fraction. Since 100 is the highest common factor of 300 and 500, when the fraction $\frac{300}{500}$ is changed to $\frac{3}{5}$, it is *reduced to lowest terms.*

Use the fundamental law of fractions to change a fraction into an equivalent fraction whose numerator is given or whose denominator is given. For example, to change $\frac{2}{5}$ to an equivalent fraction having a numerator of 14, set up the problem as $\frac{2}{5} = \frac{14}{?}$. Note that 14, the new numerator, is 7 times 2, the old numerator. Hence, the new denominator must be 7 times 5, the old denominator. The required fraction is $\frac{14}{35}$.

Use the fundamental law of fractions to compare fractions. For example, to compare $\frac{2}{5}$ and $\frac{13}{35}$, change $\frac{2}{5}$ into $\frac{14}{35}$, an equivalent fraction. Since $\frac{14}{35}$ is greater than $\frac{13}{35}$, it follows that $\frac{2}{5}$ is greater than $\frac{13}{35}$. Note that the given fractions are compared by changing them so that the resulting fractions have the same denominator, and then by comparing the numerators.

Fundamental Operations on Fractions

Adding or Subtracting Fractions (Combining Fractions)

If the fractions have the same denominator, combine them by keeping the denominator unchanged and adding or subtracting the numerators. Reduce the result to lowest terms if necessary.

$$\frac{1}{7} + \frac{8}{7} - \frac{3}{7} = \frac{1+8-3}{7} = \frac{6}{7}$$

If the fractions have different denominators, change the fractions into equivalent fractions having the same denominator by using the lowest common denominator.

The L.C.D. of $\frac{2}{5}$, $\frac{3}{10}$, and $\frac{4}{15}$ is 30. Hence,

$$\frac{2}{5} + \frac{3}{10} - \frac{4}{15} = \frac{12}{30} + \frac{9}{30} - \frac{8}{30} = \frac{12+9-8}{30} = \frac{13}{30}$$

To add or subtract mixed numbers, change each mixed number into an improper fraction and then proceed as with common fractions.

Multiplying Fractions

If none of the numerators has a factor in common with any denominator, the new numerator is found by multiplying the numerators, and the new denominator is found by multiplying the denominators.

$$\frac{1}{2}\times\frac{3}{5}\times\frac{7}{11}=\frac{1\cdot 3\cdot 7}{2\cdot 5\cdot 11}=\frac{21}{110}$$

If a numerator and a denominator have a common factor, remove that factor. After removing all such common factors, multiply the remaining factors in the numerators to obtain the new numerator, and multiply the remaining factors in the denominators to obtain the new denominator.

$$\frac{2}{\cancel{5}_1}\times\frac{\cancel{5}^1}{\cancel{7}_1}\times\frac{\cancel{14}^2}{9}=\frac{2\times 2}{9}=\frac{4}{9}$$

Dividing by Fractions

To divide by a fraction, invert the divisor and then multiply.

(*Reminder*. The divisor is the number to the right of the division sign ÷.)

$$\frac{3}{7}\div\frac{5}{7}=\frac{3}{\cancel{7}_1}\times\frac{\cancel{7}^1}{5}=\frac{3}{5}$$

$$9\div\frac{9}{10}=\frac{\cancel{9}^1}{1}\times\frac{10}{\cancel{9}_1}=10$$

$$5\frac{1}{4}\div\frac{3}{4}=\frac{\cancel{21}^7}{\cancel{4}_1}\times\frac{\cancel{4}^1}{\cancel{3}_1}=7$$

Adding or Subtracting Mixed Numbers (Without Changing to Improper Fractions)

As illustrated in the following problems, add or subtract the whole numbers, separately add or subtract the fractions, and then combine the results.

$$11\frac{3}{7}-5\frac{2}{7}=\left(11+\frac{3}{7}\right)-\left(5+\frac{2}{7}\right)$$

Subtract whole numbers: $11-5=6$

Subtract fractions: $\frac{3}{7}-\frac{2}{7}=\frac{1}{7}$

Combine results to obtain the answer, $6\frac{1}{7}$.

$$5\frac{1}{5}-3\frac{3}{5}=\left(5+\frac{1}{5}\right)-\left(3+\frac{3}{5}\right)$$

Note that $\frac{3}{5}$, the fraction in the subtrahend, is larger than $\frac{1}{5}$, the fraction in the minuend.

Hence, borrow 1, or $\frac{5}{5}$, from the whole number 5, and add $\frac{5}{5}$ as follows:

$$5\frac{1}{5} - 3\frac{3}{5} = \left(4 + \frac{5}{5} + \frac{1}{5}\right) - \left(3 + \frac{3}{5}\right)$$

$$= \left(4 + \frac{6}{5}\right) - \left(3 + \frac{3}{5}\right)$$

Subtract whole numbers: $4 - 3 = 1$

Subtract fractions: $\frac{6}{5} - \frac{3}{5} = \frac{3}{5}$

Combine results to obtain the answer, $1\frac{3}{5}$.

Multiplying or Dividing Mixed Numbers

To multiply or divide mixed numbers, change them to improper fractions and then proceed as with common fractions.

2.1 COMBINING FRACTIONS

Matching Test

In each of the following, combine the fractions as indicated. After completing this test, match your answers with the numbers in the box.

1. $\frac{1}{2} - \frac{1}{6} + \frac{3}{5}$ **2.** $\frac{1}{2} + \frac{1}{6} - \frac{3}{5}$ **3.** $\frac{1}{2} - \frac{1}{6} + \frac{5}{3}$ **4.** $\frac{5}{3} - \frac{1}{6} - \frac{1}{2}$ **5.** $1\frac{2}{3} + 2\frac{1}{6} - \frac{1}{2}$

a. $3\frac{1}{3}$	*b.* $\frac{14}{15}$	*c.* 1	*d.* $\frac{1}{15}$	*e.* 2

2.2 MULTIPLYING OR DIVIDING FRACTIONS

Matching Test

In each of the following, multiply or divide as indicated. After completing this test, match your answers with the numbers in the box.

1. $\frac{2}{3} \times \frac{7}{8} \times \frac{6}{7}$ **2.** $\frac{2}{3} \times \frac{7}{8} \div \frac{1}{6}$ **3.** $\frac{3}{2} \times \frac{8}{7} \div 6$ **4.** $\frac{3}{2} \times 6 \div \frac{9}{7}$ **5.** $1\frac{1}{2} \times 6 \div 2\frac{1}{4}$

a. $\frac{7}{2}$	*b.* 4	*c.* $\frac{1}{2}$	*d.* 7	*e.* $\frac{2}{7}$

2.3 COMPARING FRACTIONS

Matching Test

In each of the following, arrange the fractions in order of size beginning with the smallest. After completing this test, match your answers with the sets of fractions in the box.

1. $\frac{4}{15}, \frac{1}{3}, \frac{2}{5}$ **2.** $\frac{2}{5}, \frac{3}{8}, \frac{9}{20}$ **3.** $\frac{3}{5}, \frac{5}{8}, \frac{3}{4}$ **4.** $\frac{4}{5}, \frac{13}{16}, \frac{7}{10}$ **5.** $\frac{19}{24}, \frac{3}{4}, \frac{13}{15}, \frac{5}{6}$

a. $\frac{7}{10}, \frac{4}{5}, \frac{13}{16}$ *b.* $\frac{3}{5}, \frac{5}{8}, \frac{3}{4}$ *c.* $\frac{3}{8}, \frac{2}{5}, \frac{9}{20}$ *d.* $\frac{3}{4}, \frac{19}{24}, \frac{5}{6}, \frac{13}{15}$ *e.* $\frac{4}{15}, \frac{1}{3}, \frac{2}{5}$

Note. Compare the fractions in each set by changing them to equivalent fractions having the lowest common denominator. The correct equivalent fractions are as follows:

1. $\frac{4}{15}, \frac{5}{15}, \frac{6}{15}$ **2.** $\frac{16}{40}, \frac{15}{40}, \frac{18}{40}$ **3.** $\frac{24}{40}, \frac{25}{40}, \frac{30}{40}$ **4.** $\frac{64}{80}, \frac{65}{80}, \frac{56}{80}$ **5.** $\frac{95}{120}, \frac{90}{120}, \frac{104}{120}, \frac{100}{120}$

2.4 PROBLEMS THAT INCLUDE A CONVENIENT WHOLE NUMBER AND REQUIRE A FRACTIONAL ANSWER

Model Problem

A company employs 800 workers of whom $\frac{1}{4}$ are college graduates. Of those who are *not* college graduates, $\frac{1}{2}$ are high school graduates. What fraction of the total number of employees are *not* high school graduates?

(A) $\frac{3}{4}$ (B) $\frac{3}{8}$ (C) $\frac{3}{16}$ (D) $\frac{1}{8}$ (E) $\frac{1}{16}$

Solution:

Method 1: *Using Fractions Only*

Since the number of employees does not affect the answer, this method shows how the answer can be obtained without using that number.

Branching Diagram With Fractions

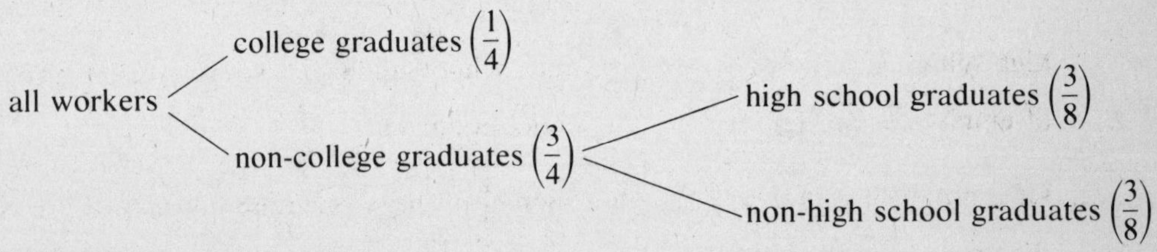

1. It is given that $\frac{1}{4}$ of the workers are college graduates. Hence, the remaining $\frac{3}{4}$ must be non-college graduates.

2. It is given that $\frac{1}{2}$ of the remaining workers are high school graduates. Hence, the fraction of workers who are high school graduates is $\frac{1}{2}\left(\frac{3}{4}\right)$, or $\frac{3}{8}$.

3. If the non-college graduates are $\frac{3}{4}$, or $\frac{6}{8}$, of all the workers and, of these, $\frac{3}{8}$ are high school graduates, then those who are *not* high school graduates must be the balance, $\frac{6}{8} - \frac{3}{8}$, or $\frac{3}{8}$.

Method 2: Using a Whole Number Together With Fractions

Since operations with fractions are tricky, it may be easier to work with a convenient whole number.

Branching Diagram With Whole Numbers

all workers (800) → college graduates (200)
all workers (800) → non-college graduates (600)
non-college graduates (600) → high school graduates (300)
non-college graduates (600) → non-high school graduates (300)

1. It is given that $\frac{1}{4}$ of the workers are college graduates. Hence, the number of college graduates is $\frac{1}{4}(800)$, or 200. The remaining 600 workers are *not* college graduates.

2. It is given that $\frac{1}{2}$ of the remaining workers are high school graduates. Hence, the number of high school graduates is $\frac{1}{2}(600)$, or 300. The balance, 300 workers, must be those that are *not* high school graduates.

3. The balance of 300 non-high school graduates $= \frac{300}{800}$, or $\frac{3}{8}$, of all the workers.

Ans. (B)

Matching Test

After completing this test, match your answers with the fractions in the box.

1. Of the 600 employees of a company, $\frac{1}{3}$ are high school graduates. Of those who are not high school graduates, $\frac{1}{4}$ are elementary school graduates. What fraction of the total number of employees are not elementary school graduates?

2. Of a group of 1,000 people, $\frac{1}{5}$ are men. $\frac{1}{4}$ of the women are not married. What fraction of the group are married women?

3. Tom is taking a trip of 1600 miles. $\frac{1}{4}$ of the entire distance was covered on the first day and $\frac{1}{6}$ of the remaining distance on the second day. What fraction of the entire trip remains to be covered?

4. Of 400 students, it was found that $\frac{1}{2}$ take a science course. $\frac{3}{8}$ of the remaining students take a foreign language. What fraction of the group is the number of students who take neither science nor a foreign language?

5. On his trip, Harry took $400. During the first week, he spent $\frac{1}{2}$ of the money; during the second week, he spent $\frac{1}{4}$ of the remainder; and during the third week, he spent $\frac{1}{5}$ of what was left. What fraction of his original sum did he still have after the third week?

a. $\frac{5}{8}$	*b.* $\frac{3}{5}$	*c.* $\frac{5}{16}$	*d.* $\frac{1}{2}$	*e.* $\frac{3}{10}$

2.5 PROBLEMS THAT REQUIRE A WHOLE-NUMBER ANSWER

Matching Test

After completing this test, match your answers with the numbers in the box.

1. A theater has as many orchestra seats as balcony seats in a total of 500 seats. If $\frac{3}{5}$ of the orchestra seats were sold and also $\frac{1}{2}$ of the balcony seats, how many seats were not sold?

2. In a district where 200 votes were cast, $\frac{3}{5}$ of the votes were for the Democratic Party candidate and $\frac{3}{5}$ of the remaining votes were for the Republican Party candidate. How many votes were for neither candidate?

3. How many soccer balls were sold by a sporting goods store if out of a total of 330 balls sold, $\frac{1}{3}$ of the total were basketballs and $\frac{5}{11}$ of the remainder were soccer balls?

4. A manufacturer of trousers calculated that of a total of 65 dozen pairs of trousers in stock, $\frac{3}{10}$ were brown. If $\frac{1}{3}$ of the remainder are gray, how many pairs of gray trousers are there in stock?

5. A tank having a capacity of 300 gallons is $\frac{3}{4}$ full. How many gallons are needed to make it $\frac{4}{5}$ full?

a. 100	*b.* 182	*c.* 225	*d.* 15	*e.* 32

2.6 PROBLEMS EXPRESSED IN FRACTIONS ONLY AND REQUIRING A FRACTIONAL ANSWER

Matching Test

After completing this test, match your answers with the fractions in the box. (*Hint.* It may be easier to use the branching process illustrated in Method 2, page 23. To use this method, choose a convenient whole number for each problem. For example, in problem 2, assume that the movie house had 200 seats.)

1. A student finished $\frac{1}{2}$ of his homework in the early afternoon after school and then $\frac{1}{4}$ of the remainder in the evening. What part of the homework still remained to be done?
2. In a movie house, the number of orchestra seats equals the number of balcony seats. For one performance, $\frac{1}{4}$ of the orchestra and $\frac{1}{2}$ of the balcony remained unsold. What part of the total number of seats were sold?
3. In three classes having equal registers, $\frac{1}{2}$ of each class made the honor roll. What part of the total register of the three classes made the honor roll?
4. During a dress sale, $\frac{1}{3}$ of the dresses were sold the first day and $\frac{1}{3}$ of the rest during the second day. What part of the original number remained unsold?
5. Of a quantity of meat purchased by a family, $\frac{1}{5}$ was eaten the first day and $\frac{3}{8}$ of the rest was eaten the next day. What part of the original quantity was eaten during the second day?

a. $\frac{5}{8}$	*b.* $\frac{3}{10}$	*c.* $\frac{3}{8}$	*d.* $\frac{1}{2}$	*e.* $\frac{4}{9}$

2.7 PROBLEMS IN COMPARING FRACTIONS

Model Problem

Select the most economical price among the following prices:

(A) 4 for 7¢ (B) 8 for 13¢ (C) 10 for 16¢ (D) 20 for 34¢ (E) 2 for 3¢

Solution:

Divide each price by the number of units to obtain the following *prices per unit:*

$$\frac{7}{4}¢ \qquad \frac{13}{8}¢ \qquad \frac{16}{10}¢ \qquad \frac{34}{20}¢ \qquad \frac{3}{2}¢$$

Change each of these fractions to an equivalent fraction having a denominator equal to the least common denominator, 40:

$$\frac{7}{4}=\frac{70}{40} \qquad \frac{13}{8}=\frac{65}{40} \qquad \frac{16}{10}=\frac{64}{40} \qquad \frac{34}{20}=\frac{68}{40} \qquad \frac{3}{2}=\frac{60}{40}$$

Since the smallest of the resulting fractions is $\frac{60}{40}$, the most economical price among the given prices is 2 for 3¢. *Ans.* (E)

Multiple-Choice Test

In each of the following, make the correct choice from the four alternatives.

1. Select the smallest of the following tool sizes:

(A) $\frac{3}{8}$ inch (B) $\frac{5}{16}$ inch (C) $\frac{19}{64}$ inch (D) $\frac{13}{32}$ inch

2. Select the most economical price among the following prices:

(A) 4 for 5¢ (B) 8 for 9¢ (C) 12 for 14¢ (D) 20 for 24¢

3. Select the winning score if the following are the records of four boys in a contest using a basketball to shoot fouls:

(A) 11 out of 15 tries (B) 14 out of 20 tries (C) 20 out of 25 tries
(D) 18 out of 24 tries

4. Select the most economical price among the following prices:

(A) 4 oz. for 7¢ (B) 8 oz. for 13¢ (C) 1 lb. for 25¢ (D) 1 lb., 4 oz. for 30¢

5. During the season, a baseball team had won (A) 4 games after 6 had been played, (B) 5 games after 9 had been played, (C) 7 games after 12 had been played, and (D) 13 games after 18 had been played. When did the team have its best record?

3. REVIEWING DECIMALS

A ***decimal fraction***, or ***decimal***, is a fraction that has a denominator of 10, of 100, of 1000, or of some other power of 10. For example, the decimal .07, if written in the form of a common fraction, is $\frac{7}{100}$.

An approximate value of a decimal is obtained by rounding off the decimal to the nearest tenth, to the nearest hundredth, etc., as called for by the problem. For example, to round off .47 to the nearest tenth, note that .47 is closer to .5 than it is to .4. Hence, .47 to the nearest tenth becomes .5. If a decimal is halfway between two possible answers, use the larger possibility. Thus, .75 to the nearest tenth becomes .8.

To estimate the answers to problems containing decimals, use approximate values. For example, to estimate the value of $.49 \times .711$, round off to $.5 \times .7$ to obtain the estimate .35.

To change a common fraction to a decimal, divide the numerator by the denominator. Continue dividing to obtain an answer to any desired number of places. For example, $\frac{3}{16} = \frac{3.0000}{16} = .1875$.

To change a decimal to a common fraction, write the decimal in fractional form and reduce the fraction to lowest terms. For example, $.35 = \frac{35}{100} = \frac{7}{20}$.

Fundamental Operations on Decimals

Adding or Subtracting Decimals

Arrange the decimals vertically so that the decimal points are in the same line. Then add or subtract as with whole numbers. Align the decimal point of the answer with the other decimal points. For example, to add 1.4, .23, and 3.75, align the addends as shown in the box.

$$\begin{array}{r} 1.4 \\ .23 \\ 3.75 \\ \hline 5.38 \end{array}$$

Multiplying or Dividing Decimals

To multiply or divide a decimal by 10, by 100, by 1000, or by any power of 10, count the number of zeros in the power of 10. Then move the decimal point that many places *to the right if multiplying* or *to the left if dividing*. For example, to multiply 1.23 by 1000, move the decimal point 3 places to the right to obtain the answer 1230. To divide 1.23 by 1000, move the decimal point 3 places to the left to obtain the answer .00123.

To multiply decimals, multiply as with whole numbers. Then mark off as many decimal places in the answer as there are decimal places in the numbers being multiplied. For example, to multiply .12 by .009, first multiply 12 by 9 to obtain 108. Since there are 2 places in .12 and 3 places in .009, mark off 5 places to obtain the answer .00108.

To divide decimals, count the number of decimal places in the divisor. Then move the decimal points of both the dividend and the divisor that many places to the right. (The purpose of this is to make the new divisor a whole number.) Divide as with whole numbers, marking off as many decimal places in the answer as there are new decimal places in the dividend. For example, to divide .048 by .12, proceed as shown in the box.

Vertical Division

$$\frac{.048}{.12} = \frac{.04.8}{.12.} = \frac{4.8}{12} = .4$$

Horizontal Division

$$\begin{array}{r} .4 \\ .12.\overline{\smash{)}\,.04.8} \\ \underline{4\ 8} \end{array}$$

TABLE OF COMMON FRACTIONS AND EQUIVALENT DECIMALS

Common Fraction	*Equivalent Decimal*	*Common Fraction*	*Equivalent Decimal*	*Common Fraction*	*Equivalent Decimal*
$\frac{1}{2}$	.5	$\frac{5}{6}$	$.83\frac{1}{3}$	$\frac{3}{5}$	.6
$\frac{1}{3}$	$.33\frac{1}{3}$	$\frac{1}{8}$	.125	$\frac{4}{5}$	.8
$\frac{2}{3}$	$.66\frac{2}{3}$	$\frac{3}{8}$	.375	$\frac{1}{7}$	$.14\frac{2}{7}$
$\frac{1}{10}$	.1	$\frac{5}{8}$	.625	$\frac{1}{9}$	$.11\frac{1}{9}$
$\frac{1}{4}$	.25	$\frac{7}{8}$	.875	$\frac{1}{12}$	$.08\frac{1}{3}$
$\frac{3}{4}$	.75	$\frac{1}{5}$	.2	$\frac{1}{16}$	$.06\frac{1}{4}$
$\frac{1}{6}$	$.16\frac{2}{3}$	$\frac{2}{5}$	.4		

3.1 DIVISION WITH DECIMALS

Model Problem

Simplify: $.6 \div \dfrac{.6}{.05}$

(A) 7.2 (B) 72 (C) 50 (D) .5 (E) .05

Solution:

1. Simplify $\dfrac{.6}{.05}$: $\dfrac{.60}{.05} = \dfrac{60}{5} = 12$

2. $.6 \div \dfrac{.6}{.05} = .6 \div 12 = \dfrac{.60}{12} = .05$ *Ans.* (E)

Matching Test

Simplify each of the following. After completing this test, match your answers with the numbers in the box.

1. $\dfrac{6}{.5}$ **2.** $\dfrac{60}{.05}$ **3.** $.6 \div \dfrac{6}{5}$ **4.** $6 \div \dfrac{.6}{5}$ **5.** $6 \div \dfrac{6}{.05}$

a. .05	*b.* 50	*c.* .5	*d.* 12	*e.* 1200

3.2 OPERATIONS WITH COMMON FRACTIONS AND DECIMALS

Model Problem

Perform the indicated operations and round off the answer to the nearest tenth:

$$\frac{1}{4} + 10(.072) - \frac{1.38}{10}$$

(A) −.4 (B) .6 (C) .7 (D) .8 (E) −.8

Solution:

To multiply by 10, move the decimal point one place to the right: $10(.072) = .72$

To divide by 10, move the decimal point one place to the left: $\dfrac{1.38}{10} = .138$

$$\frac{1}{4} + 10(.072) - \frac{1.38}{10}$$

$$= .25 + .72 - .138$$

$$= .97 - .138 = .970 - .138$$

$$= .832 = .8, \text{ to the nearest tenth} \quad \textit{Ans. } (D)$$

Matching Test

In each problem, perform the indicated operations and round off the answer to the nearest tenth. After completing this test, match your answers with the numbers in the box. (See the Table of Common Fractions and Equivalent Decimals, page 27.)

1. $\frac{3}{4} + 10(.058)$ **2.** $\frac{5.72}{10} - \frac{3}{8}$ **3.** $\frac{4}{9} \times .63$ **4.** $1.87\frac{1}{2} \div 1.12\frac{1}{2}$ **5.** $.83\frac{1}{3} \times 100(.0225)$

a. .2	*b.* 1.3	*c.* 1.7	*d.* 1.9	*e.* .3

3.3 SAVINGS ON PURCHASES

Model Problem

A 4-ounce can of tomato juice costs 10¢, while a 20-ounce can costs 35¢. (1 liquid pint = 16 ounces.) In the purchase of larger cans, how much is saved on each 8 pints?

(A) 48¢ (B) 72¢ (C) 80¢ (D) 96¢ (E) $1.20

Solution:

First find the cost of 1 pint, or 16 ounces, of tomato juice at each price.

At 10¢ for 4 ounces, 16 ounces costs 4(10¢), or 40¢.

At 35¢ for 20 ounces, 16 ounces costs $\frac{4}{5}$(35¢), or 28¢.

Thus, 1 pint of the larger can costs 12¢ less than 1 pint of the smaller can.

Hence, the saving on 8 pints = 96¢. *Ans.* (D)

Matching Test

After completing this test, match your answers with the amounts in the box.

1. If an 8-ounce can of juice costs 18¢ and a 12-ounce can costs 21¢, how much is saved in the purchase of 15 pints of the larger-size can?
2. A certain pencil is sold in one store at 3 for 13¢ while the same pencil is sold in another store at 4 for 15¢. How much is saved on 5 dozen at the cheaper price?
3. 100 tablets of a brand cost $1.45 while a package of 12 such tablets costs 27¢. How much is saved in the purchase of 300 tablets at the cheaper price?
4. 24 tea bags can be bought for 30¢ while 64 bags of the same tea are packaged to sell for 72¢. What is the saving on 6 of the larger boxes?
5. A 4-gallon can of milk costs $3.75, while a 6-gallon can costs $5.25. How much is saved in the purchase of 36 gallons by buying the larger cans?

a. $2.25	*b.* $.35	*c.* $2.40	*d.* $1.20	*e.* $.48

3.4 EARNINGS OR EXPENSES

Model Problem

A grocery clerk earns \$2.20 an hour for each hour he works before 5 P.M. and \$2.50 an hour for each hour or part of an hour that he works after 5 P.M. (The lunch hour is included among the paid hours.) What did he earn by working from 10 A.M. to 8:30 P.M.?

(A) \$22.00 (B) \$23.10 (C) \$26.25 (D) \$24.68 (E) \$24.15

Solution:

For the 7 hours worked from 10 A.M. to 5 P.M., he earned 7(\$2.20), or \$15.40.

For the remaining $3\frac{1}{2}$ hours from 5 P.M. to 8:30 P.M., he earned $3\frac{1}{2}$(\$2.50), or \$8.75.

His total earnings for the day = \$15.40 + \$8.75, or \$24.15. *Ans.* (E)

Matching Test

After completing this test, match your answers with the amounts in the box.

1. If a clerk is paid \$2.25 for each hour until 5 P.M. and \$2.75 an hour thereafter, what does he earn if he works from 9 A.M. to 9 P.M.?
2. A man worked 54 hours during a week. He is paid at the rate of \$2.20 an hour for the first 40 hours and then 50% more for each hour beyond the first 40. What did he earn for the week?
3. A worker received a total of \$444 for working 5 days a week for 6 weeks. What was his average daily wage during that time?
4. A boat uses 19 gallons of gasoline per hour. If the boat is used from 9:30 A.M. to 2:15 P.M., what is the cost of the gasoline at 36¢ a gallon?
5. On a trip, Paul began with 10 gallons of gasoline in his tank and ended with 5 gallons left. He covered 600 miles on the trip, averaging 12 miles per gallon. What was the cost of the gasoline used on the trip at 30¢ per gallon?

a. \$32.49	*b.* \$29.00	*c.* \$13.50	*d.* \$134.20	*e.* \$14.80

3.5 COSTS

Model Problem

At a school game, receipts amounted to \$272. If 250 tickets were sold to students at \$.80 each, how many tickets were sold to adults at \$1.20 each?

(A) 60 (B) 90 (C) 600 (D) 22 (E) 6

Solution:

The receipts for the student tickets amounted to 250(\$.80), or \$200.

The balance of \$272 − \$200, or \$72, was the amount received from adults.

At \$1.20 a ticket, the number of tickets sold = $\frac{72.00}{1.20} = 60$. *Ans.* (A)

Matching Test

After completing this test, match your answers with the answers in the box.

1. At a game, ticket receipts amounted to \$384. If 300 tickets were sold to students at \$.80 each, how many tickets were sold to adults at \$1.20 each?

2. At a game, receipts amounted to \$504. An equal number of tickets were sold to the adults and to the students. If the price of admission for a student was \$.75 and that for an adult was \$1.25, how many tickets were sold to the students?

3. At \$4.50 per basketball, how many dollars will $2\frac{2}{3}$ dozen cost?

4. If a purchaser buys 440 gallons of oil at 23.4¢ a gallon and 160 gallons of oil at 24.5¢ a gallon, what is his total cost in dollars?

5. The cost of an automobile trip was \$177.75. If expenses other than for gas amounted to \$120, how many gallons of gas were bought at an average cost of 38.5¢ per gallon?

a. 150	*b.* 120	*c.* 144	*d.* 252	*e.* 142.16

3.6 COMBINATION RATES

Model Problem

Soda is sold at the rate of 6 bottles per \$1.00 plus a deposit of 2¢ per bottle. How many bottles of soda can be bought for \$12.32?

(A) 60 (B) 66 (C) 69 (D) 72 (E) 75

Solution:

Every set of 6 bottles costs \$1.00 plus a deposit of \$.12. Hence, each set of 6 costs \$1.12. To find how many sets of 6 bottles can be bought, divide the total cost, \$12.32, by \$1.12. The result is 11 sets, or a total of 66 bottles. *Ans.* (B)

Matching Test

After completing this test, match your answers with the numbers in the box.

1. Soda is sold at the rate of 3 bottles for 25¢ plus a deposit of 2¢ on each bottle. How many bottles can be bought for \$2.48?

2. The cost of printing tickets is 3 for 20¢. Each ticket requires a 5¢ stamp for mailing. How many tickets are mailed if the total cost is \$2.45?

3. Books are sold at the rate of 3 for \$1.00. There is a sales tax of 4¢ per book. How many books were bought if the total cost was \$6.72?

4. Cans of beer are sold at the rate of 3 for 50¢. There is a delivery charge of 5¢ per can. How many cans may be purchased for \$3.25?

5. The purchase price of gifts is 4 for \$7.50. A tax of 25¢ is imposed on the price of each gift. How many gifts were purchased for \$42.50?

a. 18	*b.* 15	*c.* 24	*d.* 20	*e.* 21

3.7 SUCCESSIVE RATES

Model Problem

The rates of a lending library are 8¢ for the first day and 5¢ for each additional day. If the library charge for a book was 78¢, how many days was the book kept out?

(A) 12 (B) 13 (C) 14 (D) 15 (E) 16

Solution:

Deduct 8¢ from the 78¢ for the first day's charge.

Divide the remainder, 70¢, by 5 to determine the number of additional days. Since there are 14 additional days, the book was kept for 14 + 1 or 15 days. *Ans.* (D)

Matching Test

After completing this test, match your answers with the numbers in the box.

1. The library charges are 9¢ for the first day and 5¢ for each additional day. For how many days was a book borrowed if the charge was 89¢?

2. The rate for sending a parcel between two zones is 20¢ for the first pound and 15¢ for each additional pound. How many pounds does a parcel weigh if the shipping charge for the parcel is $1.85?

3. The charge for a telephone call between two cities is 50¢ for the first 3 minutes and 10¢ for each minute thereafter. How many minutes did a call take if the charge for it was $1.20?

4. The charge for sending a telegram is 45¢ for the first ten words, then 4¢ for each additional word. How many words are contained in a telegram that costs 77¢ to send?

5. The charge for a taxi ride is 30¢ for the first quarter-mile, then 10¢ for each additional quarter-mile. How many quarter-miles are there in a trip that costs $1.80?

a. 12	*b.* 16	*c.* 18	*d.* 10	*e.* 17

4. REVIEWING PERCENT AND PERCENTAGE

Reviewing Percent

Percent means hundredths. For example, 3% means 3 hundredths. Note that the percent symbol, %, combines a 1 and two zeros to signify 100.

Interchanging Percents, Decimals, and Fractions

To change a common fraction to a percent, multiply the fraction by 100 and add the % sign. For example, $\frac{3}{4} = \left(\frac{3}{4} \times 100\right)\% = 75\%$. Use the same procedure to change a whole number to a percent. For example, $3 = (3 \times 100)\% = 300\%$.

To change a percent to a common fraction or mixed number, drop the % sign and divide the number by 100. If necessary, reduce the result to lowest terms. For example,

$$4\frac{1}{2}\% = \frac{4\frac{1}{2}}{100} = \frac{9}{2} \times \frac{1}{100} = \frac{9}{200}; \; 250\% = \frac{250}{100} = 2\frac{1}{2}$$

To change a decimal to a percent, multiply by 100 by moving the decimal point two places to the right, and then add the % sign. For example, 1.25 = 125%; .155 = 15.5%.

To change a percent to a decimal, drop the % sign and divide by 100 by moving the decimal point two places to the left. For example, 390% = 3.9; 25.5% = .255.

TABLE OF COMMON FRACTIONS
AND EQUIVALENT PERCENTS

Common Fraction	*Equivalent Percent*	*Common Fraction*	*Equivalent Percent*	*Common Fraction*	*Equivalent Percent*
$\frac{1}{2}$	50%	$\frac{5}{6}$	$83\frac{1}{3}\%$	$\frac{3}{5}$	60%
$\frac{1}{3}$	$33\frac{1}{3}\%$	$\frac{1}{8}$	$12\frac{1}{2}\%$	$\frac{4}{5}$	80%
$\frac{2}{3}$	$66\frac{2}{3}\%$	$\frac{3}{8}$	$37\frac{1}{2}\%$	$\frac{1}{7}$	$14\frac{2}{7}\%$
$\frac{1}{10}$	10%	$\frac{5}{8}$	$62\frac{1}{2}\%$	$\frac{1}{9}$	$11\frac{1}{9}\%$
$\frac{1}{4}$	25%	$\frac{7}{8}$	$87\frac{1}{2}\%$	$\frac{1}{12}$	$8\frac{1}{3}\%$
$\frac{3}{4}$	75%	$\frac{1}{5}$	20%	$\frac{1}{16}$	$6\frac{1}{4}\%$
$\frac{1}{6}$	$16\frac{2}{3}\%$	$\frac{2}{5}$	40%		

Reviewing Percentage

The ***percentage*** is the answer obtained when a percent of a number is taken. The ***rate*** is the percent taken of the number. The ***base*** is the number of which a percent is taken.

Use one of the following three forms of the percentage rule to solve each of the three types of percentage problems:

Percentage Rule

(1) Percentage = Rate × Base (2) $\text{Base} = \frac{\text{Percentage}}{\text{Rate}}$ (3) $\text{Rate} = \frac{\text{Percentage}}{\text{Base}}$

To find the *percentage*, use form (1) of the rule: Percentage = Rate × Base

For example, to find 4% of 125: Percentage = .04 × 125 = 5.00

To find the *base*, use form (2) of the rule: $\text{Base} = \dfrac{\text{Percentage}}{\text{Rate}}$

For example, to find a number if 5% of the number is 24: $\text{Base} = \dfrac{24}{.05} = \dfrac{2400}{5} = 480$

To find the *rate*, use form (3) of the rule: $\text{Rate} = \dfrac{\text{Percentage}}{\text{Base}}$

For example, to find what percent 20 is of 180: $\text{Rate} = \dfrac{20}{180} = \dfrac{1}{9}$

Change the resulting fraction to a percent: $\dfrac{1}{9} = 11\dfrac{1}{9}\%$

Keep in mind. To find what percent one number is of another, first form a fraction having the first number as the numerator and the second number as the denominator. Then change the resulting fraction to a percent.

For example, to find what percent 20 is of 80, first form the fraction $\dfrac{20}{80}$.

$$\frac{20}{80} = \frac{1}{4} = \left(\frac{1}{4} \times 100\right)\% = 25\%$$

4.1 CHANGING A PERCENT TO A COMMON FRACTION

Model Problem

How many thousandths are there in 2.5%?

(A) .025 (B) 2.5 (C) 25 (D) 250 (E) 2500

Solution:

Apply the following rule: To change a percent to a common fraction, drop the % sign and divide the number by 100.

Hence, $2.5\% = \dfrac{2.5}{100}$.

Multiply the numerator and denominator by 10: $\dfrac{2.5\,(10)}{100\,(10)} = \dfrac{25}{1000}$

Since $2.5\% = \dfrac{25}{1000}$, there are 25 thousandths in 2.5%. *Ans.* (C)

Matching Test

After completing this test, match your answers with the numbers in the box.

1. How many thousandths are there in .8%?
2. How many 50ths are there in 12%?
3. What is the ratio of 80% to 16%?
4. How many 12ths are there in 75%?
5. How many 40ths are there in $37\frac{1}{2}\%$?

a. 15
b. 6
c. 9
d. 8
e. 5

4.2 FINDING WHAT PERCENT ONE NUMBER IS OF ANOTHER NUMBER

Model Problem

900 is what percent of 15,000?

(A) .6% (B) 6% (C) 60% (D) $6\frac{2}{3}\%$ (E) $66\frac{2}{3}\%$

Solution:

To find what percent one number is of another, form a fraction in which the first number is the numerator and the second one is the denominator.

The part, or fraction, that 900 is of 15,000 $= \frac{900}{15{,}000} = \frac{3}{50}$.

To change the fraction to a percent, multiply the number by 100 and add the % sign.

Thus, $\frac{3}{50} = 100\left(\frac{3}{50}\right)\%$, or 6%. *Ans.* (B)

(Note how the rule for finding the rate, Rate $= \frac{\text{Percentage}}{\text{Base}}$, was applied.)

Matching Test

After completing this test, match your answers with the percents in the box.

1. 80 is what percent of 32?

2. 8 is what percent of 320?

3. $\frac{1}{4}$ of 8 is what percent of $\frac{1}{2}$ of 32?

4. What percent of 8 is 32?

5. 32% is what percent of 80%?

a. 400%
b. $12\frac{1}{2}\%$
c. 250%
d. $2\frac{1}{2}\%$
e. 40%

4.3 FINDING A PERCENT OF A NUMBER

Model Problem

What is $\frac{1}{2}\%$ of 3600?

(A) 180 (B) 72 (C) 18 (D) 7.2 (E) 1.8

Solution:

To find a number that is some percent of a given number, change the percent to a decimal or common fraction, then multiply the fraction by the given number.

Note that $\frac{1}{2}\% = \frac{1}{2}\left(\frac{1}{100}\right) = \frac{1}{200}$. Hence, $\frac{1}{2}\%$ of $3600 = \left(\frac{1}{200}\right)3600 = 18$. *Ans.* (C)

(Note how the percentage rule, Percentage = Rate × Base, was applied.)

Matching Test

After completing this test, match your answers with the numbers in the box.

1. What is 25% of 60?
2. What is 225% of 60?
3. What is 2.5% of 60?
4. What is $\frac{1}{4}$% of 60?
5. What is .2% of 60?

a. 135
b. 15
c. .12
d. 1.5
e. .15

4.4 FINDING A NUMBER WHEN A PERCENT OF IT IS GIVEN

Model Problem

42 is $37\frac{1}{2}$% of what number?

(A) 112 (B) 28 (C) 15.75 (D) 11.2 (E) 1.12

Solution:

To find a number when a percent of it is given, change the percent to a decimal or common fraction, then divide the number by the fraction.

Since $37\frac{1}{2}\% = \frac{3}{8}$, to find the number, divide 42 by $\frac{3}{8}$.

Thus, the required number $= 42\left(\frac{8}{3}\right) = 112$. *Ans.* (A)

(Note how the rule for the base, Base $= \dfrac{\text{Percentage}}{\text{Rate}}$, was applied.)

Matching Test

After completing this test, match your answers with the numbers in the box.

1. 60 is $83\frac{1}{3}$% of what number?
2. $62\frac{1}{2}$% of what number is 60?
3. What is the number of which 60 is 120%?
4. 60 is 12% of what number?
5. 12 is 60% of what number?

a. 50
b. 500
c. 72
d. 20
e. 96

4.5 COMPARING COMMON FRACTIONS, DECIMALS, AND PERCENTS

Model Problem

Determine which of the following is greatest in value:

(A) $\frac{3}{8}$ (B) 0.3 (C) 37.2% (D) 38% (E) 0.365

Solution:

To find which is the greatest in value, make all the numbers alike in form by changing $\frac{3}{8}$ and the two percents to decimals.

1. To change $\frac{3}{8}$ to a decimal, divide 3 by 8: $8\overline{)3.000}$ = .375 Hence, $\frac{3}{8} = .375$.

2. Since "percent" means "hundredths," $37.2\% = \frac{37.2}{100} = .372$.

3. Similarly, $38\% = \frac{38}{100} = .38$.

The numbers to be compared can now be written as .375, .300, .372, .380, and .365. The greatest of these is .380; hence, 38% is the greatest in value. *Ans.* (D)

Matching Test

In each of the following, select the number having the greatest value. After completing this test, match your answers with the numbers in the box.

1. $\frac{2}{3}$, .687, 69%

2. $\frac{1}{4}$, .2499, $24\frac{1}{2}\%$

3. $\frac{.1}{.3}$, $\frac{.1}{3}$, $\frac{1}{.3}$, 310%

4. $\frac{3}{.1}$, $\frac{.3}{.1}$, .3, 310%

5. $\frac{3}{\frac{1}{2}}$, $\frac{\frac{1}{2}}{3}$, $\frac{3}{2}$, 320%

a. $\frac{1}{.3}$ $\left(\frac{1}{.3} = \frac{10}{3} = 333\frac{1}{3}\%\right)$

b. $\frac{3}{.1}$ $\left(\frac{3}{.1} = 30 = 3000\%\right)$

c. 69%

d. $\frac{3}{\frac{1}{2}}$ $\left(\frac{3}{\frac{1}{2}} = 3 \div \frac{1}{2} = 6 = 600\%\right)$

e. $\frac{1}{4}$

4.6 FINDING PERCENTS OF A NUMBER

Model Problems

(1) A manufacturer found that .2% of his production of stockings was unsuitable for marketing and had to be rejected. Of 3500 manufactured pairs of stockings, how many pairs were rejected?

(A) 3493 (B) 3430 (C) 700 (D) 70 (E) 7

Solution:

Of 3500 pairs, the number of rejected pairs = .2% of 3500

$$= \frac{.2}{100}(3500) = .2(35)$$

$$= 7 \quad \textit{Ans.}\ (E)$$

(2) A family with an income of $7500 spent 38% of the amount for rent and 45% for food. How many dollars remain for other purposes?

(A) 127.50 (B) 1275 (C) 202.50 (D) 2025 (E) 6225

Solution:

The percent of the income spent for rent and food is 38% + 45% = 83%.

The percent of the income that remains is 100% − 83% = 17%.

Hence, the number of dollars remaining is 17% × 7500 = 1275. *Ans.* (B)

After completing each test, match your answers with the numbers in the boxes.

Matching Test A

1. How many pairs of stockings were rejected in a load of 6500 pairs if .8% were not of satisfactory quality?

2. If the charge for obtaining travelers checks is $\frac{1}{2}$%, how many dollars must be paid to obtain $15,000 worth of these checks?

3. To the nearest cent, what is the number of cents in a 4% sales tax on 20 gallons of gasoline at 38.5¢ per gallon?

4. When traveling by plane, there is an overweight charge on baggage that is in excess of 40 pounds. If this charge is 1% of the one-way plane fare for each pound of baggage over 40 pounds, how many additional dollars must be paid when the fare is $80 and the baggage weighs 50 pounds?

5. A realtor who had bought two houses for $10,600 apiece sold one of them at a profit of 4% and the other at a loss of $2\frac{1}{2}$%. How many dollars of net profit did he make from the sale of both houses?

a. 75	*b.* 159	*c.* 52	*d.* 31	*e.* 8

Matching Test B

1. A family's income is $6000, of which 40% is spent for food and 27% for rent. How many dollars remain for other purposes?

2. In a city, the percent of deaths from cancer dropped from 24% to 13% of all cancer cases. In every 5000 cases, how many fewer deaths from cancer were there?

3. How many dollars will Mr. Abel save in a year if his earnings amount to $150 a week and he spends 95% of his earnings?

4. How many seniors are there in a school having 2450 students if 82% of them are not seniors?

5. A merchant purchased 40 radio sets for $7600. In order to make a profit of 15%, for how many dollars must each set be sold?

a. 218.50	*b.* 550	*c.* 390	*d.* 1980	*e.* 441

4.7 FINDING THE PERCENT OF INCREASE OR DECREASE

Model Problem

In five years the enrollment of a certain junior high school increased from 800 to 1400 pupils. What was the percent of increase in the school's enrollment during this period?

(A) $42\frac{6}{7}\%$ (B) 60% (C) 80% (D) 75% (E) 175%

Solution:

Find the percent of increase as follows:

1. Determine the increase. Since the enrollment increased from 800 to 1400 pupils, the increase was 600 pupils.

2. Divide the increase by the original enrollment.

$$\frac{600 \text{ pupils}}{800 \text{ pupils}} = \frac{3}{4}$$

3. Change the fraction $\frac{3}{4}$ to a percent. Since $\frac{3}{4} = \frac{75}{100}$, then $\frac{3}{4} = 75\%$. *Ans.* (D)

Matching Test

After completing this test, match your answers with the percents in the box.

1. Find the percent of increase in a salary that has changed from $5000 to $6750 per year.

2. Find the percent of increase in the population of a town if the population has grown from 2000 people to 10,000 people.

3. Find the percent of decrease of a stock that has dropped from $130 per share to $104 per share.

4. Find the percent of increase of the price of pads if the change in price was from 75¢ a dozen to 2 for 25¢.

5. Find the percent of profit made on a used car that was sold for $630 if the dealer bought it for $300 and then spent $150 on repairs.

a. 40%	*b.* 400%	*c.* 20%	*d.* 35%	*e.* 100%

4.8 PROBLEMS WITH COMMON FRACTIONS AND PERCENTS

Model Problems

(1) A major-league baseball team has lost 48 games out of 80 games played. During the season, a total of 162 games are to be played. If the team is to end the season winning 50% of its games, how many of the remaining games must it win?

(A) 33 (B) 39 (C) 49 (D) 59 (E) 81

Solution:

Thus far, the team has won (80−48) or 32 games.

To win 50%, or $\frac{1}{2}$, of the total of 162 games, the team must win 81 games by the end of the season.

Hence, the number of games to be won = (81 − 32) or 49 games. *Ans.* (C)

(2) In a city, it was found that 9 out of 10 of those who graduate from elementary school enter high school; of those who enter high school, 4 out of 5 graduate from high school; and of these graduates, 3 out of 4 enter college. What percent of those who graduate from elementary school enter college?

(A) 54% (B) 50% (C) 46% (D) 27% (E) 18%

Solution:

Nine-tenths of the elementary-school graduates enter high school and four-fifths of these graduate from high school. Hence, the fraction of elementary-school graduates who also graduate from high school is $\frac{4}{5} \times \frac{9}{10}$, or $\frac{18}{25}$.

Three-fourths of the high-school graduates enter college. Therefore, the fraction of elementary-school graduates who enter college is $\frac{3}{4} \times \frac{18}{25}$, or $\frac{54}{100}$. In terms of percent, $\frac{54}{100} = 54\%$. *Ans.* (A)

After completing each test, match your answers with the answers in the boxes.

Matching Test A

1. A team has won 60% of its first 25 games. If the team is to play 15 more games, how many more games must it win to end the season by winning 50% of all the games?
2. A team has lost 34 of its first 50 games. If it wins 20 of the next 30 games, what percent of the total number of games will it have won?
3. A team has thus far won 15 games and lost 20. If it wins only one-third of the next 15 games, what percent of *all* the games will it have won?
4. A student has done only 5 of 15 questions correctly. How many of the remaining 35 questions must he answer correctly to obtain a grade of 60%?
5. In a test containing 60 questions, John scored 60% on the first half of the test. How many more questions must he answer correctly to end with a grade of 75%?

a. 27	*b.* 45%	*c.* 25	*d.* 40%	*e.* 5

Matching Test B

1. In a town, it is found that 4 out of 5 graduates from high school enter college; and of those who enter college, 3 out of 8 graduate from college. What percent of those who graduate from high school graduate from college?

2. If 2 out of 5 people in a community develop a certain virus fever and of these 1 out of 10 is hospitalized, what percent of the community is hospitalized?

3. If Mr. Hale spends $\frac{1}{4}$ of his salary for food and $\frac{1}{6}$ of the remainder for rent, what percent of his salary is left for all other purposes?

4. In two classes of equal size, it is found that $\frac{1}{2}$ of the first class are boys while only $\frac{1}{4}$ of the second class are boys. What percent of both classes are boys?

5. During a television program of 30 minutes, 10 minutes were spent for commercial advertising. An actor on the program acted for $\frac{1}{2}$ the remaining time. What percent of the total time of the program did the actor occupy?

a. 4%	*b.* $37\frac{1}{2}\%$	*c.* 30%	*d.* $33\frac{1}{3}\%$	*e.* $62\frac{1}{2}\%$

4.9 PROBLEMS WITH RATIOS AND PERCENTS

Model Problems

(1) A man purchased a building lot for \$3000 and built a \$12,000 house on it. What percent of the total cost was the cost of the lot?

(A) 20% (B) 25% (C) $33\frac{1}{3}\%$ (D) 75% (E) 80%

Solution:

Find the required percent as follows:

1. The total cost of the lot and the house = \$3000 + \$12,000 = \$15,000.
2. Since the lot cost \$3000 out of the total cost of \$15,000, then the cost of the lot $= \frac{3000}{15000}$ or $\frac{1}{5}$ of the total cost.
3. Since $\frac{1}{5} = \frac{20}{100} = 20\%$, the cost of the lot = 20% of the total cost. *Ans.* (A)

(2) There are 640 pupils enrolled in a certain school. One day 32 pupils were absent. What percent of the pupils were present that day?

(A) 95% (B) 96% (C) 90% (D) 5% (E) 4%

Solution:

Find the percent of pupils present as follows:

1. The 32 absent pupils constitute $\frac{32}{640}$ of the entire school population. By reduction, this fraction is equivalent to $\frac{1}{20}$, or 5%. Thus, 5% of the pupils were absent.

2. Since all the pupils constitute 100% of the enrollment, if 5% were absent, then 100% − 5%, or 95% were present. *Ans.* (A)

(3) An article is sold at 25% above cost. What is the percent of profit based on the selling price?

(A) 20% (B) 25% (C) 30% (D) $33\frac{1}{3}$% (E) 50%

Solution:

The cost in terms of percent may be considered as 100%.

Since the selling price is the sum of the cost and the profit, the selling price = 100% + 25%, or 125%.

To find the percent of profit based on the selling price, find the ratio of the profit to the selling price as follows: $\frac{25\%}{125\%} = \frac{1}{5}$.

Expressed in percent, the ratio $\frac{1}{5} = 20\%$. *Ans.* (A)

After completing each test, match your answers with the answers in the boxes.

Matching Test A

1. In voting on a motion, 30 members of a group voted for the motion and 20 voted against it. Find the percent of the votes in favor of the motion.

2. In a mixture of alcohol and water, there are $7\frac{1}{2}$ quarts of water and 5 quarts of alcohol. Find the percent of alcohol in the mixture.

3. One of the teams in a basketball game found that its players had made 20 successful shots and 60 misses. Find the percent of successful shots made by the team.

4. A baseball team won 8 of the first 20 games and then won the next 5. Find the percent of games won.

5. In a class of 22 students, 9 are boys. If 10 students were added to the class including 3 girls, what percent of the enlarged class are boys?

a. 52%	*b.* 25%	*c.* 50%	*d.* 60%	*e.* 40%

Matching Test B

1. An article is sold at 25% below cost. What is the percent of loss based on the selling price?
2. The price of a suit was reduced by 20%. To restore the original price, what percent of increase must be made based on the reduced price?
3. A company allocates 60% of every sales dollar to buying cost, 8% to profit, and 15% to advertising cost. What percent of the sales dollar is allocated for other expenses?
4. 60% of the students in a senior class are expected to graduate from high school, and of these, 15% are expected to graduate from college. What percent of the class are expected to graduate from high school but not from college?
5. A gold alloy of a rock is estimated at 40% of the rock composition. The pure gold contained in the alloy is estimated at 10% of the alloy. What percent of the rock is estimated to be pure gold?

a. 17%	*b.* 51%	*c.* 25%	*d.* 4%	*e.* $33\frac{1}{3}\%$

4.10 SUCCESSIVE PERCENT CHANGES

Model Problem

Mr. Light's salary was increased by 20% one year and then decreased by 20% the following year. If he originally earned $10,000 a year, what has been the net change in his salary?

(A) decrease of $400 (B) decrease of $40 (C) no change (D) increase of $40
(E) increase of $400

Solution:

Method 1: Using Each Percent Separately

After the 20% increase in salary, Mr. Light earned 120% of $10,000, or $12,000.

The following year, due to the decrease of 20%, Mr. Light earned 80% of his new salary, that is, 80% of $12,000, or $9600.

If the final salary, $9600, is compared with the original salary, $10,000, it can be seen that there is a net decrease of $400.

Method 2: Multiplying the Percents Together

Because of the 20% increase, the second salary was 120% of the original salary.

Because of the 20% decrease, the final salary was 80% of the second salary. Hence, *the final salary was 80% of 120% of the original salary.* Since 80%(120%) = 96%, the final salary is 4% less than the original salary of $10,000.

Therefore, the net change is a decrease of 4% of $10,000, or $400.

Ans. (A)

Matching Test

After completing this test, match your answers with the numbers in the box.

1. A man's salary of \$5000 is first increased 10% and then decreased 10%. What is the number of dollars in the net change of salary resulting from the two changes?

2. Because of the introduction of automation, 20% of the workers in a certain plant were dismissed, while each remaining worker's output was increased by 30%. What is the net change in the original production of 2000 items?

3. A town's population of 4000 first increased by 25% and then increased by 20%. What is the net change in the population due to the two increases?

4. Taxes on a piece of property were increased 10% per year for each of three successive years. If the original yearly tax was \$4000, how many dollars was the net change in the taxes for the three-year period?

5. After selling 2500 cars, an agency increased its sales by 20% the following year and by 15% the year after that. What is the net change in the number of cars sold for the two years?

a. +950	*b.* +80	*c.* +2000	*d.* −50	*e.* +1324

4.11 SUCCESSIVE DISCOUNT AND PERCENT CHANGES

Model Problem

How much is saved in buying an article listed at \$50 if the buyer receives a single discount of 40% rather than successive discounts of 20% and 20%?

(A) \$12 (B) \$2 (C) \$1 (D) \$.20 (E) no saving

Solution:

1. *Single discount of 40%*

   In this case, the buyer pays 100% − 40%, or 60%, of the list price. Hence, his cost is 60% × \$50 = \$30.

2. *Successive discounts of 20% and 20%*

   In this case, the buyer pays 80% of the list price because of the first discount of 20%. Because of the second discount of 20%, the buyer pays 80% of the reduced price; that is, the buyer pays 80% of 80% of the original list price. Since 80% × 80% = 64%, the buyer pays 64% × \$50 = \$32.

   Hence, there is a saving of \$32 − \$30, or \$2. *Ans.* (B)

Alternate Method

1. *Single discount of 40%*

   In this case, the buyer pays 60% of the list price.

2. *Successive discounts of 20% and 20%*

   In this case, the buyer pays 80% of 80%, or 64%, of the list price.

   Hence, the buyer saves 64% − 60%, or 4%, of the list price.

   Since the list price is \$50, the buyer saves 4% of \$50, or \$2.

Matching Test

After completing this test, match your answers with the amounts in the box.

1. How much is saved in buying an article priced at $100 less 40%, rather than $100 less 20% and 20%?
2. How much is saved in buying an article priced at $200 less 20%, rather than $200 less 10% and 10%?
3. How much is saved in buying an article priced at $50 less 50%, rather than $50 less 20% and 30%?
4. A dealer pays $150 for a radio set. After marking the set at 20% above cost, he gives a discount of 20% on the selling price. How much did he lose?
5. A buyer of a TV set listed at $100 is allowed a discount of 20% but must pay a sales tax of 5% on the reduced price. How much more would he have paid if there were a discount of 15% without a sales tax?

a. $3	*b.* $6	*c.* $2	*d.* $4	*e.* $1

5. TYPES OF PERCENT PROBLEMS

5.1 TAXES

Model Problem

In a certain state, a taxpayer may deduct $2000 from his taxable income; of the remainder, he must pay 2% of the first $2500, 3% of the next $2500, and 4% of any amount thereafter. How much does a taxpayer pay on a taxable income of $8500?

(A) $145 (B) $185 (C) $195 (D) $200 (E) $225

Solution:

On the first $2000, there is no tax.

On the income between $2000 and $4500, the tax is 2% of $2500, or $50.

On the income between $4500 and $7000, the tax is 3% of $2500, or $75.

On the income between $7000 and $8500, the tax is 4% of $1500, or $60.

Hence, the total tax is $50 + $75 + $60 = $185. *Ans.* (B)

Matching Test

After completing this test, match your answers with the amounts in the box.

1. In the preceding model problem, what would the tax be if the income were $9500?
2. What is the amount of the social security tax paid in a 10-week period on a salary of $96 a week, if the tax is $3\frac{1}{4}$% of the salary?
3. What is the school tax on property assessed at $24,500 if the tax rate is $15.50 per $1000?

4. A company raised the salaries of its employees by granting a raise of 15% on the first \$3000 of salary, 10% on the next \$3000, and 5% on any part of the salary thereafter. What was the raise for a salary of \$8000?

5. What is the electricity bill for 850 kilowatt-hours of electricity if the charge is 7¢ per kilowatt-hour for the first 200 hours, 6¢ per kilowatt-hour for the next 300 hours, and 5¢ per kilowatt-hour thereafter?

a. \$850	*b.* \$379.75	*c.* \$225	*d.* \$31.20	*e.* \$49.50

5.2 PROFIT AND LOSS

Model Problem

A clothing merchant bought 60 suits for \$2655. He sold 45% of them at \$52 each and the rest at \$55 each. Find his net profit from the suits.

(A) \$619 (B) \$509 (C) \$460 (D) \$616 (E) \$564

Solution:

Since 45% of 60 is 27, the merchant sold 27 suits at \$52 apiece for \$1404.

The remaining 33 suits were sold at \$55 each for \$1815.

Thus, his total sales were the sum of \$1404 and \$1815, or \$3219.

Hence, the profit from the 60 suits equalled the difference between the total sales, \$3219, and the cost, \$2655, a difference of \$564. *Ans.* (E)

Matching Test

After completing this test, match your answers with the amounts in the box.

1. A clothing merchant bought 75 suits for \$3624. If he sold 40% of these at \$50 each and the rest at \$54 each, what was his net profit on the suits?

2. A bracelet of precious stones was bought for \$6000 and sold for \$9330. If the bracelet contained 45 carats, how much was the profit per carat?

3. A dealer mixed 10 pounds of tea costing 95¢ a pound with 12 pounds of tea costing \$1.25 a pound. If the mixture was sold for \$1.40 a pound, what was his total profit?

4. A dealer sold 20 pounds of candy at a profit of \$.28 a pound, 30 pounds of another grade of candy at a profit of \$.32 a pound, and 25 pounds of another kind at a loss of \$.12 a pound. What was his net gain?

5. A grocer's profit from the sale of $20\frac{1}{2}$ dozen eggs was \$2.20. If the eggs were sold at \$.48 a dozen, what did the eggs cost him?

a. \$12.20	*b.* \$74	*c.* \$7.64	*d.* \$306	*e.* \$6.30

5.3 COMMISSIONS AND INSURANCE

Model Problem

In addition to a salary of \$100 per week, a salesman is paid a commission of 3% on all sales over \$1000. During a certain week, his sales were \$1500. How much did he earn that week?

(A) \$101.50 (B) \$115 (C) \$125 (D) \$130 (E) \$145

Solution:

Since he receives 3% of the sales over \$1000, his commission was 3% of (\$1500 − \$1000); that is, his commission = .03(\$500), or \$15.

His earnings that week were \$115: the sum of his salary, \$100, and his commission, \$15. *Ans.* (B)

Matching Test

After completing this test, match your answers with the answers in the box.

1. What does a salesman earn during a week if, in addition to a salary of \$100, he receives a commission of 5% of sales over \$500 and his sales for that week are \$700?

2. A salesman earns a commission of 4% on all sales between \$200 and \$500 and a commission of 10% on all sales over \$500. What is the amount of his commission if his sales are \$850?

3. A vegetable merchant receives a commission of 8% for selling a farmer's tomatoes. If the merchant sells 400 boxes of tomatoes at \$6.50 a box, what is the amount of his commission?

4. What is the insurance premium on a house worth \$20,000, if it is insured for 80% of its value at \$.35 per \$100?

5. A man drives his car 15,000 miles during a certain year and estimates that the cost of insuring the car for that year was \$1.16 per 100 miles. How much had he paid for insurance?

a. \$208	*b.* \$56	*c.* \$47	*d.* \$110	*e.* \$174

5.4 INSTALLMENT BUYING

Model Problem

A home buyer paid \$20,000 for a home. He agreed to make a down payment of \$4000, to pay \$1500 at the end of each year and an interest of 6% on any unpaid balance. What will be his payment at the end of the first year?

(A) \$1596 (B) \$1620 (C) \$1740 (D) \$2460 (E) \$2700

Solution:

His unpaid balance is \$16,000, the difference between \$20,000 and the down payment of \$4000.

The interest charge on the unpaid balance will be 6% of \$16,000, or \$960.

Hence, at the end of the first year, the payment must be \$1500 plus the interest charge of \$960 for a total of \$2460. *Ans.* (D)

Matching Test

After completing this test, match your answers with the amounts in the box.

1. If a home buyer purchases a home for \$12,000, makes a down payment of \$2000, and agrees to pay \$1000 at the end of the year with interest at 6% on any unpaid balance, what will be his payment at the end of the first year?

2. The cash price of a TV set was \$250. The purchaser made a down payment of \$50 and agreed to pay \$15 in each of 16 equal installments. What was his carrying charge?

3. A refrigerator is bought on the installment plan for \$40 down and 20 monthly installments at \$10 each. A cash purchase would have been 15% less than the total paid on the installment plan. How much would have been saved by paying cash?

4. The purchaser of a used car sold for \$1400 makes a down payment of \$200 and agrees to pay the balance in 16 monthly installments. How much is each of the monthly installments?

5. A purchaser of a new car was given an allowance of \$850 on his old car. He agreed to pay \$125 per month for 24 months for the new car. What was the total cost of the new car?

a. \$75	*b.* \$36	*c.* \$3850	*d.* \$1600	*e.* \$40

6. REVIEWING COMMON UNITS OF MEASURE

Length Measures	*Time Measures*
12 inches (in.) = 1 foot (ft.)	60 seconds (sec.) = 1 minute (min.)
3 feet = 1 yard (yd.)	60 minutes = 1 hour (hr.)
36 inches = 1 yard	24 hours = 1 day
$16\frac{1}{2}$ feet = 1 rod (rd.)	7 days = 1 week
$5\frac{1}{2}$ yards = 1 rod	*Approximate Equivalents*
1760 yards = 1 mile (mi.)	30 days = 1 month
5280 feet = 1 mile	52 weeks = 1 year
6080 feet = 1 nautical mile	12 months = 1 year
	100 years = 1 century

Dry Measures	*Liquid Measures*
2 pints = 1 quart	4 gills (gi.) = 1 pint (pt.)
8 quarts = 1 peck (pk.)	16 ounces = 1 pint
4 pecks = 1 bushel (bu.)	2 pints = 1 quart (qt.)
32 quarts = 1 bushel	4 quarts = 1 gallon (gal.)
	8 pints = 1 gallon

Metric System Measures

Length:	10 millimeters (mm.) = 1 centimeter (cm.)
	100 centimeters (cm.) = 1 meter (m.)
	1000 meters (m.) = 1 kilometer (km.)
Weight:	1000 milligrams (mg.) = 1 gram (g.)
	1000 grams (g.) = 1 kilogram (kg.)
Capacity:	1000 milliliters (ml.) = 1 liter (l.) ≈ 1.056 quarts

Approximate Equivalents	*Area Measures*
2.54 centimeters = 1 inch	144 square inches (sq. in.) = 1 square foot
1 meter = 39.37 inches	9 square feet = 1 square yard
1 kilometer = 0.62 mile	160 square rods = 1 acre
1 gram = 0.035 ounce	640 acres = 1 square mile
1 kilogram = 2.2 pounds	

Weight Measures	*Volume Measures*
16 ounces (oz.) = 1 pound (lb.)	1728 cubic inches (cu. in.) = 1 cubic foot
2000 pounds = 1 ton	27 cubic feet = 1 cubic yard
2240 pounds = 1 long ton	

Fundamental Operations on Common Units of Measure

Converting Common Units of Measure

To convert from a larger unit of measure to a smaller one, determine the number of units of the smaller unit of measure needed to make one unit of the larger unit. Then *multiply* the larger measure by this number.

For example, 12 inches are needed to make 1 ft. Hence, $3\frac{1}{2}$ feet $= 12\left(3\frac{1}{2}\right)$, or 42 inches.

To convert from a smaller unit of measure to a larger one, determine the number of units of the smaller unit of measure needed to make one unit of the larger unit. Then *divide* the larger measure by this number.

For example, 12 inches are needed to make 1 foot. Hence, 42 inches $= \frac{42}{12}$, or $3\frac{1}{2}$ feet.

Adding Common Units of Measure

Arrange the units in columns, one column for each unit. Then add the columns. Convert a smaller unit to a larger unit if possible.

Add 1 yard 2 feet 5 inches and 1 yard 1 foot 4 inches.

	1 yd.	2 ft.	5 in.	
Add:	1 yd.	1 ft.	4 in.	
	2 yd.	3 ft.	9 in.	= 3 yd. 9 in. (Change 3 ft. to 1 yd.)

Ans. 3 yd. 9 in.

Subtracting Common Units of Measure

Arrange the units in columns, one column for each unit. Then subtract the columns. Borrow when the number of units in the minuend is smaller than that in the subtrahend.

> Subtract 4 pounds 8 ounces from 9 pounds 2 ounces.
>
> | | | | |
> |---|---|---|---|
> | | 9 lb. 2 oz. | 8 lb. 18 oz. | (Borrow: 1 lb. = 16 oz.) |
> | Subtract: | 4 lb. 8 oz. | 4 lb. 8 oz. | |
> | | | 4 lb. 10 oz. | |
>
> *Ans.* 4 lb. 10 oz.

Multiplying Common Units of Measure by a Number

Multiply each number of units by the multiplier. Convert a smaller unit to a larger if possible.

> Multiply 2 minutes 20 seconds by 5.
>
> | | | |
> |---|---|---|
> | 2 min. | 20 sec. | |
> | | 5 | |
> | 10 min. | 100 sec. = 11 min. 40 sec. | (Change 60 sec. to 1 min.) |
>
> *Ans.* 11 min. 40 sec.

Dividing Common Units of Measure by a Number

Divide each number of units by the divisor, beginning with the largest unit. Whenever there is a remainder, combine this with the given number of units of the next smaller unit before dividing.

> Divide 7 hours 20 minutes by 5.
>
> $$\frac{7 \text{ hr. } 20 \text{ min.}}{5} = \frac{5 \text{ hr. } 140 \text{ min.}}{5} = 1 \text{ hr. } 28 \text{ min.}$$
>
> Note that 7 hr. divided by 5 results in a remainder of 2 hours. This was converted into 120 minutes and combined with the given 20 minutes to make a total of 140 minutes.
>
> *Ans.* 1 hr. 28 min.

6.1 OPERATIONS ON COMMON UNITS

Matching Test

After completing this test, match your answers with the sets of measures in the box.

1. Add 1 lb. 6 oz., 2 lb. 7 oz., and 3 lb. 8 oz.

2. From 4 yd. 1 ft. 7 in. subtract 1 yd. 2 ft. 10 in.

3. Multiply 3 hr. 15 min. by 10.

4. Divide 1 yd. 2 ft. 6 in. by 3.

5. From the sum of 3 lb. 5 oz. and 4 lb. 12 oz., subtract 2 lb. 6 oz.

a. 1 ft. 10 in.
b. 7 lb. 5 oz.
c. 5 lb. 11 oz.
d. 2 yd. 1 ft. 9 in.
e. 1 day 8 hr. 30 min.

6.2 LENGTH UNITS

Model Problem

There are 5280 feet in a statute mile and 6080 feet in a nautical mile. A knot is a speed of 1 nautical mile per hour. A boat is sailing at 10 knots while a bicyclist on shore is traveling at 10 statute miles per hour. In one hour, how many feet farther than the bicyclist does the boat travel?

(A) 800 (B) 6000 (C) 7000 (D) 8000 (E) 80,000

Solution:

Since 10 knots is 10 nautical miles per hour, the boat covers 10(6080) or 60,800 feet in one hour. The bicycle rider in one hour covers 10(5280) feet, or 52,800 feet. Hence, the boat outdistances the rider by (60,800 – 52,800) feet or 8000 feet. *Ans.* (D)

Matching Test

After completing this test, match your answers with the numbers in the box.

1. To the nearest tenth of a mile, how many statute miles are there in 10 nautical miles? (6080 feet = 1 nautical mile, 5280 feet = 1 statute mile.)

2. How many steel plates $\frac{3}{16}$ of an inch thick are there in a pile of these plates, if the pile is a yard high?

3. How many dollars do 6 curtains cost if each curtain requires 2 yards 2 feet 6 inches of material and the price of the material is $2.50 per yard?

4. Three pieces each $7\frac{1}{2}$ inches long are cut from a moulding that is 1 yard 4 inches. How many inches of the moulding are left?

5. To the nearest centimeter, by how many centimeters does 1 meter exceed 30 inches? (1 inch = 2.54 centimeters.)

a. $42\frac{1}{2}$ *b.* $17\frac{1}{2}$ *c.* 192 *d.* 11.5 *e.* 24

6.3 TIME UNITS

Model Problem

A man's watch is losing time at the rate of $\frac{1}{2}$ minute each hour. If he sets his watch correctly at 10 A.M. and makes no further corrections thereafter, what time will the watch read at 10 A.M. the following day?

(A) 9:48 (B) 9:50 (C) 9:54 (D) 10:06 (E) 10:12

Solution:

Since the watch loses $\frac{1}{2}$ minute every hour, it will lose 1 minute every 2 hours. There is an interval of 24 hours between 10 A.M. of one day and 10 A.M. of the next day. Hence, the watch will lose 12 minutes.

If 12 minutes is subtracted from 10 o'clock, the watch will read 9:48 o'clock. ("O'clock" means "on the clock." 9:48 o'clock is the reading on the clock for either 9:48 A.M. or 9:48 P.M.) *Ans.* (A)

Matching Test

After completing this test, match your answers with the times in the box.

1. If a watch is set at 3 P.M. and is losing 1 minute every 4 hours, what time will the watch read at 11 P.M. the following day?

2. At what time must you go to bed if you wish to have $7\frac{1}{2}$ hours of sleep and arise at 6:45 A.M.?

3. Each performance at a movie house takes 2 hours and 12 minutes. If the first performance begins at 12:30 P.M., when will the fourth performance end?

4. A clock gains 5 minutes every 2 days. If it was set correctly at noon on May 1 and no further corrections were made thereafter, what time did the clock read at noon 30 days later?

5. A train scheduled to arrive at 4:15 P.M. was delayed twice, once for 1 hour and 10 minutes and the second time for 55 minutes. At what time did the train actually arrive?

a. 6:20 P.M.	*b.* 9:18 P.M.	*c.* 1:15 o'clock	*d.* 10:52 o'clock	*e.* 11:15 P.M.

6.4 LIQUID UNITS

Model Problem

How many 3-gallon cans of milk are needed to fill 480 half-pint containers?

(A) 30 (B) 20 (C) 16 (D) 10 (E) 5

Solution:

Since 1 gallon = 8 pints, then 1 gallon can be used to fill 16 half-pints.

Hence, each 3-gallon can will fill 3(16) or 48 half-pint containers.

To fill 480 half-pint containers, the number of 3-gallon cans needed is $\frac{480}{48}$, or 10. *Ans.* (D)

Matching Test

After completing this test, match your answers with the numbers in the box.

1. How many 2-gallon cans of juice are needed for 640 pint bottles if one-quarter of each bottle is to be filled with the juice?
2. 16 liquid ounces = 1 pint. To the nearest ounce, how many liquid ounces are in .95 quarts?
3. 1 British gallon is approximately equal to 1.2 American gallons. How many British gallons are there in 44.4 American gallons?
4. 1 liter = 1.056 quarts. By how many quarts does 500 liters exceed 500 quarts?
5. 1 bushel = 4 pecks and 1 peck = 8 quarts. What percent of one-half a bushel is 4 quarts?

a. 25	*b.* 30	*c.* 28	*d.* 37	*e.* 10

6.5 MISCELLANEOUS UNITS

Model Problem

1 kilometer is approximately equal to 3280 feet. To the nearest foot per second, what is the equivalent in feet per second of 100 kilometers per hour?

(A) 80 (B) 90 (C) 91 (D) 95 (E) 5467

Solution:

A rate of 100 kilometers per hour = 100(3280) or 328,000 feet per hour.

Since there are 3600 seconds in 1 hour, the number of feet per second $= \frac{328{,}000}{3600} = 91$, to the nearest foot per second. *Ans.* (C)

Matching Test

After completing this test, match your answers with the numbers in the box.

1. To the nearest foot per second, what is the equivalent of 36 kilometers per hour in feet per second?
2. 1 furlong = 220 yards. How many furlongs are there in $1\frac{1}{2}$ miles?
3. At one time, 600 Italian lire were worth 1 American dollar. At this rate, how many cents was a tip of 180 lire?
4. 10 avoirdupois ounces = 283.5 grams, and 10 troy ounces = 311 grams. By how many grams do 20 troy ounces exceed 20 avoirdupois ounces?
5. 8 cartons of soap weigh $56\frac{1}{2}$ pounds when full. If each carton weighs 7 ounces when empty, how many pounds does the soap weigh?

a. 30	*b.* 55	*c.* 33	*d.* 53	*e.* 12

7. MAXIMUM-MINIMUM PROBLEMS

7.1 SELECTION PROBLEMS WITH A MINIMUM OF TRIES

Model Problem

In a bag, there are 3 red, 4 blue, 5 brown, and 6 white marbles. If only one marble at a time can be drawn from the bag, without looking, what is the least number of draws one must make to be sure of having at least 3 marbles of any one color?

(A) 5 (B) 8 (C) 9 (D) 11 (E) 13

Solution:

On each draw, the marble may have any one of 4 colors. When the 8th marble has been drawn, it is still possible to have obtained only 2 marbles of each color. However, when the 9th marble is drawn, whatever its color, there must now be at least 3 of one color.

Ans. (C)

Matching Test

After completing this test, match your answers with the numbers in the box.

1. In the model problem, what is the least number of draws that must be made, without looking, to be certain that the drawer will have at least 2 marbles of one color?

2. There are 25 students in a class. What is the smallest number of students that can be chosen at random from this class to form a group that is sure to contain at least two students whose birthdays are in the same month?

3. A choir of freshmen and sophomores is to be formed by selecting students at random from a group of 25 freshmen and 25 sophomores. What is the smallest number of students that can be chosen at random from the group if one wants to be certain that the choir will contain at least 5 members of the same class?

4. Cards are removed from a deck of cards so that the remaining pack contains 4 number cards for each number from 1 to 10, that is, four 1's, four 2's, four 3's, and so on. If one card at a time is drawn from the pack at random, what is the smallest number of cards that must be drawn to be certain to obtain at least 3 cards having the same number?

5. A candy jar contains 100 candies made in 5 flavors. There are equal numbers of candies of each flavor, so that 20 candies are of one flavor, 20 candies are of a second flavor, etc. If one candy at a time is taken from the jar, without looking, what is the least number of candies that must be drawn to be sure that at least 3 candies are obtained that will have the same flavor?

a. 13	*b.* 11	*c.* 9	*d.* 21	*e.* 5

7.2 FINDING MAXIMUM AND MINIMUM RATIOS

Model Problem

The Park Camp can accommodate from 80 to 100 children and from 50 to 60 adults. Find the maximum and the minimum ratios of children to adults that can be accommodated.

(A) $\frac{8}{5}, \frac{5}{3}$ (B) $\frac{8}{5}, \frac{4}{3}$ (C) $\frac{2}{1}, \frac{8}{5}$ (D) $\frac{2}{1}, \frac{5}{3}$ (E) $\frac{2}{1}, \frac{4}{3}$

Solution:

To find a maximum ratio means to find the largest fraction that can be formed from a set of numerators and denominators. To form this fraction, select the largest possible numerator and the smallest possible denominator.

Thus, to obtain the maximum ratio of children to adults, divide the maximum number of children by the minimum number of adults. Hence, the maximum ratio $= \frac{100}{50} = \frac{2}{1}$.

Similarly, to find a minimum ratio, form a fraction by selecting the smallest possible numerator and the largest possible denominator.

Thus, to obtain the minimum ratio of children to adults, divide the minimum number of children by the maximum number of adults. Hence, the minimum ratio $= \frac{80}{60} = \frac{4}{3}$. *Ans.* (E)

Matching Test

After completing this test, match your answers with the answers in the box.

1. The Family Hotel can accommodate from 50 to 65 adults and from 20 to 30 children. Find the maximum and minimum ratios of adults to children.
2. The weight of a bushel of tomatoes varies from 30 to 45 pounds, while the weight of a bushel of apples varies from 50 to 60 pounds. Find the maximum and minimum ratios of the weight of a bushel of tomatoes to that of a bushel of apples.
3. During a certain period, the number of passengers per train varies from 80 to 96. When the passengers leave the train, the rate of exit varies from 6 to 10 passengers per second. Find the maximum and minimum number of seconds it takes to empty the train.
4. The weights of certain trucks may range from 1 ton to 5 tons, while the weights of certain passenger cars may range from 1500 pounds to 5000 pounds. Find the maximum and minimum ratios of the weight of a truck to that of a passenger car.
5. In a certain city, class registers vary from 20 to 35 in district *A*, while class registers vary from 30 to 40 in district *B*. Find the maximum and minimum ratios of the register of a school in district *A* to that of a school in district *B*.

a. $\frac{20}{3}, \frac{2}{5}$	*b.* $\frac{7}{6}, \frac{1}{2}$	*c.* $\frac{13}{4}, \frac{5}{3}$	*d.* $\frac{9}{10}, \frac{1}{2}$	*e.* 16, 8

7.3 FINDING MAXIMUM AND MINIMUM NUMBERS

Model Problem

Oranges sell from 6¢ to 10¢ apiece. Grapefruits sell from 8¢ to $12\frac{1}{2}$¢ apiece. If 6 oranges and 4 grapefruits are bought, find the average maximum price and the average minimum price per piece.

(A) 12.5¢, 6¢ (B) 11.5¢, 6¢ (C) 11¢, 6.8¢ (D) 11¢, 6¢ (E) 10.25¢, 8¢

Solution:

Using the maximum prices, the maximum cost of 6 oranges at 10¢ apiece and 4 grapefruits at $12\frac{1}{2}$¢ apiece is 60¢ + 50¢, or 110¢. Hence, the average maximum price for the 10 pieces of fruit is $\frac{110¢}{10}$, or 11¢ per piece.

Using the minimum prices, the minimum cost of 6 oranges at 6¢ apiece and 4 grapefruits at 8¢ apiece is 36¢ + 32¢, or 68¢. Hence, the average minimum price for the 10 pieces is $\frac{68¢}{10}$, or 6.8¢ per piece. *Ans.* (C)

Matching Test

After completing this test, match your answers with the answers in the box.

1. Oranges sell from 6¢ to 10¢ apiece while grapefruits sell from 10¢ to $17\frac{1}{2}$¢ apiece. If 4 oranges and 6 grapefruits are being bought, what are the average maximum and the average minimum prices per piece?

2. A school contains 12 classes with class sizes ranging from 25 to 30 pupils per class. One day, 10% of the pupils in these classes were absent. What was the maximum number of pupils present on that day? What was the minimum number of pupils present?

3. A basketful of apples may vary in weight from 7 to 10 pounds. If $\frac{1}{3}$ to $\frac{1}{2}$ of the apples in each basket are ripe, what are the maximum and minimum number of pounds of ripe apples contained in 18 baskets?

4. A truck carts from 10 to 15 crates of fruit on each trip to the market. If the crates vary in weight from 120 to 225 pounds per crate, what are the maximum and the minimum number of pounds carried on any trip?

5. The weight of traffic supported by a narrow bridge at noon varies from 12 to 15 tons. If the bridge is used only by small cars ranging in weight from 1200 to 1500 pounds, what are the maximum and the minimum number of such cars using the bridge at noontime?

a. 25, 16	*b.* 324, 270	*c.* 3375, 1200	*d.* 14.5¢, 8.4¢	*e.* 90, 42

7.4 USING MAXIMUMS TO FIND MINIMUMS AND MINIMUMS TO FIND MAXIMUMS

Model Problem

An oil tank has a capacity of 1000 gallons. The quantity of oil in the tank is estimated to be between 350 and 400 gallons. Find the maximum and the minimum number of gallons needed to fill the tank to 80% of its capacity.

(A) 650, 600 (B) 550, 500 (C) 450, 400 (D) 350, 300 (E) 320, 280

Solution:

80% of the capacity of the tank in gallons = .80(1000) = 800.

Using the estimated minimum number of gallons already in the tank, the maximum number of gallons needed = 800 − 350, or 450.

Using the estimated maximum number of gallons already in the tank, the minimum number of gallons needed = 800 − 400, or 400. *Ans.* (C)

Matching Test

After completing this test, match your answers with the pairs of numbers in the box.

1. An oil tank has a capacity of 500 gallons and the number of gallons in the tank is estimated to vary from 150 to 180 gallons. What are the maximum and minimum numbers of gallons needed to fill the tank to 60% of its capacity?

2. A container whose weight is 2 pounds weighs 12 pounds when filled with orange juice. If the weight of the juice in an orange varies from 4 to 5 ounces, what are the maximum and minimum number of oranges needed to fill the container with juice?

3. Books whose cost may vary from $1.50 to $2 per book are sold at a price that may vary from $2.50 to $4. If 10 books are sold, what are the maximum and minimum number of dollars profit that may be made?

4. Suits that are bought at costs ranging from $25 to $30 per suit are sold at prices that range from $35 to $50 per suit. For 6 suits, what are the maximum and minimum number of dollars profit that may be made?

5. The marked prices of 10 dresses range from $15 to $20 per dress. In a sale, these prices are reduced at discounts ranging from 10% to 20% of the marked price. What are the maximum and minimum reductions on the dresses in dollars?

a. 150, 30	*b.* 25, 5	*c.* 150, 120	*d.* 40, 15	*e.* 40, 32

8. PRACTICE PROBLEMS

1. An estimate of the value of $\frac{69 \times 2014}{97}$ is

(A) 140 (B) 150 (C) 1400 (D) 1500 (E) 1300 **[1.1]**

2. To balance an object, seven 8-pound and ten 6-pound weights were used. Had only four 8-pound weights been available, how many more 6-pound weights would have been needed?

(A) 4 (B) 5 (C) 6 (D) 13 (E) 14 **[1.2]**

3. Three fractions are being added. The lowest common denominator of the first two fractions is 24 and the lowest common denominator of the last two fractions is 36. What is the least common denominator for all three fractions?

(A) 6 (B) 12 (C) 36 (D) 72 (E) 144 **[1.3]**

4. If 25 flower pots are placed in a row so that the centers of their bases are 8 inches apart, how many feet are there in the distance from the center of the first flower pot to the center of the last pot?

(A) 16 (B) $16\frac{2}{3}$ (C) $17\frac{1}{3}$ (D) 192 (E) 200 **[1.4]**

5. Find the missing term in the following set: 25, 30, 37, 46, 57, (?)

(A) 60 (B) 66 (C) 68 (D) 70 (E) 85 **[1.5]**

6. Combine: $\frac{1}{2} + \frac{3}{5} - \frac{5}{6}$

(A) $\frac{1}{15}$ (B) $\frac{4}{15}$ (C) $\frac{11}{15}$ (D) $\frac{14}{15}$ (E) $\frac{29}{15}$ **[2, 2.1]**

7. Multiply and divide as indicated: $2\frac{1}{2} \times \frac{1}{6} \div 1\frac{1}{4}$

(A) $\frac{2}{15}$ (B) $\frac{1}{3}$ (C) $\frac{25}{48}$ (D) $\frac{1}{6}$ (E) $\frac{5}{6}$ **[2, 2.2]**

8. Arrange the following fractions in order of size beginning with the smallest: $\frac{7}{24}, \frac{5}{16}, \frac{13}{48}$

(A) $\frac{13}{48}, \frac{5}{16}, \frac{7}{24}$ (B) $\frac{13}{48}, \frac{7}{24}, \frac{5}{16}$ (C) $\frac{5}{16}, \frac{13}{48}, \frac{7}{24}$ (D) $\frac{7}{24}, \frac{13}{48}, \frac{5}{16}$

(E) $\frac{7}{24}, \frac{5}{16}, \frac{13}{48}$ **[2, 2.3]**

9. A man has a monthly income of \$600. During a certain month, he spent $\frac{1}{3}$ of this income for food and $\frac{2}{5}$ of the remainder for rent. What fraction of the income was spent for rent?

(A) $\frac{2}{15}$ (B) $\frac{4}{15}$ (C) $\frac{2}{5}$ (D) $\frac{4}{5}$ (E) $\frac{11}{15}$ **[2.4]**

10. The plane fare for a husband and wife was \$140. If the husband paid full fare for himself and $\frac{3}{4}$ of the full fare for his wife, how many dollars were paid for the husband's ticket?

(A) 20 (B) 60 (C) 75 (D) 80 (E) 105 **[2, 2.5]**

11. In a school, $\frac{1}{3}$ of the student body are freshmen and $\frac{1}{4}$ are sophomores. One-half of each of these classes are members of athletic teams. What part of the entire student body are the students in these two classes who are members of athletic teams?

(A) $\frac{1}{12}$ (B) $\frac{7}{24}$ (C) $\frac{11}{12}$ (D) $\frac{17}{24}$ (E) $\frac{23}{24}$ **[2, 2.6]**

12. Select the smallest of the following tool sizes: $\frac{1}{2}$ inch, $\frac{15}{32}$ inch, $\frac{33}{64}$ inch, $\frac{7}{16}$ inch, $\frac{3}{8}$ inch

(A) $\frac{1}{2}$ (B) $\frac{15}{32}$ (C) $\frac{33}{64}$ (D) $\frac{7}{16}$ (E) $\frac{3}{8}$ **[2.7]**

13. Simplify: $6 \div \frac{.6}{5}$

(A) .5 (B) 5 (C) 50 (D) .72 (E) 500 **[3.1]**

14. Combine and express your answer to the nearest tenth: $2\frac{5}{8} + 10(.47)$

(A) 2.7 (B) 3.1 (C) 7.3 (D) 5.0 (E) 49.6 **[3.2]**

15. A 5-gallon can of milk costs \$4.75, while a 6-gallon can costs \$5.25. How much is saved in the purchase of 60 gallons by buying the larger cans?

(A) \$1.25 (B) \$1.75 (C) \$2.25 (D) \$4.50 (E) \$3.75 **[3.3]**

16. John estimates that his car uses gasoline at an average rate of 15 miles per gallon on the road. He is going to take a trip of 2400 miles. How much money should he set aside for the purchase of gasoline that costs 40¢ per gallon?

(A) \$6.40 (B) \$36 (C) \$48 (D) \$64 (E) \$640 **[3.4]**

17. If golf balls are priced at \$9.00 a dozen and the tax is 10% of the selling price, what will be the total cost of $2\frac{1}{3}$ dozen?

(A) \$19.80 (B) \$21.90 (C) \$22 (D) \$23.10 (E) \$21.21 **[3.5]**

18. The toll charge on a bridge is 50¢ for every car and 75¢ for every two occupants. At one time, a group of cars containing exactly two occupants each passed over the bridge. If the tolls they paid totaled \$16.25, how many persons crossed the bridge in these cars?

(A) 13 (B) 16 (C) 26 (D) 32 (E) 39 **[3.6]**

19. The local rate for sending a parcel is 5¢ for the first pound and 3¢ for each additional 2 pounds. How many pounds does a parcel weigh if the cost for local delivery is 32¢?

(A) 9 (B) 17 (C) 18 (D) 19 (E) 21 **[3.7]**

20. How many 24ths are there in $83\frac{1}{3}\%$?

(A) .2 (B) 2 (C) 20 (D) 21 (E) 28.8 **[4.1]**

21. .8 is what percent of 3.2?

(A) .25% (B) 2.5% (C) 25% (D) 250% (E) 25.6% **[4.2]**

22. What is 102% of 60?

(A) 6.12 (B) 61.2 (C) 612 (D) 6120 (E) 60.12 **[4.3]**

23. The decimal .56 is .7% of what number?

(A) .392 (B) .8 (C) 8 (D) 80 (E) 800 **[4.4]**

24. Of the following numbers, select the number that is the greatest in value: $\frac{2}{.3}, \frac{3}{.2}, \frac{.3}{2}, \frac{.3}{.2}$, 320%

(A) $\frac{2}{.3}$ (B) $\frac{3}{.2}$ (C) $\frac{.3}{2}$ (D) $\frac{.3}{.2}$ (E) 320% **[4.5]**

25. Of $750,000 collected in taxes, 22% was set aside for a school fund. How many thousands of dollars are there in the school fund?

(A) 16.5 (B) 165 (C) 1650 (D) 165,000 (E) 151.5 **[4.6]**

26. Each day, the number of microbes in a jar increases by 30%. How many thousands of microbes are there today if there were 260,000 yesterday?

(A) 78 (B) 182 (C) 200 (D) 338 (E) 267.8 **[4.7]**

27. Find the percent of decrease in the net profits of a company if, during the year, the net profit dropped from $12,500 to $11,125.

(A) 1.1 (B) 11 (C) 12 (D) 12.4 (E) 89 **[4.7]**

28. In a test, one-fourth of Tom's answers to the first 40 questions were incorrect. If he answers one-fifth of the next 40 questions incorrectly, what grade will he receive on the test?

(A) $22\frac{1}{2}$% (B) 45% (C) $47\frac{1}{2}$% (D) $52\frac{1}{2}$% (E) $77\frac{1}{2}$% **[4.8]**

29. In a school club, there are twice as many sophomores as juniors and three times as many juniors as seniors. What percent of the club are sophomores?

(A) 10% (B) 30% (C) $33\frac{1}{3}$% (D) 60% (E) $66\frac{2}{3}$% **[4.8]**

30. One of the teams in a school consists of 9 seniors, 8 juniors, and 7 sophomores. What percent of this team are not seniors?

(A) $37\frac{1}{2}$% (B) 60% (C) $62\frac{1}{2}$% (D) 75% (E) 80% **[4.9]**

31. In a school, the first-year students comprise 35% of the school population, while the second-year students comprise 30% of it. If 60% of the first-year students and 50% of the second-year students take mathematics, what percent of the school population is the group of first- and second-year students who are taking mathematics?

(A) 20% (B) $35\frac{1}{2}$% (C) 36% (D) 55% (E) 55.4% **[4.9]**

32. A certain automobile depreciates in value at the rate of 25% per year. If the original value of the car is $3200, what is the net change in the value of the car over a two-year period?

(A) −$200 (B) −$1400 (C) −$1500 (D) −$1600 (E) −$1800 **[4.10]**

33. How much is saved if an article priced at $100 less 20%, 20%, and 20% is bought instead for $100 less 60%?

(A) $0 (B) $4 (C) $11 (D) $11.20 (E) $12.20 **[4.11]**

34. In a certain state, the gasoline tax is $12\frac{1}{2}$% of the price of gasoline. When the tax is added to the price, the gasoline sells for 39¢ per gallon. What is the amount of the tax collected by the state for every 1200 gallons sold?

(A) $48 (B) $52 (C) $58.50 (D) $520 (E) $585 **[5.1]**

35. A grocer's profit on 20 dozen eggs was $5.20. If the eggs were sold at $.56 a dozen, what did the eggs cost him?

(A) $6.00 (B) $7.20 (C) $11.20 (D) $16.40 (E) $106.80 **[5.2]**

36. A house is insured for \$15,000 at a cost of \$.45 per \$100 for one year. If a 3-year insurance policy costs $2\frac{1}{2}$ times as much as a 1-year policy, what will it cost to insure the house for 3 years?

(A) \$33.75 (B) \$135 (C) \$168.75 (D) \$202.50 (E) \$206.25 **[5.3]**

37. The down payment on a set of furniture amounted to 30% of the price. The remainder was paid in 16 equal installments of \$56 each. What was the total cost of the furniture?

(A) \$627.20 (B) \$896 (C) \$1104.80 (D) \$1280 (E) \$12,800 **[5.4]**

38. From twice 2 hr. 20 min., subtract one-half of 3 hr. 20 min.

(A) 2 hr. (B) $2\frac{1}{3}$ hr. (C) $2\frac{2}{3}$ hr. (D) 3 hr. (E) $3\frac{1}{3}$ hr. **[6, 6.1]**

39. To the nearest foot, by how many feet does 100 meters exceed 100 yards? (1 meter = 39.37 inches.)

(A) 2.8 (B) 28 (C) 31 (D) 34 (E) 337 **[6.2]**

40. A plane left at 11:30 A.M. on a scheduled trip of 2 hours and 40 minutes. If there was a delay of 15 minutes en route, at what time did the plane actually arrive at its destination?

(A) 1:25 P.M. (B) 2:10 P.M. (C) 2:25 P.M. (D) 2:40 P.M. (E) 3:25 P.M. **[6.3]**

41. How many gills are there in $\frac{3}{4}$ of a gallon? (1 gallon = 4 quarts, 1 quart = 2 pints, 1 pint = 2 cups, and 1 cup = 2 gills.)

(A) 12 (B) 18 (C) 24 (D) 32 (E) 48 **[6.4]**

42. How many pounds heavier is $\frac{1}{8}$ of a long ton compared to 12% of a short ton? (1 long ton = 2240 lb., 1 short ton = 2000 lb.)

(A) 20 (B) 24 (C) 30 (D) 36 (E) 40 **[6.5]**

43. A group of 7 dresses is on special sale in a dress shop. Each dress is of one distinct color, so that the dresses are of 7 different colors. What is the minimum number of dresses that must be bought unseen in order to obtain at least 2 dresses having the same color?

(A) 6 (B) 7 (C) 8 (D) 14 (E) 15 **[7.1]**

44. If x may vary from 2 to 3, and y may vary from 3 to 5, find the maximum and minimum ratios of x^2 to y^2.

(A) $1, \frac{9}{25}$ (B) $1, \frac{2}{5}$ (C) $1, \frac{4}{25}$ (D) $\frac{9}{25}, 1$ (E) $\frac{4}{9}, \frac{9}{25}$ **[7.2]**

45. Before it is cooked, a meat roast may weigh from 3 to 5 pounds. When cooked for a certain time, the roast may lose from 30% to 50% of its weight. Find the maximum and minimum number of pounds in the total weight of 4 roasts after cooking.

(A) 10, 6 (B) 14, 6 (C) 14, 8.4 (D) 10, 8.4 (E) 6, 14 **[7.3]**

46. In a class of 120 seniors, the dropout rate may vary from 5% to 15%. Find the maximum and the minimum numbers of seniors in this class who graduate.

(A) 18, 6 (B) 114, 102 (C) 92, 114 (D) 138, 126 (E) 114, 82 **[7.4]**

II. REVIEWING MONOMIALS AND POLYNOMIALS

1. REVIEWING ALGEBRAIC EXPRESSIONS

In algebra, a letter represents a number. For example, the letter n represents 9 in the equations $5n = 45$, $n + 5 = 14$, and $\sqrt{n} = 3$. A letter used to represent a number is called a ***literal number.***

By means of letters and symbols, lengthy verbal statements can be replaced by short algebraic statements. For example, the verbal statement "The sum of twice a number and five times a number is the equivalent of twenty less than eight times the same number" may be represented by the algebraic statement $2n + 5n = 8n - 20$.

Fundamental Operations in Algebra

The four fundamental operations of algebra are addition, subtraction, multiplication, and division. The student should become thoroughly familiar with the symbols, the terminology, and the laws or principles governing each of these operations.

ADDITION

In algebra, as in arithmetic, the symbol for addition is +, ***addends*** are the numbers being added, and their ***sum*** is the answer obtained.

For example, if x and 3 are addends, their sum is $x + 3$.

Commutative Law of Addition. **Interchanging addends does not change their sum.**

In arithmetic, since $3 + 4 = 7$ and $4 + 3 = 7$, then $3 + 4 = 4 + 3$. Also, $3 + 4 + 5 = 3 + 5 + 4 = 5 + 3 + 4$.

In algebra, $a + b = b + a$. Also, $a + b + c = a + c + b = c + a + b$. Hence, addends may be rearranged in any order.

Associative Law of Addition. **Changing the order in which addends are added does not change their sum.**

In arithmetic, $3 + 4 + 5 = 3 + (4 + 5) = (3 + 4) + 5 = 12$.

In algebra, $a + b + c = a + (b + c) = (a + b) + c$.

The sum of any number and 0 is the same number. For example, $157 + 0 = 157$. Since the addition of 0 does not change the identity of any number, 0 is called the ***identity element for addition.***

Additive inverses, or ***opposites,*** are numbers whose sum is 0. For example, since the sum of $(+5)$ and (-5) is 0, $(+5)$ and (-5) are additive inverses, or opposites, of each other. They are also called ***negatives*** of each other.

SUBTRACTION

In algebra, as in arithmetic, the symbol for subtraction is $-$, the ***minuend*** is the number from which we subtract, the ***subtrahend*** is the number being subtracted, and the ***difference*** is the answer obtained.

For example, if x is the minuend and 3 is the subtrahend, the difference is $x - 3$.

The commutative and associative laws do not apply to subtraction. For example, $x - 3 \neq 3 - x$. ($\neq$ means "is not equal to.") In fact, $(x - 3)$ and $(3 - x)$ are opposites. Also, $7 - (4 - 3) \neq (7 - 4) - 3$.

Multiplication

In algebra, the symbols for multiplication are $\times$, $\cdot$, or ( ). Multiplication can also be represented without any operational symbol, as in the product xy.

For example, $3 \times a = 3(a) = 3 \cdot a = 3a$.

Factors are the numbers being multiplied. Their ***product*** is the answer obtained. In a product, each factor is the ***coefficient*** of the remaining factors. For example, in the product $3b$, 3 and b are factors, 3 is the ***numerical coefficient*** of b, and b is the ***literal coefficient*** of 3.

Commutative Law of Multiplication. **Interchanging factors does not change their product.**

In arithmetic, since $3 \times 4 = 12$ and $4 \times 3 = 12$, then $3 \times 4 = 4 \times 3$. Also, $3 \times 4 \times 5 = 3 \times 5 \times 4 = 5 \times 3 \times 4$.

In algebra, $ab = ba$. Also, $abc = acb = cab$. Hence, factors may be rearranged in any order.

Associative Law of Multiplication. **Changing the order in which factors are multiplied does not change their product.**

In arithmetic, $3 \times 4 \times 5 = 3 \times (4 \times 5) = (3 \times 4) \times 5 = 60$.

In algebra, $a \times b \times c = a(bc) = (ab)c$.

The product of any number and 1 is the same number. For example, $157 \times 1 = 157$. Since multiplication by 1 does not change the identity of any number, 1 is called the ***identity element for multiplication***.

Multiplicative inverses, or ***reciprocals***, are numbers whose product is 1. For example, since the product of $\frac{3}{5}$ and $\frac{5}{3}$ is 1, then $\frac{3}{5}$ and $\frac{5}{3}$ are multiplicative inverses, or reciprocals, of each other.

Division

In algebra, as in arithmetic, the symbols for division are $\div$, :, and the fraction line. Also, the ***dividend*** is the number being divided, the ***divisor*** is the number by which we divide, and the ***quotient*** is the answer obtained.

For example, if x is the dividend and 3 is the divisor, the quotient is $\frac{x}{3}$.

The commutative and associative laws do not apply to division. For example, $\frac{x}{3} \neq \frac{3}{x}$ for any value of x except 3. In fact, $\frac{x}{3}$ and $\frac{3}{x}$ are reciprocals of each other, if $x \neq 0$. Also, $12 \div (6 \div 2) \neq (12 \div 6) \div 2$.

> *Note.* Division by zero is impossible. An expression such as $\frac{5}{0}$ is meaningless or undefined. However, 0 divided by any other number is 0. For example, $\frac{0}{5} = 0$.

Fundamental Operations With Powers

Power operation is repeated multiplication. For example, $2^5 = 32$ means $2 \cdot 2 \cdot 2 \cdot 2 \cdot 2 = 32$. Here, 5 is the ***exponent*** of the ***base*** 2. Also, 32 is the fifth ***power*** of the base 2.

Exponent	
Base	= Power

TABLE OF POWERS

(This table should be memorized.)

Exponent

Base	1	2	3	4	5
1	1	1	1	1	1
2	2	4	8	16	32
3	3	9	27	81	243
4	4	16	64	256	1024
5	5	25	125	625	3125
10	10	100	1000	10,000	100,000

Operations on Powers

To multiply powers having the same base, keep the base unchanged and add exponents: $x^2x^3y^2y^4 = x^5y^6$.

To divide unequal powers having the same base, first arrange them as a fraction. Keep the base unchanged and subtract the smaller exponent from the larger exponent.

If the larger exponent is in the numerator	If the larger exponent is in the denominator
$\frac{x^5}{x^2} = x^3$	$\frac{x^2}{x^5} = \frac{1}{x^3}$

Note. If powers are equal, their quotient is 1: $\frac{x^3}{x^3} = 1$

To find the power of a power of a base, keep the base and multiply exponents: $(x^2)^3 = x^6$.

Powers of Negative Numbers

An odd power of a negative number is negative, while an even power of a negative number is positive.

For example, $(-2)^3 = (-2)(-2)(-2) = -8$, but $(-2)^4 = (-2)(-2)(-2)(-2) = +16$.

Terminology of Algebraic Expressions

A ***term*** is a number or the product of numbers. An ***expression*** consists of one term alone or of two or more terms being added or subtracted, that is, connected by plus or minus signs.

For example, $3xy$ is a term, while $x + y - 3$ is an expression of three terms.

A ***monomial*** is an expression of one term only.

For example, 7, x, $7x$, $\frac{1}{7}x^2$, and $\frac{1}{7}x^2y^3z$ are monomials.

A ***polynomial*** is a monomial or the sum of two or more monomials.

A polynomial of two terms is a ***binomial***; if a polynomial has three terms, it is a ***trinomial.***

For example, $3x + y$ is a binomial, and $x + y - 3$ is a trinomial.

A ***factor of a term*** is any of the numbers, letters, or expressions that are multiplied together to form the term.

For example, the factors of the term $3xy$ are 3, x, and y, while the factors of $5(a + b)$ are 5 and $(a + b)$.

The ***literal factors of a term*** are those factors represented by letters.

For example, the letters x and y are the literal factors of the term $3xy$.

Like terms are terms that have the same literal factors, each with the same exponent. ***Unlike terms*** are terms whose literal factors differ or have different exponents.

For example, $3a^2bc^3$ and $5a^2bc^3$ are like terms, while $3ab^2$ and $5a^2b$ are unlike terms.

The factors of a term may be either numerical or literal coefficients. The ***numerical coefficient*** is the numerical factor, while the ***literal coefficient*** is the product of the literal factors.

For example, in $8ab^2c$, 8 is the numerical coefficient of ab^2c, while ab^2c is the literal coefficient of 8.

1.1 COMBINING ODD AND EVEN POWERS OF POSITIVE AND NEGATIVE NUMBERS

Model Problem

The value of $(-1)^5 + (-1)^4 + (-2)^5 + (-3)^3 + (-10)^2$ is

(A) 39 (B) 41 (C) 95 (D) 105 (E) 161

Solution:

Since an even power of a negative number is positive, $(-1)^4 = +1$ and $(-10)^2 = +100$.

Since an odd power of a negative number is negative, $(-1)^5 = -1$, $(-2)^5 = -32$, and $(-3)^3 = -27$.

Hence, to evaluate $(-1)^5 + (-1)^4 + (-2)^5 + (-3)^3 + (-10)^2$, substitute as follows:

$$(-1) + 1 + (-32) + (-27) + 100 = 41 \quad \textit{Ans.}\ (B)$$

Matching Test

Evaluate each expression. After completing this test, match your answers with the numbers in the box. (Use the table of powers to check your work.)

1. $2^5 + (-2)^4 + 2^3$ **2.** $3^4 + (-3)^3 + (-3)^2$ **3.** $4^4 + (-4)^3 - (-4)^2$ **4.** $5^3 - (-5)^2 + 5^2$
5. $10^3 + (-10)^2 - 10^2$

a. 125	*b.* 56	*c.* 176	*d.* 1000	*e.* 63

1.2 FINDING EQUIVALENT EXPRESSIONS WITHOUT EXPONENTS

Model Problem

The expression m^3 means

(A) $m + m + m$ (B) $3 + m$ (C) $3m$ (D) $m \times m \times m$ (E) none of these

Solution:

The expression m^3 means that m is used as a factor three times; that is, $m^3 = m \times m \times m$.
Ans. (D)

Wrong Choices:

Choices (A) and (C) are equivalent, since $3m$ means that m is used as an addend three times; that is, $3m = m + m + m$.

In choice (B), $3 + m$ means that 3 is added to m. The expressions $3 + m$, $3m$, and m^3 have different meanings.

Matching Test

Express each of the following by using only the letter m and the operational signs. After completing this test, match your answers with the expressions in the box.

1. $(m^2)^3$ **2.** $3m^2$ **3.** $(m^3)^2$ **4.** $2m^3$ **5.** $(2m)^3$

a. $(mm)(mm)(mm)$	*b.* $(m + m)(m + m)(m + m)$	*c.* $(mmm)(mmm)$
d. $mm + mm + mm$	*e.* $mmm + mmm$	

1.3 MULTIPLICATION OF POWERS WITH FRACTIONAL COEFFICIENTS

Model Problem

The product of $\frac{1}{2}x^3$ and $\frac{1}{2}x^3$ is

(A) x^6 (B) $\frac{1}{4}x^6$ (C) x^9 (D) $\frac{1}{4}x^9$ (E) none of these

Solution:

Find the product of $\frac{1}{2}x^3$ and $\frac{1}{2}x^3$ as follows:

1. Multiply the numerical coefficients: $\left(\frac{1}{2}\right)\left(\frac{1}{2}\right) = \frac{1}{4}$
2. Since x^3 and x^3 are powers having the same base, multiply x^3 and x^3 by keeping the base and adding their exponents: $x^3 \cdot x^3 = x^6$
3. Multiply the results: $\frac{1}{4}x^6$

Ans. (B)

Wrong Choices:

(A) In this choice, the coefficients $\frac{1}{2}$ and $\frac{1}{2}$ were added. They should be multiplied.

(C) In this choice, the coefficients were added and the exponents were multiplied. Instead, the coefficients should be multiplied and the exponents added.

(D) In this choice, the exponents were multiplied. They should be added.

Matching Test

Find the product of each set of expressions. After completing this test, match your answers with the expressions in the box.

1. $\frac{1}{3}x^2$ and $\frac{1}{3}x^2$ **2.** $\frac{1}{2}x^3$ and $\frac{1}{3}x^2$ **3.** $\frac{1}{2}x^2$ and $\frac{1}{2}x^2$ **4.** $\frac{1}{2}x^3$, $\frac{1}{2}x^3$, and $\frac{1}{2}x^3$ **5.** $\frac{1}{3}x^2$, $\frac{1}{3}x^2$, and $\frac{1}{3}x^2$

a. $\frac{1}{6}x^5$	*b.* $\frac{1}{8}x^9$	*c.* $\frac{1}{27}x^6$	*d.* $\frac{1}{9}x^4$	*e.* $\frac{1}{4}x^4$

1.4 MULTIPLICATION OF POWERS WITH INTEGRAL COEFFICIENTS

Model Problem

If $(-3a^2)$ is multiplied by $(-6a^3)$, the product is

(A) $9a^5$ (B) $-9a^5$ (C) $9a^6$ (D) $18a^5$ (E) $18a^6$

Solution:

The product of $(-3a^2)$ and $(-6a^3)$ is found as follows:

1. Multiply the numerical coefficients: $(-3)(-6) = 18$
2. Since a^2 and a^3 are powers having the same base, multiply a^2 and a^3 by keeping the base and adding their exponents: $a^2 \cdot a^3 = a^5$
3. Multiply the results: $18a^5$

Ans. (D)

Matching Test

Determine the product of each set of expressions. After completing this test, match your answers with the expressions in the box.

1. $(3a^2)$ and $(6a^4)$ **2.** $(2a^3)$ and $(-6a^4)$ **3.** $(-2a^3)$ and $(-4a^6)$ **4.** $(-3a^2)$ and $(4a^6)$ **5.** $(-4a^2)$ and $(6a^3)$

a. $-12a^7$	*b.* $-24a^5$	*c.* $18a^6$	*d.* $-12a^8$	*e.* $8a^9$

1.5 FINDING A POWER OF A POWER

Model Problem

The expression $(2x^2)^3$ is equivalent to

(A) $6x^5$ (B) $6x^6$ (C) $8x^5$ (D) $8x^6$ (E) none of these

Solution:

In the expression $(2x^2)^3$, $2x^2$ is being cubed. To cube $2x^2$, use it as a factor three times. Hence, $(2x^2)^3 = (2x^2)(2x^2)(2x^2) = (2)(2)(2)x^2x^2x^2 = 8x^6$. *Ans.* (D)

Wrong Choices:

In choices (A) and (C), the exponent of x is 5. This would be true if x^2 were multiplied by x^3. This is not the same as $(x^2)^3$.

In choice (B), the coefficient of x^6 is 6. The product of (2)(2)(2) is not 6 but 8.

Matching Test

Match each expression with its equivalent in the box.

1. $(2x^3)^2$ **2.** $(3x^2)^2$ **3.** $(2x^2)^3$ **4.** $(3x^2)^3$ **5.** $(2x^3)^3$

a. $27x^6$	*b.* $4x^6$	*c.* $8x^6$	*d.* $9x^4$	*e.* $8x^9$

1.6 ALGEBRAIC REPRESENTATION INVOLVING COMMON UNITS OF MEASURE

Model Problem

Represent the number of hours in x days and y hours.

(A) $24x + y$ (B) $x + 24y$ (C) $24(x + y)$ (D) $\frac{x}{24} + y$ (E) $x + \frac{y}{24}$

Solution:

1. Since each day contains 24 hours, the number of hours in x days equals 24 times x hours, or $24x$ hours. For example, in 10 days there are 24(10) hours, or 240 hours.
2. The total number of hours in x days and y hours is the sum of $24x$ and y, that is, $24x + y$.

Ans. (A)

After completing each test, match your answers with the expressions in the boxes.

Matching Test A

1. Represent the number of minutes in x hours and y minutes.
2. Represent the number of hours in x days and y minutes.
3. Represent the number of days in x weeks and y days.
4. Represent the number of days in x weeks and y hours.
5. Represent the number of seconds in y minutes and x seconds.

a. $7x + 7$
b. $24x + \frac{y}{60}$
c. $60y + x$
d. $60x + y$
e. $7x + \frac{y}{24}$

Matching Test B

1. Represent the number of inches in a yards and b inches.
2. Represent the number of inches in a yards, b feet, and c inches.
3. Represent the number of cents in q quarters and d dimes.
4. Represent the number of nickels equivalent to q quarters and d dimes.
5. Represent the number of cents in q quarters, n nickels, and c cents.

a. $25q + 5n + c$
b. $36a + 12b + c$
c. $5q + 2d$
d. $25q + 10d$
e. $36a + b$

1.7 ALGEBRAIC REPRESENTATION IN VERBAL PROBLEMS

Model Problem

If Tom makes \$$x$ a day and spends \$$y$ a day, how many dollars will he save in 10 days?

(A) $10(x - y)$ (B) $10x - y$ (C) $10(y - x)$ (D) $\frac{x - y}{10}$ (E) $\frac{y - x}{10}$

Solution:

Since he makes \$$x$ a day and spends \$$y$ a day, he saves $(x - y)$ dollars each day.

Hence, in 10 days, he will save $10(x - y)$ dollars.

For example, if he makes \$20 a day and spends \$15 a day, then, in 10 days, he will save $10(20 - 15)$ dollars, or \$50. *Ans.* (A)

Matching Test

After completing this test, match your answers with the expressions in the box.

1. If a man makes \$$y$ a day and spends \$10 a day, how many dollars will he save in x days?
2. If a man makes \$$x$ in 10 days and spends \$$y$ each of those days, how many dollars will he save in the 10 days?
3. If a man makes \$$x$ a day and spends \$$y$ in 10 days, how many dollars will he save in the 10 days?
4. If a man makes \$$x$ in 10 days and spends \$$y$ in the 10 days, how much was saved per day?
5. If a man makes \$$y$ a day and spends \$10 each day, how many days will it take to save \$$x$?

a. $x - 10y$	*b*. $\frac{x - y}{10}$	*c*. $\frac{x}{y - 10}$	*d*. $x(y - 10)$	*e*. $10x - y$

1.8 ALGEBRAIC REPRESENTATION OF SUCCESSIVE OPERATIONS

Model Problem

If we take any number, multiply by 3, add 6, multiply by 4, subtract 12 times the original number, the result will always equal 24. An algebraic expression which may be used to prove this is:

(A) $4(3x) + 6 - 12x$ (B) $4 \cdot 3(x + 6) - 12x$ (C) $4(3x + 24) - 12x$
(D) $4(3x + 6) - 12x$ (E) $x(3 + 6 \cdot 4) - 12x$

Solution:

If a number, x, is multiplied by 3, we obtain $3x$.

If 6 is added to the new number, $3x$, we obtain $3x + 6$.

If the new number, $3x + 6$, is multiplied by 4, we obtain $4(3x + 6)$.

If, from the new number $4(3x + 6)$, we subtract 12 times the original number, we obtain $4(3x + 6) - 12x$. *Ans.* (D)

Matching Test

In each of the following, find the algebraic equation that may be used to prove the statement. After completing this test, match your answers with the equations in the box.

1. Take a number, add 1, multiply by 2, and subtract twice the original number; the result will always equal 2.

2. Take a number, subtract 2, multiply by 3, and add 6; the result will always be triple the original number.

3. Take a number, add 2, multiply by 4, and subtract three times the original number; the result will always be 8 more than the original number.

4. Take a number, double it, add 4, multiply by 2, and subtract 8; the result will always be 4 times the original number.

5. Take a number, add 1, multiply by 2, add 3 more than the original number, and subtract 3 times the original number; the result will always be 5.

a. $2(2x + 4) - 8 = 4x$	*b*. $2(x + 1) - 2x = 2$	*c*. $2(x + 1) + (x + 3) - 3x = 5$
d. $4(x + 2) - 3x = x + 8$	*e*. $3(x - 2) + 6 = 3x$	

2. EVALUATING EXPRESSIONS

Order of Operations

1. Substitute the value given for each letter.
2. If the expression contains parentheses, evaluate within the parentheses.
3. Do powers and roots.
4. Do multiplications and divisions in order from left to right.
5. Do additions and subtractions in order from left to right.

Example 1	*Example* 2
Evaluate $3 + 3a^3$ when $a = 3$.	Evaluate $(2 + 2a)^2$ when $a = 2$.
Solution:	*Solution:*
Substitute 3 for a.	Substitute 2 for a.
$3 + 3 \cdot 3^3$	$(2 + 2 \cdot 2)^2$
Take the power of 3.	To evaluate within the parentheses, multiply, then add.
$3 + 3 \cdot 27$	$(2 + 4)^2$
Multiply before adding.	6^2
$3 + 81$	Take the power of 6.
Add.	36
84	

In Example 2, note how the presence of parentheses may change the order of operations.

2.1 EVALUATING MONOMIALS CONTAINING ONE VARIABLE

Model Problem

Which of the following is greatest in value when $x = 2$?

(A) $3x^2$ (B) $\frac{48}{x^3}$ (C) $6x + 1$ (D) $(3x)^2$ (E) $(3 + x)^2$

Solution:

When $x = 2$, $x^2 = (2)(2) = 4$ and $x^3 = (2)(2)(2) = 8$.

To find which of the five expressions is greatest in value when $x = 2$, substitute 2 for x, 4 for x^2, and 8 for x^3; then evaluate each expression.

(A) $3x^2 = 3(4) = 12$ (B) $\frac{48}{x^3} = \frac{48}{8} = 6$ (C) $6x + 1 = 6(2) + 1 = 12 + 1 = 13$

(D) $(3x)^2 = (3 \cdot 2)^2 = 6^2 = 36$ (E) $(3 + x)^2 = (3 + 2)^2 = 5^2 = 25$

Ans. (D)

Matching Test

After completing this test, match your answers with the numbers in the box.

1. Evaluate $3x^2$ when $x = -1$.
2. Evaluate $(3x)^2$ when $x = 2$.
3. Evaluate $2x^3$ when $x = 1$.
4. Evaluate $(2x)^3$ when $x = 10$.
5. Evaluate $(x + 3)^2$ when $x = 0$.

a. 8000
b. 36
c. 9
d. 3
e. 2

2.2 EVALUATING MONOMIALS CONTAINING TWO VARIABLES

Model Problem

If $a = 5$ and $b = -3$, what is the value of $2ab^2$?

(A) 60 (B) −90 (C) 90 (D) −450 (E) 900

Solution:

Find the value of $2ab^2$ by substituting 5 for a and −3 for b.

$$2(5)(-3)^2$$

To square (−3), multiply −3 by −3. The product is +9.

$$2(5)(9)$$
$$(10)(9)$$
$$90 \quad \textit{Ans. } \text{(C)}$$

Matching Test

After completing this test, match your answers with the numbers in the box.

1. Evaluate $2ab^3$ when $a = 2$, $b = 1$.
2. Evaluate $2a^3b$ when $a = 2$, $b = -1$.
3. Evaluate $3ab^2$ when $a = -2$, $b = 1$.
4. Evaluate $3a^2b$ when $a = -2$, $b = -1$.
5. Evaluate a^3b^2 when $a = -2$, $b = 1$.

a.	−12
b.	−16
c.	− 6
d.	4
e.	− 8

2.3 EVALUATING TRINOMIALS CONTAINING ONE VARIABLE

Model Problem

Find the value of $3a^2 - 2a + 1$ when $a = -1$.

(A) 0 (B) 2 (C) −4 (D) 6 (E) 12

Solution:

To find the value of $3a^2 - 2a + 1$, substitute −1 for a.

$$3(-1)^2 - 2(-1) + 1 \qquad (-1)^2 = (-1)(-1) = 1$$
$$3(1) + 2 + 1$$
$$3 + 2 + 1$$
$$6 \quad \textit{Ans. } \text{(D)}$$

Matching Test

After completing this test, match your answers with the numbers in the box.

1. Evaluate $3a^2 - 2a + 1$ when $a = -2$.
2. Evaluate $3a^2 + 2a + 1$ when $a = -2$.
3. Evaluate $a^2 + 3a - 2$ when $a = -2$.
4. Evaluate $3a^2 - 2a - 1$ when $a = -1$.
5. Evaluate $2a^2 - 3a + 1$ when $a = -1$.

a.	4
b.	6
c.	17
d.	9
e.	−4

2.4 EVALUATING ONE-VARIABLE EXPRESSIONS CONTAINING PARENTHESES

Model Problem

If $x = 8$, the value of $x - 5(x + 2)$ is

(A) -42 (B) -22 (C) 30 (D) 42 (E) -15

Solution:

To find the value of $x - 5(x + 2)$ when $x = 8$, first substitute for x.

$$8 - 5(8 + 2)$$

Combine the numbers in parentheses.

$$8 - 5 \cdot 10$$

Multiply before subtracting.

$$8 - 50 = -42 \quad \textit{Ans.}\ \text{(A)}$$

Matching Test

Evaluate each expression when $x = 8$. After completing this test, match your answers with the numbers in the box.

1. $x + 5(x + 2)$ **2.** $x - 5(x - 2)$ **3.** $(x - 5)(x - 2)$ **4.** $(5 - x)(x + 2)$ **5.** $5x - (x + 2)$

a. -22	*b.* -30	*c.* 58	*d.* 18	*e.* 30

2.5 EVALUATING EXPRESSIONS CONTAINING PARENTHESES AND HAVING MORE THAN ONE VARIABLE

Model Problem

If $a = -1$, $b = 2$, and $c = -3$, find the value of $a + b(a + c)$.

(A) 3 (B) -4 (C) 5 (D) 7 (E) -9

Solution:

Find the value of the given expression by substituting -1 for a, 2 for b, and -3 for c.

$$a + b(a + c)$$

$$-1 + 2(-1 - 3)$$

Evaluate within parentheses.

$$-1 + 2(-4)$$

Multiply before adding.

$$-1 - 8$$

Add.

$$-9 \quad \textit{Ans.}\ \text{(E)}$$

After completing each test, match your answers with the numbers in the boxes.

Matching Test A

Find the value of $a + b(a + c)$ for the given values of a, b, and c.

1. $a = 1,\ b = 2,\ c = -3$	*a.* 4
2. $a = 1,\ b = -2,\ c = -3$	*b.* 5
3. $a = 1,\ b = -2,\ c = 3$	*c.* 2
4. $a = -3,\ b = -1,\ c = -2$	*d.* −7
5. $a = 3,\ b = 1,\ c = -2$	*e.* −3

Matching Test B

In each of the following, $a = -1$, $b = 2$, and $c = -3$. Find the value of each expression.

1. $a + b(a - c)$ **2.** $a - b(a - c)$ **3.** $a - c(a - b)$ **4.** $b + c(a - b)$ **5.** $(a + c)(a + b)$

a. −4 *b.* −5 *c.* 11 *d.* −10 *e.* 3

3. SIMPLIFYING EXPRESSIONS CONTAINING PARENTHESES

Parentheses, ( ), are used to group terms into a single term. For example, $3(x + y)$ is a single term representing three times the sum of x and y.

Other symbols of grouping are brackets, $[a + b]$, braces, $\{a + b\}$, and the vinculum, or bar, $\overline{a + b}$.

To simplify an expression containing parentheses, remove the parentheses and then combine like terms.

To remove parentheses:

1. If a plus sign precedes the parentheses, do not change the sign of any of the enclosed terms.
2. If a minus sign precedes the parentheses, change the sign of each enclosed term.
3. Apply the ***Distributive Law of Multiplication:*** **To multiply a polynomial by a monomial, multiply each term of the polynomial by the monomial.** (Think: "Distribute" the multiplication over each of the terms.)

   For example, $5(x + y - z) = 5x + 5y - 5z$. Also, $\frac{x + y - z}{5} = \frac{1}{5}(x + y - z) = \frac{x}{5} + \frac{y}{5} - \frac{z}{5}$.
4. Where more than one set of grouping symbols is used, remove one set at a time, beginning with the innermost set.

$$5x - [4 - 3(2 - x)] = 5x - [4 - 6 + 3x]$$
$$= 5x - 4 + 6 - 3x = 2x + 2$$

$$a^2 + a + 1 + 2a^2 - 2a - 1$$
$$= a^2 + 2a^2 + a - 2a + 1 - 1$$
$$= 3a^2 - a$$

To simplify after removing parentheses, combine like terms (that is, add or subtract terms having the same literal coefficient). To combine like terms, add or subtract their numerical coefficients and keep their common literal coefficients.

3.1 SIMPLIFYING ONE-VARIABLE EXPRESSIONS CONTAINING PARENTHESES

Model Problem

Perform the indicated operations and combine like terms: $3p - 2p(p - 4)$

(A) $11p - 2p^2$ (B) $-2p^2 - 5p$ (C) $p^2 - 4p$ (D) $9p^2$ (E) $4 - p$

Solution:

$$3p - 2p(p - 4)$$

Remove parentheses.

$$3p - 2p^2 + 8p$$

Collect like terms.

$$3p + 8p - 2p^2$$

Combine like terms.

$$11p - 2p^2 \quad \textit{Ans.} \text{ (A)}$$

Matching Test

Simplify each expression. After completing this test, match your answers with the expressions in the box.

1. $3p + 2p(p + 4)$ **2.** $3p - 2p(p + 4)$ **3.** $3p(p - 2) - 2p$ **4.** $2p(p - 3) + 2p$

5. $2 + p(p - 3) + 2p$

a. $-2p^2 - 5p$	*b.* $2p^2 - 4p$	*c.* $3p^2 - 8p$	*d.* $2p^2 + 11p$	*e.* $p^2 - p + 2$

3.2 SIMPLIFYING TWO-VARIABLE EXPRESSIONS CONTAINING PARENTHESES

Model Problem

Perform the indicated operations and write the result in simplest form:

$$3x + 2(x - 4y) - (x + y)$$

(A) $-5xy$ (B) $5x - 7y$ (C) $5x - 9y$ (D) $4x - 7y$ (E) $4x - 9y$

Solution:

$$3x + 2(x - 4y) - (x + y)$$

Remove parentheses.

$$3x + 2x - 8y - x - y$$

Collect like terms.

$$3x + 2x - x - 8y - y$$

Combine like terms.

$$4x - 9y \quad \textit{Ans.} \text{ (E)}$$

Matching Test

Simply each expression. After completing this test, match your answers with the expressions in the box.

1. $3x + 2(x - 4y) + (x + y)$

2. $3x - 2(x - 4y) - (x + y)$

3. $3x - 2y - (x - 4y) + 2(x + y)$

4. $3(x + y) - 2(3x - 2y) - (x - 4y)$

5. $3(x - y) - 2(x - 4y) + 3(3x - 2y)$

a. $-4x + 11y$
b. $10x - y$
c. $4x + 4y$
d. $6x - 7y$
e. $7y$

4. OPERATIONS ON POLYNOMIALS

Arranging the Terms of a Polynomial

The terms of a polynomial may be arranged in descending or ascending order by arranging them so that the exponents of the same letter decrease or increase in successive terms. For example, $4a + a^3 + a^2$ may be written as $a^3 + a^2 + 4a$, in which the exponents 3, 2, and 1 decrease, or as $4a + a^2 + a^3$, in which the exponents increase.

Adding or Subtracting Polynomials

To add or subtract polynomials, first arrange them in order by placing like terms in the same column. Then add or subtract like terms. (Generally, the descending order is used.)

$$\begin{array}{rrrrr} & x^4 & & -x^2 & +2 \\ & 2x^4 & -x^3 & +x^2 & \\ \hline \text{Add:} & 3x^4 & -x^3 & & +2 \end{array}$$

For example, to find the sum of $-x^2 + x^4 + 2$ and $-x^3 + 2x^4 + x^2$, first arrange the terms of both polynomials in descending order and then place like terms in the same columns, as shown above.

4.1 ADDING TWO EXPRESSIONS

Model Problem

The sum of $3x^2 - 4$ and $4x^2 + 5x + 4$ is

(A) $7x^2 + 5x$ (B) $7x^2 + 5x + 8$ (C) $7x^2 + x + 4$ (D) $4x^2 + 8x$ (E) $x^2 + 5x + 8$

Solution:

To find the sum of the two expressions, first arrange them in order by placing like terms in the same column. Then add like terms.

$$\begin{array}{rrr} 4x^2 & +5x & +4 \\ 3x^2 & & -4 \\ \hline 7x^2 & +5x & \end{array}$$

Ans. (A)

Matching Test

Find the sum of each pair of expressions. After completing this test, match your answers with the expressions in the box.

1. $3x^2 - 4,\ 4x^2 + 5x - 4$	*a.* $7x^2 - 5x$
2. $3x^2 - 4,\ 4x^2 - 5x + 4$	*b.* $8x^2 - 4x$
3. $3x^2 - 4,\ -4x^2 + 5x + 4$	*c.* $7x^2 + 5x - 8$
4. $3x^2 - 4,\ -4x + 5x^2 + 4$	*d.* $-x^2 + 4x + 1$
5. $3x^2 - 4,\ -4x^2 + 5 + 4x$	*e.* $-x^2 + 5x$

4.2 ADDING THREE EXPRESSIONS

Model Problem

What is the sum of $a + 2b + 3c$, $4a - 5b + 5c$, and $3a - b + 2c$?

(A) $8a - 4b + 10c$ (B) $8a + 4b + 10c$ (C) $7a - 4b + 10c$ (D) $7a + 4b + 10c$
(E) $8a - 3b + 10c$

Solution:

To find the sum of the polynomials, first arrange them in order by placing like terms in the same column. Then add like terms.

$$\begin{array}{r} a + 2b + 3c \\ 4a - 5b + 5c \\ 3a - b + 2c \\ \hline 8a - 4b + 10c \end{array} \quad \textit{Ans.}\ (A)$$

Matching Test

Find the sum of each set of expressions. After completing this test, match your answers with the expressions in the box.

1. $a + 2b + 3c,\ 3a + b + 2c,\ 4a + 5b + 5c$	*a.* $-a + 6b + c$
2. $a + 2b + 3c,\ 3a - 2b + c,\ 4a - 5b - 5c$	*b.* $-7a + 8b - c$
3. $a + 2b + 3c,\ 2a - b + 3c,\ 5b - 4a - 5c$	*c.* $8a - 5b - c$
4. $a + 2b + 3c,\ 3a - c - 2b,\ 5b - 5a - 4c$	*d.* $8a + 8b + 10c$
5. $a + 2b + 3c,\ 2b + c - 3a,\ 4b - 5c - 5a$	*e.* $-a + 5b - 2c$

4.3 SUBTRACTION INVOLVING ONE VARIABLE

Model Problem

Subtract $x^2 + x - 1$ from $3x^2 - 5x$.

(A) $4x^2 - 6x + 1$ (B) $2x^2 + 6x + 1$ (C) $2x^2 - 6x + 1$ (D) $4x^2 - 4x + 1$
(E) $2x^2 - 6x - 1$

Solution:

To subtract the expression $x^2 + x - 1$, change the sign of each term and then add the result to $3x^2 - 5x$.

$$\begin{array}{rr} & 3x^2 - 5x \\ \text{Subtract:} & \underline{x^2 + x - 1} \end{array} \rightarrow \begin{array}{rr} & 3x^2 - 5x \\ \text{Add:} & \underline{-x^2 - x + 1} \\ & 2x^2 - 6x + 1 \end{array} \quad \textit{Ans. (C)}$$

Matching Test

After completing this test, match your answers with the expressions in the box.

1. Subtract $x^2 - x + 1$ from $3x^2 - 5x$.	*a.* $-5x + 8$
2. Subtract $3x^2 - x - 1$ from $3x^2 - 5x$.	*b.* $-4x + 1$
3. Subtract $3x^2 - 8x$ from $3x^2 - 5x$.	*c.* $2x^2 - 4x - 1$
4. Subtract $3x^2 - 8$ from $3x^2 - 5x$.	*d.* $3x$
5. Subtract $3x - x^2 + 3$ from $3x^2 - 5x$.	*e.* $4x^2 - 8x - 3$

4.4 SUBTRACTION INVOLVING TWO VARIABLES

Model Problem

Subtract $x + y - 2$ from $2x - 4y + 5$.

(A) $3x - 3y + 3$ (B) $-x + 5y - 7$ (C) $x + 5y + 7$ (D) $x - 5y - 7$ (E) $x - 5y + 7$

Solution:

To subtract the expression $x + y - 2$, change the sign of each term and then add the result to $2x - 4y + 5$.

$$\begin{array}{rr} & 2x - 4y + 5 \\ \text{Subtract:} & \underline{x + y - 2} \end{array} \rightarrow \begin{array}{rr} & 2x - 4y + 5 \\ \text{Add:} & \underline{-x - y + 2} \\ & x - 5y + 7 \end{array} \quad \textit{Ans. (E)}$$

Matching Test

After completing this test, match your answers with the expressions in the box.

1. Subtract $x + y - 2$ from $2x + 4y - 5$.	*a.* $4x - 3y - 2$
2. Subtract $x + y - 2$ from $2x - 5y + 4$.	*b.* $-x - 3y + 6$
3. Subtract $x + y - 2$ from $5x - 2y - 4$.	*c.* $x - 6y + 6$
4. Subtract $x + y - 2$ from $5y - 4x - 2$.	*d.* $x + 3y - 3$
5. Subtract $x + y - 2$ from $4 - 2y$.	*e.* $-5x + 4y$

Multiplying Polynomials

To multiply polynomials, multiply each term of the multiplicand by each term of the multiplier. Then combine like terms. For example, to multiply $(a-b+c)$ by $(a+b-c)$, first multiply $(a-b+c)$ by a, then by b, and then by $-c$, arranging the three products in rows so that like terms appear in the same columns.

$$\begin{array}{lrrrrr}
 & a & -\ b & +\ c & & \\
 & a & +\ b & -\ c & & \\
\hline
 & a^2 & -\ ab & +\ ac & & \\
 & & ab & & -\ b^2 + & bc \\
\text{Add:} & & & -\ ac & & +\ bc - c^2 \\
\hline
 & a^2 & & & -\ b^2 & +\ 2bc - c^2
\end{array}$$

To multiply two binomials by inspection, use the procedure illustrated in set 4.5.

4.5 MULTIPLYING BINOMIALS BY INSPECTION: FINDING THE AREA OF A RECTANGLE

Model Problem

The length of a rectangle is $3x+2$ and the width is $x-5$. Express the area of the rectangle as a trinomial in x.

(A) $3x^2+3x-10$ (B) $3x^2-10$ (C) $3x^2-13x-10$ (D) $3x^2+13x-10$ (E) $8x-6$

Solution:

To find the area of the rectangle, multiply the length $(3x+2)$ by the width $(x-5)$.

1. Multiply the first terms: $(3x)(x)=3x^2$

2. Add the inner and outer products:

$$\overbrace{(3x+2)\qquad(x-5)}^{-15x}\ \text{(outer)},\quad \underbrace{2\qquad x}_{+2x}\ \text{(inner)} \longrightarrow -15x+2x=-13x$$

3. Multiply the last terms: $(+2)(-5)=-10$

4. Combine the results to obtain the answer. *Ans.* (C)

Matching Test

In each of the following, write the expression for the area of the rectangle. After completing this test, match your answers with the expressions in the box.

1. The length of the rectangle is $3x+2$ and the width is $x+5$.

2. The length of the rectangle is $3x-2$ and the width is $x+5$.

3. The length of the rectangle is $3x-2$ and the width is $x-5$.

4. The length of the rectangle is $2x-3$ and the width is $x+5$.

5. The length of the rectangle is $2x+3$ and the width is $x-5$.

a. $3x^2+17x+10$
b. $2x^2-7x-15$
c. $3x^2+13x-10$
d. $2x^2+7x-15$
e. $3x^2-17x+10$

4.6 SQUARING A BINOMIAL

Model Problems

(1) The expression $(x + 5)^2$ is equal to

(A) $x^2 + 25$ (B) $x^2 + 5x + 25$ (C) $x^2 + 10x + 25$ (D) $x^2 + 10$ (E) $x^2 + 5x + 10$

Solution:

Since the expression $(x + 5)^2$ is the product of $x + 5$ and $x + 5$, determine the correct choice by multiplying.

$$\begin{array}{r} x + 5 \\ x + 5 \\ \hline x^2 + 5x \qquad \\ 5x + 25 \\ \hline x^2 + 10x + 25 \end{array}$$

Ans. (C)

(2) If one side of a square is represented by $x - 2$, the area is represented by

(A) $x^2 + 4$ (B) $4x - 8$ (C) $x^2 - 4x + 4$ (D) $2x - 4$ (E) $x^2 - 4$

Solution:

To find the area of a square, square a side. Hence, if a square has a side of length $(x - 2)$, then its area is $(x - 2)^2$.

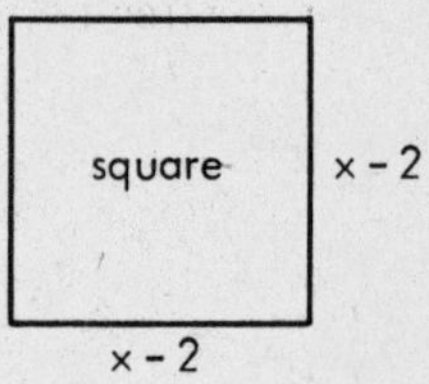

Since $(x - 2)^2 = (x - 2)(x - 2)$, multiply $(x - 2)$ by $(x - 2)$.

$$\begin{array}{r} x - 2 \\ x - 2 \\ \hline x^2 - 2x \qquad \\ - 2x + 4 \\ \hline x^2 - 4x + 4 \end{array}$$

Ans. (C)

After completing each test, match your answers with the expressions in the boxes.

Matching Test A

Square each binomial.

1. $(x - 5)^2$ **2.** $(5 - x)^2$ **3.** $(5 - 2x)^2$ **4.** $(5x - 2)^2$ **5.** $(5x + 2)^2$

a. $25x^2 + 20x + 4$ *b.* $25 - 20x + 4x^2$ *c.* $25x^2 - 20x + 4$ *d.* $x^2 - 10x + 25$
e. $25 - 10x + x^2$

Matching Test B

1. Represent the area of a square whose side is $x + 2$.

2. Represent the area of a square whose side is $2x - 2$.

3. Represent the area of a square whose side is $5x - 2$.

4. Represent the side of a square whose area is $4x^2 + 4x + 1$.

5. Represent the side of a square whose area is $9x^2 + 60x + 100$.

a. $4x^2 - 8x + 4$
b. $3x + 10$
c. $x^2 + 4x + 4$
d. $25x^2 - 20x + 4$
e. $2x + 1$

4.7 RELATING $(x + y)^2$, $(x - y)^2$, AND xy

Model Problem

If $x^2 + y^2 = 34$ and $xy = 15$, find $(x + y)^2$ and $(x - y)^2$.

(A) 49, 19 (B) 34, 4 (C) 94, −26 (D) 64, 4 (E) 49, 4

Solution:

$$(x + y)^2 = x^2 + 2xy + y^2 = x^2 + y^2 + 2xy \qquad (x - y)^2 = x^2 - 2xy + y^2 = x^2 + y^2 - 2xy$$

Substitute 34 for $x^2 + y^2$ and 15 for xy.

$$(x + y)^2 = 34 + 2(15) = 34 + 30 = 64 \qquad (x - y)^2 = 34 - 2(15) = 34 - 30 = 4$$

Ans. (D)

Matching Test

After completing this test, match your answers with the numbers in the box.

1. If $x^2 + y^2 = 52$ and $xy = 24$, find $(x + y)^2$.
2. If $(x + y)^2 = 81$ and $xy = 18$, find $x^2 + y^2$.
3. If $(x - y)^2 = 49$ and $xy = 18$, find $x^2 + y^2$.
4. If $(x + y)^2 = 169$ and $x^2 + y^2 = 89$, find xy.
5. If $x^2 + y^2 = 116$ and $xy = 40$, find $(x - y)^2$.

a.	45
b.	40
c.	85
d.	100
e.	36

4.8 SIMPLIFYING AN EXPRESSION CONTAINING THE SQUARE OF A BINOMIAL

Model Problem

Remove parentheses and combine similar terms: $(a + 2)^2 - 2(a - 2)$

(A) $a^2 - 2a + 8$ (B) $a^2 + 2a + 4$ (C) $a^2 - 2a$ (D) $a^2 + 2a + 2$ (E) $a^2 + 2a + 8$

Solution:

Remove parentheses by replacing $(a + 2)^2$ with $a^2 + 4a + 4$ and $-2(a - 2)$ with $-2a + 4$.

$$(a + 2)^2 - 2(a - 2)$$
$$a^2 + 4a + 4 - 2a + 4$$

$$\begin{array}{r} a + 2 \\ \underline{a + 2} \\ a^2 + 2a \quad\quad \\ \underline{+ 2a + 4} \\ a^2 + 4a + 4 \end{array}$$

Collect like terms.

$$a^2 + 4a - 2a + 4 + 4$$

Combine.

$$a^2 + 2a + 8 \quad \textit{Ans. } (E)$$

Matching Test

Simplify each expression. After completing this test, match your answers with the expressions in the box.

1. $(a+2)^2 - 3(a-2)$	*a.* $-a^2 - 6a + 2$
2. $(a-2)^2 + 3(a-2)$	*b.* $a^2 - a - 2$
3. $3(a-2)^2 - 2(a-3)$	*c.* $3a^2 - 14a + 18$
4. $(2-a)^2 - 3(2-a)$	*d.* $a^2 - a - 2$
5. $2(3-a) - (a+2)^2$	*e.* $a^2 + a + 10$

4.9 MULTIPLYING A TRINOMIAL BY A BINOMIAL: FINDING THE VOLUME OF A RECTANGULAR SOLID

Model Problem

The volume of a rectangular solid is found by multiplying the area of the base by the height. If the area of the base is $x^2 - x + 1$ and the height is $x + 1$, express the volume of the solid as a polynomial.

(A) $x^3 + 1$ (B) $x^3 - 2x^2 - 2x + 1$ (C) $x^3 - 1$ (D) $x^3 - 2x^2 + 1$ (E) $x^3 - x^2 + 1$

Solution:

Find the expression for the volume by multiplying the area of the base, $x^2 - x + 1$, by the height, $x + 1$, as follows:

$$\begin{array}{rrrr} x^2 & -\,x & +\,1 & \\ & x & +\,1 & \\ \hline x^3 - x^2 & +\,x & & \\ x^2 & -\,x & +\,1 & \\ \hline x^3 & & +\,1 & \end{array}$$

Ans. (A)

Matching Test

In each of the following, the two given expressions represent, respectively, the area of the base and the height of a rectangular solid. Write the expression for the volume. After completing this test, match your answers with the expressions in the box.

1. $x^2 + x + 1,\ x + 1$	*a.* $x^3 + 1$
2. $x^2 + 2x + 1,\ x + 1$	*b.* $x^3 - 1$
3. $x^2 + x + 1,\ x - 1$	*c.* $x^3 + 3x^2 + 4x + 4$
4. $x^2 - x + 1,\ x + 1$	*d.* $x^3 + 3x^2 + 3x + 1$
5. $x^2 + x + 2,\ x + 2$	*e.* $x^3 + 2x^2 + 2x + 1$

Dividing Polynomials

Dividing a Polynomial by a Monomial

To divide a polynomial by a monomial, divide each term of the polynomial by the monomial.

$$\frac{2a^3 + 4a^2 + 6a}{2a} = \frac{2a^3}{2a} + \frac{4a^2}{2a} + \frac{6a}{2a}$$
$$= a^2 + 2a + 3$$

$$\frac{2a^3 + 4a^2 + 6a}{2a} = \frac{2a(a^2 + 2a + 3)}{2a}$$
$$= a^2 + 2a + 3$$

This result can also be obtained by factoring.

For another illustration of this method, see set 4.10.

Dividing a Polynomial by Another Polynomial

The procedure for dividing one polynomial by another is similar to long division in arithmetic. For illustrations of this method, see sets 4.12 and 4.13.

4.10 DIVIDING A BINOMIAL BY A MONOMIAL (FACTORING METHOD)

Model Problem

If $5x - 15x^2$ is divided by $5x$, the result is

(A) $1 - 3x$ (B) $3x - 1$ (C) $-3x$ (D) $-2x$ (E) $1 - 15x^2$

Solution:

Divide $5x - 15x^2$ by $5x$ as follows:

1. Express the division in the form of a fraction: $\frac{5x - 15x^2}{5x}$
2. Factor $5x - 15x^2$. Since $5x$ is a common factor of both terms, use $5x$ as one of the factors. The other factor is $(1 - 3x)$. $\frac{\overset{1}{\cancel{5x}}(1 - 3x)}{\underset{1}{\cancel{5x}}}$
3. Divide out (cancel) the common factor $5x$: $1 - 3x$ *Ans.* (A)

Matching Test

After completing this test, match your answers with the expressions in the box.

1. Divide $5x + 15x^2$ by $5x$.	*a.* $3 - x$
2. Divide $15x - 5x^2$ by $5x$.	*b.* $1 + 3x$
3. Divide $15x - 6x^2$ by $3x$.	*c.* $3 - x$
4. Divide $18x^2y - 6xy$ by $6xy$.	*d.* $5 - 2x$
5. Divide $18x^3y^2 - 6x^4y^2$ by $6x^3y^2$.	*e.* $3x - 1$

4.11 DIVIDING A BINOMIAL BY A MONOMIAL: FINDING A DIMENSION OF A RECTANGLE

Model Problem

What is the length of a rectangle whose width is $4n$ and whose area is $20n^2 + 12n$?

(A) $5n^2 + 3n$ (B) $80n^3 + 48n^2$ (C) $20n^2 + 3$ (D) $5n + 3$ (E) $8n$

Solution:

The length of a rectangle can be found by dividing the area of the rectangle by its width.

Hence, length $= \dfrac{\text{area}}{\text{width}} = \dfrac{20n^2 + 12n}{4n}$.

Divide each term of $20n^2 + 12n$ by $4n$ to obtain the length $5n + 3$. *Ans.* (D)

Matching Test

In the following table, find the expression for the missing dimension of the rectangle. After completing this test, match your answers with the expressions in the box.

	Length	*Width*	*Area*
1.	?	$2n$	$20n^2 + 12n$
2.	?	$4n$	$20n^2 - 12n$
3.	$4n$	?	$28n^2 - 8n$
4.	$10n$	?	$10n^2 + 20n$
5.	$2n$	?	$12n - 18n^2$

a. $7n - 2$
b. $n + 2$
c. $5n - 3$
d. $10n + 6$
e. $6 - 9n$

4.12 FINDING A BINOMIAL FACTOR BY DIVISION

Model Problem

The trinomial $x^2 - 3x - 18$ is exactly divisible by

(A) $x + 2$ (B) $x + 3$ (C) $x + 6$ (D) $x - 9$ (E) $x + 9$

Solution:

To show that the trinomial $x^2 - 3x - 18$ is exactly divisible by $x + 3$, divide the trinomial by $x + 3$.

$$\begin{array}{r|l} & x \quad - 6 \\ \hline x + 3 & x^2 - 3x - 18 \\ & x^2 + 3x \\ \hline & \quad - 6x - 18 \\ & \quad - 6x - 18 \\ \hline \end{array}$$

1. Arrange the problem in the form of long division.
2. Divide x^2 by x to obtain x, the first term of the quotient.
3. Multiply the divisor, $x + 3$, by x to obtain $x^2 + 3x$.
4. By subtracting like terms, $-6x$ is obtained. Combine $-6x$ with the next place, -18.
5. Divide $-6x$ by x to obtain -6, the second term of the quotient.
6. Multiply the divisor, $x + 3$, by -6 to obtain $-6x - 18$.
7. By subtracting like terms, a remainder of 0 is obtained. Hence, it must follow that $x^2 - 3x - 18$ is exactly divisible by $x + 3$. *Ans.* (B)

Wrong Choices:

In choices (A), (C), (D), and (E), if the division process is used, a remainder is obtained. Hence, in each of the remaining choices there is no exact divisibility.

Matching Test

Match each expression with the binomial in the box by which it is exactly divisible.

1. $x^2 + 3x - 18$ **2.** $x^2 - 7x - 18$ **3.** $x^2 - 7x - 30$ **4.** $x^2 - 11x + 30$ **5.** $x^2 + 17x + 30$

a. $x + 2$	*b.* $x - 10$	*c.* $x - 9$	*d.* $x - 6$	*e.* $x + 6$

4.13 FINDING A REMAINDER

Model Problem

If $x^2 - 5x - 8$ is divided by $x - 2$, the remainder is

(A) 4 (B) −2 (C) −14 (D) 14 (E) 2

Solution:

To find the remainder, divide $x^2 - 5x - 8$ by $x - 2$.

$$\begin{array}{r|l} & x \quad -3 \\ \hline x - 2 & x^2 - 5x - 8 \\ & x^2 - 2x \\ \hline & \quad - 3x - 8 \\ & \quad - 3x + 6 \\ \hline & \qquad\quad - 14 \end{array}$$

1. Arrange the problem in the form of long division.
2. Divide x^2 by x to obtain x, the first term of the quotient.
3. Multiply the divisor, $x - 2$, by x to obtain $x^2 - 2x$.
4. By subtracting like terms, $-3x$ is obtained. Combine $-3x$ with the next place, -8.
5. Divide $-3x$ by x to obtain -3, the second term of the quotient.
6. Multiply the divisor, $x - 2$, by -3 to obtain $-3x + 6$.
7. By subtracting like terms, -14 is obtained. No further steps are needed. -14 is the remainder.

Ans. (C)

Matching Test

After completing the indicated divisions, match the remainders obtained with the numbers in the box.

1. Divide $x^2 - 5x - 8$ by $x + 2$.
2. Divide $x^2 - 5x + 8$ by $x - 2$.
3. Divide $x^2 + 5x + 8$ by $x - 1$.
4. Divide $2x^2 + 5x + 8$ by $x + 1$.
5. Divide $2x^2 - 5x - 8$ by $2x - 1$.

a. −10
b. 6
c. 14
d. 5
e. 2

5. FACTORING POLYNOMIALS

5.1 FACTORING A POLYNOMIAL HAVING A COMMON MONOMIAL FACTOR

A common monomial factor of a polynomial is a monomial that is a factor of each term of the polynomial. The highest common monomial factor of a polynomial is the product of all the common monomial factors. For example, $14x + 28x^2$ has the common monomial factors 7, 2, and x, but its *highest* common monomial factor is $14x$.

Model Problem

Factor: $21st^5 - 35s^2t^4 + 7st^3$

(A) $7st(3t^4 - 5st^3 + t^2)$ (B) $7st^3(3t^2 - 5st + 1)$ (C) $7st^3(3t^2 - 5st)$

(D) $7st^2(3t^3 - 5st^2 + t)$ (E) $7t^3(3st^2 - 5s^2t + s)$

Solution:

Since $21st^5 - 35s^2t^4 + 7st^3$ has a common monomial factor, factor as follows:

1. Use the highest common monomial factor, $7st^3$, as one factor.
2. Find the other factor by dividing each term of the polynomial by the highest common monomial factor. Obtain $3t^2 - 5st + 1$ in this way.

Hence, $21st^5 - 35s^2t^4 + 7st^3 = 7st^3(3t^2 - 5st + 1)$. *Ans.* (B)

Matching Test

Factor each expression. After completing this test, match your answers with the expressions in the box.

1. $6st^5 + 9s^2t^4 - 12s^3t^3$	*a.* $6s^3t^3(2t^3 - 3s^2t + 4s^3)$
2. $12s^2t^5 - 20s^3t^4 - 24s^4t^3$	*b.* $4s^2t^3(3t^2 - 5st - 6s^2)$
3. $12s^3t^6 - 18s^5t^4 + 24s^6t^3$	*c.* $8s^2t^4(2t^2 + 3st + 6s^2)$
4. $16s^2t^6 + 24s^3t^5 + 48s^4t^4$	*d.* $12st^3(3t^4 - 2s^2t^2 + 4s^4)$
5. $36st^7 - 24s^3t^5 + 48s^5t^3$	*e.* $3st^3(2t^2 + 3st - 4s^2)$

5.2 FACTORING THE DIFFERENCE OF TWO SQUARES: $x^2 - y^2 = (x + y)(x - y)$

An expression such as $x^2 - y^2$, which is the difference of two squares, has two factors. One factor of $x^2 - y^2$ is $x + y$, the sum of the principal, or positive, square roots; and the other factor is $x - y$, their difference.

Model Problem

Factor: $9a^2 - b^2$

(A) $9(a - b)(a + b)$ (B) $(9a - b)(a + b)$ (C) $(9a + b)(a - b)$

(D) $(3a + b)(3a - b)$ (E) none of these

Solution:

Since $9a^2 - b^2$ is the difference of two squares, factor as follows:

1. Obtain the principal square root of each square: $\sqrt{9a^2} = 3a$, $\sqrt{b^2} = b$
2. One factor is the sum of the principal square roots; the other is their difference. Hence, $9a^2 - b^2 = (3a + b)(3a - b)$. *Ans.* (D)

Matching Test

Factor each expression. After completing this test, match your answers with the expressions in the box.

1. $4b^2 - 9a^2$

2. $9b^2 - 4a^2$

3. $9 - 4a^2b^2$

4. $4a^2b^2 - 9$

5. $9a^2 - \dfrac{b^2}{4}$

a. $(3b - 2a)(3b + 2a)$
b. $\left(3a - \dfrac{b}{2}\right)\left(3a + \dfrac{b}{2}\right)$
c. $(2b - 3a)(2b + 3a)$
d. $(3 - 2ab)(3 + 2ab)$
e. $(2ab - 3)(2ab + 3)$

5.3 FACTORING A TRINOMIAL OF THE FORM $x^2 + bx + c$

If factorable, a trinomial of the form $x^2 + bx + c$ has two binomial factors. These binomial factors can be found as follows:

1. The first term in each binomial is x.
2. To find the two last terms, select two numbers whose product is c, the last term of the trinomial, and whose sum is b, the coefficient of x in the middle term.

The following model problem illustrates how this procedure is applied:

Model Problem

Find the factors of $x^2 - x - 42$.

(A) $(x - 7)(x + 6)$ (B) $(x + 7)(x - 6)$ (C) $(x - 14)(x + 3)$
(D) $(x - 3)(x + 14)$ (E) $(x - 14)(x - 3)$

Solution:

To factor a trinomial such as $x^2 - x - 42$, whose first term, x^2, has the coefficient 1, proceed as follows:

1. Use x and x, the factors of x^2, as the first terms of the binomial factors.
2. Find two numbers whose product is -42, the last term, and whose sum is -1, the numerical coefficient of the middle term. These numbers are -7 and $+6$. Use these numbers, -7 and $+6$, as the last terms of the required binomial factors.
3. Form binomial factors from the results obtained in steps 1 and 2.

$$(x - 7)(x + 6)$$

Ans. (A)

Matching Test

Factor each expression. After completing this test, match your answers with the expressions in the box.

1. $x^2 - x - 12$ **2.** $x^2 + x - 12$ **3.** $x^2 - 7x + 12$ **4.** $x^2 - 8x + 12$ **5.** $x^2 + 4x - 12$

a. $(x+4)(x-3)$	*b.* $(x-4)(x-3)$	*c.* $(x-6)(x-2)$
d. $(x+6)(x-2)$	*e.* $(x-4)(x+3)$	

5.4 FACTORING A PERFECT-SQUARE TRINOMIAL

A perfect-square trinomial is a trinomial containing two terms that are perfect squares and a third term that is double the product of the square roots of the perfect-square terms.

For example, $x^2 - 6x + 9$ and $x^2 + 6x + 9$ are perfect square trinomials. If factored, $x^2 - 6x + 9 = (x-3)^2$ and $x^2 + 6x + 9 = (x+3)^2$.

Model Problem

Factor: $49 - 14x + x^2$

(A) $(7-x)(7+x)$ (B) $7(1-x)^2$ (C) $7(x-1)^2$ (D) $(7-x)^2$ (E) $(7+x)^2$

Solution:

The expression $49 - 14x + x^2$ is a perfect-square trinomial. Note that the end terms, 49 and x^2, are perfect squares. Note also that the middle term, $-14x$, is double the product of 7 and $-x$, which are respective square roots of the end terms.

Hence, $49 - 14x + x^2$ is the square of $7 - x$.

Ans. (D)

Matching Test

Factor each expression. After completing each test, match your answers with the expressions in the box.

1. $9x^2 - 12x + 4$ **2.** $4x^2 + 12x + 9$ **3.** $4x^4 + 12x^2 + 9$ **4.** $9x^4 - 12x^2 + 4$

5. $9 - 12x^2 + 4x^4$

a. $(2x^2+3)^2$	*b.* $(3x^2-2)^2$	*c.* $(3x-2)^2$	*d.* $(3-2x^2)^2$	*e.* $(2x+3)^2$

5.5 FACTORING A POLYNOMIAL COMPLETELY

A polynomial is factored completely only if the polynomial factors cannot be factored further. For example, to factor $8a^2 - 8b^2$ completely, first remove the highest common factor, 8, to obtain $8(a^2 - b^2)$; then factor this result into $8(a+b)(a-b)$.

Model Problem

If $3x^2 - 48$ is factored completely, the result is

(A) $3(x^2 - 16)$ (B) $(3x + 12)(x - 4)$ (C) $(3x - 12)(x + 4)$ (D) $3(x - 4)^2$
(E) $3(x + 4)(x - 4)$

Solution:

To factor the expression $3x^2 - 48$ completely, first note that 3 is a common factor of its two terms. Hence, 3 is a factor of $3x^2 - 48$. The remaining factor is $x^2 - 16$.

Since $x^2 - 16$ is the difference of two squares, its factors are $(x + 4)$ and $(x - 4)$.

Hence, $3x^2 - 48 = 3(x + 4)(x - 4)$. *Ans.* (E)

Matching Test

Factor each expression *completely*. After completing each test, match your answers with the expressions in the box.

1. $4x^2 - 36$	*a.* $9(1 - 2x)(1 + 2x)$
2. $9x^2 - 36$	*b.* $9(x + 2)(x - 2)$
3. $4 - 36x^2$	*c.* $9(1 - 2x^2)(1 + 2x^2)$
4. $9 - 36x^2$	*d.* $4(1 - 3x)(1 + 3x)$
5. $9 - 36x^4$	*e.* $4(x + 3)(x - 3)$

6. REVIEWING ALGEBRAIC FRACTIONS

Reducing a Fraction to Lowest Terms

A fraction is reduced to lowest terms when its numerator and denominator have no common factor except 1.

For example, $\frac{2x}{5x}$ is reduced to $\frac{2}{5}$ by dividing its numerator and denominator by the common factor, x. The resulting fraction, $\frac{2}{5}$, is in lowest terms.

To reduce a fraction to lowest terms, apply the following principle:

The value of a fraction is not changed if its numerator and denominator are divided by the same number (except zero).

For example, $\frac{2(a + b)}{3(b + a)} = \frac{2}{3}$, since $(a + b)$ and $(b + a)$ have the same value.

However, $\frac{2(a - b)}{3(b - a)} = -\frac{2}{3}$. Here, the two binomial factors are negatives of each other and may be eliminated, leaving a quotient of -1. Thus, $\frac{2\overset{(-1)}{\cancel{(a - b)}}}{3\cancel{(b - a)}} = -\frac{2}{3}$.

6.1 ADDING OR SUBTRACTING FRACTIONS (COMBINING FRACTIONS)

To add or subtract fractions having the same denominator, combine them by keeping the denominator unchanged and adding or subtracting the numerators.

$$\frac{1}{b}+\frac{8}{b}-\frac{a}{b}=\frac{1+8-a}{b}=\frac{9-a}{b}$$

To add or subtract fractions having different denominators, change them into equivalent fractions having the same denominator, using the lowest common denominator for the purpose. Then apply the procedure for adding or subtracting fractions having the same denominator.

For example, $6b$ is the lowest common denominator of $\frac{4}{3}$, $\frac{1}{b}$, and $\frac{1}{6}$.

$$\frac{4}{3}+\frac{1}{b}-\frac{1}{6}=\frac{8b}{6b}+\frac{6}{6b}-\frac{b}{6b}=\frac{8b+6-b}{6b}=\frac{7b+6}{6b}$$

Model Problems

(1) Combine into one fraction: $\frac{1}{a}+\frac{3}{5}$

(A) $\frac{4}{a+5}$ (B) $\frac{4}{5a}$ (C) $\frac{5+3a}{5a}$ (D) $\frac{8}{5a}$ (E) $\frac{8}{5}$

Solution:

Since the denominators are a and 5, the lowest common denominator is $5a$.

Change each fraction into an equivalent fraction whose denominator is the lowest common denominator, $5a$:

$$\frac{1}{a}+\frac{3}{5}$$

$$\frac{5}{5a}+\frac{3a}{5a}$$

Since the denominators are now the same, combine into one fraction by adding numerators and keeping the denominator:

$$\frac{5+3a}{5a} \quad \textit{Ans. (C)}$$

(2) Combine into one fraction: $1-\frac{b}{a}$

(A) $\frac{a-b}{a}$ (B) $\frac{1-b}{a}$ (C) $\frac{1-b}{1-a}$ (D) $\frac{b-a}{a}$ (E) $\frac{1-ab}{a}$

Solution:

Substitute $\frac{a}{a}$ for 1: $\frac{a}{a}-\frac{b}{a}$

Since the two fractions now have the same denominator, keep the denominator and combine the numerators:

$$\frac{a-b}{a} \quad \textit{Ans. (A)}$$

In the following tests, combine the two terms of each expression into one fraction. After completing these tests, match your answers with the fractions in the boxes.

Matching Test A

1. $a+\frac{2}{3}$ **2.** $\frac{1}{a}+\frac{2}{3}$ **3.** $\frac{1}{a}+\frac{3}{2}$ **4.** $a+\frac{3}{2}$ **5.** $\frac{3}{a}+\frac{1}{2}$

a. $\frac{2+3a}{2a}$	*b.* $\frac{2a+3}{2}$	*c.* $\frac{3a+2}{3}$	*d.* $\frac{6+a}{2a}$	*e.* $\frac{3+2a}{3a}$

Matching Test B

1. $\frac{a}{b}+1$ **2.** $\frac{b}{a}-1$ **3.** $\frac{1}{a}+b$ **4.** $\frac{1}{b}+a$ **5.** $1-\frac{a}{b}$

a. $\frac{b-a}{b}$	*b.* $\frac{b-a}{a}$	*c.* $\frac{1+ab}{b}$	*d.* $\frac{1+ab}{a}$	*e.* $\frac{a+b}{b}$

6.2 MULTIPLYING FRACTIONS

To multiply fractions whose numerators contain no factors in common with any of the denominators, multiply the numerators to obtain the numerator of the answer, and multiply the denominators to obtain the denominator of the answer. For example, $\frac{3}{a}\cdot\frac{b}{5}=\frac{3b}{5a}$.

To multiply fractions in which the numerators and denominators contain common factors, first divide out all such common factors. Then multiply the remaining factors in the numerators to obtain the numerator of the answer, and multiply the remaining factors in the denominators to obtain the denominator of the answer. For example, $\frac{3}{2a}\cdot\frac{a}{c}=\frac{3}{\underset{2}{\cancel{2a}}}\cdot\frac{\overset{1}{\cancel{a}}}{c}=\frac{3}{2c}$.

Model Problem

Multiply: $\frac{3b}{a}\cdot\frac{9c}{b}$

(A) $\frac{27bc}{ab}$ (B) $\frac{27c}{a}$ (C) $\frac{12bc}{ab}$ (D) $\frac{12c}{a}$ (E) none of these

Solution:

Cancel the common factor b: $\frac{\overset{3}{\cancel{3b}}}{a}\cdot\frac{9c}{\underset{1}{\cancel{b}}}$

Multiply the numerators to obtain the numerator of the answer, and multiply the denominators to obtain its denominator:

$$\frac{3(9c)}{1(a)}=\frac{27c}{a} \quad \textit{Ans. (B)}$$

Matching Test

Multiply the given fractions. After completing this test, match your answers with the fractions in the box.

1. $\frac{2b}{a}\cdot\frac{4c}{b}$ **2.** $\frac{2b}{a}\cdot\frac{c}{4b}$ **3.** $\frac{b}{2a}\cdot\frac{4c}{b}$ **4.** $\frac{a}{2b}\cdot\frac{4c}{b}$ **5.** $\frac{a}{2b}\cdot\frac{4b}{c}$

a. $\frac{c}{2a}$	*b.* $\frac{2a}{c}$	*c.* $\frac{2c}{a}$	*d.* $\frac{8c}{a}$	*e.* $\frac{2ac}{b^2}$

6.3 DIVIDING FRACTIONS

To divide by a fraction, invert the fraction and then multiply.

For example, $\frac{3}{a}\div\frac{a}{c}=\frac{3}{a}\cdot\frac{c}{a}=\frac{3c}{a^2}$.

Model Problem

Divide: $\frac{a+b}{a-b}\div\frac{b+a}{b-a}$

(A) 1 (B) -1 (C) $\frac{(a+b)^2}{(a-b)^2}$ (D) $-\frac{(a+b)^2}{(a-b)^2}$ (E) $\frac{a^2+b^2}{a^2-b^2}$

Solution:

To divide by a fraction, invert the fraction and multiply.

Thus, invert $\frac{b+a}{b-a}$ and multiply: $\frac{a+b}{a-b}\cdot\frac{b-a}{b+a}$

Since $(a-b)$ and $(b-a)$ are opposites, substitute $-(a-b)$ for $(b-a)$.

Since $(a+b)$ and $(b+a)$ are equals, substitute $(a+b)$ for $(b+a)$.

Cancel common factors: $$\frac{\overset{1}{\cancel{(a+b)}}}{\underset{1}{\cancel{(a-b)}}}\cdot\frac{\overset{(-1)}{-\cancel{(a-b)}}}{\underset{1}{\cancel{(a+b)}}}$$

$1(-1)=-1$ *Ans.* (B)

Matching Test

Divide the first fraction by the second. After completing this test, match your answers with the expressions in the box.

1. $\frac{a-b}{a+b}\div\frac{b-a}{b+a}$ **2.** $\frac{a+b}{a-b}\div\frac{b+a}{a-b}$ **3.** $\frac{a-b}{b}\div\frac{a-b}{a}$

4. $\frac{a+b}{a}\div\frac{b+a}{b}$ **5.** $\frac{a-b}{a}\div\frac{b-a}{b}$

a. $\frac{a}{b}$	*b.* -1	*c.* $\frac{b}{a}$	*d.* 1	*e.* $-\frac{b}{a}$

6.4 SIMPLIFYING COMPLEX FRACTIONS

A complex fraction is a fraction that contains at least one other fraction within it. For example,

$\dfrac{2+\frac{1}{2}}{3}$ and $\dfrac{\frac{1}{4}+3}{\frac{1}{2}}$ are complex fractions.

The following methods may be used to simplify a complex fraction:

Method 1:
1. Combine the terms of the numerator.
2. Combine the terms of the denominator.
3. Divide the new numerator by the new denominator.

For example, $\dfrac{\frac{1}{2}+\frac{1}{4}}{\frac{1}{2}-\frac{1}{4}}=\dfrac{\frac{3}{4}}{\frac{1}{4}}=\frac{3}{4}\div\frac{1}{4}=\frac{3}{\cancel{4}_1}\cdot\frac{\cancel{4}^1}{1}=3.$

Method 2:
1. Find the lowest common denominator of the fractions contained within the complex fraction.
2. Multiply both the numerator and the denominator by the lowest common denominator.

For example, simplify $\dfrac{\frac{1}{2}+\frac{1}{4}}{\frac{1}{2}-\frac{1}{4}}$ by multiplying the numerator and the denominator by the lowest common denominator, 4, as follows:

$$\frac{4\left(\frac{1}{2}+\frac{1}{4}\right)}{4\left(\frac{1}{2}-\frac{1}{4}\right)}=\frac{2+1}{2-1}=\frac{3}{1}=3$$

Model Problem

Simplify: $\dfrac{\frac{1}{2}-\frac{1}{3}}{\frac{2}{3}-\frac{3}{5}}$

(A) $\frac{2}{5}$ (B) $\frac{25}{38}$ (C) $\frac{5}{2}$ (D) 3 (E) $\frac{3}{2}$

Solution:

Method 1

Simplify the numerator: $\frac{1}{2}-\frac{1}{3}=\frac{3}{6}-\frac{2}{6}=\frac{1}{6}$

Simplify the denominator: $\frac{2}{3}-\frac{3}{5}=\frac{10}{15}-\frac{9}{15}=\frac{1}{15}$

Divide the numerator by the denominator: $\frac{1}{6}\div\frac{1}{15}=\frac{1}{\cancel{6}_2}\times\frac{\cancel{15}^5}{1}=\frac{5}{2}$

Method 2

Find the lowest common denominator of all the fractions within the complex fraction.

The lowest common denominator of $\frac{1}{2}, \frac{1}{3}, \frac{2}{3}$, and $\frac{3}{5}$ is 30.

Multiply both the numerator and the denominator by the L.C.D. and reduce if necessary:

$$\frac{30\left(\frac{1}{2}-\frac{1}{3}\right)}{30\left(\frac{2}{3}-\frac{3}{5}\right)} = \frac{15-10}{20-18} = \frac{5}{2}$$

Ans. (C)

After completing each test, match your answers with the answers in the boxes.

Matching Test A

Simplify each complex fraction.

1. $\dfrac{\frac{1}{2}-\frac{1}{3}}{1-\frac{2}{3}}$ **2.** $\dfrac{\frac{1}{2}+\frac{1}{3}}{1-\frac{2}{3}}$ **3.** $\dfrac{\frac{1}{2}+\frac{1}{3}}{\frac{1}{6}+\frac{2}{3}}$ **4.** $\dfrac{1+\frac{1}{3}}{\frac{1}{6}+\frac{2}{3}}$ **5.** $\dfrac{1+\frac{1}{4}}{\frac{5}{6}-\frac{2}{3}}$

a. 1	*b.* $\frac{8}{5}$ or $1\frac{3}{5}$	*c.* $\frac{1}{2}$	*d.* $\frac{15}{2}$ or $7\frac{1}{2}$	*e.* $\frac{5}{2}$ or $2\frac{1}{2}$

Matching Test B

Evaluate the fraction $\dfrac{\frac{1}{x}+\frac{1}{y}}{\frac{x}{y}}$ for the given values of x and y.

1. $x=2,\ y=3$ **2.** $x=2,\ y=4$ **3.** $x=3,\ y=4$ **4.** $x=1,\ y=5$ **5.** $x=5,\ y=1$

a. 6	*b.* $\frac{5}{4}$	*c.* $\frac{7}{9}$	*d.* $\frac{3}{2}$	*e.* $\frac{6}{25}$

Matching Test C

Evaluate the fraction $\frac{a+b}{c}$ for the given values of a, b, and c.

1. $a=\frac{1}{2},\ b=\frac{1}{3},\ c=\frac{1}{4}$	*a.* $\frac{7}{4}$
2. $a=\frac{1}{2},\ b=\frac{1}{3},\ c=\frac{1}{3}$	*b.* $\frac{10}{3}$
3. $a=\frac{1}{2},\ b=\frac{1}{3},\ c=\frac{1}{2}$	*c.* $\frac{7}{2}$
4. $a=\frac{1}{3},\ b=\frac{1}{4},\ c=\frac{1}{6}$	*d.* $\frac{5}{2}$
5. $a=\frac{1}{3},\ b=\frac{1}{4},\ c=\frac{1}{3}$	*e.* $\frac{5}{3}$

Matching Test D

Simplify each complex fraction.

1. $\dfrac{a+\frac{1}{2}}{b}$ **2.** $\dfrac{a+\frac{1}{b}}{2}$ **3.** $\dfrac{\frac{1}{a}+2}{b}$ **4.** $\dfrac{\frac{1}{a}+b}{2}$ **5.** $\dfrac{a}{\frac{1}{b}+2}$

a. $\frac{1+2a}{ab}$ *b.* $\frac{1+ab}{2a}$ *c.* $\frac{ab+1}{2b}$ *d.* $\frac{2a+1}{2b}$ *e.* $\frac{ab}{1+2b}$

7. PRACTICE PROBLEMS

1. Find the value of $(-1)^2+(-1)^3+(-2)^2+(-2)^3$.
(A) -4 (B) 4 (C) -2 (D) 2 (E) -14 **[1.1]**

2. The expression $(3m)^2$ is equivalent to
(A) $3mm$ (B) $(m+m+m)(m+m+m)$ (C) $3(m+m+m)$ (D) $3m+3m$
(E) $mm+mm+mm$ **[1.2]**

3. The product of $\frac{1}{3}x^3$, $\frac{1}{2}x^2$, and $\frac{1}{2}x^3$ is

(A) $\frac{1}{12}x^{18}$ (B) $\frac{4}{3}x^{18}$ (C) $\frac{4}{3}x^8$ (D) $\frac{1}{12}x^8$ (E) $\frac{1}{3}x^{18}$ **[1.3]**

4. The product of $(3a^4)$ and $(-2a^6)$ is
(A) a^{10} (B) $6a^{10}$ (C) $-6a^{10}$ (D) $6a^{24}$ (E) $-6a^{24}$ **[1.4]**

5. The expression $(3x^3)^2$ is equivalent to
(A) $6x^6$ (B) $9x^5$ (C) $27x^6$ (D) $9x^9$ (E) $9x^6$ **[1.5]**

6. Represent the number of minutes in x hours and y seconds.

(A) $\frac{x}{60}+60y$ (B) $60x+\frac{y}{60}$ (C) $60x+y$ (D) $\frac{x+y}{60}$ (E) $60(x+y)$ **[1.6]**

7. Represent the number of feet in a yards, b feet, and c inches.

(A) $3a + b + \frac{c}{12}$ (B) $\frac{a}{3} + b + 12c$ (C) $3a + b + 12c$ (D) $36a + 12b + c$

(E) $\frac{a}{36} + \frac{b}{12} + c$ **[1.6]**

8. A man earned x dollars in 10 days and spent y dollars during each of those days. How many dollars did he save per day?

(A) $\frac{x - y}{10}$ (B) $x - \frac{y}{10}$ (C) $\frac{y - x}{10}$ (D) $\frac{x}{10} - y$ (E) $10(x - 10y)$ **[1.7]**

9. Take a number, multiply it by 3, add 2, multiply by 2, and subtract 6 times the original number; the result will always be 4. An algebraic equation that may be used to prove this is:

(A) $3n + 2(2 - 6n) = 4$ (B) $3n + 2(2) - 6n = 4$ (C) $2(3n + 2) - 6n = 4$
(D) $2(3n + 2 - 6n) = 4$ (E) $3(n + 2) + 2 - 6n = 4$ **[1.8]**

10. Evaluate $(x - 2)^3$ when $x = 5$.

(A) 27 (B) 9 (C) -3 (D) 117 (E) -1000 **[2.1]**

11. Evaluate a^2b^3 when $a = 2$ and $b = -1$.

(A) 4 (B) -32 (C) -4 (D) -64 (E) 32 **[2.2]**

12. Evaluate $2a^2 - a - 3$ when $a = -1$.

(A) 4 (B) -4 (C) 1 (D) 0 (E) -2. **[2.3]**

13. Evaluate $5 + x(x + 2)$ when $x = 8$.

(A) 130 (B) 85 (C) 71 (D) 122 (E) 133 **[2.4]**

14. Find the value of $a + b(a + c)$ when $a = -2$, $b = -1$, and $c = -3$.

(A) -7 (B) 15 (C) -8 (D) 3 (E) 8 **[2.5]**

15. Simplify: $2 + p(p - 3) + 2p$

(A) $p^2 - p + 2$ (B) $p^2 + 2p - 1$ (C) $-3p^2 + 2p + 2$ (D) $p^2 + p + 2$
(E) $p^2 + 5p + 2$ **[3.1]**

16. Simplify: $3(x - y) + 2[(x - 4y) + (3x - 2y)]$

(A) $8x + 9y$ (B) $11x - 9y$ (C) $8x - 13y$ (D) $11x + 15y$ (E) $11x - 15y$ **[3.2]**

17. Find the sum of $3x^2 + 4$ and $-4x^2 + 5x + 4$.

(A) $x^2 + 5x + 8$ (B) $-x^2 + 5x + 8$ (C) $-x^2 \pm 9x + 4$ (D) $x^2 + 9x + 4$
(E) $-7x^2 + 5x + 8$ **[4.1]**

18. Find the sum of $a + 2b + 3c$, $2c + b - 3a$, and $4b - 5a + 5c$.

(A) $7a - 2b + 5c$ (B) $-7a + 7b + 10c$ (C) $-9a + 7b + 10c$ (D) $-7a - 7b + 10c$
(E) $-7a + 7b + 8c$ **[4.2]**

19. Subtract $3 - 5x - 2x^2$ from $3x^2 - 5x$.

(A) $x^2 - 10x + 3$ (B) $5x^2 + 3$ (C) $3 - 5x^2$ (D) $5x^2 - 3$ (E) $2x^2$ **[4.3]**

20. Subtract $x + y - 2$ from $2y + 5x$.

(A) $4x + y + 2$ (B) $-4x - y - 2$ (C) $x + 4y + 2$ (D) $4x + y - 2$
(E) $4x - y - 2$ **[4.4]**

21. Write an expression for the area of a rectangle whose length is $2x + 3$ and whose width is $3x - 5$.

(A) $6x^2 + 19x - 15$ (B) $5x^2 - x - 15$ (C) $6x^2 - x - 15$ (D) $6x^2 + x - 15$
(E) $10x - 4$ **[4.5]**

22. The expression $(2x - 5)^2$ is equal to
(A) $4x^2 + 20x - 25$ (B) $4x^2 + 25$ (C) $4x^2 - 25$ (D) $4x^2 - 20x + 25$
(E) $4x^2 - 10x + 25$ **[4.6]**

23. If $(x - y)^2 = 4$ and $x^2 + y^2 = 74$, find xy.
(A) 36 (B) 35 (C) 70 (D) 72 (E) 29 **[4.7]**

24. Simplify: $3(1 - 2a) + (2a - 1)^2$
(A) $4a^2 - 6a + 4$ (B) $4a^2 - 10a + 4$ (C) $-2a + 1$ (D) $4a^2 - 6a + 2$
(E) $4a^2 - 2a + 4$ **[4.8]**

25. Write an expression for the volume of a rectangular solid whose base is $2x^2 + 3x + 4$ and whose height is $x - 3$.
(A) $2x^3 + 3x^2 - 12$ (B) $2x^3 - 3x^2 - 12$ (C) $2x^3 + 3x^2 + 5x - 12$
(D) $2x^3 - 3x^2 - 5x - 12$ (E) $2x^3 + 3x^2 - 5x - 12$ **[4.9]**

26. Divide $18x^3y - 12xy^3$ by $6xy$.
(A) $3x^2y - 2xy^2$ (B) $3x - 2y$ (C) $3x^3 - 2y^3$ (D) $3x^2 - 2y^2$ (E) $6x^2y^2$ **[4.10]**

27. What is the length of a rectangle whose width is $4n$ and whose area is $20n - 24n^2$?
(A) $5n - 6n^2$ (B) $5 - 6n$ (C) $5n - 6$ (D) $5 - 6n^2$ (E) $5n^2 - 6$ **[4.11]**

28. The trinomial $x^2 + 3x - 18$ is exactly divisible by
(A) $x + 6$ (B) $x + 3$ (C) $x - 6$ (D) $x - 9$ (E) $x + 9$ **[4.12]**

29. If $2x^2 + 5x - 8$ is divided by $2x + 3$, what is the remainder?
(A) -11 (B) 11 (C) -20 (D) 20 (E) -5 **[4.13]**

30. Factor completely: $36s^2t^6 - 24s^3t^5 + 48s^4t^4$
(A) $4s^2t^4(9t^2 - 6st^2 + 12s^2)$ (B) $12s^2t^4(3t^2 - 2st + 4s^2)$ (C) $6s^2t^4(6t^2 - 4st + 8s^2)$
(D) $6s^2t^3(6t^3 - 4st^2 + 8s^2t)$ (E) $12s^2t^2(3t^4 - 2st^3 + 4s^2t^2)$ **[5.1]**

31. Factor: $9a^2b^2 - 4$
(A) $(9ab - 4)(ab + 1)$ (B) $(3ab - 2)(3ab - 2)$ (C) $(3a - 2b)(3a + 2b)$
(D) $(3ab + 2)(3ab - 2)$ (E) $(9ab - 1)(ab + 4)$ **[5.2]**

32. Factor: $x^2 - x - 20$
(A) $(x + 5)(x - 4)$ (B) $(x - 5)(x + 4)$ (C) $(x - 10)(x + 2)$ (D) $(x - 20)(x + 1)$
(E) $(x + 10)(x - 2)$ **[5.3]**

33. Factor: $9x^4 - 12x^2 + 4$
(A) $(3x^4 - 2)^2$ (B) $(3x^2 - 4)^2$ (C) $(9x^2 - 2)(x^2 - 2)$ (D) $(3x^2 - 2)^2$
(E) $(9x^2 - 4)(x^2 - 1)$ **[5.4]**

34. Factor completely: $8x^4 - 72x^2$
(A) $8x^2(x + 3)(x - 3)$ (B) $8(x^2 - 3)(x^2 + 3)$ (C) $8x^2(x^2 - 9)$
(D) $2x^2(2x - 6)(2x + 6)$ (E) $8(x^2 - 3x)(x^2 + 3x)$ **[5.5]**

35. Combine into one fraction: $\frac{a}{3} + \frac{3}{2}$

(A) $\frac{2a + 3}{6}$ (B) $\frac{a + 3}{5}$ (C) $\frac{2a + 9}{6}$ (D) $\frac{a + 3}{6}$ (E) $\frac{2a + 9}{5}$ **[6.1]**

36. Combine into one fraction: $1 + \frac{a}{b}$

(A) $\frac{b + a}{b}$ (B) $\frac{1 + a}{b}$ (C) $\frac{1 + ab}{b}$ (D) $\frac{1 + ab}{ab}$ (E) $\frac{1}{ab}$ **[6.1]**

37. Multiply: $\frac{2ab}{b} \cdot \frac{b}{4c}$

(A) $\frac{8ac}{b^2}$ (B) $\frac{b}{2c}$ (C) $\frac{ab}{2c}$ (D) $\frac{ab}{2b}$ (E) $\frac{a}{2c}$ **[6.2]**

38. Divide: $\frac{a+b}{a} \div \frac{a-b}{a}$

(A) -1 (B) $\frac{a+b}{a-b}$ (C) $\frac{a-b}{a+b}$ (D) 0 (E) $\frac{a^2-b^2}{a^2}$ **[6.3]**

39. Simplify: $\dfrac{2-\frac{1}{4}}{\frac{5}{6}+\frac{2}{3}}$

(A) $\frac{7}{6}$ (B) $\frac{21}{8}$ (C) $\frac{6}{7}$ (D) $\frac{1}{6}$ (E) $\frac{3}{8}$ **[6.4]**

40. Evaluate the fraction $\dfrac{\frac{1}{x}+\frac{1}{y}}{\frac{x}{y}}$ if $x = 3$ and $y = 5$.

(A) $\frac{5}{24}$ (B) $\frac{13}{45}$ (C) $\frac{8}{9}$ (D) $\frac{8}{25}$ (E) $\frac{1}{9}$ **[6.4]**

41. Evaluate the fraction $\frac{a+b}{c}$ if $a = \frac{1}{2}$, $b = \frac{3}{4}$, and $c = \frac{1}{4}$.

(A) $\frac{3}{2}$ (B) $\frac{1}{5}$ (C) $\frac{16}{5}$ (D) $\frac{5}{16}$ (E) 5 **[6.4]**

42. Simplify: $\dfrac{\frac{1}{a}+\frac{1}{b}}{2}$

(A) $\frac{ab}{2a+b}$ (B) $\frac{a+b}{2ab}$ (C) $\frac{ab}{a+2b}$ (D) $\frac{2a+2b}{ab}$ (E) $\frac{2ab}{a+b}$ **[6.4]**

III. REVIEWING THE SOLUTION OF FIRST-DEGREE EQUATIONS

The equations to be solved in this chapter contain a single unknown having one value. Solving these equations means finding the root of the equation. A ***root of an equation*** is a value that, when substituted for the unknown, makes the two sides of the equation equal.

Inverse Operations in the Solution of Equations

Addition and subtraction are inverse operations. If a number is added to an unknown, solve the equation by means of subtraction. If a number is subtracted from an unknown, solve the equation by means of addition.

For example, solve $x + 5 = 15$ by subtracting 5 from each side; solve $x - 5 = 15$ by adding 5 to each side.

Multiplication and division are inverse operations. If the unknown is multiplied by a number, solve the equation by means of division. If the unknown is divided by any number other than zero, solve the equation by means of multiplication.

For example, solve $5x = 15$ by dividing each side by 5; solve $\frac{x}{5} = 15$ by multiplying each side by 5.

To solve an equation, use the inverse operations to "undo" the operational connections among the terms, except where this would result in division by zero. Bear in mind that the same operation, using equal numbers, must be performed on each side of the equation.

For example, to solve $2x - 3 = 7$, add 3 to each side to obtain $2x = 10$; then divide each side by 2 to obtain $x = 5$. Generally, undo addition and subtraction before undoing multiplication and division.

Transposing a term is a procedure for removing the term from one side of an equation and placing its opposite on the other side. Opposite terms differ only in sign.

For example, to solve $2x - 3 = 7$, transpose -3 by dropping it from the left side and adding its opposite, $+3$, to the right side, thus obtaining $2x = 7 + 3$.

1. SOLVING EQUATIONS HAVING INTEGRAL COEFFICIENTS

Use inverse operations to combine the terms containing the unknown into one term. Isolate this single term from other terms. Then divide both sides of the equation by the coefficient of the unknown.

For example, by combining terms containing x, the equation $2x + 3x - 2 = x + 10$ becomes $4x - 2 = 10$. Since 2 is being subtracted on the left, the inverse operation to apply is addition: add 2 to each side to obtain $4x = 12$. Since $4x$ is a product, the inverse operation is division: divide each side by 4, the coefficient of x, to obtain $x = 3$.

1.1 EQUATIONS WITH THE UNKNOWN ON ONE SIDE

Model Problem

Solve for x: $5 - 3x = 17$

(A) -4 (B) 4 (C) $7\frac{1}{3}$ (D) $-7\frac{1}{3}$ (E) 15

Solution:

Solve the given equation for x as follows:

$$\begin{array}{lrcl} & 5 - 3x & = & 17 \\ \text{Subtract 5 from each side:} & -5 \quad & = & -5 \\ \hline & -3x & = & 12 \end{array}$$

Divide each side by -3: $\frac{-3x}{-3} = \frac{12}{-3}$, $x = -4$ *Ans.* (A)

> *Note.* By transposing, $5 - 3x = 17$ becomes $-3x = 17 - 5$. Since $-3x = 12$, $x = -4$.

Matching Test

Solve each equation for x. After completing this test, match your answers with the numbers in the box.

1. $3 + 2x = 9$ **2.** $3 - 2x = 9$ **3.** $2 + 3x = 9$ **4.** $9x - 2 = 3$ **5.** $2 = 3 - 9x$

a. -3	*b.* $\frac{7}{3}$	*c.* 3	*d.* $\frac{1}{9}$	*e.* $\frac{5}{9}$

1.2 EQUATIONS WITH THE UNKNOWN ON BOTH SIDES

Model Problem

Solve for y: $9y - 5 = 7 + 8y$

(A) 2 (B) -2 (C) $2\frac{2}{9}$ (D) 12 (E) -12

Solution:

To solve the given equation for y, subtract $8y$ from each side and add 5 to each side.

$$\begin{array}{rcl} 9y - 5 & = & 7 + 8y \\ -8y + 5 & = & 5 - 8y \\ \hline y & = & 12 \end{array} \qquad \textit{Ans.}\ \text{(D)}$$

> *Note.* By transposing, $9y - 5 = 7 + 8y$ becomes $9y - 8y = 7 + 5$. Hence, $y = 12$.

Matching Test

Solve each equation for y. After completing this test, match your answers with the numbers in the box.

1. $9y - 5 = 7 + 6y$ **2.** $9y - 5 = 7 - 6y$ **3.** $9y + 5 = 7 + 6y$ **4.** $5y + 9 = 7 + 6y$
5. $5y + 9 = 6y - 7$

a. $\frac{4}{5}$	*b.* 16	*c.* 4	*d.* $\frac{2}{3}$	*e.* 2

1.3 DETERMINING THE VALUE OF A CONSTANT

Model Problem

If 3 is a root of the equation $x^2 - 5x + k = 0$, what is the value of k?

(A) -6 (B) 6 (C) 9 (D) -9 (E) 24

Solution:

Find the value of k by applying the following principle:

A root of an equation is a value of the unknown that satisfies the equation. Hence, if 3 is a root of the equation, then x must have this value.

$$x^2 - 5x + k = 0$$

Substitute 3 for x: $3^2 - 5(3) + k = 0$

$$9 - 15 + k = 0$$

$$-6 + k = 0$$

Add 6 to each side.

$$k = 6 \quad \textit{Ans.}\ (B)$$

After completing each test, match your answers with the numbers in the boxes.

Matching Test A

Find the value of k if 2 is a root of each equation.

1. $x^2 + 5x + k = 0$ **2.** $x^2 - 5x - k = 0$ **3.** $x^2 - 5kx = 0$ **4.** $kx^2 - 5x = 0$ **5.** $kx^2 + 5x = 0$

a. -6	*b.* $-2\frac{1}{2}$	*c.* $\frac{2}{5}$	*d.* $2\frac{1}{2}$	*e.* -14

Matching Test B

Find the value of k if each given number is a root of the equation $x^2 - 5x + k = 0$.

1. -4 **2.** -2 **3.** 6 **4.** 2 **5.** 4

a. 6	*b.* -36	*c.* 4	*d.* -6	*e.* -14

1.4 CHECKING A ROOT BY SUBSTITUTION ON EACH SIDE OF AN EQUATION

Model Problem

If the equation $9x - 4 = 7x + 6$ is solved and then checked by substitution, the number to which each member of the equation will reduce in the check will be

(A) 5 (B) 41 (C) 0 (D) −49 (E) 49

Solution:

To solve the given equation, subtract $7x$ and add 4 to each side.

$$\begin{array}{rcr} 9x - 4 &=& 7x + 6 \\ -7x + 4 &=& -7x + 4 \\ \hline 2x &=& 10 \end{array}$$

Divide each side by 2: $\frac{2x}{2} = \frac{10}{2}$, $x = 5$

To check for $x = 5$, substitute 5 for x in the given equation.

$$9x - 4 = 7x + 6$$
$$9(5) - 4 \stackrel{?}{=} 7(5) + 6$$
$$45 - 4 \stackrel{?}{=} 35 + 6$$
$$41 = 41$$

Each member of the equation reduces to 41. *Ans.* (B)

> *Note.* By transposing, $9x - 4 = 7x + 6$ becomes $9x - 7x = 6 + 4$. Hence, $2x = 10$ and $x = 5$.

Matching Test

Solve each equation for x; check this value by substitution. After completing this test, *match the number to which each side of the equation will reduce* with one of the numbers in the box.

1. $9x + 4 = 7x + 6$

2. $9x - 4 = 7x - 6$

3. $9x + 4 = 7x - 6$

4. $4x + 9 = 7x - 6$

5. $4x - 9x = 7x - 6$

a. −41
b. 29
c. 13
d. $-\frac{5}{2}$
e. −13

2. SOLVING EQUATIONS CONTAINING PARENTHESES

Applying the distributive law, remove the parentheses first. Then use inverse operations.

For example, to solve $3(x+2) = 12$, remove parentheses to obtain $3x + 6 = 12$. Then use inverse operations to obtain $3x = 6$ and $x = 2$.

2.1 EQUATIONS WITH ONE SET OF PARENTHESES

Model Problem

Solve for y: $3(y-2) = 12$

(A) -2 (B) 2 (C) 6 (D) -6 (E) 15

Solution:

To solve the given equation for y, first remove parentheses by replacing $3(y-2)$ with $3y - 6$.

Add 6 to each side:

$$\begin{aligned} 3y - 6 &= 12 \\ +6 &= +6 \\ \hline 3y &= 18 \end{aligned}$$

Divide each side by 3: $\dfrac{3y}{3} = \dfrac{18}{3}$, $y = 6$ *Ans.* (C)

After completing each test, match your answers with the numbers in the boxes.

Matching Test A

Solve each equation for y.

1. $3(y+2) = 12$ **2.** $3(y+5) = 12$ **3.** $2(y-5) = 12$ **4.** $2(y-3) = 12$ **5.** $2(y+3) = 12$

a. 3 *b.* -1 *c.* 9 *d.* 2 *e.* 11

Matching Test B

Solve each equation for x.

1. $5x - 3(x-2) = 12$

2. $5x + 3(x-2) = 12$

3. $5(x+2) - 3x = 12$

4. $5(x-3) - 3 = 12$

5. $3(x-5) - 2x = 12$

a. $2\frac{1}{4}$

b. 1

c. 27

d. 3

e. 6

2.2 EQUATIONS WITH TWO SETS OF PARENTHESES

Model Problem

Solve for x: $4(x-1)-3(x-2)=10$

(A) -8 (B) 2 (C) 8 (D) 12 (E) 20

Solution:

To solve the given equation for x, first remove parentheses by replacing $4(x-1)$ with $4x-4$ and $-3(x-2)$ with $-3x+6$.

$$4x-4-3x+6=10$$

Collect like terms on the left.

$$x+2=10$$

Subtract 2 from each side.

$$x=8 \quad \textit{Ans.}\ \text{(C)}$$

Matching Test

Solve each equation for x. After completing this test, match your answers with the numbers in the box.

1. $4(x-1)-3(x-2)=10$	*a.* 3
2. $4(x+1)-3(x-2)=10$	*b.* 0
3. $4(x-1)-3(x+2)=10$	*c.* 20
4. $3(x-1)+4(x-2)=10$	*d.* 8
5. $3(1-x)+4(2+x)=10$	*e.* -1

3. SOLVING EQUATIONS CONTAINING FRACTIONAL COEFFICIENTS

Clear the equation of fractions as follows:

If the fractions in the equation have the same denominator, multiply each side of the equation by the denominator.

For example, to solve $\frac{x}{3}+\frac{2}{3}=1$, multiply each side by 3 to obtain $x+2=3$. Solving, $x=1$.

If the fractions in the equation have different denominators, multiply each side of the equation by the lowest common denominator.

For example, to solve $\frac{x}{3}+\frac{2}{9}=1$, multiply each side by 9 to obtain $3x+2=9$. Hence, $3x=7$ and $x=2\frac{1}{3}$.

3.1 EQUATIONS WITH FRACTIONAL COEFFICIENTS

Model Problems

(1) Solve for y: $\frac{2y}{3} = 6$

(A) $5\frac{1}{3}$ (B) 9 (C) 6 (D) 16 (E) 12

Solution:

Solve the given equation for y as follows:

$$\frac{2y}{3} = 6$$

Multiply each side by 3: $3\left(\frac{2y}{3}\right) = 3(6)$

$$2y = 18$$

Divide each side by 2: $\frac{2y}{2} = \frac{18}{2}$, $y = 9$ *Ans.* (B)

> *Note.* Each side of the equation $\frac{2y}{3} = 6$ may be multiplied by $\frac{3}{2}$, the reciprocal of $\frac{2}{3}$.
>
> $$\frac{\overset{1}{\cancel{3}}}{\underset{1}{\cancel{2}}}\left(\frac{\overset{y}{\cancel{2y}}}{\underset{1}{\cancel{3}}}\right) = \frac{3}{\underset{1}{\cancel{2}}}(\overset{3}{\cancel{6}}), \quad y = 9$$

(2) Solve for d: $\frac{2}{3}d + 4 = 6$

(A) 15 (B) $\frac{4}{3}$ (C) -3 (D) 4 (E) 3

Solution:

Solve the given equation for d as follows:

$$\frac{2}{3}d + 4 = 6$$

Subtract 4 from each side: $\quad - 4 = -4$

$$\frac{2}{3}d = 2$$

Multiply each side by 3.

$$2d = 6$$

Divide each side by 2.

$$d = 3 \quad \textit{Ans. } (E)$$

After completing each test, match your answers with the numbers in the boxes.

Matching Test A

Solve each equation for y.

1. $\frac{3y}{2} = 6$ **2.** $\frac{3y}{2} = 18$ **3.** $\frac{2y}{3} = 18$ **4.** $\frac{3y}{4} = 18$ **5.** $\frac{3y}{4} = 12$

a. 16	*b.* 4	*c.* 24	*d.* 12	*e.* 27

Matching Test B

Solve each equation for d.

1. $\frac{2}{3}d - 4 = 6$ **2.** $\frac{3}{2}d - 4 = 8$ **3.** $\frac{3}{4}d - 4 = 8$ **4.** $\frac{4}{3}d + 4 = 8$ **5.** $\frac{4}{5}d + 4 = 8$

a. 8	*b.* 5	*c.* 15	*d.* 3	*e.* 16

3.2 FRACTIONAL EQUATIONS WITH THE UNKNOWN ON ONE SIDE

Model Problem

Solve for n: $\frac{n}{2} + \frac{n}{3} = 25$

(A) 5 (B) 30 (C) 36 (D) 25 (E) 150

Solution:

To solve the given equation for n, first clear the equation of fractions. To do this, multiply both sides by the lowest common denominator, 6.

$$\frac{n}{2} + \frac{n}{3} = 25$$

$$6\left(\frac{n}{2} + \frac{n}{3}\right) = 6(25)$$

$$3n + 2n = 150$$

$$5n = 150$$

Divide each side by 5: $\frac{5n}{5} = \frac{150}{5}$, $n = 30$ *Ans.* (B)

After completing each test, match your answers with the numbers in the boxes.

Matching Test A

Solve each equation for n.

1. $\frac{n}{2}+\frac{n}{3}=35$ **2.** $\frac{n}{3}+\frac{n}{4}=35$ **3.** $\frac{n}{2}+\frac{n}{4}=6$ **4.** $\frac{n}{2}-\frac{n}{4}=6$ **5.** $\frac{n}{2}-\frac{n}{3}=6$

a. 8	*b*. 24	*c*. 36	*d*. 60	*e*. 42

Matching Test B

Solve each equation for n.

1. $\frac{n}{2}+\frac{n}{3}-\frac{n}{4}=35$ **2.** $\frac{n}{2}+\frac{n}{4}-\frac{n}{3}=35$ **3.** $\frac{n}{3}+\frac{n}{4}-\frac{n}{2}=35$ **4.** $\frac{2n}{3}-\frac{n}{4}=\frac{5}{2}$ **5.** $\frac{3n}{4}-\frac{2n}{3}=\frac{5}{2}$

a. 420	*b*. 6	*c*. 30	*d*. 84	*e*. 60

3.3 FRACTIONAL EQUATIONS WITH THE UNKNOWN ON BOTH SIDES

Model Problem

Solve for x: $\frac{2x}{3}+3=\frac{x}{3}$

(A) 6 (B) −6 (C) 9 (D) −9 (E) 12

Solution:

To solve the given equation, first clear the equation of fractions.

Multiply each term by 3: $\overset{1}{\cancel{3}}\left(\frac{2x}{\underset{1}{\cancel{3}}}\right)+3(3)=\overset{1}{\cancel{3}}\left(\frac{x}{\underset{1}{\cancel{3}}}\right)$

$$2x+9=x$$

Subtract $x+9$ from each side:

$$\begin{aligned} x+9 &= x+9 \\ \hline x &= -9 \end{aligned}$$ *Ans.* (D)

Matching Test

Solve each equation for x. After completing this test, match your answers with the numbers in the box.

1. $\frac{2x}{3}-3=\frac{x}{3}$ **2.** $\frac{3x}{2}-3=\frac{x}{2}$ **3.** $\frac{3x}{2}-5=\frac{x}{2}$ **4.** $\frac{3x}{5}+2=\frac{x}{5}$ **5.** $\frac{5x}{3}=2-\frac{x}{3}$

a. 3	*b*. 5	*c*. 1	*d*. 9	*e*. −5

4. SOLVING EQUATIONS CONTAINING DECIMAL COEFFICIENTS

Clear the equation of decimals by multiplying both sides of the equation by the denominator of the decimal having the greatest number of decimal places.

For example, if $.32x = .2x + 6$, multiply each side by 100, the denominator of .32, that is, of $\frac{32}{100}$.

4.1 EQUATIONS WITH DECIMAL COEFFICIENTS

Model Problems

(1) Solve for x: $0.06x = 0.468$

(A) .078 (B) .408 (C) 780 (D) 78 (E) 7.8

Solution:

To solve the given equation for x, first eliminate the decimals. If each side is multiplied by 1000, then each decimal point must be moved three places to the right, thus clearing the equation of decimals.

$$0.06x = 0.468$$

$$60x = 468$$

Divide each side by 60: $$\frac{60x}{60} = \frac{468}{60}$$

$$x = 7.8, \text{ or } 7\frac{4}{5} \quad \textit{Ans. } (E)$$

(2) Solve for n: $1.1n = 7.6 - .42n$

(A) .2 (B) 5 (C) .5 (D) 50 (E) 608

Solution:

To solve the given equation for n, first eliminate the decimals by multiplying each term on both sides by 100. This will move each decimal point two places to the right and clear the equation of decimals.

$$1.1n = 7.6 - .42n$$

$$110n = 760 - 42n$$

Add $42n$ to each side: $$\underline{+\ 42n = \qquad +\ 42n}$$

$$152n = 760$$

Divide each side by 152: $$\frac{152n}{152} = \frac{760}{152}, \quad n = 5 \quad \textit{Ans. } (B)$$

After completing each test, match your answers with the numbers in the boxes.

Matching Test A

Solve each equation for x.

1. $.4x = 1.6$ **2.** $.4x = 16$ **3.** $.04x = 16$ **4.** $1.6x = .4$ **5.** $.16x = 4$

a. 4	*b.* 400	*c.* .25	*d.* 40	*e.* 25

Matching Test B

Solve each equation for n.

1. $.5n = 2.1 - .2n$	*a.* 30
2. $.5n = 21 - .2n$	*b.* 3
3. $.05n = .021 - .02n$	*c.* 70
4. $.5n = 21 + .2n$	*d.* 7
5. $.05n = .21 + .02n$	*e.* .3

4.2 EQUATIONS WITH INTEGRAL COEFFICIENTS AND DECIMALS

Model Problem

Solve for x: $4x + 2.7 = 0.6 - 3x$

(A) $-.3$ (B) $.3$ (C) $-3\frac{1}{3}$ (D) 3 (E) -91

Solution:

To solve the given equation for x, first eliminate the decimals by multiplying each term on both sides by 10. This will move each decimal point one place to the right and clear the equation of decimals.

$$4x + 2.7 = 0.6 - 3x$$

$$40x + 27 = 6 - 30x$$

Add $30x - 27$ to each side: $$\underline{30x - 27 = -27 + 30x}$$
$$70x = -21$$

Divide each side by 70: $$\frac{70x}{70} = \frac{-21}{70}$$

$$x = -\frac{3}{10} = -.3 \quad \textit{Ans.} \text{ (A)}$$

Matching Test

Solve each equation for x. After solving each test, match your answers with the numbers in the box.

1. $5x + 1.8 = 3.4 - 3x$	*a.* .08
2. $5x + 3.4 = 1.8 - 3x$	*b.* .2
3. $5x + 3.4 = 1.8 + 3x$	*c.* 2.6
4. $5x - 3.4 = 1.8 + 3x$	*d.* $-.8$
5. $3x + .34 = .18 + 5x$	*e.* $-.2$

5. SOLVING EQUATIONS CONTAINING PERCENT COEFFICIENTS

To solve an equation containing a percent coefficient, change the percent to a decimal or to a fraction, and then proceed as previously shown for solving equations containing fractional or decimal coefficients.

For example, if $12\frac{1}{2}\%x = 5$, change the equation to $\frac{1}{8}x = 5$. If $3\frac{1}{2}\%x = 10.5$, change the equation to $.035x = 10.5$.

5.1 EQUATIONS WITH PERCENT COEFFICIENTS

Model Problem

If 75% of x is 24, find the value of x.

(A) 96 (B) 36 (C) $23\frac{1}{4}$ (D) 18 (E) 32

Solution:

Since $75\% = \frac{3}{4}$, then "75% of x is 24" may be written as the equation $\frac{3x}{4} = 24$.

Solve the equation $\frac{3x}{4} = 24$ for x.

Multiply each side by 4.

$$4\left(\frac{3x}{4}\right) = 4(24)$$

$$3x = 96$$

Divide each side by 3.

$$x = 32 \quad \textit{Ans.} \text{ (E)}$$

Matching Test

Find the value of x. After completing this test, match your answers with the numbers in the box.

1. 20% of x is 4 **2.** 2% of x is 4 **3.** 200% of x is 4 **4.** $4\%x = 2$ **5.** $40\%x = 2$

a. 50	*b.* 200	*c.* 20	*d.* 5	*e.* 2

5.2 VERBAL PERCENT PROBLEMS

Model Problem

In March, a pair of skis sold for \$24. This amount was 80% of the original price. What was the original price?

(A) \$30 (B) \$35 (C) \$40 (D) \$32 (E) \$19.20

Solution:

Let $p =$ the original price in dollars of a pair of skis.
80% of the original price is \$24.

Hence, 80% of $p = 24$

80% can be written as .80, as .8, or as $\frac{4}{5}$. (See Note below.)

$$.8p = 24$$

Multiply each side by 10 to eliminate the decimal point.

$$8p = 240$$

Divide each side by 8: $\frac{8p}{8} = \frac{240}{8}$, $p = 30$

Hence, the original price was \$30. *Ans.* (A)

> *Note.* To solve $\frac{4}{5}p = 24$, multiply each side by $\frac{5}{4}$ to obtain
>
> $$\frac{\not{5}^{1}}{\not{4}_{1}}\left(\frac{\not{4}^{1}}{\not{5}_{1}}p\right) = \frac{5}{\not{4}_{1}}(\not{24}^{6})$$
>
> Hence, $p = 30$.

Matching Test

Find the original price of a pair of skis by using the given information. After completing this test, match your answers with the numbers in the box.

1. The selling price, \$36, was 80% of the original price.
2. The selling price, \$36, was 75% of the original price.
3. The selling price, \$36, was 120% of the original price.
4. The selling price, \$30, was the amount paid after the purchaser was allowed a discount of 25% of the original price.
5. The selling price, \$30, was the amount paid after the purchaser was allowed a discount of 40% of the original price.

a. 40	*b.* 30	*c.* 50	*d.* 48	*e.* 45

6. SOLVING LITERAL EQUATIONS

Literal equations are equations containing two or more letters, each letter representing a variable. All formulas are literal equations.

To solve a literal equation for one of the variables, follow the same procedures that are required to solve any equation for an unknown.

For example, to solve $S = P + C$ for P, subtract C from each side to obtain $S - C = P$. To solve $D = RT$ for R, divide each side by T to obtain $\frac{D}{T} = R$.

6.1 TRANSFORMING LITERAL EQUATIONS AND FORMULAS

Model Problems

(1) Solve for x: $y = mx + b$

(A) $y - \frac{b}{m}$ (B) $\frac{y}{m} - b$ (C) $\frac{y-m}{b}$ (D) $\frac{b-y}{m}$ (E) $\frac{y-b}{m}$

Solution:

Solve the given equation for x as follows:
Subtract b from each side (or transpose b).

$$\begin{aligned} y &= mx + b \\ -b &= \quad\ - b \\ \hline y - b &= mx \end{aligned}$$

Divide each side by m: $\frac{y-b}{m} = \frac{mx}{m}$

$$\frac{y-b}{m} = x \quad \textit{Ans. } \text{(E)}$$

(2) Solve for x: $ax + b = c - b$

(A) $\frac{c-2b}{a}$ (B) $\frac{2b-c}{a}$ (C) $c - \frac{2b}{a}$ (D) $\frac{2b}{a} - c$ (E) $c - 2b - a$

Solution:

Solve the given equation for x as follows:

Subtract b from each side:

$$\begin{aligned} ax + b &= c - b \\ -b &= \quad - b \\ \hline ax \quad &= c - 2b \end{aligned}$$

Divide each side by a: $\frac{ax}{a} = \frac{c-2b}{a}$

$$x = \frac{c-2b}{a} \quad \textit{Ans. } \text{(A)}$$

(3) Solve for x: $ax - cx = b - ax$

(A) $b + c - 2a$ (B) $\frac{b}{c-2a}$ (C) $\frac{b}{2a-c}$ (D) $\frac{b}{2a} - c$ (E) $\frac{b}{2c} - a$

Solution:

All terms in x must be brought to one side. To do this, add ax to both sides.

$$\begin{aligned} ax - cx &= b - ax \\ \underline{ax \qquad} &= \underline{\qquad ax} \\ 2ax - cx &= b \end{aligned}$$

Factor x.

$$x(2a - c) = b$$

Divide each side by $(2a - c)$.

$$x = \frac{b}{2a - c} \quad \textit{Ans.}\ \text{(C)}$$

(4) Given the formula $V = lwh$. Express l in terms of V, w, and h.

(A) $l = \frac{wh}{V}$ (B) $l = \frac{Vh}{w}$ (C) $l = \frac{V}{wh}$ (D) $l = \frac{Vw}{h}$ (E) $l = V - wh$

Solution:

To solve the formula $V = lwh$ for l, divide each side by wh.

$$\frac{V}{wh} = \frac{lwh}{wh}$$

$$\frac{V}{wh} = l \quad \textit{Ans.}\ \text{(C)}$$

After completing each test, match your answers with the answers in the boxes.

Matching Test A

Solve each equation for x.

1. $y = ax + b$ **2.** $y = bx + a$ **3.** $y + bx = a$ **4.** $by = ax$ **5.** $bx - ay = 0$

a. $\frac{ay}{b}$	*b.* $\frac{y - a}{b}$	*c.* $\frac{by}{a}$	*d.* $\frac{a - y}{b}$	*e.* $\frac{y - b}{a}$

Matching Test B

Solve each equation for x.

1. $ax - b = c + b$

2. $ax - c = c + b$

3. $cx - a = a - b$

4. $bx + cx - a = a$

5. $bx + cx = a - bx$

a. $\frac{2a}{b + c}$
b. $\frac{2b + c}{a}$
c. $\frac{a}{2b + c}$
d. $\frac{2a - b}{c}$
e. $\frac{b + 2c}{a}$

Matching Test C

Transform each equation or formula as indicated.

1. Solve $RT = D$ for T.
2. Solve $V = lwh$ for h.
3. Solve $A = \frac{1}{2}bh$ for b.
4. Solve $p = 2l + 2w$ for w.
5. Solve $F = \frac{9}{5}C + 32$ for C.

a. $h = \frac{V}{lw}$

b. $C = \frac{5}{9}(F - 32)$

c. $w = \frac{p - 2l}{2}$

d. $T = \frac{D}{R}$

e. $b = \frac{2A}{h}$

Matching Test D

Express x in terms of y and z.

1. $z = 3xy$ **2.** $z = 2\pi x + y$ **3.** $z = 2\pi xy$ **4.** $z = \frac{1}{4}xy$ **5.** $z = \pi xy + \pi y$

a. $x = \frac{z - \pi y}{\pi y}$ *b.* $x = \frac{z - y}{2\pi}$ *c.* $x = \frac{4z}{y}$ *d.* $x = \frac{z}{3y}$ *e.* $x = \frac{z}{2\pi y}$

6.2 TRANSFORMING ONE LITERAL EQUATION INTO ANOTHER

Model Problem

The equation $b - a = a - b$ is true if

(A) $a + b = 1$ (B) $a - b = 1$ (C) $ab = 1$ (D) $\frac{a}{b} = 1$ (E) $b - a = 1$

Solution:

Test each choice to see which of the five equations can be transformed into $b - a = a - b$. It can thus be shown that the correct choice is (D). If $\frac{a}{b} = 1$, then $a = b$, and both $b - a$ and $a - b$ would be equal to 0. Hence, $b - a = a - b$. *Ans.* (D)

Alternate Method

Since $b - a = a - b$, then $b + b = a + a$ by transposition. Hence, $2b = 2a$, $b = a$, and $\frac{a}{b} = 1$.

Matching Test

Transform each equation into one of the equations in the box.

1. $1-\frac{1}{b}=\frac{a}{b}$ **2.** $\frac{1}{a}-b=0$ **3.** $\frac{a}{b}=\frac{1}{b}-1$ **4.** $1=\frac{1}{b}+\frac{a}{b}$ **5.** $\frac{b}{a}+\frac{1}{a}=1$

a. $a+b=1$	*b.* $a=b+1$	*c.* $b=a+1$	*d.* $ab=1$	*e.* $b-1=a$

6.3 LITERAL EQUATIONS CONTAINING FRACTIONS

Model Problem

Solve for x: $x-\frac{1}{c}=\frac{x}{c}$

(A) $\frac{1}{c-1}$ (B) $\frac{1}{1-c}$ (C) $\frac{1}{c+1}$ (D) $\frac{1}{c}-1$ (E) $1-\frac{1}{c}$

Solution:

To solve the given equation for x, first clear the equation of fractions. To do this, multiply each term of the equation by the common denominator, c.

$$cx-\overset{1}{\cancel{c}}\left(\frac{1}{\underset{1}{\cancel{c}}}\right)=\overset{1}{\cancel{c}}\left(\frac{x}{\underset{1}{\cancel{c}}}\right)$$

$$cx-1=x$$

Collect the x-terms on one side. To do this, subtract x and add 1 to each side.

$$cx-x=1$$

Factor the left side.

$$x(c-1)=1$$

Divide each side by $c-1$: $\frac{x(c-1)}{c-1}=\frac{1}{c-1}$

$$x=\frac{1}{c-1} \quad \textit{Ans. (A)}$$

Matching Test

Solve each equation for x. After completing this test, match your answers with the expressions in the box.

1. $x+\frac{x}{c}=\frac{1}{c}$ **2.** $x+\frac{x}{c}=c$ **3.** $\frac{x}{c}-c=x$ **4.** $x+\frac{1}{c}=c$ **5.** $x+c=\frac{1}{c}$

a. $\frac{c^2}{c+1}$	*b.* $\frac{c^2-1}{c}$	*c.* $\frac{1}{c+1}$	*d.* $\frac{1-c^2}{c}$	*e.* $\frac{c^2}{1-c}$

7. SOLVING EQUATIONS CONSISTING OF EQUAL FRACTIONS OR RATIOS

An equation that consists of equal fractions or ratios is called a ***proportion.***

To solve a proportion, first clear the equation of fractions. This can be done by ***cross-multiplication.*** For example, if $\frac{x+2}{10} = \frac{2}{5}$, cross-multiply to obtain $5(x+2) = 2(10)$. Another method is to multiply both sides of the equation by the lowest common denominator. Thus, in the example just given, multiply each side by the lowest common denominator, 10, to obtain $x + 2 = 4$.

7.1 EQUATIONS CONSISTING OF EQUAL FRACTIONS

Model Problem

Solve for m: $\frac{3+m}{2-m} = \frac{7}{3}$

(A) 2 (B) $1\frac{1}{4}$ (C) -2 (D) $-\frac{1}{2}$ (E) $\frac{1}{2}$

Solution:

To solve the given equation for m, cross-multiply to clear the equation of fractions.

$$\frac{3+m}{2-m} = \frac{7}{3}$$

$$3(3+m) = 7(2-m)$$

Remove parentheses.

$$9 + 3m = 14 - 7m$$

Add $-9 + 7m$ to each side:

$$\frac{-9 + 7m = -9 + 7m}{10m = 5}$$

Divide each side by 10:

$$\frac{10m}{10} = \frac{5}{10}$$

$$m = \frac{1}{2} \quad \textit{Ans. } (E)$$

> *Note.* By transposing, $9 + 3m = 14 - 7m$ becomes $3m + 7m = 14 - 9$. Hence, $10m = 5$ and $m = \frac{1}{2}$.

Matching Test

Solve each equation for m. After completing this test, match your answers with the numbers in the box.

1. $\frac{2-m}{3-m} = \frac{4}{3}$ **2.** $\frac{2-m}{4-m} = \frac{3}{4}$ **3.** $\frac{m-2}{m+4} = \frac{3}{4}$ **4.** $\frac{m+4}{3} = \frac{m+2}{2}$ **5.** $\frac{m-2}{2} = \frac{m+4}{3}$

a. 2	*b.* 20	*c.* 14	*d.* 6	*e.* −4

7.2 LITERAL EQUATIONS CONSISTING OF EQUAL FRACTIONS

Model Problem

Solve for x: $\frac{a}{b}=\frac{c}{x}$

(A) $bc-a$ (B) $b(c-a)$ (C) $\frac{bc}{a}$ (D) $\frac{ba}{c}$ (E) $\frac{b}{ac}$

Solution:

To solve the given equation for x, cross-multiply to clear the equation of fractions.

$$\frac{a}{b}=\frac{c}{x}$$

$$ax=bc$$

Divide each side by a: $\frac{ax}{a}=\frac{bc}{a}$; $x=\frac{bc}{a}$ *Ans.* (C)

Matching Test

Solve each equation for x. After completing this test, match your answers with the numbers in the box.

1. $\frac{a}{c}=\frac{b}{x}$ **2.** $\frac{c}{a}=\frac{b}{x}$ **3.** $\frac{b}{a}=\frac{c}{x}$ **4.** $\frac{3b}{a}=\frac{2a}{x}$ **5.** $\frac{3b}{2a}=\frac{2b}{x}$

a. $\frac{2a^2}{3b}$	*b.* $\frac{ac}{b}$	*c.* $\frac{bc}{a}$	*d.* $\frac{ab}{c}$	*e.* $\frac{4a}{3}$

8. EVALUATING VARIABLES IN A LITERAL EQUATION

To find the value of a variable in a literal equation when specific numerical values are assigned to the other variables, follow the same order of operations as that described in Chapter II, set 2, for evaluating expressions:

1. Substitute the value given for each letter.
2. If the equation contains parentheses, evaluate within the parentheses.
3. Do powers and roots.
4. Do multiplications and divisions in order from left to right.
5. Do additions and subtractions in order from left to right.

Model Problem

If $m=\frac{12}{2-x}$, find the value of m when $x=-2$.

(A) 12 (B) -12 (C) -3 (D) 3 (E) none of these

Solution:

To find the value of m when $x=-2$, substitute -2 for x in the given equation.

$$m=\frac{12}{2-x}=\frac{12}{2-(-2)}=\frac{12}{2+2}=\frac{12}{4}=3 \quad \textit{Ans. (D)}$$

Matching Test

After completing this test, match your answers with the numbers in the box.

If $m = \dfrac{12}{2 - x}$,

1. find m when $x = -4$.
2. find m when $x = 4$.
3. find m when $x = \frac{1}{2}$.
4. find x when $m = 1$.
5. find x when $m = -1$.

a. 14
b. 2
c. −6
d. −10
e. 8

8.1 EVALUATING A VARIABLE IN AN AREA FORMULA

Model Problem

A formula for the lateral area, S, of a right circular cylinder is $S = 2\pi rh$, where r is the radius of the base and h is the altitude of the cylinder. Find the lateral area of a right circular cylinder if the radius of the base is 10 and the altitude is 14. $\left(\text{Use } \pi = \frac{22}{7}.\right)$

(A) 220 (B) 440 (C) 880 (D) 1760 (E) 8.8

Solution:

Find S in the formula $S = 2\pi rh$ by substituting 10 for r, 14 for h, and $\frac{22}{7}$ for π.

$$S = 2\left(\frac{22}{\not{7}_{1}}\right)(10)(\not{14}^{\,2})$$

$$= 2(22)(10)(2) = 880 \quad \textit{Ans. } (C)$$

Matching Test

After completing this test, match your answers with the numbers in the box.

Given the formula $S = 2\pi rh$: $\left(\text{Use } \pi = \frac{22}{7}.\right)$

1. Find S when $r = 20$ and $h = 7$.
2. Find S when $r = 5$ and $h = 14$.
3. Find S when $r = 2\frac{1}{2}$ and $h = 7$.
4. Find r when $S = 220$ and $h = 5$.
5. Find r when $S = 330$ and $h = 21$.

a. 7
b. $2\frac{1}{2}$
c. 110
d. 440
e. 880

8.2 EVALUATING A VARIABLE IN AN INTEREST FORMULA

Model Problem

Using the formula $A = p + prt$, find t when $A = \$1650$, $p = \$1500$, and $r = 4\%$.

(A) 90 (B) 9 (C) $\frac{2}{5}$ (D) $2\frac{1}{2}$ (E) 25

Solution:

To find t, substitute 1650 for A, 1500 for p, and .04 for r in the given formula.

$$A = p + prt$$
$$1650 = 1500 + (1500)(.04)t$$
$$1650 = 1500 + 60t$$

Subtract 1500 from each side.

$$150 = 60t$$

Divide each side by 60:

$$\frac{150}{60} = \frac{60t}{60}$$

$$2\frac{1}{2} = t \quad \textit{Ans.}\ \text{(D)}$$

Matching Test

After completing this test, match your answers with the numbers in the box.

Given the formula $A = p + prt$:

1. Find A when $p = 1500$, $r = 4\%$, $t = 2$.

2. Find A when $p = 2500$, $r = 2\%$, $t = 2$.

3. Find p when $A = 2790$, $r = 2\frac{1}{2}\%$, $t = 5$.

4. Find t when $A = 1620$, $p = 1500$, $r = 4\%$.

5. Find r when $A = 2400$, $p = 2000$, $t = 4$.

a. 5%
b. 1620
c. 2
d. 2600
e. 2480

8.3 EVALUATING A VARIABLE IN A TEMPERATURE FORMULA

Model Problem

Given the formula $C = \frac{5}{9}(F - 32)$. Find C when $F = 5$.

(A) $-27\frac{5}{9}$ (B) $-26\frac{4}{9}$ (C) -25 (D) -15 (E) 15

Solution:

Find C in the formula $C = \frac{5}{9}(F - 32)$ by substituting 5 for F.

$$C = \frac{5}{9}(5 - 32)$$

$$= \frac{5}{\cancel{9}_{1}}(-\cancel{27}^{(-3)}) = -15 \quad \textit{Ans.}\ \text{(D)}$$

Matching Test

After completing this test, match your answers with the numbers in the box.

Given the formula $C = \frac{5}{9}(F - 32)$:

1. Find C when $F = -4$.

2. Find C when $F = 41$.

3. Find C when $F = -40$.

4. Find F when $C = 0$.

5. Find F when $C = 100$.

a. 32
b. −20
c. 5
d. 212
e. −40

9. PRACTICE PROBLEMS

1. Solve for x: $2 - 3x = 9$

(A) $-\frac{7}{3}$ (B) $\frac{7}{3}$ (C) $-\frac{11}{3}$ (D) $\frac{11}{3}$ (E) -9 **[1.1]**

2. Solve for y: $9 - 5y = 7 - 6y$

(A) 2 (B) $-\frac{2}{11}$ (C) -2 (D) $\frac{2}{11}$ (E) 16 **[1.2]**

3. Find the value of k if 2 is a root of the equation $x^2 + (k+5)x = 0$.

(A) -3 (B) -4 (C) $-4\frac{1}{2}$ (D) -7 (E) -14 **[1.3]**

4. Solve the equation $4x - 9x = 6 - 7x$ for x, then check by substitution. The number to which each side of the equation reduces is

(A) 3 (B) -3 (C) -15 (D) 15 (E) -5 **[1.4]**

5. Solve for y: $5(y+2) = 12$

(A) $\frac{22}{5}$ (B) 2 (C) $\frac{2}{5}$ (D) $-\frac{2}{5}$ (E) $\frac{5}{2}$ **[2.1]**

6. Solve for y: $9y - 5(y-2) = 14$

(A) 1 (B) 2 (C) $3\frac{1}{2}$ (D) 4 (E) 6 **[2.1]**

7. Solve for x: $3(1-x) - 4(2-x) = 10$

(A) 15 (B) -15 (C) $-\frac{15}{7}$ (D) $-\frac{5}{7}$ (E) 21 **[2.2]**

8. Solve for y: $\frac{4y}{3} = 12$

(A) 24 (B) 16 (C) 9 (D) 6 (E) 11 **[3.1]**

9. Solve for d: $\frac{3}{5}d - 4 = 8$

(A) 12 (B) $\frac{20}{3}$ (C) 20 (D) 24 (E) -20 **[3.1]**

10. Solve for n: $\frac{n}{3} - \frac{n}{4} = 6$

(A) 60 (B) 24 (C) 48 (D) 72 (E) 14 **[3.2]**

11. Solve for n: $\frac{3n}{4} - \frac{2n}{3} = \frac{5}{3}$

(A) 40 (B) 20 (C) 36 (D) 24 (E) −20 **[3.2]**

12. Solve for x: $\frac{5x}{3} = 2 + \frac{x}{3}$

(A) $\frac{3}{2}$ (B) $\frac{2}{3}$ (C) $\frac{1}{2}$ (D) 3 (E) 2 **[3.3]**

13. Solve for x: $.016x = .04$

(A) 25 (B) 4 (C) 2.5 (D) 40 (E) .25 **[4.1]**

14. Solve for n: $.5n = .21 + .2n$

(A) 7 (B) .07 (C) .7 (D) .3 (E) 3 **[4.1]**

15. Solve for x: $3x - .34 = .18 + 5x$

(A) −.26 (B) .26 (C) −2.6 (D) 2.6 (E) −.065 **[4.2]**

16. Solve for x: $400\%x = 2$

(A) .5 (B) .005 (C) .05 (D) .8 (E) 200 **[5.1]**

17. The selling price of $30 was the amount paid for a pair of skis after the original price had been increased by 25%. Find the number of dollars in the original price of the pair of skis.

(A) $22\frac{1}{2}$ (B) 20 (C) 24 (D) $37\frac{1}{2}$ (E) 25 **[5.2]**

18. Solve for x: $by + ax = c$

(A) $\frac{c - y}{a + b}$ (B) $\frac{by - c}{a}$ (C) $\frac{bcy}{a}$ (D) $\frac{c - by}{a}$ (E) $\frac{c + by}{a}$ **[6.1]**

19. Solve for x: $2ax + cx = b - ax$

(A) $\frac{b}{3a + c}$ (B) $\frac{b}{a + c}$ (C) $\frac{b - a}{2a + c}$ (D) $\frac{b - c}{3a}$ (E) $\frac{b}{3ac}$ **[6.1]**

20. If $\frac{a + b}{a} = 7$, find the value of $\frac{a}{b}$.

(A) $\frac{1}{6}$ (B) 6 (C) 8 (D) $\frac{1}{8}$ (E) $\frac{6}{7}$ **[6.1]**

21. If $3x - 2y = 0$, find the value of $\frac{x}{y}$.

(A) $\frac{2}{3}$ (B) $\frac{3}{2}$ (C) 1 (D) $-\frac{2}{3}$ (E) $-\frac{3}{2}$ **[6.1]**

22. Solve for B: $V = \frac{1}{3}Bh$

(A) $\frac{h}{3V}$ (B) $\frac{3h}{V}$ (C) $\frac{V}{3h}$ (D) $\frac{3V}{h}$ (E) $3Vh$ **[6.1]**

23. Solve $z = 3x + 3y$ for x.

(A) $\frac{z - 3y}{3}$ (B) $\frac{y - 3z}{3}$ (C) $\frac{y - z}{3}$ (D) $z - \frac{y}{3}$ (E) $\frac{z + 3y}{3}$ **[6.1]**

24. The equation $b - \frac{1}{a} = 1$ can be transformed into

(A) $b = \frac{1}{1-a}$ (B) $a = \frac{1}{b-1}$ (C) $b = \frac{1}{a}$ (D) $\frac{a}{b} = \frac{1}{b} - 1$ (E) $a = \frac{1}{b+1}$ **[6.2]**

25. Solve for x: $x - \frac{1}{c} = c$

(A) $\frac{c^2+1}{c}$ (B) $\frac{c^2-1}{c}$ (C) $c - \frac{1}{c}$ (D) $\frac{1}{c} - c$ (E) $\frac{c}{c^2+1}$ **[6.3]**

26. Solve for m: $\frac{3}{m+2} = \frac{2}{m-4}$

(A) 22 (B) 10 (C) 16 (D) $\frac{16}{5}$ (E) -16 **[7.1]**

27. Solve for x: $\frac{2a}{b} = \frac{3a}{x}$

(A) $\frac{3b}{2}$ (B) $\frac{2b}{3}$ (C) $\frac{3}{2b}$ (D) $\frac{2}{3b}$ (E) $\frac{3a^2}{b}$ **[7.2]**

28. If $m = \frac{12}{2-x}$, find the value of x when $m = 24$.

(A) $1\frac{1}{4}$ (B) $1\frac{1}{2}$ (C) 1 (D) -1 (E) $\frac{2}{3}$ **[8]**

29. If $S = 2\pi rh$, find h when $S = 352$ and $r = 14$. $\left(\text{Use } \pi = \frac{22}{7}.\right)$

(A) 8 (B) 4 (C) 2 (D) 16 (E) 12 **[8.1]**

30. If $A = p + prt$, find p when $A = 2730$, $r = 3\%$, and $t = 10$.
(A) 910 (B) 1820 (C) 2100 (D) 210 (E) 2200 **[8.2]**

31. If $C = \frac{5}{9}(F - 32)$, find F when $C = -5$.

(A) 41 (B) 23 (C) -41 (D) -23 (E) $-20\frac{5}{9}$ **[8.3]**

IV. REVIEWING PROBLEM-SOLVING

The four steps in problem-solving are:

1. *Representation.* Represent the unknowns by letters.
2. *Translation.* Obtain equations containing the letters by translating the relations given in the problem.
3. *Solution.* Find the values of the unknowns by solving the equations.
4. *Verification.* Check the values found for the unknowns by substituting these values in the original statements of the problem.

1. NUMBER PROBLEMS

If the number problem has one unknown. Represent the unknown by n or x and obtain an equation from the given relations in the problem.

Example: "If a number is increased by 5, the result is 20. What is the number?"

Solution: Represent the number by x and obtain the equation $x + 5 = 20$. Solving, $x = 15$.

If the number problem has two unknowns. Two methods of solution may be used:

Method 1. Represent both unknowns in terms of a single letter by using one of the given relations. Then obtain a single equation by using a second relation.

Method 2. Represent both unknowns by using two different letters. Then obtain two equations using two relations. (The two equations thus obtained are called ***simultaneous equations.***)

Example: "The sum of two numbers is 12 and one is twice the other. Find the two numbers."

Solution:

Method 1. Represent the numbers by using x and $2x$ and obtain the equation $x + 2x = 12$.

Method 2. Represent the unknowns by x and y and obtain the equations $y = 2x$ and $x + y = 12$.

1.1 REPRESENTING THE UNKNOWNS

Model Problem

The sum of two numbers is 10. Let s represent the smaller number. Write in simplest form the expression for the larger number increased by 9.

(A) $s - 19$ (B) $19 - s$ (C) $s + 19$ (D) $s - 1$ (E) $1 - s$

Solution:

If the sum of two numbers is 10, either one of the numbers is 10 minus the other number. Since the smaller number $= s$, the larger number $= 10 - s$.

Hence, the larger number increased by $9 = 10 - s + 9$

$= 19 - s$ *Ans.* (B)

Matching Test

Write the required expression in simplest form. After completing this test, match your answers with the expressions in the box.

1. The sum of two numbers is 10. If one of the numbers is represented by x, write the expression for the other number decreased by 2.

2. The sum of two numbers is 10. If one of the numbers is represented by x, write the expression for twice the other number.

3. The sum of two numbers is represented by x. If one of the numbers is 10, write the expression for the other number increased by 9.

4. The sum of two numbers is represented by x. If one of the numbers is 10, write the expression for 5 more than twice the other number.

5. The sum of two numbers is represented by $2x + 6$. If one of the numbers is 10, write the expression for one-half of the other number.

a. $20 - 2x$	*b*. $8 - x$	*c*. $x - 1$	*d*. $2x - 15$	*e*. $x - 2$

1.2 SOLVING A NUMBER PROBLEM BY USING ONE UNKNOWN

Model Problems

(1) In a collection of 60 coins, the number of half dollars is one-third the number of dimes and the number of quarters is 10 less than twice the number of dimes. Find the value of the collection.

(A) \$13.60 (B) \$14.00 (C) \$19.85 (D) \$21.00 (E) \$25.50

Solution:

Let $x =$ the number of dimes.

The number of half-dollars equals one-third the number of dimes.

Hence, $\frac{1}{3}x =$ the number of half-dollars.

The number of quarters equals 10 less than twice the number of dimes.

Hence, $2x - 10 =$ the number of quarters.

The total number of coins in the collection equals 60.

$$\text{Hence, } x + \frac{1}{3}x + 2x - 10 = 60$$

Since $x + \frac{1}{3}x + 2x = 3\frac{1}{3}x = \frac{10x}{3}$, then $\frac{10x}{3} - 10 = 60$

Add 10 to each side.

$$\frac{10x}{3} = 70$$

Multiply each side by 3.

$$10x = 210$$

Divide each side by 10.

$$x = 21$$

The number of half-dollars $= \frac{1}{3}x = \frac{1}{3}(21) = 7$.

The number of quarters $= 2x - 10 = 2(21) - 10 = 42 - 10 = 32$.

Since there are 21 dimes, 7 half-dollars, and 32 quarters, the value of the collection of coins is \$2.10 + \$3.50 + \$8.00, or \$13.60. *Ans.* (A)

(2) The total weight of a satellite to be put into orbit is 650 pounds. This weight is made up entirely of the container, the regulating devices, and the recording equipment. The container weighs two-thirds as much as the regulating devices, and the recording equipment weighs 50 pounds less than the container. Find the number of pounds in the weight of the container.

(A) 150 (B) 200 (C) 250 (D) 350 (E) 450

Solution:

Since the container weighs two-thirds as much as the regulating devices,

let $3x =$ the weight of the regulating devices in pounds,
and $2x =$ the weight of the container in pounds.

Since the recording equipment weighs 50 pounds less than the container,

$2x - 50 =$ the weight of the recording equipment in pounds.

Since the total weight of the satellite is 650 pounds,

$$2x + 3x + (2x - 50) = 650$$

$$7x - 50 = 650$$

Add 50 to each side:

$$+50 = +50$$

$$7x = 700$$

Divide each side by 7.

$$x = 100$$

The number of pounds in the weight of the container $= 2x = 200$. *Ans.* (B)

(3) A citrus-fruit grower has a 120-acre grove completely planted with three different kinds of trees as follows: 30 more acres with tangerine trees than with grapefruit trees and 3 times as many acres with orange trees as with grapefruit trees. Find the number of acres of orange trees that he planted.

(A) 18 (B) 46 (C) 48 (D) 58 (E) 54

Solution:

Let $x =$ the number of acres of grapefruit trees.
Then $x + 30 =$ the number of acres of tangerine trees,
and $3x =$ the number of acres of orange trees.

The total number of acres planted equals 120.

Hence, $x + 3x + x + 30 = 120$

Collect like terms.

$$5x + 30 = 120$$

Subtract 30 from each side.

$$5x = 90$$

Divide each side by 5.

$$x = 18$$

Since $x = 18$, then $x + 30 = 48$ and $3x = 54$.

Hence, 54 acres of orange trees were planted. *Ans.* (E)

Matching Test

After completing this test, match your answers with the numbers in the box.

1. A dozen eggs cost twice as much as a pound of chicken. What is the price in cents of a dozen eggs if the cost of $\frac{1}{2}$ dozen eggs and 2 pounds of chicken is 96 cents?
2. A tie and a pair of socks cost $1.60 together. If the tie costs $1 more than the pair of socks, what is the cost of the tie in cents?
3. A, B, C, and D are consecutive points on a line. CD is three times the length of AB. If $AC = 14$ inches and $BD = 20$ inches, what is the number of inches in BC?
4. For every 2 coins that Harry has, Mary has 3 and Joe has 5. If together they have 90 coins, how many does Mary have?
5. There were 200 people at a party. The adults outnumbered the children by 20, and among the children the girls outnumbered the boys by 8. How many girls were there at the party?

a. 27	*b.* 130	*c.* 64	*d.* 11	*e.* 49

1.3 SOLVING A NUMBER PROBLEM BY USING TWO UNKNOWNS

Model Problem

The difference between two numbers is 10. If the smaller number is increased by six times the larger number, the result is 4. Find the larger number.

(A) 2 (B) −2 (C) 8 (D) −8 (E) $9\frac{1}{3}$

Solution:

Let l = the larger number,
and s = the smaller number.

The smaller number increased by 6 times the larger number is 4: $s + 6l = 4$
The difference of two numbers is 10: $l - s = 10$
Add: $7l = 14$

$l = 2$

Ans. (A)

Matching Test

After completing this test, match your answers with the pairs of numbers in the box.

1. Find two numbers whose difference is 10 and whose sum is 40.
2. Find two numbers whose sum is 40 if the larger exceeds the smaller by 2.
3. Find two numbers such that the larger exceeds the smaller by 10 and is also three times as large as the smaller.
4. Find two numbers such that the larger is three times the smaller and the sum of the larger and twice the smaller is 45.
5. Find two numbers such that the difference between the larger and twice the smaller is 10 and their sum is 25.

a. 21, 19	*b.* 25, 15	*c.* 15, 5	*d.* 27, 9	*e.* 20, 5

2. CONSECUTIVE-INTEGER PROBLEMS

Integers may be positive, zero, or negative, such as +5, 0, and −5. Except for zero, which has no sign, integers combine a whole number with a positive or negative sign.

In a set of consecutive integers such as 3, 4, 5 or −3, −2, −1, note that the members of such a set *increase by 1* from left to right.

In a set of consecutive odd or consecutive even integers such as 3, 5, 7 or 4, 6, 8, note that the members of each set *increase by 2* from left to right.

Representation of Consecutive Integers

Use n, $n+1$, $n+2$, . . . , or, if there is an odd number of such integers, use n for the middle member. For example, if the problem begins, "The sum of 5 consecutive integers is 25," represent the members of the set by $n-2$, $n-1$, n, $n+1$, and $n+2$, and obtain the equation $5n=25$. Since $n=5$, the required integers are 3, 4, 5, 6, and 7.

Representation of Consecutive Odd or Consecutive Even Integers

Use n, $n+2$, $n+4$, . . . , or, if there is an odd number of integers, use n for the middle member. For example, if the problem begins, "The sum of 3 consecutive even integers is 24," represent the members of the set by $n-2$, n, and $n+2$, and obtain the equation $3n=24$. Since $n=8$, the required integers are 6, 8, and 10.

2.1 REPRESENTING THE UNKNOWNS

Model Problem

If n represents the first of three consecutive integers, then the sum of these integers is

(A) $n+1$ (B) $3n$ (C) $3n+3$ (D) $3n+2$ (E) $3n+6$

Solution:

In a set of consecutive integers, each integer is one more than the previous one.

Hence, if n = the first consecutive integer,
then $n+1$ = the second consecutive integer,
and $n+2$ = the third consecutive integer.

The sum of the three consecutive integers equals $n+n+1+n+2=3n+3$. *Ans.* (C)

Matching Test

n represents one of the integers in a set of consecutive integers. Write an expression that can be used to represent the sum of the set. After completing this test, match your answers with the expressions in the box.

1. n is the last of a set of three consecutive integers.

2. n is the middle number of a set of three consecutive integers.

3. n is the first of a set of four consecutive integers.

4. n is the first of a set of five consecutive integers.

5. n is the middle number of a set of five consecutive integers.

a. $5n+10$
b. $4n+6$
c. $3n$
d. $5n$
e. $3n-3$

2.2 TRANSLATING A RELATIONSHIP INTO AN EQUATION

Model Problem

The product of two consecutive even integers is 168. If n represents the smaller even integer, then an equation that can be used to find the integers is

(A) $n + n + 2 = 168$ (B) $n + n + 1 = 168$ (C) $n(n + 1) = 168$
(D) $n(n + 2) = 168$ (E) $n^2 + 2 = 168$

Solution:

Consecutive even integers, such as 100, 102, 104, . . . , are such that any one of them is 2 more than the preceding one.

Hence, if n = the first consecutive even integer,
then $(n + 2)$ = the second consecutive even integer.

The statement "The product of two consecutive even integers is 168" becomes the equation $n(n + 2) = 168$. *Ans.* (D)

Wrong Choices:

In choices (B) and (C), the numbers differ by one. This is true for consecutive integers but not for consecutive even integers.

Choice (A) would be correct for the statement "The *sum* of two consecutive even integers is 168."

Matching Test

n is the first of two consecutive even integers. Write an equation that can be used to find these integers. After completing this test, match your answers with the equations in the box.

1. Their product is 120.

2. Their sum is 102.

3. The square of their sum is 100.

4. The sum of their squares is 100.

5. The difference of their squares is 100.

a. $2n + 2 = 102$

b. $4n + 4 = 100$

c. $n(n + 2) = 120$ or $n^2 + 2n = 120$

d. $(2n + 2)^2 = 100$ or $4n^2 + 8n + 4 = 100$

e. $n^2 + (n + 2)^2 = 100$ or $2n^2 + 4n + 4 = 100$

2.3 SOLVING CONSECUTIVE-INTEGER PROBLEMS

Model Problem

Find four consecutive odd integers such that the sum of the first three integers is six more than the last integer.

(A) 3, 4, 5, 6 (B) 3, 5, 7, 9 (C) 5, 7, 9, 11 (D) 7, 9, 11, 13 (E) 9, 11, 13, 15

Solution:

Let $n =$ the first consecutive odd integer.

Then $n + 2 =$ the second consecutive odd integer,
$n + 4 =$ the third consecutive odd integer,
and $n + 6 =$ the fourth consecutive odd integer.

Since the sum of the first three consecutive odd integers is six more than the last number,

$$n + (n + 2) + (n + 4) = (n + 6) + 6$$

Eliminate parentheses and combine like terms.

$$3n + 6 = n + 12$$

Subtract $n + 6$ from each side: $\quad n + 6 = n + 6$

$$2n = 6$$

Divide each side by 2.

$$n = 3$$

Therefore, the four consecutive odd integers are 3, 5, 7, and 9.

Ans. (B)

> *Note.* By transposing, $3n + 6 = n + 12$ becomes $3n - n = 12 - 6$. Hence, $2n = 6$ and $n = 3$.

Matching Test

After completing this test, match your answers with the sets of numbers in the box.

Find three consecutive odd integers such that

1. their sum is 33.
2. the sum of the first two is 3 more than the last.
3. three times the first is 1 less than twice the last.
4. the sum of the last two is 5 less than three times the first.
5. the sum of the first two is -8.

a. 7, 9, 11
b. 11, 13, 15
c. $-5, -3, -1$
d. 9, 11, 13
e. 5, 7, 9

3. AGE PROBLEMS

A person's *future age n* years hence is calculated by *adding n* years to his present age. A person's *past age n* years ago is calculated by *subtracting n* years from his present age. For example, in 15 years, a person who is now 25 years old will be (25 + 15), or 40 years old, whereas 15 years ago he was (25 − 15), or 10 years old.

3.1 REPRESENTING THE UNKNOWNS

Model Problem

Mr. Cole was 30 years old 5 years ago. How old will he be y years hence?

(A) $35 - y$ (B) $35 + y$ (C) $25 - y$ (D) $25 + y$ (E) $y - 35$

Solution:

Since he was 30 years old 5 years ago, his present age is $(30 + 5)$ or 35 years.

In y years from now, or y years hence, his age in years will be $35 + y$. *Ans.* (B)

Matching Test

After completing this test, match your answers with the expressions in the box.

1. Represent the age of a person 25 years hence if his present age is y years.
2. Represent the age of a person y years ago if his present age is 25 years.
3. Represent the age of a person 25 years ago if his present age is y years.
4. Represent the age of a person y years ago if he will be 20 years old in 5 years.
5. Represent the age of a person 5 years hence if he was 25 years old y years ago.

a. $y + 30$	*b.* $y - 25$	*c.* $15 - y$	*d.* $25 - y$	*e.* $y + 25$

4. AVERAGES

To find the ***average*** of a set of n numbers, divide their sum by n. For example, the average of the 5 marks 80%, 81%, 83%, 85%, and 86% is 83%—their sum, 415%, divided by 5.

To find the ***weighted average*** of separate averages, weight is given to each average according to the number of items in the average. For example, if a person averages 35 miles per hour for 6 hours, and then 45 miles per hour for 4 hours, find the weighted average speed as follows: The first distance traveled is 35(6), or 210 miles. The second distance traveled is 45(4), or 180 miles. Hence, the entire distance traveled is $(210 + 180)$, or 390 miles. Since the time is 10 hours, the weighted average speed for the total distance is $\frac{390}{10}$, or 39 miles per hour.

4.1 AVERAGING COMMON FRACTIONS, DECIMALS, AND PERCENTS

Model Problems

(1) Find the average of $\frac{1}{2}$ and $\frac{2}{3}$.

(A) $\frac{5}{6}$ (B) $\frac{3}{10}$ (C) $\frac{7}{3}$ (D) $\frac{7}{6}$ (E) $\frac{7}{12}$

Solution:

To find the average of two numbers, first add the two numbers and then take one-half of their sum.

Hence, if the numbers are $\frac{1}{2}$ and $\frac{2}{3}$, their average is $\frac{1}{2}\left(\frac{1}{2}+\frac{2}{3}\right)$

1. Add: $\frac{1}{2}+\frac{2}{3}=\frac{3}{6}+\frac{4}{6}=\frac{7}{6}$

2. Take one-half of their sum: $\frac{1}{2}\left(\frac{7}{6}\right)=\frac{7}{12}$ *Ans.* (E)

(2) The average of $\frac{2}{5}$, 0.6, and 8% is

(A) $\frac{4}{5}$ (B) $\frac{9}{25}$ (C) 0.4 (D) 6% (E) 4%

Solution:

The three given numbers consist of a common fraction, a decimal, and a percent. To find their average, first change their forms so that the numbers are all of the same kind. For example, they can all be written as decimals:

$$\begin{aligned} \frac{2}{5} &= 0.4 \\ 0.6 &= 0.6 \\ 8\% &= 0.08 \\ \hline \text{sum} &= 1.08 \end{aligned}$$

To find the average of three numbers, add the numbers and divide their sum by 3.

The sum of the given numbers is 1.08. Hence, their average $= \frac{1.08}{3}$

$$= 0.36 = \frac{36}{100} = \frac{9}{25}$$

Ans. (B)

Matching Test

Find the average of each of the given sets of numbers. After completing this test, match your answers with the numbers in the box.

1. $\frac{1}{3}, \frac{2}{3}$ **2.** $\frac{1}{2}, \frac{1}{4}, \frac{1}{8}$ **3.** $2\frac{1}{2}, 4\frac{1}{4}, 7\frac{1}{8}$ **4.** 2.5, 4.4, 7.8, 10.1 **5.** 2.5%, 4.4%, 7.8%, 10.1%

a. 6.2	*b.* $\frac{1}{2}$	*c.* $4\frac{5}{8}$	*d.* $\frac{7}{24}$	*e.* .062 or 6.2%

4.2 AVERAGING QUANTITIES THAT ARE EXPRESSED ALGEBRAICALLY

Model Problem

Find the average of the numbers represented by $x-8$ and $3x+2$.

(A) $2x-6$ (B) $4x-6$ (C) $4x-3$ (D) $2x+3$ (E) $2x-3$

Solution:

The average of two numbers is found by adding the numbers and then dividing their sum by 2. Hence, the first step is to add:

$$\text{Add:}\quad \begin{array}{r} x-8 \\ 3x+2 \\ \hline 4x-6 \end{array}$$

Divide the sum by 2: $\dfrac{4x-6}{2} = \dfrac{\overset{1}{\cancel{2}}(2x-3)}{\underset{1}{\cancel{2}}} = 2x - 3$ *Ans.* (E)

Matching Test

Find the average of each set of numbers. After completing this test, match your answers with the expressions in the box.

1. $x + 8,\ 3x + 2$	*a.* $4x - 8y$
2. $x - 8y,\ 3x - 4y$	*b.* $x + 3y$
3. $x,\ 8y,\ 3x,\ 4y$	*c.* $4x + 3$
4. $x,\ 2x,\ 9x + 9$	*d.* $2x - 6y$
5. $x,\ 2x - 15y,\ 9x - 9y$	*e.* $2x + 5$

4.3 AVERAGING CONSECUTIVE INTEGERS

Model Problem

The average of two positive consecutive integers is $\frac{17}{2}$. Find the smaller of these two integers.

(A) 6 (B) 7 (C) 8 (D) 9 (E) 10

Solution:

The average of two numbers is found by adding the two numbers and then dividing their sum by 2.

Hence, if n and $n + 1$ are used to represent two consecutive integers, then their average $= \frac{2n+1}{2}$.

It is given that the average is $\frac{17}{2}$. Hence: $\frac{2n+1}{2} = \frac{17}{2}$

$$2n + 1 = 17$$

Subtract 1 from each side:

$$\begin{array}{rcl} 2n + 1 &=& 17 \\ -1 &=& -1 \\ \hline 2n &=& 16 \end{array}$$

Divide each side by 2.

$n = 8$ *Ans.* (C)

Matching Test

After completing this test, match your answers with the numbers in the box.

1. Find the average of 16 and 17.
2. Find the average of 16, 17, 18, and 19.
3. Find the smaller of two positive consecutive integers if their average is $\frac{9}{2}$.
4. Find the larger of two negative consecutive integers if their average is $-\frac{9}{2}$.
5. Find the largest of three positive consecutive integers if their average is 10.

a. 11
b. $\frac{35}{2}$
c. -4
d. $\frac{33}{2}$
e. 4

4.4 GIVEN THE AVERAGE OF A SET OF NUMBERS, TO FIND ONE MEMBER OF THE SET

Model Problem

If the average of two numbers is $2x + 1$ and one of the numbers is x, the other number is

(A) $x + 1$ (B) $2x - 1$ (C) $3x + 1$ (D) $3x + 2$ (E) $2x + 1$

Solution:

The average of two numbers is found by adding the two numbers and then dividing their sum by 2.

If x and y are used to represent the two numbers, then $\frac{x+y}{2}$ = the average of the two numbers.

Hence, $\frac{x+y}{2} = 2x + 1$

Multiply each side by 2.

$$x + y = 4x + 2$$

Subtract x from each side.

$$y = 3x + 2 \quad \textit{Ans.} \text{ (D)}$$

> *Note.* The correct choice can be found by testing each of the choices. If the average of x and $3x + 2$ is tested, then their average $= \frac{x + 3x + 2}{2} = \frac{4x + 2}{2} = 2x + 1$.

Matching Test

In each of the given sets of numbers, one number in the set is missing. If the average of the set is $2x$, find the missing number. After completing this test, match your answers with the numbers in the box.

1. $x + 2$, ? **2.** $2x - 1$, ? **3.** x, $2x$, ? **4.** x, $x + 2$, ? **5.** x, $2x$, $\frac{x}{2}$, ?

a. $3x$	*b.* $4x - 2$	*c.* $3x - 2$	*d.* $4\frac{1}{2}x$	*e.* $2x + 1$

4.5 COMBINING AVERAGES

Model Problem

Mr. Ford drove for 6 hours at an average speed of 30 miles per hour. He then drove for 9 hours, averaging 50 miles per hour. Find his average speed in miles per hour for the entire trip.

(A) 40 (B) 70 (C) 45 (D) 42 (E) 38

Solution:

1. The entire time spent traveling equals $6 + 9$ or 15 hours.
2. During the first part of the trip, Mr. Ford covered a distance of 30(6) or 180 miles. During the second part, he traveled 50(9) or 450 miles. Hence, the distance traveled for the entire trip equals $180 + 450$ or 630 miles.
3. The average rate for the entire trip is found by dividing the entire distance traveled by the entire time spent traveling. Hence, the average rate for the entire trip equals $\frac{630}{15}$ or 42 miles per hour. *Ans.* (D)

Matching Test

After completing this test, match your answers with the numbers in the box.

1. In a class, the average mark for 25 girls is 80 and the average mark for 15 boys is 72. Find the average mark for the entire group.
2. In a senior class, the average age of 6 seniors is 15.5 years and the average age of 34 others is 17.5 years. Find the average age in years for the entire group.
3. During a trip that lasted 14 hours, Mr. Brown averaged 34 miles per hour for the first 6 hours and 48 miles per hour thereafter. Find his average speed in miles per hour for the entire trip.
4. On a state turnpike, Mr. Cole paid 3 cents a mile for the first 50 miles and then 2 cents a mile for the remaining 150 miles. Find his average cost in cents per mile for the entire trip.
5. A clothier sold 100 suits at an average price of $60 per suit. If the average for 20 suits was $50 per suit and the average for 30 suits was $70 per suit, find the average price in dollars for the remaining suits.

a. 17.2	*b.* 77	*c.* 58	*d.* $2\frac{1}{4}$	*e.* 42

5. PERIMETER PROBLEMS

The ***perimeter of a polygon*** is the sum of its sides. Hence, the perimeter of a rectangle whose length is l and whose width is w is found by using the formula $p = 2l + 2w$. Also, the perimeter of a regular polygon is the product of n, the number of sides, and s, the length of one side: that is, $p = ns$.

5.1 REPRESENTING THE LENGTH OF A RECTANGLE IN TERMS OF ITS WIDTH

Model Problem

A tennis court is 6 feet longer than twice its width w. Express in terms of w the number of feet in the length of the court.

(A) $2w - 6$ (B) $2(w + 6)$ (C) $2(w - 6)$ (D) $2w + 6$ (E) $6(w + 2)$

Solution:

The length of the court is 6 feet longer than twice the width. Hence,

if w = the number of feet in the width,
then $2w + 6$ = the number of feet in the length. *Ans.* (D)

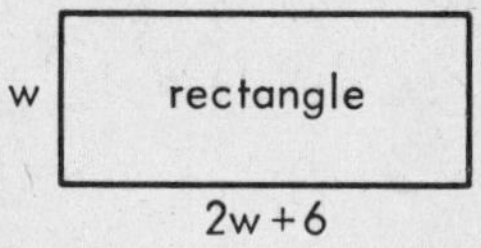

Matching Test

In each of the following, let w represent the number of feet in the width of a rectangular court. Write an expression for the number of feet in the length of the court. After completing this test, match your answers with the expressions in the box.

1. The length is 6 feet shorter than twice the width.
2. The length equals one-half the width less 6 feet.
3. The court is 8 feet longer than one-third its width.
4. The length is twice the sum of the width increased by 6 feet.
5. The length is one-half the difference between 20 feet and twice the width.

a. $2w - 6$
b. $\frac{1}{2}(20 - 2w)$
c. $\frac{w}{2} - 6$
d. $2(w + 6)$
e. $\frac{w}{3} + 8$

5.2 REPRESENTING THE PERIMETER OF A RECTANGLE, GIVEN THE DIMENSIONS

Model Problem

What is the perimeter of a rectangle whose width is represented by w and whose length is represented by $6w - 4$?

(A) $7w - 4$ (B) $12w - 8$ (C) $7w - 8$ (D) $14w - 4$ (E) $14w - 8$

Solution:

Find the perimeter of the rectangle by applying the following rule:

The perimeter of a rectangle equals twice the length added to twice the width.

Since w = the width and $6w - 4$ = the length,

$$\text{perimeter} = 2w + 2(6w - 4)$$

Replace $2(6w - 4)$ with $12w - 8$.

$$\text{perimeter} = 2w + 12w - 8$$

Collect the like terms containing w.

$$\text{perimeter} = 14w - 8 \quad \textit{Ans. } (E)$$

Matching Test

After completing this test, match your answers with the expressions in the box.

1. The length and width of a rectangle are represented by $6x + 4$ and $2x$, respectively. Represent the perimeter.
2. The length and width of a rectangle are represented by $6x - 4$ and $x + 2$. Express the perimeter.
3. The perimeter of a rectangle is represented by $12x + 4$. If the length is represented by $3x + 2$, represent the width.
4. The adjacent sides of a parallelogram are represented by $7x - 10$ and $x + 3$. Represent the perimeter.
5. The adjacent sides of a parallelogram are represented by $7x + 10$ and $5\frac{1}{2}x$. Represent the perimeter.

a. $16x + 8$	*b.* $25x + 20$	*c.* $14x - 4$	*d.* $16x - 14$	*e.* $3x$

5.3 REPRESENTING THE PERIMETER OF A TRIANGLE, GIVEN THE SIDES

Model Problem

The sides of a triangle are represented by $7x - 3$, $2x + 5$, and $4x - 2$. Express in simplest form the perimeter of the triangle in terms of x.

(A) $13x$ (B) $12x$ (C) $13x + 6$ (D) $13x - 10$ (E) $12x + 4$

Solution:

The perimeter of a triangle is found by adding its three sides.

To find the sum of the sides, arrange the three expressions in order, placing like terms in the same column. Then add like terms.

$$\begin{array}{r} 7x - 3 \\ 2x + 5 \\ \underline{4x - 2} \\ 13x \qquad \end{array} \qquad \textit{Ans. } (A)$$

Matching Test

After completing this test, match your answers with the expressions in the box.

1. The sides of a triangle are represented by $7x + 3$, $2x + 5$, and $4x + 2$. Express the perimeter.
2. The sides of a triangle are represented by $3x + 7$, $5x - 2$, and $2x - 4$. Express the perimeter.
3. Two sides of a triangle are represented by $7x - 3$ and $2x - 5$. If the perimeter is represented by $10x + 13$, express the remaining side.
4. Each of the equal sides of an isosceles triangle is represented by $2x - 1$ and the remaining side by $x + 8$. Express the perimeter.
5. Each of the equal sides of an isosceles triangle is represented by $3x + 5$. If the perimeter is represented by $10x + 5$, express the remaining side.

a. $x + 21$	*b.* $10x + 1$	*c.* $4x - 5$	*d.* $5x + 6$	*e.* $13x + 10$

5.4 REPRESENTING THE PERIMETER OF A REGULAR POLYGON, GIVEN A SIDE

Model Problem

Write an expression for the perimeter of a square whose side is represented by $\frac{1}{4}s - 7$.

(A) $s - 7$ (B) $\frac{1}{2}s - 14$ (C) $\frac{1}{2}s - 7$ (D) $s - 28$ (E) $2s - 28$

Solution:

Find the expression for the perimeter of the given square by applying the following rule:

The perimeter of a square equals four times a side.

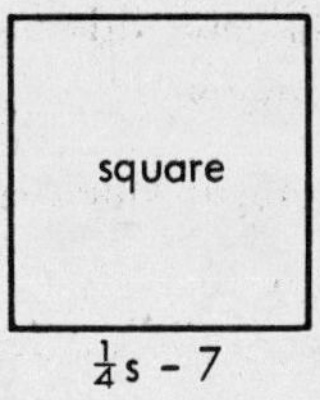

Hence, the perimeter $= 4\left(\frac{1}{4}s - 7\right)$

$= s - 28$ *Ans.* (D)

After completing each test, match your answers with the expressions in the boxes.

Matching Test A

Each expression represents a side of a square. Write an expression for the perimeter of the square.

1. $\frac{1}{4}s + 7$ **2.** $\frac{1}{2}s - 7$ **3.** $\frac{3}{2}s + 5$ **4.** $s - \frac{7}{2}$ **5.** $8 - 2s$

a. $2s - 28$	*b.* $32 - 8s$	*c.* $s + 28$	*d.* $6s + 20$	*e.* $4s - 14$

Matching Test B

Each expression represents a side of the given figure. Write an expression for the perimeter of the figure.

1. Equilateral triangle: $\frac{s}{3} - 7$

2. Equilateral triangle: $3s - 2\frac{1}{3}$

3. Regular hexagon: $\frac{s}{3} - 7$

4. Regular hexagon: $\frac{5s}{6} - \frac{7}{3}$

5. Regular octagon: $\frac{s}{4} + 2$

a. $9s - 7$
b. $2s + 16$
c. $s - 21$
d. $2s - 42$
e. $5s - 14$

5.5 REPRESENTING A SIDE OF A REGULAR POLYGON, GIVEN THE PERIMETER

Model Problem

If the perimeter of an equilateral triangle is represented by $15x - 12$, find one side of the triangle.

(A) $15x - 4$ (B) $5x - 12$ (C) $45x - 36$ (D) $3x - 4$ (E) $5x - 4$

Solution:

Since an equilateral triangle is a triangle having three equal sides, the perimeter of an equilateral triangle equals three times the length of a side.

Hence, to find the length of any side, divide the perimeter by three.

$$\text{each side} = \frac{\text{perimeter}}{3}$$

Substitute $15x - 12$ for the perimeter.

$$\text{each side} = \frac{15x - 12}{3} = 5x - 4 \quad \textit{Ans.}\ \text{(E)}$$

Matching Test

Each expression represents the perimeter of the given figure. Write an expression for a side of the figure. After completing this test, match your answers with the expressions in the box.

1. Equilateral triangle: $12x - 15$
2. Square: $12x - 12$
3. Square: $10x + 16$
4. Regular hexagon: $12x + 12$
5. Regular hexagon: $6x - 20$

a. $2x + 2$
b. $4x - 5$
c. $2\frac{1}{2}x + 4$
d. $3x - 3$
e. $x - 3\frac{1}{3}$

How to Check Your Answer in a Verbal Problem

Some of the solutions to the model problems in the following pages contain an extra step: a procedure for checking your answer. This may seem to be unnecessary, since it is natural to assume that an answer is correct if it exactly matches one of the multiple choices. Bear in mind, however, that even if your answer agrees with one of the choices, it may still be wrong. The choice that matches your answer may be the result of an error that is commonly made and, hence, it may very well have been included among the multiple choices for that reason. Therefore, if time permits, it is advisable to check your answer. In a verbal problem, this is done by checking your answer against the original statement, or wording, of the problem. *Do not check your equation, since the equation may be wrong!*

5.6 FINDING THE SIDES OF A RECTANGLE

Model Problem

A rectangular playground has a perimeter of 510 feet. If the length is 45 feet less than three times the width, find the number of feet in the width of the playground.

(A) $52\frac{1}{2}$ (B) 75 (C) 111 (D) $138\frac{3}{4}$ (E) 255

Solution:

Let w = the number of feet in the width of the rectangle.

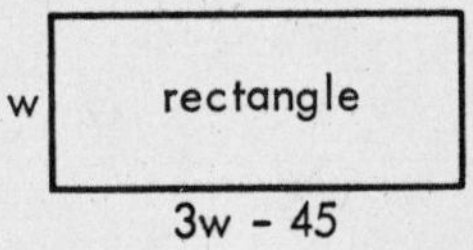

Then $3w - 45$ = the number of feet in the length of the rectangle.

The perimeter of a rectangle equals the sum of twice the length and twice the width.

Hence, since the perimeter equals 510, then $2w + 2(3w - 45) = 510$

Clear of parentheses.

$$2w + 6w - 90 = 510$$

Collect w-terms and add 90 to each side.

$$8w = 600$$

Divide each side by 8:

$$\frac{8w}{8} = \frac{600}{8}$$

$$w = 75$$

Hence, the width of the playground is 75 feet. *Ans.* (B)

Check in the original problem.

The length $= 3w - 45 = 3(75) - 45 = 180$.

1. The length equals three times the width less 45 feet.

   Three times the width equals 3(75 feet), or 225 feet.

   $$180 \text{ feet} \stackrel{?}{=} 225 \text{ feet} - 45 \text{ feet}$$

   $$180 \text{ feet} = 180 \text{ feet}$$

2. The perimeter equals 510 feet.

   Twice the length equals 360 feet. Twice the width equals 150 feet.

   $$360 \text{ feet} + 150 \text{ feet} \stackrel{?}{=} 510 \text{ feet}$$

   $$510 \text{ feet} = 510 \text{ feet}$$

Matching Test

Find the sides of each given figure. After completing this test, match your answers with the pairs of values in the box.

1. The perimeter of a rectangle is 60. The length is triple the width.

2. The perimeter of a rectangle is 60. The length is 10 more than the width.

3. The perimeter of a rectangle is 68. The length is 10 more than twice the width.

4. The perimeter of a rectangle is 52. The length is 2 more than one-third the width.

5. The perimeter of a parallelogram is 52. One of the sides is 2 more than one-half an adjacent side.

a. 16, 10	*b.* 26, 8	*c.* 20, 10	*d.* $22\frac{1}{2}, 7\frac{1}{2}$	*e.* 8, 18

5.7 FINDING THE SIDES OF A TRIANGLE

Model Problem

The perimeter of a triangle is 63 inches. The longest side exceeds the shortest side by 6 inches. The sum of the lengths of the longest and the shortest side is twice the length of the third side. Find the number of inches in the length of the shortest side of the triangle.

(A) 18 (B) 24 (C) 21 (D) 12 (E) 15

Solution:

Let $x =$ the number of inches in the shortest side.
Then $x + 6 =$ the number of inches in the longest side. (The longest side exceeds the shortest side by 6 inches.)
Also, $x + 3 =$ the number of inches in the third side.

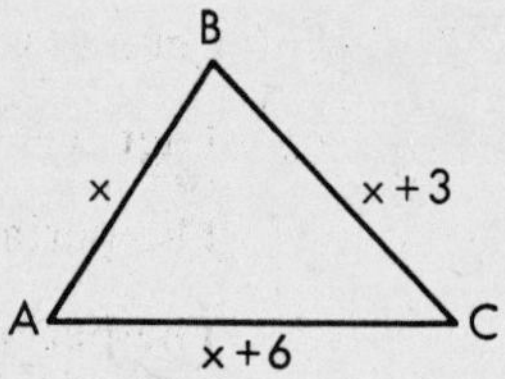

(The third side $= \dfrac{x + (x + 6)}{2} = \dfrac{2x + 6}{2} = x + 3$.)

Since the perimeter, 63 inches, is the sum of the three sides,

$$x + (x + 6) + (x + 3) = 63$$

$$3x + 9 = 63$$

Subtract 9 from each side:

$$\begin{array}{r} -9 = -9 \\ \hline 3x \quad = 54 \end{array}$$

Divide each side by 3:

$$\frac{3x}{3} = \frac{54}{3}$$

$$x = 18$$

Hence, the length of the shortest side of the triangle is 18 inches. *Ans.* (A)

Matching Test

Find the sides of each given triangle. After completing this test, match your answers with the sets of values in the box.

1. The perimeter is 51. The second side is 3 less than the first and the remaining side is 3 less than the second.

2. The perimeter is 63. Two of the sides are equal and the remaining side is 3 more than either of them.

3. The perimeter is 46. Each of two equal sides is 8 more than the remaining side.

4. The perimeter is 36. The sides are in the ratio 2:3:4.

5. The perimeter is 85. Two sides are in the ratio 2:3 and the remaining side is 5 less than their sum.

a. 18, 18, 10	*b.* 8, 12, 16	*c.* 20, 20, 23	*d.* 20, 17, 14	*e.* 18, 27, 40

5.8 RELATING THE PERIMETERS OF A SQUARE AND AN EQUILATERAL TRIANGLE

Model Problem

One side of a square exceeds one side of an equilateral triangle by 6. The perimeter of the square exceeds the perimeter of the triangle by 34. Find the length of one side of the square.

(A) 4 (B) 10 (C) 12 (D) 16 (E) 18

Solution:

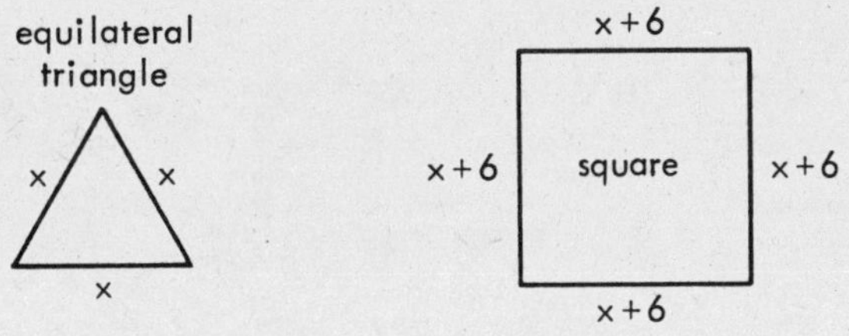

Let $x =$ the length of a side of the equilateral triangle.

Then $x + 6 =$ the length of a side of the square.

The perimeter of the square $= 4(x + 6) = 4x + 24$.

The perimeter of the equilateral triangle $= 3x$.

The perimeter of the square exceeds the perimeter of the equilateral triangle by 34.

$$\text{Hence, } 4x + 24 = 3x + 34$$

$$\text{Subtract } 3x + 24 \text{ from each side: } \quad \underline{3x + 24 = 3x + 24}$$

$$x \quad = \quad 10$$

Since $x = 10$, then $x + 6 = 16$.

Hence, the length of a side of the triangle is 10 and the length of a side of the square is 16.

Ans. (D)

Check in the original problem.

One side of a square exceeds one side of an equilateral triangle by 6.

$$16 \stackrel{?}{=} 10 + 6$$
$$16 = 16$$

The perimeter of the square exceeds the perimeter of the equilateral triangle by 34.

$$4(16) \stackrel{?}{=} 3(10) + 34$$
$$64 \stackrel{?}{=} 30 + 34$$
$$64 = 64$$

Matching Test

Find a side of the given square and find a side of the given equilateral triangle. After completing this test, match your answers with the pairs of values in the box.

1. The side of the square exceeds the side of the equilateral triangle by 5. The perimeter of the square exceeds the perimeter of the triangle by 34.
2. The side of the square exceeds the side of the equilateral triangle by 6. The sum of the perimeters equals 108.

3. The side of the square exceeds the side of the equilateral triangle by 7. The perimeter of the square is twice the perimeter of the equilateral triangle.

4. The side of the square exceeds the side of the equilateral triangle by 8. The perimeter of the triangle exceeds one-half the perimeter of the square by 10.

5. The side of the square exceeds the side of the equilateral triangle by 2. The ratio of the perimeter of the square to the perimeter of the equilateral triangle is 3:2.

a. 34, 26	*b.* 21, 14	*c.* 18, 16	*d.* 19, 14	*e.* 18, 12

5.9 RELATING PERIMETERS OF REGULAR POLYGONS

Model Problem

The length of a rectangle exceeds the width by 2. A side of a square is equal to the width of the rectangle. A side of an equilateral triangle is equal to the length of the rectangle. If the perimeter of the square is equal to the perimeter of the equilateral triangle, find the perimeter of the rectangle.

(A) 6 (B) 12 (C) 18 (D) 24 (E) 28

Solution:

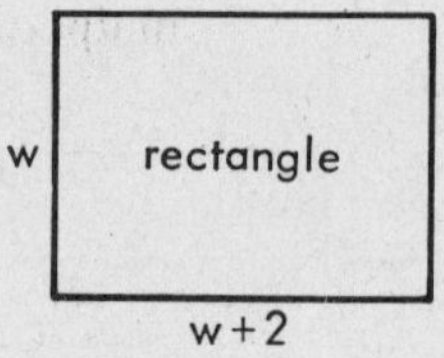

(a) Let w = the width of the rectangle.

1. The length of the rectangle exceeds its width by 2.

   Hence, $w + 2$ = the length of the rectangle.

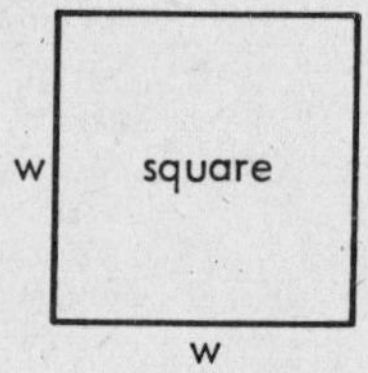

2. Each side of the square equals the width of the rectangle.

   Hence, w = each side of the square.

   The perimeter of the square = $4w$.

3. Each side of the equilateral triangle equals the length of the rectangle.

   Hence, $w + 2$ = each side of the equilateral triangle.

   The perimeter of the equilateral triangle $= 3(w + 2)$

   $= 3w + 6$

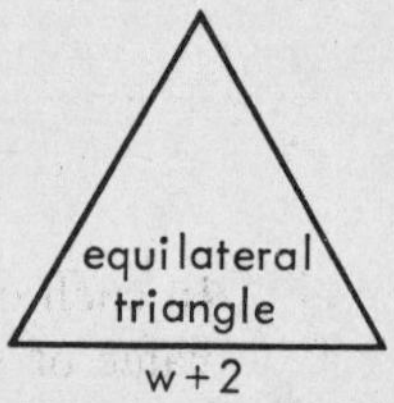

(b) The perimeter of the square equals the perimeter of the equilateral triangle.

$$\begin{aligned} \text{Hence,}\quad 4w &= 3w + 6 \\ \text{Subtract } 3w \text{ from each side:}\quad -3w &= -3w \\ w &= 6 \end{aligned}$$

Since $w = 6$, then $w + 2 = 8$.

The perimeter of the rectangle equals the sum of twice the length and twice the width.

Hence, the perimeter of the rectangle $= 2(8) + 2(6)$

$= 16 + 12 = 28$ *Ans.* (E)

After completing each test, match your answers with the numbers in the boxes.

Matching Test A

Find the perimeter of the shaded figure.

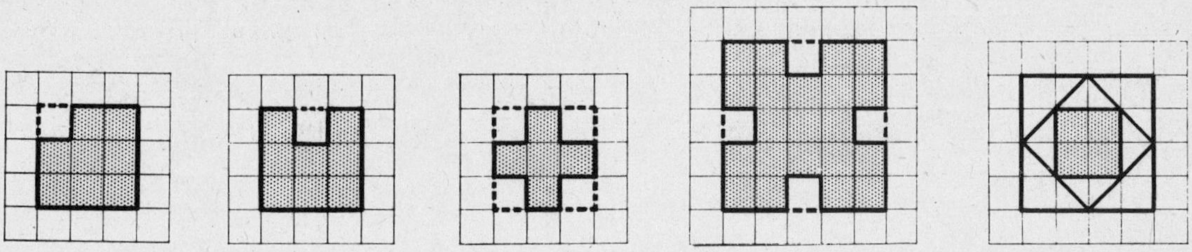

1. **2.** **3.** **4.** **5.**

1. A square of side 1 unit is cut from a square of side 3 units.
2. A square of side 2 units is cut from a square of side 6 units.
3. Squares of side 3 units are cut from a square of side 9 units.
4. Squares of side 2 units are cut from a square of side 10 units.
5. The outermost square has a side of 12 units. Two new squares are formed by joining midpoints successively as shown.

a. 36	*b.* 56	*c.* 28	*d.* 12	*e.* 24

Matching Test B

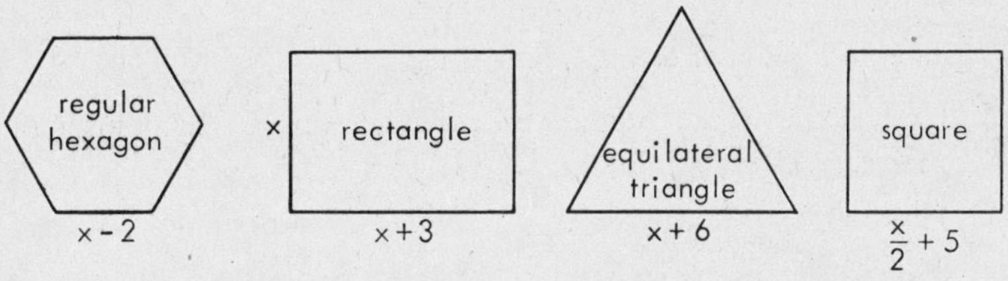

Fig. 1 Fig. 2 Fig. 3 Fig. 4

In each of the following, use the representation shown in the two indicated figures to find the value of x.

1. The perimeter of the rectangle (Fig. 2) equals the perimeter of the square (Fig. 4).
2. The perimeter of the square (Fig. 4) equals the perimeter of the equilateral triangle (Fig. 3).
3. The perimeter of the rectangle (Fig. 2) equals the perimeter of the equilateral triangle (Fig. 3).
4. The perimeter of the regular hexagon (Fig. 1) equals the perimeter of the square (Fig. 4).
5. The perimeter of the regular hexagon (Fig. 1) equals the perimeter of the rectangle (Fig. 2).

a. 2	*b.* 9	*c.* 8	*d.* 12	*e.* 7

6. VALUE, COST, AND MIXTURE PROBLEMS

Value problem formula: $T = NV \begin{cases} T = \text{the total value of a number of units} \\ N = \text{the number of units} \\ V = \text{the value of each unit} \end{cases}$

For example, if the value of each of 5 pens is 15¢, then the total value is 5(15¢), or 75¢.

Cost problem formula: $C = NP \begin{cases} C = \text{the total cost of a number of units} \\ N = \text{the number of units} \\ P = \text{the price of each unit} \end{cases}$

For example, if the price of each of 5 cards is 25¢, then the total cost is 5(25¢), or 125¢.

Solution problem formula: $Q = RS \begin{cases} Q = \text{the quantity or number of units of an element in a solution} \\ S = \text{the number of units of the solution} \\ R = \text{the rate or percent of the element in the solution} \end{cases}$

For example, if the number of quarts of a solution is 40 and the percent of salt in the solution is 20%, then the quantity of salt in the solution is 20% of 40 quarts, or 8 quarts.

Note that the formulas $T = NV$, $C = NP$, and $Q = RS$ are similar in structure. In each formula, two quantities are being multiplied to obtain a product.

Types of Mixture Problems

In mixture problems, many sorts of mixtures are made. Some examples are:

1. Different kinds of coins or stamps are assembled into a single collection.
2. Different grades of coffee or tea are mixed to form a special blend.
3. Different types of nuts or candies are mixed to form a specific assortment.
4. Tickets of admission to a game or performance are sold in certain quantities at different prices.
5. Numbers of items are bought at different prices.
6. (*a*) Solutions are combined to form a special mixture. (*b*) A solution is strengthened by adding more of the dissolved substance or weakened (diluted) by adding pure water.

6.1 SOLVING MIXTURE PROBLEMS BY USING ARITHMETIC ONLY

Model Problem

A merchant has two blends of tea. In the first blend, 2 pounds of 60¢ tea are mixed with every pound of 90¢ tea. In the second, 2 pounds of 90¢ tea are mixed with every pound of 60¢ tea. What would be his additional profit if he sold 30 pounds of the first blend at the average price per pound of the second blend?

(A) \$1.50 (B) \$2.40 (C) \$3.00 (D) \$3.60 (E) \$4.50

Solution:

In the first mixture, the value of 3 pounds is \$1.20 + \$.90, or \$2.10.

In the second mixture, the value of 3 pounds is \$1.80 + \$.60, or \$2.40.

The average price for each pound of the first mixture is $\frac{\$2.10}{3}$, or \$.70 per pound; that for each pound of the second mixture is $\frac{\$2.40}{3}$, or \$.80 per pound.

If the first blend, which averages 70¢ per pound, is sold instead at 80¢ per pound, the average price of the second blend, it follows that the merchant would make an added profit of 10¢ on each pound. Hence, on 30 pounds, he would make an added profit of \$3.00. *Ans.* (C)

Matching Test

After completing this test, match your answers with the values in the box.

1. What is the value per pound of a mixture of 30 pounds of tea at 60¢ a pound and 10 pounds of tea at 80¢ a pound?
2. What was the selling price per pound of a mixture of 10 pounds of tea at 60¢ a pound and 15 pounds of tea at 80¢ a pound if the mixture was sold at an increase of 25% above these prices?
3. What was the selling price per pound of a mixture of 20 pounds of tea at 60¢ a pound and 5 pounds of tea at 80¢ a pound if the mixture was sold at a reduction of 25% below these prices?
4. Instead of mixing 2 pounds of tea at 80¢ a pound with every pound of tea at 60¢ a pound, a merchant mixed 2 pounds of tea at 60¢ a pound with every pound of tea at 80¢ a pound. What would be the added profit if 12 pounds of the second mixture were sold at the average price per pound of the first mixture?
5. A dealer mixed 11 pounds of tea worth 40¢ a pound, 3 pounds of tea worth 60¢ a pound, and 6 pounds of tea worth 80¢ a pound. What was the average worth per pound of the mixture?

a. 80¢	*b.* 48¢	*c.* 55¢	*d.* 90¢	*e.* 65¢

6.2 COMBINING VALUES IN COST AND MIXTURE PROBLEMS: USING ONE UNKNOWN

Model Problem

Mrs. Newton spent \$27.50 for a dress, a hat, and a pair of shoes. She spent three times as much for the dress as for the hat and \$5.00 more for the shoes than for the hat. How much did she pay for the hat?

(A) \$4.50 (B) \$5.00 (C) \$9.50 (D) \$13.50 (E) \$15.00

Solution:

Let $h =$ the price in dollars of the hat.

Then $3h =$ the price in dollars of the dress.

Also, $h + 5 =$ the price in dollars of the pair of shoes.

The total cost of the hat, dress, and shoes equals \$27.50.

Hence, $h + 3h + h + 5 = 27.50$

Collect the terms containing h and subtract 5 from each side.

$$5h = 22.50$$

Divide each side by 5: $\frac{5h}{5} = \frac{22.50}{5}$

$$h = 4.50$$

Hence, the price of the hat was \$4.50. *Ans.* (A)

Check in the original problem.

1. The total cost equals \$27.50. The price of the dress $= 3h = 13.50$ and the price of the pair of shoes $= h + 5 = 9.50$.

$$\$4.50 + \$13.50 + \$9.50 \stackrel{?}{=} \$27.50$$

$$\$27.50 = \$27.50$$

2. The cost of the dress equals three times that of the hat.

$$\$13.50 \stackrel{?}{=} 3(\$4.50)$$

$$\$13.50 = \$13.50$$

3. The cost of the shoes equals \$5 more than that of the hat.

$$\$9.50 \stackrel{?}{=} \$4.50 + \$5.00$$

$$\$9.50 = \$9.50$$

Matching Test

After completing this test, match your answers with the numbers in the box.

1. A collection of 18 coins consists of pennies and nickels. What is its value in cents if there are five times as many pennies as nickels?

2. A collection of 20 stamps consists of 4¢ and 7¢ stamps. What is its value in cents if the number of 4¢ stamps is one-fourth the number of 7¢ stamps?

3. A collection of 24 coins consists of nickels, dimes, and quarters. There are three times as many dimes as nickels and three more quarters than dimes. What is the total value of the collection in cents?

4. In a collection of coins, the ratio of nickels to dimes is 2:3. If there are 20 coins in all, what is the total value in cents?

5. In a sale, a total of 15 hats and dresses were sold at \$7 for each hat and \$20 for each dress. If the number of dresses sold is two-thirds the number of hats sold, what is the sales total in dollars?

a. 128	*b.* 183	*c.* 30	*d.* 405	*e.* 160

6.3 COMBINING VALUES IN COST AND MIXTURE PROBLEMS: USING TWO UNKNOWNS

Model Problems

(1) Tickets for a high school football game cost 50¢ each if purchased at the advance sale before the day of the game and 75¢ each if bought at the gate on the day of the game. For a particular game, 1200 tickets were sold and the receipts were $712.50. How many tickets were bought at the gate on the day of this game?

(A) 450 (B) 500 (C) 600 (D) 700 (E) 750

Solution:

	(¢) Price	× Number of tickets	= (¢) Cost
Gate tickets	75	x	$75x$
Advance tickets	50	y	$50y$

The total number of tickets sold was 1200.

(1) $x + y = 1200$

The total receipts for the tickets were $712.50.

(2) $75x + 50y = 71\ 250$

Multiply each side of equation (1) by 50: $50x + 50y = 60\ 000$

Combine with equation (2) by subtraction: $25x = 11\ 250$

Divide each side by 25: $\frac{25x}{25} = \frac{11\ 250}{25}$

$x = 450$

Hence, the number of tickets bought at the gate was 450. *Ans.* (A)

(2) A wholesale merchant has two kinds of tea, one priced at 75 cents per pound and the other at 85 cents per pound. How many pounds of the 75¢ tea should he use to make a mixture of 500 pounds to sell at 79 cents per pound?

(A) 150 (B) 200 (C) 300 (D) 350 (E) 400

Solution:

Let x = the number of pounds of the 75¢ tea
and y = the number of pounds of the 85¢ tea.

	(¢) Price per lb.	× Number of lb.	= (¢) Total value
75¢ tea	75	x	$75x$
85¢ tea	85	y	$85y$
Mixture	79	500	39 500

Since the weight of the mixture is the total weight of both kinds of tea,

$$x + y = 500$$

Since the value of the mixture is the total value of both kinds of tea,

$$75x + 85y = 39\ 500$$

Multiply each side of the first equation by 85 and then subtract.

$$\begin{array}{rrl} \text{Multiply by 85: } x + y = 500 \rightarrow & 85x + 85y &= 42\ 500 \\ \text{Subtract: } & \underline{75x + 85y} &= \underline{39\ 500} \\ & 10x &= \ \ 3\ 000 \end{array}$$

Divide each side by 10.

$$x = 300$$

Hence, 300 pounds of the 75¢ tea should be used. *Ans.* (C)

(3) A purse contained 28 coins in nickels and dimes. If the total value was \$2.20, how many nickels were there?

(A) 8 (B) 10 (C) 12 (D) 16 (E) 20

Solution:

	(¢) Value of coin	× Number of coins =	(¢) Total value
Dimes	10	d	$10d$
Nickels	5	n	$5n$

The total value of the coins equals the sum of the value of the dimes and the value of the nickels.

Hence, $10d + 5n = 220$

The number of coins equals the sum of the number of dimes and the number of nickels.

Hence, $d + n = 28$

Divide each side of the first equation by 5 and then subtract as follows:

$$\begin{array}{rrl} & 2d + n &= 44 \\ \text{Subtract: } & \underline{d + n} &= \underline{28} \\ & d &= 16 \end{array}$$

Since $d + n = 28$, $n = 12$ *Ans.* (C)

Matching Test

After completing this test, match your answers with the pairs of numbers in the box.

1. There are 24 coins in a collection of nickels and dimes. If the collection is worth \$1.60, how many nickels and how many dimes are there?
2. Tickets for a game were sold for either 75¢ each or for 50¢ each. If the receipts for 24 tickets were \$14.50, how many tickets were sold at each price?
3. A stamp collection consists of 4¢ and 6¢ stamps. If there are 24 stamps in all and their value is \$1.18, how many stamps of each type are there?

4. Rose bushes are sold at \$1.50 per dozen and at \$2.00 per dozen. If 24 dozen are purchased for \$45, how many dozens are bought at each price?

5. A total of 24 pads and pencils are bought for \$4.10. If each pencil costs 10¢ and each pad costs 20¢, how many of each are bought?

a. 6, 18	*b.* 13, 11	*c.* 16, 8	*d.* 7, 17	*e.* 10, 14

6.4 DRY AND WET MIXTURE PROBLEMS: FINDING THE AMOUNTS TO BE ADDED TO OBTAIN A DESIRED RESULT

Model Problems

(1) A dealer has 400 gallons of low-octane gasoline worth 29 cents a gallon. How many gallons of high-octane gasoline worth 35 cents a gallon should he blend in with this low-octane gasoline so as to produce a mixture worth 31 cents a gallon?

(A) 200 (B) 240 (C) 300 (D) 400 (E) 600

Solution:

	(¢) Price ×	Number of gallons =	(¢) Total cost
Low-octane gas	29	400	29(400)
High-octane gas	35	x	$35x$
Mixture	31	$x + 400$	$31(x + 400)$

The total cost of the mixture equals the sum of the costs of the low-octane gas and the high-octane gas.

Hence, $29(400) + 35x = 31(x + 400)$

$$11\,600 + 35x = 31x + 12\,400$$

Subtract: $11\,600 + 31x = 31x + 11\,600$

$$4x = 800$$

Divide each side by 4: $\frac{4x}{4} = \frac{800}{4}$

$$x = 200$$

Hence, 200 gallons of high-octane gas must be added. *Ans.* (A)

Check in the original problem as follows:

The cost of the mixture equals the sum of the costs of the low-octane gas and the high-octane gas.

1. The cost of 400 gallons of low-octane gas equals 400(\$.29), or \$116.
2. The cost of 200 gallons of high-octane gas equals 200(\$.35), or \$70.
3. The cost of 600 gallons of the mixture equals 600(\$.31), or \$186.

$$\$116 + \$70 \stackrel{?}{=} \$186$$
$$\$186 = \$186$$

(2) How many cubic centimeters of pure acid must be added to 1000 cubic centimeters of a solution which is 75% acid to make a solution which is 90% acid?

(A) 600 (B) $666\frac{2}{3}$ (C) 1200 (D) 1250 (E) 1500

Solution:

	Percent of acid in solution ×	(cu. cm.) Solution =	(cu. cm.) Quantity of acid
Original solution	.75	1000	.75(1000)
Pure acid added	1.00	x	x
Resulting solution	.90	$(1000 + x)$	$.90(1000 + x)$

The acid in the resulting solution is the sum of the acid in the original solution and the pure acid added.

Hence, $.90(1000 + x) = .75(1000) + x$

Multiply by 100 to clear the equation of decimals.

$$90(1000 + x) = 75(1000) + 100x$$

$$90\,000 + 90x = 75\,000 + 100x$$

Subtract $(75\,000 + 90x)$ from each side:

$$\frac{75\,000 + 90x = 75\,000 + 90x}{15\,000 = 10x}$$

Divide each side by 10:

$$\frac{15\,000}{10} = \frac{10x}{10}$$

$$1500 = x$$

Hence, 1500 cubic centimeters of pure acid must be added. *Ans.* (E)

Matching Test

After completing this test, match your answers with the numbers in the box.

1. How many quarters must be added to 18 nickels so that the average value of a coin in the collection is 10 cents?
2. How many pounds of cashews at \$.75 a pound must be added to 18 pounds of almonds at \$1.00 a pound to obtain a mixture worth \$.85 a pound?
3. How many gallons of high-octane gasoline worth 40¢ a gallon must be mixed with 18 gallons of low-octane gasoline worth 30¢ a gallon to produce a mixture worth 34¢ a gallon?
4. How many cubic centimeters of an 85% acid solution must be mixed with 18 cubic centimeters of a 70% acid solution in order to make a 75% acid solution?
5. How many pints of pure acid must be added to 18 pints of a 72% acid solution to increase the strength of the solution to 76% acid?

a. 27 *b.* 9 *c.* 12 *d.* 3 *e.* 6

6.5 SOLVING WET MIXTURE PROBLEMS BY USING ARITHMETIC ONLY

Model Problem

A 20-ounce solution of salt and water contains 4 ounces of salt. What is the percent of salt in the new solution if 8 ounces of water are evaporated?

(A) $14\frac{2}{7}\%$ (B) $16\frac{2}{3}\%$ (C) 20% (D) 25% (E) $33\frac{1}{3}\%$

Solution:

The new solution contains 20 − 8 or 12 ounces of salt and water. Of these 12 ounces, 4 ounces are salt.

Hence, the salt is $\frac{4}{12}$, or $\frac{1}{3}$, of the new solution.

The percent of salt in the new solution is $33\frac{1}{3}\%$. *Ans.* (E)

Matching Test

In each of the following, a 20-ounce solution containing 3 ounces of salt is changed as indicated. Find the percent of the salt in the resulting solution. After completing this test, match your answers with the percents in the box.

1. 5 ounces of water are evaporated.

2. 5 ounces of water are added.

3. 5 ounces of pure salt are added.

4. 5 ounces of water are evaporated and replaced by 5 ounces of salt.

5. 5 ounces of water are evaporated and 1 ounce of salt is added.

a. 40%
b. 32%
c. 12%
d. 20%
e. 25%

6.6 STRENGTHENING OR DILUTING A SOLUTION

Model Problem

A chemist has 10 cubic centimeters of a 10% solution of acid. He wishes to add enough water to produce an 8% solution. How many cubic centimeters of water must he add?

(A) $2\frac{1}{2}$ (B) $3\frac{1}{3}$ (C) 4 (D) 30 (E) 40

Solution:

Let x = the number of cubic centimeters of water to be added.

	Percent acid ×	(cu. cm.) Solution =	(cu. cm.) Quantity of acid
Original solution	10% = .10	10	10(.10) = 1
Water added	0%	x	0
Final solution	8% = .08	$(x + 10)$	$.08(x + 10)$

Since the quantity of acid in the final solution equals the quantity of acid in the original solution,

$$.08(x+10)=1$$

Multiply each side by 100.

$$8x+80=100$$

$$8x=20, \quad x=2\frac{1}{2} \quad \textit{Ans. (A)}$$

Matching Test

After completing this test, match your answers with the numbers in the box.

A chemist has 30 cubic centimeters (c.c.) of a 12% acid solution:

1. How many c.c. of water should he add to dilute the solution to 4% acid?

2. How many c.c. of water should he add to dilute the solution to 10% acid?

3. How many c.c. of pure acid should he add to strengthen the solution to 34% acid?

4. How many c.c. of pure acid should he add to strengthen the solution to 45% acid?

5. How many c.c. of water would have to be evaporated to obtain a 20% acid solution?

a. 18	*b.* 10	*c.* 60	*d.* 6	*e.* 12

7. INTEREST PROBLEMS

Interest problem formula: $I=PR$ $\begin{cases} I = \text{the annual interest} \\ P = \text{the principal, or investment} \\ R = \text{the annual rate of interest} \end{cases}$

Other Forms	
$P=\frac{I}{R}$	$R=\frac{I}{P}$

For example, the annual interest on \$2000 at 6% per year is .06(\$2000), or \$120.

In the formula $A=P+PR$, A is the amount due at the end of a year when the principal, P, is invested at an annual rate of interest R. For example, the amount due on \$2000 at 6% per year is \$2000 + .06(\$2000), or \$2120.

7.1 USING THE SIMPLE-INTEREST FORMULA

Model Problems

(1) Mr. Brown deposited a sum of money in a bank at 6% annual interest. If his deposit grew to \$371 after one year, how many dollars had he originally deposited?

(A) 300 (B) 325 (C) 350 (D) 360 (E) 365

Solution:

The simple-interest formula for the amount due at the end of one year is $A = P + PR$, where A is the amount due, P is the original deposit or principal, and R is the annual rate of interest.

$$A = P + PR$$

Substitute \$371 for P and .06 for R.

$$371 = P + .06P$$
$$371 = 1.06P$$

Multiply by 100.

$$37\,100 = 106P$$
$$350 = P \quad \textit{Ans.}\ \text{(C)}$$

(2) Find the simple interest on \$3600 for 4 months at 4% annual interest.

(A) \$4.80 (B) \$36 (C) \$48 (D) \$60 (E) \$72

Solution:

The simple-interest formula for interest due is $I = PRT$, where I is the interest earned, P is the original deposit or principal, R is the annual rate of interest, and T is the number of years or the fraction of a year that the money is on deposit.

$$I = PRT$$

Substitute 3600 for P, .04 for R, and $\frac{4}{12}$, or $\frac{1}{3}$, for T.

$$I = 3600(.04)\left(\frac{1}{3}\right)$$
$$= 48 \quad \textit{Ans.}\ \text{(C)}$$

Matching Test

After completing this test, match your answers with the numbers in the box.

1. How many dollars had been deposited if, after one year, the amount due at 6% annual interest was \$530?
2. How many dollars had been deposited if, after one year, the amount due at 5% annual interest was \$630?
3. Find the simple interest on \$4000 for 2 months at 6% annual interest.
4. Find the simple interest on \$600 for 5 months at 6% annual interest.
5. If the amount deposited in a bank is \$600 and the annual rate of interest is 4%, how many dollars will be due the depositor at the end of 1 year and 5 months?

a. 600	*b.* 500	*c.* 634	*d.* 40	*e.* 15

7.2 FINDING THE AMOUNTS INVESTED AT DIFFERENT RATES, GIVEN THE TOTAL INVESTMENT AND THE TOTAL INCOME

Model Problem

An investment of \$2700, part at 3% and the remainder at 4%, earns a yearly income of \$93. Find the amount invested at 3%.

(A) \$600 (B) \$1200 (C) \$1500 (D) \$1800 (E) \$2100

Solution:

Let x = the number of dollars invested at 3%
and y = the number of dollars invested at 4%.

	Rate of interest	× (\$) Principal	= (\$) Annual interest
1st investment	.03	x	$.03x$
2nd investment	.04	y	$.04y$

The sum of the two investments is \$2,700: $x + y = 2700$
The yearly income is \$93: $.03x + .04y = 93$

Multiply each side of the first equation by 4 and each side of the second equation by 100.

$$\begin{aligned} 4x + 4y &= 10\,800 \\ 3x + 4y &= 9\,300 \\ \text{Subtract: } x &= 1\,500 \end{aligned}$$

Hence, the amount invested at 3% was \$1500. *Ans.* (C)

Matching Test

Find the amount invested at each of the given rates. After completing this test, match your answers with the pairs of amounts in the box.

1. A total of \$3000 is invested, part at 4% and the rest at 5%. The total annual interest is \$125.
2. A total of \$4000 is invested, part at 4% and the rest at 5%. The total annual interest is \$165.
3. A total of \$4000 is invested, part at 4% and the rest at 6%. The total annual interest is \$176.
4. A total of \$4000 is invested. On one investment, a gain of 5% is made, but on the remainder, there is a loss of 6%. The net annual gain is \$35.
5. A total of \$6000 is invested. A sum of \$1000 is invested at 3%, a second sum is invested at 4%, and the rest at 5%. The total annual interest is \$260.

a. \$2000 at 4% \$3000 at 5%	*b.* \$2500 at 4% \$ 500 at 5%	*c.* \$3500 at 4% \$ 500 at 5%
d. \$3200 at 4% \$ 800 at 6%	*e.* \$2500 at 5% \$1500 at a loss of 6%	

7.3 FINDING THE AMOUNTS INVESTED AT DIFFERENT RATES, GIVEN THE TOTAL INVESTMENT AND A RELATION BETWEEN THE INCOMES

Model Problem

A man has two investments that total \$7500. The annual interest on one investment at 4% is \$165 more than the interest on the other at 5%. How much is invested at 4%?

(A) \$1500 (B) \$2500 (C) \$4000 (D) \$5000 (E) \$6000

Solution:

Let $x =$ the number of dollars invested at 4%
and $y =$ the number of dollars invested at 5%.

	Rate of interest	× (\$) Principal	= (\$) Annual interest
1st investment	.04	x	$.04x$
2nd investment	.05	y	$.05y$

The sum of the two investments is \$7500.

Hence, $x + y = 7500$

The interest on the 4% investment is \$165 more than the interest on the 5% investment.

Hence, $.04x - .05y = 165$

Multiply each side of the first equation by 5 and each side of the second equation by 100.

$$\begin{aligned} x + y = 7500 &\rightarrow 5x + 5y = 37\,500 \\ .04x - .05y = 165 &\rightarrow 4x - 5y = 16\,500 \\ \text{Add:} &\quad 9x \quad = 54\,000 \end{aligned}$$

Divide each side by 9:

$$\frac{9x}{9} = \frac{54\,000}{9}$$

$$x = 6000$$

Hence, the amount invested at 4% is \$6000. *Ans.* (E)

Matching Test

Find the amount invested at each of the given rates. After completing this test, match your answers with the pairs of amounts in the box.

1. A total of \$7500 is invested, the first part at 4% and the second at 5%. The annual interest on the first is \$75 more than that on the second.
2. A total of \$7500 is invested, the first part at 5% and the second at 4%. The annual interest on the first is \$150 more than that on the second.
3. A total of \$5000 is invested, the first part at 4% and the second at 5%. The annual interest on the first is \$70 less than that on the second.
4. Two investments are made, the first at 4% and the second at 5%. The amount of the first investment is \$3000 more than that of the second and the annual interest on the first is \$110 more than that of the second.

5. Two investments are made, the first at 5% and the second at 4%. The amount of the first investment is \$1500 less than that of the second and the annual interest on the first is \$50 less than that of the second.

a. \$5000 at 5% \$2500 at 4%	*b.* \$1000 at 5% \$2500 at 4%	*c.* \$2000 at 4% \$3000 at 5%
d. \$4000 at 4% \$1000 at 5%	*e.* \$5000 at 4% \$2500 at 5%	

7.4 PROBLEMS WITH MORE THAN TWO RATES OF INTEREST

Model Problem

A man has \$2400 invested at 5%. How much additional money must he invest at 2% in order that his total annual income may equal 3% of the entire investment?

(A) \$1200 (B) \$2400 (C) \$3600 (D) \$4000 (E) \$4800

Solution:

Let $P =$ the number of dollars to be invested at 2%.

	Rate of interest ×	(\$) Principal =	(\$) Annual interest
1st investment	.05	2400	.05(2400)
2nd investment	.02	P	$.02P$
Total investment	.03	$P + 2400$	$.03(P + 2400)$

The total annual income equals 3% of the entire investment.

Hence, $.05(2400) + .02P = .03(P + 2400)$

Remove parentheses.

$$120 + .02P = .03P + 72$$

$$120 - 72 = .03P - .02P$$

$$48 = .01P$$

Multiply by 100.

$$4800 = P \quad \textit{Ans.}\ \text{(E)}$$

Matching Test

After completing this test, match your answers with the numbers in the box.

1. A man has \$2400 invested at 6%. How many more dollars must he invest at 2% so that his total annual income is 3% of his entire investment?

2. A man has \$2400 invested at 6%. How many more dollars must he invest at 3% so that his total annual income is 5% of his entire investment?

3. A man has invested \$2000 at 6%. How many more dollars must he invest at 3% so that his total annual income is 4% of his entire investment?

4. A man has \$2000 invested at 5%. How many more dollars must he invest at 2% so that his total annual income is 4% of his entire investment?

5. A man has \$2000 invested at 5% and also \$1200 invested at 3%. How many more dollars must he invest at 3% so that his total annual income is 4% of his entire investment?

a. 1200	*b.* 4000	*c.* 1000	*d.* 800	*e.* 7200

8. MOTION PROBLEMS

Motion problem formula: $D = RT \begin{cases} D = \text{the distance traveled} \\ R = \text{the rate or speed} \\ T = \text{the time spent traveling} \end{cases}$

Other Forms	
$R = \dfrac{D}{T}$	$T = \dfrac{D}{R}$

When the formula $D = RT$ is applied to motion problems, the units must be in agreement. For example, if the rate, or speed, is 1 mile per *minute* and the time is 5 *hours*, first convert 1 mile per minute to 60 miles per hour. Then the distance traveled is 60×5, or 300 miles.

Unless otherwise stated, a speed may be taken to mean (1) a uniform, or fixed, speed or (2) an average speed. For example, a distance of 120 miles may be covered in 3 hours by traveling at a uniform rate of 40 miles during each hour or by traveling at a changing rate such as 35 miles during the first hour, 40 miles during the second hour, and 45 miles during the third hour. In the case of the changing rate, the average speed is 40 miles per hour.

> *Note.* A rate of 60 miles per hour is equivalent to a rate of 88 feet per second. When converting from miles per hour to feet per second, or conversely, this relationship will be found useful.

8.1 USING ARITHMETIC TO FIND THE SPEED

Model Problem

A jet airplane travels 1.65 miles in 10 seconds. What is its speed in miles per hour?

(A) 594 mph (B) 660 mph (C) 990 mph (D) 1188 mph (E) 1650 mph

Solution:

1. If the plane travels 1.65 miles in 10 seconds, then, in 1 minute, it will go 6 times as far; that is, the plane will travel 1.65×6 or 9.9 miles per minute.
2. If the plane travels 9.9 miles per minute, then, in 1 hour, it will go 60 times as far; that is, the plane will travel 9.9×60 or 594 miles per hour.

Ans. (A)

Alternate Method

10 seconds is $\frac{1}{360}$ of an hour. Hence, in 1 hour, the plane will go 1.65×360 or 594 miles.

Matching Test

Find the speed of the plane in miles per hour for the given distance and time. After completing this test, match your answers with the rates in the box.

1. The jet plane travels 1.75 miles in 10 seconds.

2. The jet plane travels 1.85 miles in 10 seconds.

3. The jet plane travels 4.85 miles in 20 seconds.

4. The jet plane travels 6.65 miles in 30 seconds.

5. The jet plane travels 10.25 miles in 45 seconds.

	mph
a.	798
b.	666
c.	820
d.	630
e.	873

8.2 USING A FORMULA TO FIND THE SPEED

Model Problems

(1) If a plane travels 540 miles in an hour and a half, what is its average rate in miles per hour?

(A) 360 mph (B) 450 mph (C) 540 mph (D) 720 mph (E) 810 mph

Solution:

The average speed is found by dividing the distance by the time. If D is the distance in miles; T, the time in hours; and R, the rate in miles per hour, then

$$R = \frac{D}{T}$$

Substitute 540 for D and $1\frac{1}{2}$ for T.

$$R = \frac{540}{1\frac{1}{2}}$$

$$= 540 \div \frac{3}{2}$$

$$= \overset{180}{\cancel{540}} \cdot \frac{2}{\underset{1}{\cancel{3}}} = 360 \quad \textit{Ans.}\ \text{(A)}$$

Note. If each side is divided by T, the distance formula $D = RT$ becomes $R = \frac{D}{T}$.

(2) A meteorologist observed that a weather balloon had moved a distance of 12 miles in 15 minutes. What was the average speed of the balloon in miles per hour?

(A) 3 mph (B) 6 mph (C) 12 mph (D) 24 mph (E) 48 mph

Solution:

To find the rate, divide the distance traveled by the time:

$$R = \frac{D}{T}$$

Since 15 minutes $= \frac{1}{4}$ hour, T (in hours) $= \frac{1}{4}$.

$$R = \frac{12}{\frac{1}{4}}$$

$$= 12(4) = 48 \quad \textit{Ans.}\ (E)$$

After completing each test, match your answers with the numbers in the boxes.

Matching Test A

Find the average speed in miles per hour required to fly a distance of

1. 2 miles in 10 seconds.
2. 2 miles in 30 seconds.
3. 20 miles in 4 minutes.
4. 240 miles in 40 minutes.
5. 480 miles in $1\frac{1}{2}$ hours.

	mph
a.	720
b.	300
c.	360
d.	320
e.	240

Matching Test B

1. How many miles per hour does a weather balloon average if it moves 16 miles in 48 minutes?
2. How many miles per hour did Mr. Adams average on a trip of 154 miles if it took him 4 hours, including a half-hour stop for lunch?
3. On a trip, a man traveled for 4 hours at 50 mph. Returning over the same route, he took 3 hours and 20 minutes. How many miles per hour did he average on the return journey?
4. A train left Washington at 8:00 A.M. and arrived at a city 375 miles away at 3:30 P.M. How many miles per hour did the train average between the two cities?
5. A student left his home at 8:00 A.M. and arrived at school, 27 miles away, at exactly 8:36 A.M. How many miles per hour did he average from home to school?

a. 44	*b.* 50	*c.* 45	*d.* 60	*e.* 20

8.3 CONVERTING SPEEDS

Model Problem

A speed of 60 mph is equivalent to . . . feet per second.

(A) 44 (B) 60 (C) 88 (D) 90 (E) 152

Solution:

Since 1 mile = 5280 feet, a rate of 60 mph = 60(5280) feet per hour.

Since 1 hour = 3600 seconds, 60 mph = $\frac{60(5280)}{3600}$ or 88 feet per second. *Ans.* (C)

Matching Test

After completing this test, match your answers with the numbers in the box.

1. A speed of 15 mph is equivalent to . . . feet per second.

2. A speed of 66 feet per second is equivalent to . . . mph.

3. A runner at a track meet ran 132 feet in 5 seconds. His speed during that interval was . . . mph.

4. An airplane flying at 300 mph travels a distance of . . . feet in $\frac{1}{10}$ of a second.

5. A boy scout on a hike traveled 88 feet in 10 seconds. His speed during that interval was . . . mph.

a. 18	*b.* 44	*c.* 45	*d.* 22	*e.* 6

8.4 USING A FORMULA TO FIND THE TIME

Model Problem

How many *minutes* will it take a plane to fly 650 miles if its average speed is 520 mph?

(A) $1\frac{1}{4}$ (B) 75 (C) 80 (D) 90 (E) 125

Solution:

The time (in hours) is found by dividing the distance (in miles) by the average speed (in miles per hour).

If D is the distance in miles; R, the average speed in miles per hour; and T, the time in hours, then

$$T = \frac{D}{R}$$

Substitute 650 for D and 520 for R.

$$T = \frac{650}{520} = \frac{\overset{1}{\cancel{130}}(5)}{\underset{1}{\cancel{130}}(4)} = \frac{5}{4}$$

Hence, the time required is $1\frac{1}{4}$ hours.

Since the number of minutes must be found, change $1\frac{1}{4}$ hours into minutes by multiplying $1\frac{1}{4}$ by 60. The time in minutes is 75. *Ans.* (B)

> *Note.* By dividing each side by R, the distance formula $D = RT$ becomes $T = \frac{D}{R}$.

Matching Test

After completing this test, match your answers with the numbers in the box.

Find how many minutes it takes an airplane to fly a distance of

1. 390 miles at 520 mph.	*a.* 62
2. 385 miles at 550 mph.	*b.* 42
3. 560 miles at 420 mph.	*c.* 48
4. 544 miles at 680 mph.	*d.* 45
5. 465 miles at 450 mph.	*e.* 80

8.5 PROBLEMS WITH MIXED TIME UNITS

Model Problem

A train triggers a bell when it is $\frac{1}{2}$ mile from a crossing. If the train travels at 45 mph, how many seconds after the bell is triggered will it take to reach the crossing?

(A) 30 (B) 40 (C) 45 (D) 48 (E) 60

Solution:

To find the time, divide the distance traveled by the rate, or speed:

$$T = \frac{D}{R}$$

Since the distance is $\frac{1}{2}$ mile and the rate is 45 mph, substitute $\frac{1}{2}$ for D and 45 for R:

$$T \text{ (in hours)} = \frac{\frac{1}{2}}{45} = \frac{1}{90}$$

Since 1 hour equals 3600 seconds, $\frac{1}{90}$ hour equals 40 seconds. *Ans.* (B)

Matching Test

After completing this test, match your answers with the numbers in the box.

1. A train triggers a bell when it is $\frac{3}{4}$ mile from a crossing. If the train travels at 30 mph, how many seconds after the bell is triggered will it take to reach the crossing?
2. How many seconds longer will it take to go 2 miles at 30 mph rather than at 40 mph?
3. How many minutes longer will it take to go 10 miles at 20 mph rather than at 50 mph?
4. To be on time, a train must travel a distance of 2 miles at 45 mph. If the train travels that distance at 40 mph, how many seconds late will it be on arrival?
5. A train travels a distance of 150 miles from city A to city B at the rate of 40 mph. If the train leaves city A at 9:40 A.M. and is scheduled to arrive in city B at 1 P.M., how many minutes late will it be?

a. 60	*b*. 18	*c*. 20	*d*. 25	*e*. 90

8.6 USING A PROPORTION WHEN THE SPEED IS CONSTANT

Model Problem

A train travels 150 miles in 3 hours and 20 minutes. At the same rate, how many miles will it travel in 5 hours and 20 minutes?

(A) 200 (B) 240 (C) 250 (D) 300 (E) 320

Solution:

Since the speed does not change, the ratio of the distances is equal to the ratio of the times. Hence, a proportion can be formed as follows:

Let $x =$ the distance traveled in 5 hours and 20 minutes, or in 320 minutes.

Since 3 hours and 20 minutes = 200 minutes, $\frac{x}{150} = \frac{320}{200}$

$$x = \left(\frac{320}{200}\right)150$$

$$= 240 \quad \textit{Ans.} \text{ (B)}$$

Matching Test

After completing this test, match your answers with the numbers in the box.

1. A train covers 150 miles in 3 hours. At the same rate, how many miles will it travel in 4 hours and 18 minutes?
2. A train covers 156 miles in 3 hours and 15 minutes. How many minutes will it require to go 264 miles at the same speed?
3. A plane travels 675 miles in an hour and a half. At the same speed, how many miles will it travel in 34 minutes?

4. A plane requires 1 hour and 12 minutes to go 600 miles. At the same speed, how many minutes will it require to travel 700 miles?

5. A projectile takes 54 minutes to go 4500 miles. At that rate, how many minutes will be required to go 6000 miles?

a. 330	*b*. 72	*c*. 84	*d*. 255	*e*. 215

8.7 TRAVELING A ROUND TRIP, DISTANCE KNOWN

Model Problem

A steamboat made a round trip between two ports 24 miles apart. If the average speed for the entire trip was 4.8 mph and the steamboat took $4\frac{1}{2}$ hours to travel downstream, how many hours did it take to go upstream?

(A) $\frac{1}{2}$ (B) $4\frac{1}{2}$ (C) $5\frac{1}{2}$ (D) 6 (E) 9

Solution:

To find the time required for the entire trip, divide the total distance by the average speed.

$$T = \frac{D}{R}$$

The round trip was 2(24) or 48 miles. The average rate was 4.8 mph. Hence, $D = 48$ and $R = 4.8$.

$$T = \frac{48}{4.8} = 10$$

Hence, the round trip took 10 hours.

Since the time required to go downstream was $4\frac{1}{2}$ hours, the time taken to return upstream was $10 - 4\frac{1}{2}$ or $5\frac{1}{2}$ hours. *Ans.* (C)

Matching Test

After completing this test, match your answers with the numbers in the box.

1. The distance between two ports is 34 miles. A boat averages 8.5 mph during the round trip between the two ports, taking 5 hours for the trip going. How many hours were required for the return trip?

2. A canoeist goes $1\frac{1}{2}$ miles downstream in 14 minutes and returns the same distance in 16 minutes. How many miles per hour did he average for the entire trip?

3. A boat goes downstream a distance of $12\frac{1}{2}$ miles in 1 hour and returns the same distance in $1\frac{1}{2}$ hours. How many miles per hour does it average for the entire trip?

4. On a round trip between two towns, a bus travels 8 miles each way and averages 20 mph for the entire trip. If the trip one way required 20 minutes, how many minutes were spent going the other way?

5. A taxi travels 12 miles in 40 minutes going one way and makes the return trip over the same distance in 50 minutes. How many miles per hour did the taxi average for the round trip?

a. 28	*b*. 10	*c*. 16	*d*. 3	*e*. 6

8.8 TRAVELING A ROUND TRIP, DISTANCE NOT GIVEN

Model Problem

Tom's father drove from home to his summer camp in 4 hours. Returning over the same route, he took one hour longer because heavy traffic forced him to drive an average of 8 mph slower. What was his average rate going to camp?

(A) 30 mph (B) 32 mph (C) 36 mph (D) 40 mph (E) 45 mph

Solution:

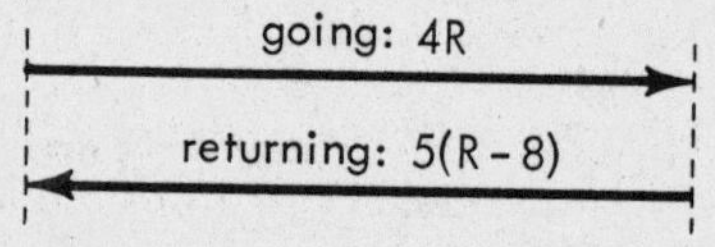

Let R = the rate in miles per hour of the trip from home to camp.

	(miles per hour) Average rate	× (hours) Time	= (miles) Distance
Going	R	4	$4R$
Returning	$R - 8$	5	$5(R - 8)$

Since the return route was the same, the distances going and returning are equal.

$$\text{Hence, } 4R = 5(R - 8)$$

$$4R = 5R - 40$$

$$40 = R \quad \textit{Ans.}\ \text{(D)}$$

Matching Test

After completing this test, match your answers with the numbers in the box.

A motorist makes a round trip from home to camp and back again, returning along the same route:

1. The time going was 4 hours. On the return trip, the rate was 4 mph slower and the time 1 hour longer. Find the average rate going to camp.

2. The time going was 3 hours. On the return trip, the rate was 3 mph slower and the time 1 hour longer. Find the average rate going to camp.

3. The time going was 3 hours. On the return trip, the rate was 10 mph slower and the time 2 hours longer. Find the average rate going to camp.

4. The rate going was 40 mph. On the return trip, the rate was 30 mph and the time 2 hours longer. How many hours did it take to go to camp?

5. The rate going was 20 mph. On the return trip, the rate was 15 mph and the time 3 hours longer. How many hours did it take to go to camp?

a. 25	*b*. 6	*c*. 12	*d*. 20	*e*. 9

8.9 USING A FORMULA IN A ROUND-TRIP PROBLEM

Model Problem

A motorist drove 100 miles from city A to city B at an average speed of 20 mph. His average return rate over the same distance was 30 mph. What was his average speed for the entire trip?

(A) 12 mph (B) 24 mph (C) 25 mph (D) 27 mph (E) 36 mph

Solution:

Let r_1 = the rate going, in miles per hour,
and r_2 = the return rate, in miles per hour.

The average rate for the entire trip may be found by applying the following formula:

$$\text{rate for the entire trip} = \frac{2r_1r_2}{r_1 + r_2}$$

Substitute 20 for r_1 and 30 for r_2.

$$= \frac{2(20)(30)}{20 + 30}$$

$$= \frac{2(20)(30)}{50}$$

$$= 24 \quad \textit{Ans.}\ \text{(B)}$$

(*Note.* In this problem, the distance of the trip is immaterial.)

Matching Test

After completing this test, match your answers with the numbers in the box.

1. How many miles per hour was the rate for the entire trip, if the rate going was 42 mph and the rate returning over the same route was 30 mph?

2. How many miles per hour was the speed for the entire trip of 167 miles, if the rate going was 60 mph and the return rate over the same distance was 40 mph?

3. How many miles per minute did a plane average for the entire trip, if the rate going was 20 miles per minute and the rate for the return trip over the same distance was 15 miles per minute?

4. How many feet per second did a runner average for an entire race if his speed one way was 25 feet per second and his return speed over the same distance was 15 feet per second?

5. A motorist made a round trip, returning along the same route, and found that his average speed for the entire trip was $13\frac{1}{3}$ mph. If his rate going was 20 mph, how many miles per hour did he average on the return trip?

a. $18\frac{3}{4}$	*b.* 10	*c.* $17\frac{1}{7}$	*d.* 35	*e.* 48

8.10 TWO-STAGE TRIPS INVOLVING AVERAGE SPEEDS

Model Problem

For the first 80 miles of a trip of 210 miles, Jack averaged 40 mph. If he wishes to average 30 mph for the entire trip, how many miles per hour should his speed be for the rest of the trip?

(A) 15 (B) 20 (C) 26 (D) 30 (E) 24

Solution:

If the average speed for the entire trip of 210 miles is to be 30 mph, then the entire time of travel will be $\frac{210}{30}$ or 7 hours.

For the first 80 miles, the time of travel was $\frac{80}{40}$ or 2 hours.

Hence, for the remaining 130 miles, Jack has $7 - 2$ or 5 hours to spend.

To go 130 miles in 5 hours, the rate is $\frac{130}{5}$ or 26 mph. *Ans.* (C)

Matching Test

After completing this test, match your answers with the numbers in the box.

1. For the first 45 miles, a motorist travels at 45 mph. For the next 120 miles, he travels at 30 mph. How many miles per hour did he average for the entire trip?

2. For the first 45 miles, a bus takes 3 hours. For the next 100 miles, the bus averages 50 mph. How many miles per hour was the average speed for the entire trip?

3. On a trip of 204 miles, a motorist covered the first 60 miles at 30 mph. At how many miles per hour must he travel for the rest of the trip if the average speed for the entire trip is to be 34 mph?

4. A man travels 45 miles at an average rate of 20 mph. At how many miles per hour must he travel for the next 90 miles so that the entire trip will have taken $4\frac{1}{2}$ hours?

5. A car travels 120 miles at 20 mph, the next 120 miles at 24 mph, and the last 120 miles at 30 mph. How many miles per hour did the car average for the entire trip?

a. 36	*b.* 29	*c.* 40	*d.* 24	*e.* 33

8.11 OVERTAKING A SLOWER VEHICLE

Model Problem

A train leaves a city traveling at an average rate of 60 mph. Three hours later a plane leaves the same city traveling in the same direction at a rate of 180 mph. How many hours will it take the plane to overtake the train?

(A) $\frac{1}{2}$ (B) 1 (C) $1\frac{1}{3}$ (D) $1\frac{1}{2}$ (E) 2

Solution:

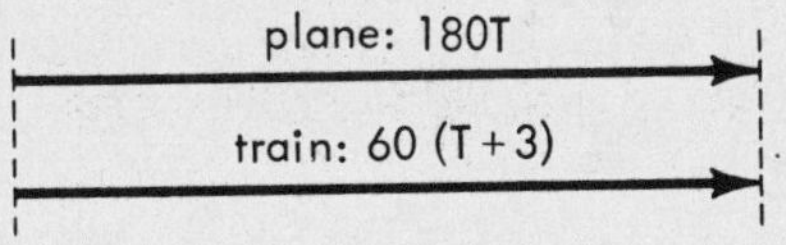

Let T = the number of hours it takes the plane to overtake the train.

	(miles per hour) Average rate	× (hours) Time	= (miles) Distance
Plane	180	T	$180T$
Train	60	$T+3$	$60(T+3)$

The distances traveled by the train and the overtaking plane are equal.

Hence, $180T = 60(T+3)$

$$180T = 60T + 180$$

Subtract $60T$ from each side: $-60T = -60T$

$$120T = 180$$

Divide each side by 120: $\frac{120T}{120} = \frac{180}{120}$

$$T = 1\frac{1}{2}$$

Hence, the plane takes $1\frac{1}{2}$ hours to overtake the train. *Ans.* (D)

Matching Test

After completing this test, match your answers with the numbers in the box.

In each of the following, a slower vehicle is overtaken by a faster one. Both vehicles leave from the same city.

1. A train traveling at 60 mph starts out 2 hours earlier than a plane traveling at 180 mph. How many hours does it take the plane to overtake the train?

2. A train traveling at 60 mph starts out 8 hours earlier than a plane traveling at 180 mph. How many hours does it take the plane to overtake the train?

3. A train traveling at 50 mph started out 10 hours earlier than a plane that overtook it. If the plane overtook the train in 5 hours, at how many miles per hour did the plane travel?

4. A train that had started out 8 hours earlier than a plane was overtaken by the plane in 2 hours. If the plane traveled at 200 mph, at how many miles per hour did the train travel?

5. A plane traveling at 200 mph was overtaken by another plane traveling at 300 mph. It took the faster plane 5 hours to overtake the slower one. How many hours earlier than the faster plane did the slower plane leave?

a. 40	*b.* 150	*c.* $2\frac{1}{2}$	*d.* 4	*e.* 1

8.12 TRAVELING IN OPPOSITE DIRECTIONS TO MEET

Model Problem

A train leaves Albany for New York City and travels at an average speed of 40 mph. At the same time, a train leaves New York City for Albany and travels at an average speed of 60 mph. The distance between New York City and Albany is 150 miles. How far from New York City will the trains be when they pass each other?

(A) 30 mi. (B) 60 mi. (C) 72 mi. (D) 75 mi. (E) 90 mi.

Solution:

Since the two trains start out at the same time, they will have traveled the same number of hours when they pass each other.

Hence, let t = the number of hours traveled by each train.

	(miles per hour) Rate	× (hours) Time	= (miles) Distance
Train from New York City	60	t	$60t$
Train from Albany	40	t	$40t$

Since the sum of the two distances traveled equals 150 miles,

$$60t + 40t = 150$$

$$100t = 150$$

Divide each side by 100.

$$t = 1\frac{1}{2}$$

Hence, the train from New York City will have traveled $1\frac{1}{2}$ hours at 60 mph. Therefore, the distance from New York City equals 90 miles. *Ans.* (E)

Matching Test

After completing this test, match your answers with the numbers in the box.

Towns A and B are 300 miles apart. A train at A starts out for B at the same time that a train at B starts out for A.

1. The speeds of the trains are 60 mph and 40 mph. In how many hours will they pass each other?
2. The trains pass each other in 5 hours. If the speed of the faster train is 35 mph, how many miles per hour does the slower train travel?
3. The trains pass each other in 6 hours. If the rate of the slower train is 22 mph, how many miles per hour does the faster train travel?
4. The trains pass each other in 4 hours. If the faster train is twice as fast as the slower one, how many miles per hour does the faster train travel?
5. The trains pass each other in 4 hours. If the rate of the faster train is 5 mph faster, how many miles per hour does the slower train travel?

a. 28	*b*. 50	*c*. 25	*d*. 3	*e*. 35

8.13 TRAVELING FROM THE SAME POINT IN OPPOSITE DIRECTIONS

Model Problems

(1) Two automobiles leave a certain place at the same time and travel in opposite directions along a straight road. The speed of one is three-fourths that of the other; and in 5 hours they are 315 miles apart. What is the speed of the slower automobile?

(A) $20\frac{1}{4}$ mph (B) 24 mph (C) 27 mph (D) 32 mph (E) 36 mph

Solution:

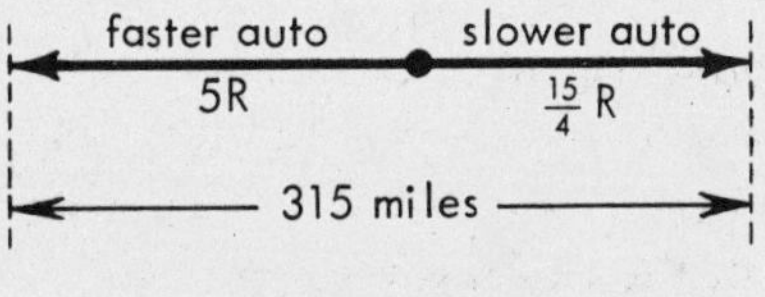

Let R = the rate in miles per hour of the faster automobile.

	(miles per hour) Average rate	× (hours) Time	= (miles) Distance
Faster auto	R	5	$5R$
Slower auto	$\frac{3}{4}R$	5	$\frac{15}{4}R$

The sum of the two distances is 315 miles.

$$\text{Hence, } 5R + \frac{15}{4}R = 315$$

Multiply each side by 4.

$$20R + 15R = 1260$$
$$35R = 1260$$
$$R = 36, \quad \frac{3}{4}R = 27$$

Hence, the speed of the slower automobile is 27 mph. *Ans.* (C)

(2) Two trains start from the same point at the same time and travel in opposite directions, one going 12 mph faster than the other. After 6 hours they are 648 miles apart. Find the speed of the faster train.

(A) 36 mph (B) 42 mph (C) 48 mph (D) 54 mph (E) 60 mph

Solution:

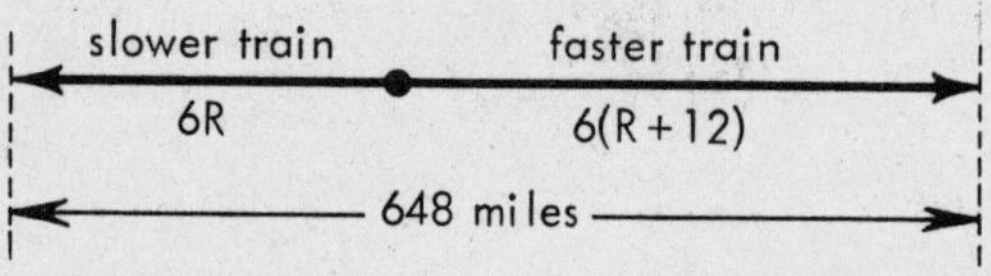

Let R = the rate of the slower train in miles per hour.

	(miles per hour) Average rate ×	(hours) Time =	(miles) Distance
Slower train	R	6	$6R$
Faster train	$R + 12$	6	$6(R + 12)$

The sum of the two distances traveled is 648 miles.

$$\text{Hence,} \quad 6R + 6(R + 12) = 648$$

Clear of parentheses.

$$6R + 6R + 72 = 648$$

Collect like terms and subtract 72 from each side.

$$12R = 576$$

Divide each side by 12.

$$R = 48$$

Since $R = 48$, then $R + 12 = 60$.

Hence, the speed of the faster train is 60 mph. *Ans.* (E)

Check in the original problem.

One train goes 12 mph faster than the other.

$$60 \stackrel{?}{=} 48 + 12$$
$$60 = 60$$

After 6 hours, the trains are 648 miles apart. In 6 hours, the faster train traveled 6(60), or 360 miles; while the slower train traveled 6(48), or 288 miles.

$$360 + 288 \stackrel{?}{=} 648$$
$$648 = 648$$

Matching Test

After completing this test, match your answers with the pairs of numbers in the box.

Two cars leave a certain place at the same time and travel in opposite directions along a straight road. At the end of 3 hours, they are 180 miles apart. Find how many miles per hour each car is traveling if

1. one car travels twice as fast as the other.	*a.* 36, 24
2. one car travels three times as fast as the other.	*b.* 34, 26
3. the speed of one car is two-thirds that of the other.	*c.* 45, 15
4. one car is going 8 mph faster than the other.	*d.* 40, 20
5. one car is going 6 mph slower than the other.	*e.* 33, 27

8.14 USING RATIOS OF TIME, RATE, AND DISTANCE

Model Problem

Comparing two trips that he had taken, a traveler found that on the second trip he had gone twice as fast and taken three times as long to reach his destination. How many times as much distance was traveled during the second trip as during the first?

(A) 5 (B) 6 (C) 8 (D) 9 (E) 10

Solution:

The ratio of the speeds, or rates, equals 2. (He went twice as fast.)

The ratio of the times spent traveling equals 3. (He took three times as long.)

Since the ratio of the distances traveled equals the product of the rate-ratio and the time-ratio, the ratio of the distances $= 2(3) = 6$.

Hence, the traveler went 6 times as far. *Ans.* (B)

Note. The ratio of the distances traveled may be shown to equal the product of the rate-ratio and the time-ratio as follows:

1. If a man first travels a distance of d_1 miles in t_1 hours at a speed of r_1 miles per hour, then $d_1 = r_1 t_1$.
2. If the man then travels a distance of d_2 miles in t_2 hours at a speed of r_2 miles per hour, then $d_2 = r_2 t_2$.
3. Since $d_2 = r_2 t_2$ and $d_1 = r_1 t_1$,

$$\frac{d_2}{d_1} = \frac{r_2 t_2}{r_1 t_1} = \left(\frac{r_2}{r_1}\right)\left(\frac{t_2}{t_1}\right)$$

Hence, the distance-ratio equals the product of the rate-ratio and the time-ratio.

Matching Test

After completing this test, match your answers with the answers in the box.

1. A motorist drives at half the speed and takes 6 times as long to complete his trip as he had done on an earlier journey. What is the ratio of the new distance to the distance he had previously covered?

2. If the speed of a car is doubled and the distance is quadrupled, what is the ratio of the new traveling time to the old?

3. A troop of soldiers marched 70 miles in 3 days. If they must now march 140 miles in 5 days, how many times as fast must they march?

4. At 30 mph, Mr. Cole can get to his office in 45 minutes. If he travels three-halves as fast, how many minutes will he save getting to his office?

5. A trip of 250 miles is covered in 5 hours. If the same distance is to be covered in five-sixths as much time, how many miles per hour faster must one travel?

a. $\frac{6}{5}$	*b.* $\frac{3}{1}$	*c.* 10	*d.* 15	*e.* $\frac{2}{1}$

9. COMBINATION PROBLEMS

In this type of problem, similar quantities, such as two prices or speeds or hourly rates of pay, are combined in two ways. In the model problem below, note how a problem of this type is readily solved by means of ***simultaneous equations***.

When a letter is used, remember that the letter may represent only a number. For example, the letter h may represent the number of hats purchased, the number of dollars in the price of the hat, or the number of ounces in the weight of the hat, but h may never stand for *hats*.

9.1 USING SIMULTANEOUS EQUATIONS TO SOLVE COMBINATION PROBLEMS

Model Problem

A housewife bought 4 quarts of milk and two pounds of butter for \$2.42. Several days later with no change in prices, she paid \$1.49 for 3 quarts of milk and one pound of butter. What was the cost of a quart of milk?

(A) 21¢ (B) 28¢ (C) 35¢ (D) 42¢ (E) 65¢

Solution:

Let m = the price in cents of a quart of milk
and b = the price in cents of a pound of butter.

Since 4 quarts of milk and 2 pounds of butter cost \$2.42, $4m + 2b = 242$

Since 3 quarts of milk and 1 pound of butter cost \$1.49, $3m + b = 149$

Multiply each side of the second equation by 2 and subtract as follows:

$$6m + 2b = 298$$

Subtract: $4m + 2b = 242$

$$2m = 56$$

$$m = 28$$

Hence, a quart of milk cost 28¢. *Ans.* (B)

Matching Test

After completing this test, match your answers with the pairs of numbers in the box.

1. A man paid \$39 for 5 shirts and 4 hats. At the same prices, 4 shirts and 5 hats would cost \$42. How many dollars did he pay for each shirt and for each hat?
2. A man paid \$36 for a number of shirts at \$5 apiece and a number of hats at \$8 apiece. Had he bought 1 more hat and twice as many shirts, the cost would have been \$64. How many shirts and how many hats did he buy?
3. A distance of 65 miles was covered by traveling for 5 hours at one speed and for 10 hours at another. Had the first speed been used for 10 hours and the second speed for 5 hours, the distance covered would have been 55 miles. How many miles per hour were the two speeds?
4. A motorist traveled a distance of 170 miles by driving at 30 mph for a period of time and then at 40 mph for another period of time. Had the speeds been 10 mph faster, the distance covered would have been 220 miles. For how many hours did he travel at each speed?
5. John worked for 3 hours and Henry worked for 5 hours. The sum of their combined earnings was \$37. Had each worked 2 hours longer, they would have earned \$55. How many dollars per hour did each earn?

a. 4, 5	*b*. 3, 5	*c*. 4, 2	*d*. 3, 6	*e*. 3, 2

10. DIGIT PROBLEMS

The two-digit number 37 equals 3 tens + 7. In general, if t and u represent, respectively, the tens digit and the units digit of a two-digit number, *the number* $= 10t + u$. If the digits are reversed, that is, interchanged, 37 becomes 73. Since reversing the digits makes t the units digit and u the tens digit, *the number with the digits reversed* $= 10u + t$.

10.1 EXPRESSING RELATIONS BETWEEN NUMBERS AND THEIR DIGITS

Model Problems

Using the letters t and u to represent, respectively, the tens and units digit of a two-digit number, write the equations that may be used to solve the following problems: (*Note*. Only the equations are required, not the solutions.)

(1) The sum of the digits of a two-digit number is 7. If 9 is added to the number, the result is a number with the digits of the original number reversed. Find the original number.

(A) $tu = 7$, $10t + u + 9 = 10u + t$

(B) $t + u = 7$, $10u + t + 9 = 10t + u$

(C) $t + u = 7$, $10t + u + 10u + t = 9$

(D) $t + u = 7$, $10t + u + 9 = 10u + t$

(E) $t + 7 = u$, $10t + u + 9 = 10u + t$

Solution:

Letting t and u represent the tens and units digits,

$10t + u =$ the original number,
$10u + t =$ the new number formed by reversing the digits.

Since the sum of the digits is 7,

$$t + u = 7$$

Since the result of adding 9 to the number is the number with the digits reversed,

$$10t + u + 9 = 10u + t$$

Ans. (D)

(2) The sum of the tens digit and twice the units digit of a two-digit number is 14. If the number is divided by the units digit, the result is 16. Find the number.

(A) $t + 2u = 14$ $\quad \dfrac{10t + u}{u} = 16$

(B) $2t + u = 14$ $\quad \dfrac{10t + u}{u} = 16$

(C) $t + 2u = 14$ $\quad \dfrac{t + u}{u} = 16$

(D) $t + u = 14$ $\quad \dfrac{10t + u}{t} = 16$

(E) $2(t + u) = 14$ $\quad \dfrac{10t + u}{u} = 16$

Solution:

The original number is $10t + u$.

The sum of the tens digit and twice the units digit equals 14:

$$t + 2u = 14$$

The number divided by the units digit is 16:

$$\frac{10t + u}{u} = 16$$

Ans. (A)

For each problem in tests *A* and *B*, let t and u represent, respectively, the tens digit and the units digit of the number. Then write the two equations that may be used to find the number. (The solution of the equations is not required.) After completing each test, match your answers with the pairs of equations in the boxes.

Matching Test A

1. The tens digit of a two-digit number is 5 less than twice the units digit. If the digits are reversed, the sum of the resulting number and the original number is 77. Find the original number.

2. The units digit of a two-digit number is 5 less than 3 times the tens digit. If the digits are reversed, the resulting number exceeds the original number by 9. Find the original number.

3. The digits of a two-digit number are such that the ratio of the tens digit to the units digit is 3:4. If the digits are reversed, the resulting number is 25 less than twice the original number. Find the original number.

4. The digits of a two-digit number are such that the sum of the tens digit and twice the units digit is 11. If the digits are reversed, the resulting number is 1 more than 6 times the sum of the digits. Find the original number.

5. A two-digit number is 1 less than 5 times the sum of the digits. The number obtained by reversing the digits is 6 less than 7 times the sum of the digits. Find the original number.

a. $t = 2u - 5$	*b.* $u = 3t - 5$	*c.* $10t + u = 5(t + u) - 1$
$(10t + u) + (10u + t) = 77$	$(10u + t) - (10t + u) = 9$	$10u + t = 7(t + u) - 6$
d. $\frac{t}{u} = \frac{3}{4}$	*e.* $t + 2u = 11$	
$10u + t = 2(10t + u) - 25$	$10u + t = 6(t + u) + 1$	

Note. The five problems of test A have the same solution. In each case, the original two-digit number is 34.

Matching Test B

1. The digits of a two-digit number are such that the ratio of the tens digit to the units digit is 3:2. The number that results when the digits are reversed is 27 less than the original number. Find the original number.

2. The difference between the tens digit and the units digit of a two-digit number is 3. If the number is divided by the sum of the digits, the quotient is 6 and the remainder is 6. Find the number.

3. The tens digit of a two-digit number is 6 more than one-half the units digit. If the digits are reversed, the new number is 21 more than one-half the original number. Find the original number.

4. If the tens digit of a two-digit number is increased by 3, the units digit is one-half the result. The sum of the number and the new number formed by reversing the digits is 165. Find the original number.

5. The digits of a two-digit number are such that the sum of one-third the tens digit and one-half the units digit is 6. The number with the digits reversed exceeds one-third the original number by 37. Find the original number.

a. $u = \frac{1}{2}(t + 3)$	*b.* $t = \frac{u}{2} + 6$	
$(10t + u) + (10u + t) = 165$	$10u + t = \frac{1}{2}(10t + u) + 21$	
c. $\frac{t}{3} + \frac{u}{2} = 6$	*d.* $\frac{t}{u} = \frac{3}{2}$	*e.* $t - u = 3$
$(10u + t) - \frac{10t + u}{3} = 37$	$10u + t = (10t + u) - 27$	$\frac{10t + u}{t + u} = 6 + \frac{6}{t + u}$

Note. The five problems of test B have the same solution. In each case, the original two-digit number is 96.

11. WORK PROBLEMS

In 3 days, a machine that requires 5 days to do a complete job can finish $\frac{3}{5}$ of the job. In general, in n days, a machine that requires p days to do a complete job will finish $\frac{n}{p}$ of the job.

In work problems, unless otherwise stated, it is assumed that workers and machines operate at a fixed rate of work.

When machines or persons work together, add the parts of the job completed by each to obtain the amount of work done. For example, if Ann takes 4 hours to do a job and works for 1 hour, and Bess takes 8 hours to do the same job and works for 4 hours, together they will finish $\frac{1}{4}+\frac{4}{8}$, or $\frac{3}{4}$, of the job.

11.1 FINDING THE PART OF THE JOB COMPLETED

Model Problem

John takes x days to complete a job. Express in terms of x the part of the job he could complete in 2 days.

(A) $\frac{x+2}{x}$ (B) $\frac{2}{x+2}$ (C) $\frac{x}{x+2}$ (D) $\frac{2}{x}$ (E) $\frac{x}{2}$

Solution:

If John takes x days to complete a job, then, at that rate, he is able to do $\frac{1}{x}$ of the job in one day. For example, if he takes 5 days to do a job, then he is able to do $\frac{1}{5}$ of the job in one day.

Hence, in 2 days, John can do twice as much work as he can do in one day; that is, he can complete $2\left(\frac{1}{x}\right)$, or $\frac{2}{x}$, of the job. *Ans.* (D)

Matching Test

After completing this test, match your answers with the numbers in the box.

> *Hint.* In problem **3**, add $\frac{5}{12}$ and $\frac{5}{18}$; in **4**, double $\frac{3}{12}$; in **5**, add $\frac{1}{2}$, $\frac{1}{3}$, and $\frac{1}{6}$, and then multiply by $\frac{1}{3}$.

1. What part of a job can John do in 3 days if it takes him 10 days to complete the job?

2. If Tom works twice as fast as John and John needs 10 days to complete a certain job, what part of the job can Tom complete in 3 days?

3. If each works alone, John takes 12 hours and Tom takes 18 hours to complete a certain job. What part of the job can be done if John and Tom work together for 5 hours?

4. A pool receives its water from two inlet pipes. Each pipe, if used alone, can fill the entire pool in 12 hours. If the two pipes are used at the same time, what part of the pool can be filled in 3 hours?

5. A pool receives its water from three inlet pipes of different capacities. If used alone, the first pipe can fill the pool in 2 hours, the second pipe can fill it in 3 hours, and the third, in 6 hours. What part of the pool can be filled in 20 minutes if the three pipes are used together?

a. $\frac{3}{5}$	*b.* $\frac{3}{10}$	*c.* $\frac{1}{3}$	*d.* $\frac{25}{36}$	*e.* $\frac{1}{2}$

11.2 WORKING TOGETHER AT DIFFERENT RATES

Model Problem

Ann requires 3 hours and Bess requires 6 hours to typewrite a certain report. If the girls work together, how many hours will it take them to typewrite the report?

(A) 2 (B) 3 (C) 4 (D) $4\frac{1}{2}$ (E) 9

Solution:

If Ann requires 3 hours and Bess, 6 hours, then, in 1 hour, Ann will do $\frac{1}{3}$ and Bess will do $\frac{1}{6}$ of the report.

In x hours, Ann will do $\frac{x}{3}$ and Bess will do $\frac{x}{6}$ of the report.

The part done by Ann + the part done by Bess = 1.

$$\frac{x}{3} + \frac{x}{6} = 1$$

$$\frac{2x}{6} + \frac{x}{6} = 1$$

$$\frac{3x}{6} = 1$$

Hence, $\frac{x}{2} = 1$ and $x = 2$. *Ans.* (A)

Matching Test

After completing this test, match your answers with the numbers in the box.

1. One typist can typewrite a report in 6 hours and another typist can do it in 12 hours. If the two typists worked together, how many hours would they require to do the report?

2. One man can paint a sign in 2 hours and his friend can paint it in 3 hours. If the two men work together, how many hours will they need to paint the sign?

3. One farmer can plow a field in 21 days and another farmer can plow it in 28 days. If the two farmers work together, how many days will they require to plow the field?

4. One inlet pipe can fill a pool in 20 minutes and a second pipe, if used alone, can fill it in 60 minutes. If the two pipes are turned on at the same time, how many minutes will it take to fill the pool?

5. Each of three outlet pipes, when used alone, can drain a full pool in 15 minutes. If the pool is full and all three pipes are opened simultaneously, how many minutes will it take to drain the pool?

a. 15	*b.* 12	*c.* 5	*d.* $1\frac{1}{5}$	*e.* 4

12. FRACTION PROBLEMS

If 3 is added to the numerator and 2 is subtracted from the denominator of the fraction $\frac{1}{10}$, the new fraction, $\frac{4}{8}$, has a value of $\frac{1}{2}$. In general, if n represents the numerator and d represents the denominator of a fraction, and 3 is added to the numerator while 2 is subtracted from the denominator, the new fraction can be expressed as $\frac{n+3}{d-2}$.

12.1 EXPRESSING RELATIONS BETWEEN FRACTIONS AND THEIR TERMS

Model Problem

In the following problem, use x to represent the denominator of the original fraction. Then write an equation that may be used to solve the problem. (Only the equation is required, not the solution.)

The numerator of a fraction is 5 less than the denominator. If the numerator is doubled and the denominator is increased by 7, the value of the resulting fraction is $\frac{2}{3}$. Find the original fraction.

(A) $\frac{2x}{x+7}=\frac{2}{3}$ (B) $\frac{2x-10}{x-7}=\frac{2}{3}$ (C) $\frac{2x-10}{x+7}=\frac{2}{3}$ (D) $\frac{2x-5}{x+7}=\frac{2}{3}$ (E) $\frac{2x}{x+12}=\frac{2}{3}$

Solution:

Let $x =$ the denominator of the original fraction.

Then $x - 5 =$ the numerator of the original fraction.

Since the numerator is doubled, the new numerator $= 2(x-5) = 2x - 10$.

Since the denominator is increased by 7, the new denominator $= x + 7$.

The value of the resulting fraction is $\frac{2}{3}$. Hence, $\frac{2x-10}{x+7}=\frac{2}{3}$. *Ans.* (C)

Matching Test

For each problem in this test, let x represent the denominator of the original fraction. Then write an equation that may be used to find the fraction. (The solution of the equation is not required.) After completing this test, match your answers with the equations in the box.

1. The numerator is 5 less than the denominator. If the numerator is increased by 2 and the denominator remains unchanged, the resulting fraction equals $\frac{3}{4}$. Find the original fraction.

2. The numerator is 5 less than the denominator. If the denominator is doubled and the numerator remains unchanged, the resulting fraction equals $\frac{2}{5}$. Find the original fraction.

3. The numerator is 5 more than the denominator. If the numerator is increased by 5 and the denominator is tripled, the resulting fraction equals $\frac{2}{3}$. Find the original fraction.

4. The numerator is 5 more than the denominator. If the numerator is doubled and the denominator is multiplied by 5, the ratio of the new numerator to the new denominator is 4:5. Find the original fraction.

5. The numerator is 5 less than twice the denominator. If the numerator is doubled and the denominator is increased by 17, the value of the resulting fraction is $\frac{2}{7}$. Find the original fraction.

a. $\frac{2(x+5)}{5x} = \frac{4}{5}$ *b.* $\frac{x-5}{2x} = \frac{2}{5}$ *c.* $\frac{2(2x-5)}{x+17} = \frac{2}{7}$ *d.* $\frac{(x-5)+2}{x} = \frac{3}{4}$ *e.* $\frac{(x+5)+5}{3x} = \frac{2}{3}$

Note. You may wish to solve each problem completely. The required fractions are: **1.** $\frac{7}{12}$ **2.** $\frac{20}{25}$ **3.** $\frac{15}{10}$ **4.** $\frac{10}{5}$ **5.** $\frac{3}{4}$

12.2 SOLVING FRACTION PROBLEMS

Model Problem

A certain fraction is equivalent to $\frac{2}{5}$. If the numerator of this fraction is decreased by 2 and its denominator is increased by 1, the resulting fraction is equivalent to $\frac{1}{3}$. Find the denominator of the original fraction.

(A) 14 (B) 21 (C) 35 (D) 40 (E) 42

Solution:

Since the fraction is equivalent to $\frac{2}{5}$, the terms of the fraction may be represented by $2x$ and $5x$.

Let $2x$ = the numerator of the fraction
and $5x$ = the denominator of the fraction.

Since the numerator is decreased by 2, the new numerator $= 2x - 2$.

Since the denominator is increased by 1, the new denominator $= 5x + 1$.

The new fraction equals $\frac{1}{3}$. Hence, $\frac{2x-2}{5x+1} = \frac{1}{3}$

Cross-multiply to clear of fractions.

$$3(2x-2) = 5x + 1$$

Clear of parentheses.

$$\begin{aligned} 6x - 6 &= 5x + 1 \\ -5x + 6 &= -5x + 6 \\ \hline x &= 7 \end{aligned}$$

Subtract $5x$ and add 6 to each side:

Hence, the denominator of the original fraction $= 5x = 35$. *Ans.* (C)

Check in the original problem.

The numerator of the original fraction $= 2x = 14$.

A certain fraction is equivalent to $\frac{2}{5}$.

$$\frac{14}{35} \stackrel{?}{=} \frac{2}{5}$$

$$\frac{2}{5} = \frac{2}{5}$$

If the numerator is decreased by 2 and the denominator is increased by 1, the resulting fraction is equivalent to $\frac{1}{3}$.

$$\frac{14-2}{35+1} \stackrel{?}{=} \frac{1}{3}$$

$$\frac{12}{36} \stackrel{?}{=} \frac{1}{3}$$

$$\frac{1}{3} = \frac{1}{3}$$

Matching Test

Find the original fraction. After completing this test, match your answers with the fractions in the box.

1. A fraction is equivalent to $\frac{2}{5}$. If the numerator is increased by 5 and the denominator is unchanged, the result is equal to $\frac{3}{5}$.

2. A fraction is equivalent to $\frac{3}{5}$. If both the numerator and the denominator are increased by 5, the result equals $\frac{7}{10}$.

3. A fraction is equivalent to $\frac{4}{5}$. If the numerator is decreased by 4 and the denominator is increased by 4, the result equals $\frac{1}{2}$.

4. A fraction is equivalent to $\frac{3}{7}$. If the numerator is doubled and the denominator is increased by 20, the result equals $\frac{2}{3}$.

5. A fraction is equivalent to $\frac{3}{7}$. If the numerator is increased by 6 and the denominator is increased by 2 and then doubled, the result equals $\frac{3}{8}$.

a. $\frac{16}{20}$	*b.* $\frac{9}{15}$	*c.* $\frac{10}{25}$	*d.* $\frac{6}{14}$	*e.* $\frac{30}{70}$

13. PRACTICE PROBLEMS

1. The sum of two numbers is $6x + 2$. If one of the numbers is 10, write in simplest form the expression for the sum of twice the other number and 8.
(A) $12x - 8$ (B) $12x + 8$ (C) $12x + 10$ (D) $6x + 8$ (E) $12x + 4$ **[1.1]**

2. John had a collection of 20 coins. The number of half-dollars was 2 more than the number of quarters, while the number of dimes was 6 more than the number of quarters. Find the value of the collection.
(A) \$4.50 (B) \$4.75 (C) \$5.00 (D) \$5.50 (E) \$6.00 **[1.2]**

3. Find two numbers such that their sum is 14 and three times the smaller decreased by 12 equals the larger.
(A) $7\frac{1}{2}, 6\frac{1}{2}$ (B) $8\frac{1}{2}, 5\frac{1}{2}$ (C) 8, 6 (D) $9\frac{1}{2}, 4\frac{1}{2}$ (E) 9, 5 **[1.3]**

4. If n represents the middle number of a set of nine consecutive numbers, write an expression that can be used to represent the sum of the set.
(A) $9n + 18$ (B) $9n - 18$ (C) $9n$ (D) $9n + 36$ (E) $9n - 36$ **[2.1]**

5. If n is the first of two consecutive even integers, write an equation that can be used to find the integers if the sum of the squares of twice the first and the second is 100.
(A) $(2n)^2 + (n + 2)^2 = 100$ (B) $2n^2 + (n + 2)^2 = 100$ (C) $(3n + 2)^2 = 100$
(D) $n^2 + (n + 2)^2 = 100$ (E) $(2n)^2 + (n + 1)^2 = 100$ **[2.2]**

6. Find three consecutive odd integers such that the sum of the first two is four times the last.
(A) $-11, -9, -7$ (B) 7, 9, 11 (C) $-7, -5, -3$ (D) 3, 5, 7 (E) $-7, -6, -5$ **[2.3]**

7. Represent the age of a person 5 years ago if he was 20 years old y years ago.
(A) $y + 15$ (B) $y + 20$ (C) $y + 25$ (D) $15 - y$ (E) $25 - y$ **[3.1]**

8. Find the average of $\frac{1}{2}$, $\frac{1}{4}$, and $-\frac{1}{8}$.

(A) $\frac{1}{24}$ (B) $\frac{5}{24}$ (C) $\frac{5}{8}$ (D) $\frac{1}{8}$ (E) $\frac{5}{12}$ **[4.1]**

9. Find the average of x, $2x$, $15y$, and $9x + 9y$.
(A) $3x + 6y$ (B) $3x + 9y$ (C) $4x + 6y$ (D) $3x + 6y$ (E) $4x + 8y$ **[4.2]**

10. Find the largest of three negative consecutive integers if their average is -10.
(A) -10 (B) -9 (C) -11 (D) -8 (E) -12 **[4.3]**

11. If the average of a set of 5 numbers is $2x$, and four of the numbers are x, $2x$, $3x$, and 12, find the fifth number.
(A) $4x + 6$ (B) $6x - 12$ (C) $6x + 12$ (D) $4x - 12$ (E) $4x + 12$ **[4.4]**

12. Henry averaged 40 mph for 2 hours and 15 minutes, and 50 mph for 1 hour and 30 minutes. Find his average speed for the entire trip.
(A) 44 (B) 40 (C) 42 (D) 45 (E) 48 **[4.5]**

13. The length of a rectangle is two-thirds of the sum of three times the width and 20. If w represents the width of the rectangle, which of the following expressions represents its length?
(A) $\frac{2}{3}(3w) + 20$ (B) $2w + 20$ (C) $3w + \frac{40}{3}$ (D) $\frac{w + 40}{3}$ (E) $\frac{2}{3}(3w + 20)$ **[5.1]**

14. One of the adjacent sides of a parallelogram is represented by $5x - 7$, and the perimeter by $20x - 18$. Represent the other adjacent side.
(A) $5x + 2$ (B) $10x - 4$ (C) $10x + 4$ (D) $5x - 2$ (E) $15x - 11$ **[5.2]**

15. The base of an isosceles triangle is represented by $3x + 5$. If the perimeter is represented by $13x - 5$, write an expression representing each of the equal legs.
(A) $5x - 5$ (B) $5x + 5$ (C) $7x - 5$ (D) $7x - 10$ (E) $10x - 10$ **[5.3]**

16. If $\frac{1}{2}(3s - 6)$ represents the side of a square, find an expression for the perimeter.
(A) $3s - 12$ (B) $6s - 12$ (C) $6s - 6$ (D) $12s - 24$ (E) $12s - 12$ **[5.4]**

17. If $\frac{2}{3}s + \frac{1}{2}$ represents the side of a regular hexagon, then an expression for the perimeter is
(A) $8s + 6$ (B) $2s + 1\frac{1}{2}$ (C) $4s + 6$ (D) $4s + 3$ (E) $2s + 3$ **[5.4]**

18. If $12x - 48$ represents the perimeter of a regular octagon, find an expression for the side.
(A) $\frac{3x}{2} - 12$ (B) $3x - 12$ (C) $\frac{3x}{2} - 6$ (D) $2x - 8$ (E) $\frac{3x - 6}{2}$ **[5.5]**

19. Find the sides of a parallelogram having a perimeter of 64 if one of the sides is 4 more than one-third an adjacent side.
(A) 20, 12 (B) 22, 10 (C) 27, 13 (D) 24, 12 (E) 21, 11 **[5.6]**

20. Find the sides of a triangle if the perimeter is 52, two sides have a ratio of 3:2, and the remaining side is 10 more than their difference.
(A) 24, 16, 12 (B) 21, 14, 17 (C) 24, 16, 18 (D) 18, 12, 16 (E) 22, 18, 12 **[5.7]**

21. The side of a square exceeds the side of an equilateral triangle by 10. If each side of the equilateral triangle is increased by 14, the perimeters of the new triangle and the square will then be equal. Find a side of the square and a side of the equilateral triangle.
(A) $12\frac{1}{2}$, $2\frac{1}{2}$ (B) 13, 3 (C) 11, 1 (D) 12, 2 (E) $11\frac{1}{2}$, $1\frac{1}{2}$ **[5.8]**

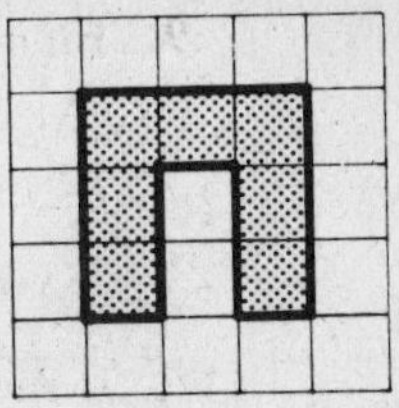

22. The shaded figure at the right is the result of cutting a rectangle out of a square of side 6 units. Find the perimeter of the figure.
(A) 24 (B) 26 (C) 28 (D) 32 (E) 36 **[5.9]**

23. A side of a regular hexagon and a side of a square are represented, respectively, by $x - 2$ and $\frac{x}{2} + 6$. If the perimeter of the hexagon equals the perimeter of the square, find the value of x.
(A) 9 (B) 10 (C) 8 (D) 7.5 (E) 8.5 **[5.9]**

24. A dealer made a tea mixture by blending 14 pounds of tea at 40¢ a pound, 4 pounds of tea at 60¢ a pound, and 2 pounds of tea at 80¢ a pound. The average price per pound of this mixture was then increased by 25%. Find the selling price per pound of the mixture.
(A) 45¢ (B) 50¢ (C) 60¢ (D) 55¢ (E) 75¢ **[6.1]**

25. Mr. Cobb bought a total of 12 items consisting of some hats at $5 apiece, some ties at $2 apiece, and a few pairs of shoes at $10 a pair. He bought twice as many hats as ties, and as many pairs of shoes as the number of hats and ties combined. How much did he pay for all the items?
(A) $54 (B) $64 (C) $96 (D) $78 (E) $84 **[6.2]**

26. A purse contained 25 coins in dimes and quarters. If the total value of the coins was $5.20, how many quarters were there?
(A) 7 (B) 18 (C) 16 (D) 20 (E) none of these **[6.3]**

27. How many quarts of water must be added to 18 quarts of a 32% acid solution to reduce the strength of the solution to 12% acid?
(A) 16 (B) 20 (C) 24 (D) 30 (E) 36 **[6.4]**

28. Find the percent of salt in a mixture of two solutions if one solution weighing 10 ounces has 7 ounces of salt and the other solution weighing 20 ounces has 3 ounces of salt.
(A) $33\frac{1}{3}$% (B) 25% (C) 30% (D) 36% (E) 40% **[6.5]**

29. A chemist has 30 cubic centimeters of a 12% acid solution. How many cubic centimeters of water must he evaporate and replace with the same quantity of pure acid to obtain a 62% acid solution?
(A) 10 (B) 12 (C) 15 (D) 16 (E) 18 **[6.6]**

30. Mr. Gold borrowed $150 from a private lender for 6 months at an annual rate of interest of 60%. How many dollars in interest did he owe at the end of the 6-month period?
(A) $4.50 (B) $9 (C) $18 (D) $27 (E) $45 **[7.1]**

31. A sum of $5600 was invested in two parts. One investment yielded a gain of $3\frac{1}{2}$%, but the other investment sustained a loss of 2%. If the net annual gain was $108, what was the amount of the investment that yielded the gain?
(A) $3500 (B) $4000 (C) $3600 (D) $3800 (E) $4050 **[7.2]**

32. Two investments are made, the first at 3% and the second at 5%. The amount of the first investment is $500 more than that of the second and the annual interest on the first is $45 less than that of the second. How much was invested at 3%?
(A) $2700 (B) $3000 (C) $3200 (D) $3500 (E) $3600 **[7.3]**

33. Mr. Fanning has \$3000 invested at 3% and \$1250 invested at 6%. How many more dollars must he invest at 6% so that his total annual income is 4% of his entire investment?
(A) \$250 (B) \$200 (C) \$350 (D) \$500 (E) \$2500 **[7.4]**

34. If a jet plane travels 7.23 miles in 36 seconds, find its speed in miles per hour.
(A) 675 (B) 720 (C) 723 (D) 725 (E) 867.6 **[8.1]**

35. Find the average speed in miles per hour required to fly 3.4 miles in 36 seconds.
(A) 340 (B) 360 (C) 410 (D) 425 (E) 510 **[8.2]**

36. A train that ordinarily travels a distance of 60 miles in $1\frac{1}{2}$ hours must now take 15 minutes less to cover the same distance. How many miles per hour faster must it now average to make the new trip?
(A) 4 (B) 5 (C) 6 (D) 12 (E) 8 **[8.2]**

37. A projectile is moving at 2250 mph. How many feet does the projectile travel in $\frac{1}{100}$ of a second?
(A) 33 (B) 30 (C) 35 (D) 38 (E) 44 **[8.3]**

38. How many minutes does it take to fly 520 miles at 400 mph?
(A) 60 (B) 72 (C) 75 (D) 78 (E) 84 **[8.4]**

39. A bus travels a distance of 1300 miles at an average speed of 26 mph. How many hours of traveling are saved if the trip is made instead by jet plane at a speed of 650 mph?
(A) 48 (B) 72 (C) 24 (D) 7.2 (E) 4.8 **[8.5]**

40. A missile takes 45 minutes to go 5400 miles. How many miles farther will it travel in 50 minutes at the same speed?
(A) 500 (B) 720 (C) 600 (D) 900 (E) 6000 **[8.6]**

41. A river boat makes a round trip between two ports that are 18 miles apart. If its average speed for the entire trip is 4.8 mph and the boat takes $3\frac{1}{2}$ hours to go downstream, how many hours does it take to go upstream?
(A) 4.5 (B) 6 (C) 7.5 (D) 8 (E) 4 **[8.7]**

42. John drove to camp at a rate of 60 mph. On the return trip, his rate was 10 mph less and he took 1 hour longer to travel the same route. How many hours did it take to go to camp?
(A) $4\frac{1}{2}$ (B) 5 (C) $5\frac{1}{2}$ (D) 6 (E) 8 **[8.8]**

43. A motorist made a round trip at an average rate of 25 mph. If his rate going was 20 mph, how many miles per hour did he average on the return trip?
(A) 27 (B) 30 (C) 32 (D) $33\frac{1}{3}$ (E) 36 **[8.9]**

44. For the first 100 miles, a motorist traveled at a speed of 25 mph. After traveling a further distance of 240 miles, he found that his average rate for the entire trip was 34 mph. What was the average rate in miles per hour at which he traveled the distance of 240 miles?
(A) 40 (B) 43 (C) 45 (D) 48 (E) 50 **[8.10]**

45. A boy scout leaves camp and walks along a road at the rate of 4 mph. After a time, he is overtaken by a boy traveling on a bicycle at 12 mph, who had also started from camp. If the bicyclist took 3 hours to reach the boy scout, how many hours earlier than the cyclist had the slower boy started out?
(A) 4 (B) 5 (C) 6 (D) 8 (E) 9 **[8.11]**

46. Two trains travel toward each other from points 300 miles apart. After they pass each other, the trains continue traveling until they are 20 miles apart. If the rates of the trains are 55 mph and 25 mph, how many hours after starting out did the trains complete their trips?

(A) $2\frac{1}{2}$ (B) $3\frac{1}{2}$ (C) $4\frac{1}{2}$ (D) 5 (E) 4 **[8.12]**

47. Two automobiles leave from the same point at the same time and travel in opposite directions. At the end of 4 hours, they are 180 miles apart. Find the rate of the faster car in miles per hour if the faster car traveled twice as fast as the slower one.

(A) 15 (B) 30 (C) 20 (D) 24 (E) 36 **[8.13]**

48. To get to work, a man ordinarily takes 30 minutes to go 20 miles. If he leaves 10 minutes late, how many miles per hour faster must he travel in order to get to work at his usual time?

(A) 18 (B) 24 (C) 48 (D) 20 (E) 60 **[8.14]**

49. A man paid $72 for some shirts and hats priced at $6 per shirt and $10 per hat. If each shirt had cost $3 less and each hat had cost $2 more, his purchase would have cost $57. How many shirts did he buy?

(A) 7 (B) 6 (C) 4 (D) 8 (E) 9 **[9.1]**

50. If t and u represent, respectively, the tens and units digit of a two-digit number, write the equations that may be used to solve the following problem:

The digits of a two-digit number are such that the ratio of the tens digit to the units digit is 3:4. When the digits are reversed, the resulting number is 18 more than the original number. Find the original number.

(A) $\frac{t}{u}=\frac{3}{4}$

$10u+t+18=10t+u$

(B) $\frac{t}{u}=\frac{3}{4}$

$10u+t=18$

(C) $\frac{t}{u}=\frac{3}{4}$

$10u+t=10t+u+18$

(D) $\frac{t}{u}=\frac{4}{3}$

$10u+t=10t+u+18$

(E) $t-u=\frac{3}{4}$

$(10u+t)+(10t+u)=18$ **[10.1]**

51. A two-digit number is three times the sum of its digits. The number obtained by reversing the digits is 8 times the sum of the digits. Find the original number.

(A) 45 (B) 36 (C) 27 (D) 54 (E) 72 **[10.1]**

52. An inlet pipe can fill a pool in 4 hours and an outlet pipe can drain the full pool in 6 hours. If the two pipes are opened at the same time, what part of the pool will be filled in $1\frac{1}{2}$ hours?

(A) $\frac{1}{16}$ (B) $\frac{3}{10}$ (C) $\frac{1}{6}$ (D) $\frac{1}{12}$ (E) $\frac{1}{8}$ **[11.1]**

53. A pool receives its water from two inlet pipes and is drained through a single outlet pipe. Each inlet pipe, when used alone, can fill the pool in 20 minutes, and the outlet pipe can drain the pool in 12 minutes. If the three pipes are opened at the same time, how many minutes will it take to fill the pool?

(A) 30 (B) 60 (C) 40 (D) 80 (E) 120 **[11.2]**

54. Let x represent the numerator of the fraction sought in the following problem, and write an equation that may be used to find the fraction:
The denominator of a fraction is twice the numerator. If the numerator is increased by 10 and the denominator is increased by 5, the resulting fraction is $\frac{4}{5}$. Find the original fraction.

(A) $\frac{x+10}{2x+5}=\frac{4}{5}$ (B) $\frac{2x+10}{x+5}=\frac{4}{5}$ (C) $\frac{x+5}{2x+10}=\frac{4}{5}$ (D) $\frac{2x+5}{x+10}=\frac{4}{5}$

(E) $\frac{x+10}{2(2x+5)}=\frac{4}{5}$ **[12.1]**

55. A certain fraction is equivalent to $\frac{2}{7}$. If the numerator is doubled and the denominator is decreased by 1, the value of the new fraction is $\frac{3}{5}$. Find the original fraction.

(A) $\frac{6}{21}$ (B) $\frac{4}{14}$ (C) $\frac{8}{28}$ (D) $\frac{3}{11}$ (E) $\frac{9}{31}$ **[12.2]**

56. The numerator of a fraction is one-half of the denominator. If the numerator is increased by 2 and the denominator is decreased by 2, the value of the resulting fraction is $\frac{3}{4}$. Find the original fraction.

(A) $\frac{6}{12}$ (B) $\frac{8}{16}$ (C) $\frac{16}{26}$ (D) $\frac{7}{14}$ (E) $\frac{28}{42}$ **[12.2]**

V. REVIEWING RATIOS, PROPORTIONS, AND VARIATION

1. REVIEWING RATIOS

The ***ratio*** of one quantity to another expressed in the same unit of measure is the first quantity divided by the second. For example, the ratio of 15 ft. to 5 ft. = 15 ft. ÷ 5 ft. = $\frac{15}{5} = 3$.

A ratio may be expressed as follows:

1. by using a colon, 2 : 3
2. by using the word "to," 2 to 3
3. as a common fraction, $\frac{2}{3}$
4. as a decimal fraction, $.66\frac{2}{3}$
5. as a percent, $66\frac{2}{3}\%$

To simplify a ratio, eliminate any fraction contained in the ratio and reduce to lowest terms. For example, the ratio of 2.4 to 1.8 = $\frac{2.4}{1.8} = \frac{24}{18} = \frac{4}{3}$.

The ratio of three or more quantities may be expressed in the form of a ***continued ratio***. For example, "the ratio of $3 to $5 to $7" is the continued ratio 3:5:7. The ratio 3:5:7 is a combination of the ratios 3:5, 3:7, and 5:7.

Understanding Scales

A scale of 7 feet per inch means that each inch on a map, blueprint, or other scaled diagram represents 7 feet. For example, if the length of a room is represented by a 3-inch line on a diagram having a scale of 7 feet per inch, then the actual length of the room is 7(3), or 21 feet.

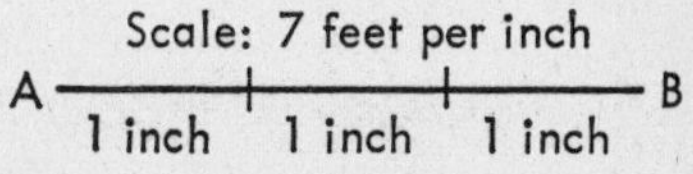

Line AB represents a length of 21 feet.

A scale may be expressed as a ratio. This is done by expressing the scale using only one unit of measure. For example, the scale of 7 feet per inch may be expressed as 84 inches per inch, or as a ratio of 84:1.

1.1 FINDING A RATIO

Model Problems

(1) The ratio of the number of degrees in a right angle to the number of degrees in a straight angle is

(A) 1:4 (B) 1:2 (C) 4:1 (D) 2:1 (E) 1:3

Solution:

To obtain the ratio of the number of degrees in a right angle to the number of degrees in a straight angle, divide the first quantity by the second quantity.

$$\frac{\text{Number of degrees in a right angle}}{\text{Number of degrees in a straight angle}} = \frac{90}{180}$$

$$= \frac{1}{2} \text{ or } 1:2 \qquad \textit{Ans.}\ \text{(B)}$$

(2) A team played seven-fifths as many games as it won. Find the ratio of the number of games won to the number of games lost.

(A) $\frac{5}{2}$ (B) $\frac{2}{5}$ (C) $\frac{5}{3}$ (D) $\frac{3}{5}$ (E) $\frac{5}{7}$

Solution:

Since the number of games played is $\frac{7}{5}$ of the number of games won,

let $7x$ = the number of games played
and $5x$ = the number of games won.

Then $7x - 5x$, or $2x$ = the number of games lost.

The ratio of games won to games lost $= \frac{5x}{2x} = \frac{5}{2}$. *Ans.* (A)

(3) Express the ratio of 30 seconds to 2 minutes to 1 hour in lowest terms.

(A) 1:2:60 (B) $\frac{1}{2}$:2:120 (C) 1:4:120 (D) 1:4:60 (E) 1:2:120

Solution:

To find the ratio of the three quantities, change them to the same unit of measure. If this is done, then the ratio 30 sec. to 2 min. to 1 hr. = 30 sec.: 120 sec.: 3600 sec.

Remove the unit of measure and divide each number by the highest common factor, 30.

30:120:3600 = 1:4:120 *Ans.* (C)

After completing each test, match your answers with the ratios or fractions in the boxes.

Matching Test A

Find the ratios of the given quantities.

1. \$2.50 to \$1.50 to 50¢.

2. $2\frac{1}{2}$ days to 2 days to 12 hours.

3. $2\frac{1}{4}$ years to $1\frac{1}{2}$ years to 9 months.

4. 120 mi. per hr. to 80 mi. per hr. to 1 mi. per min.

5. 15 sec. to $\frac{1}{2}$ min. to 1 min. 15 sec.

a. 6:4:3
b. 3:2:1
c. 5:3:1
d. 5:4:1
e. 1:2:5

Matching Test B

1. What is the ratio of 2^3 to 3^2?

2. 1 furlong is 220 yards. What is the ratio of $1\frac{1}{2}$ furlongs to a mile?

3. A man can do a job in 4 hours. If he works on the job from 9:15 A.M. to 10:45 A.M., what is the ratio of the work completed to the work that still remains to be done?

4. A man who earns \$6000 a year finds that his food costs average \$125 a month. What is the ratio of the cost of his food to his earnings?

5. There are twice as many boys as girls in a school. If one-half of the boys and all of the girls voted, what is the ratio of non-voters to voters in the school?

a. 3:5	*b.* 1:2	*c.* 3:16	*d.* 1:4	*e.* 8:9

Matching Test C

1. A line segment is divided into two parts in the ratio of 3:4. What part of the entire segment is the shorter part?

2. A line segment is divided into three parts in the ratio of 2:3:4. What part of the entire segment is the largest part?

3. A team's ratio of games won to games lost is 7:5. What part of the games played is the games won?

4. Sand and gravel are mixed in the ratio of 2:3. What part of the mixture is the gravel?

5. In a salad dressing, oil and vinegar are mixed in the ratio of 2:5. What part of the dressing is the vinegar?

a. $\frac{4}{9}$	*b.* $\frac{5}{7}$	*c.* $\frac{3}{7}$	*d.* $\frac{3}{5}$	*e.* $\frac{7}{12}$

Matching Test D

Determine the ratio of *the number of games won to the number of games lost.*

1. A team won three-quarters as many games as it lost.
2. A team played five-fourths as many games as it lost.
3. A team played seven-fourths as many games as it won.
4. A team won three-fifths as many games as it played.
5. A team lost five-eighths as many games as it played.

a. $\frac{1}{4}$
b. $\frac{3}{5}$
c. $\frac{3}{4}$
d. $\frac{4}{3}$
e. $\frac{3}{2}$

1.2 FINDING A RATIO: GEOMETRIC PROBLEMS

Model Problem

In the figure, each side of the equilateral triangle is divided into three equal segments. What is the ratio of the shaded parts to the unshaded parts?

(A) 4:9 (B) 5:4 (C) 9:4 (D) 1:2 (E) 4:5

Solution:

As shown, the entire triangle is divided into three equal parallelograms and three equal triangles.

Each triangle is one-half the size of each parallelogram.

The shaded parts are two equal parallelograms.

The unshaded section consists of one parallelogram and 3 triangles.

Hence, the unshaded part equals $2\frac{1}{2}$ parallelograms.

Thus, the ratio of the shaded part to the unshaded part $= 2{:}2\frac{1}{2}$, or 4:5. *Ans.* (E)

After completing each test, match your answers with the ratios in the boxes.

Matching Test A

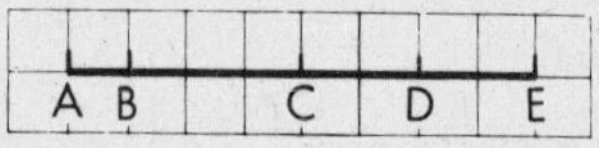

Fig. 1

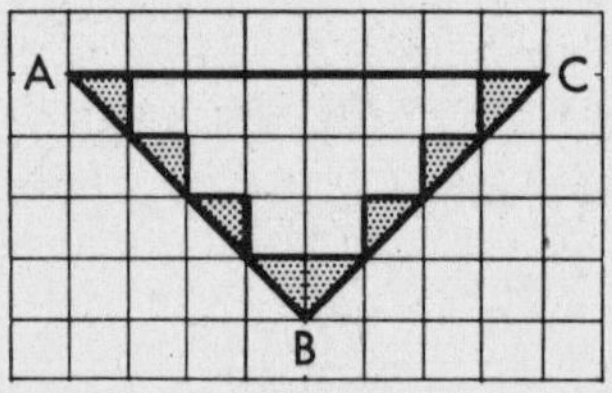

Fig. 2

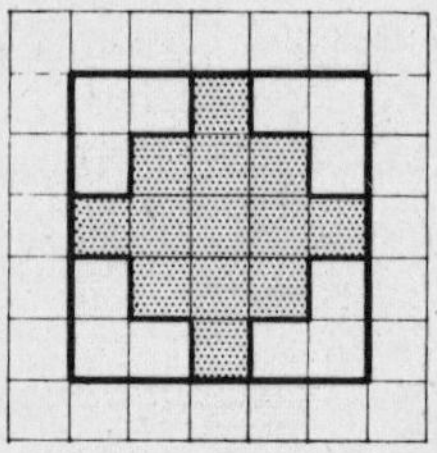

Fig. 3

1. (Fig. 1) What is the ratio of *BC* to *AE*?

2. (Fig. 1) What is the ratio of *CD* to *AC*?

3. (Fig. 1) What is the ratio of *AB*:*BC*:*DE*?

4. (Fig. 2) What is the ratio of the shaded part to triangle *ABC*?

5. (Fig. 3) What is the ratio of the shaded part to the unshaded part in the square?

a. 1:4
b. 1:3:2
c. 13:12
d. 3:8
e. 1:2

Matching Test B

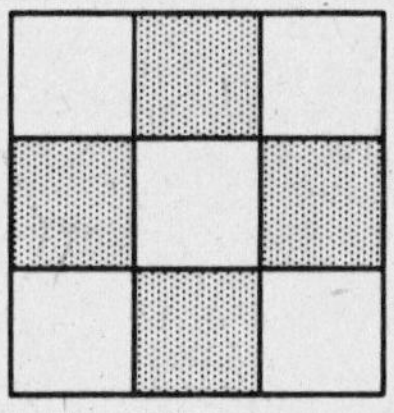

Fig. 1

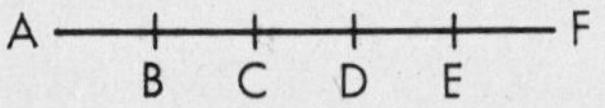

Fig. 2

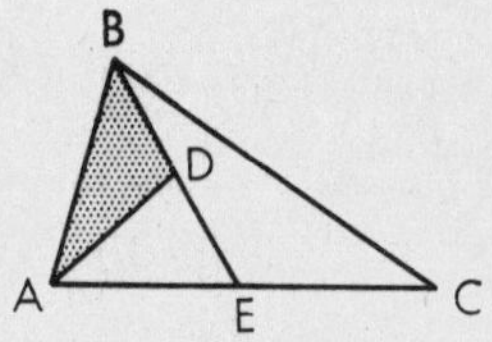

Fig. 3

1. (Fig. 1) What is the ratio of the shaded to the unshaded parts if the sides of the square have been trisected?

2. (Fig. 2) What is the ratio of BD to CF if line AF has been divided into 5 equal parts?

3. (Fig. 3) BE and AD are medians. What is the ratio of the shaded part to the unshaded part? (*Hint.* A median divides a triangle into two triangles of equal areas.)

4. If the sides of a triangle are in the ratio 3:5:7, what is the ratio of the smallest side to the sum of the remaining sides?

5. If the angles of a quadrilateral are in the ratio 9:7:3:2, what is the ratio of the largest angle to the sum of the remaining angles?

a. 1:3	*b.* 3:4	*c.* 1:4	*d.* 2:3	*e.* 4:5

1.3 PROBLEMS INVOLVING RATIOS

Model Problems

(1) A line segment 30 inches long is divided into two parts which are in the ratio of 2 to 3. Find the number of inches in the shorter part.

(A) 10 (B) 8 (C) 6 (D) 12 (E) 18

Solution:

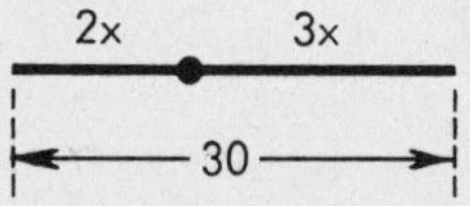

Since the line segment is divided in the ratio 2:3,

let $2x$ = the number of inches in the shorter segment

and $3x$ = the number of inches in the longer segment.

The entire line segment is 30 inches.

$$\text{Hence, } 2x + 3x = 30$$

Combine like terms.

$$5x = 30$$

Divide each side by 5.

$$x = 6$$

$$2x = 12$$

Therefore, the shorter segment is 12 inches. *Ans.* (D)

(2) The perimeter of a triangle is 31 inches and one of its sides is 10 inches. If the other two sides are in the ratio 3:4, how many inches are there in the longer of these two sides?

(A) 3 (B) 9 (C) 10 (D) 12 (E) 21

Solution:

Since the perimeter is 31 inches and one side is 10 inches, the sum of the other two sides is 21 inches.

Since these two sides are in the ratio 3:4, let $3x$ and $4x$ represent the numbers of inches in these sides.

$$3x + 4x = 21$$
$$7x = 21, \quad x = 3$$

Since $4x = 12$, the number of inches in the longer side is 12. *Ans.* (D)

(3) Three men invest money in an enterprise with the agreement that all profits are to be divided in the ratio 3:7:8. If the profits at the end of the first year are $2700, what is the largest share?

(A) $450 (B) $1050 (C) $1200 (D) $1500 (E) $1800

Solution:

Since the profits are to be divided in the ratio 3:7:8,

let $3x$ = the share in dollars of the first partner,
$7x$ = the share in dollars of the second partner,
$8x$ = the share in dollars of the third partner.

The total of the profits equals $2700.

Hence, $3x + 7x + 8x = 2700$
$$18x = 2700, \quad x = 150$$

If $x = 150$, then $8x = 1200$.

Hence, the largest share of the profits is $1200. *Ans.* (C)

(4) A cake recipe requires $1\frac{1}{2}$ teaspoons of yeast to every 9 ounces of flour. How many teaspoons of yeast are needed if 15 ounces of flour are used?

(A) $\frac{9}{10}$ (B) $1\frac{2}{3}$ (C) $1\frac{1}{2}$ (D) 2 (E) $2\frac{1}{2}$

Solution:

Let x = the number of teaspoons of yeast needed.

Form a proportion (see set 2 of this chapter) in which the ratio of the numbers of teaspoons of yeast equals the ratio of the numbers of ounces of flour.

x teaspoons are needed for 15 ounces of flour.

$1\frac{1}{2}$ teaspoons are needed for 9 ounces of flour.

Hence, $\dfrac{x}{1\frac{1}{2}} = \dfrac{15}{9}$

$$x = \left(1\frac{1}{2}\right)\left(\frac{15}{9}\right) = \frac{3}{2}\left(\frac{5}{3}\right) = \frac{5}{2}, \text{ or } 2\frac{1}{2} \quad \textit{Ans. (E)}$$

(5) An alloy of silver and copper contains 39 ounces of silver. If there are 15 ounces of silver for every 25 ounces of copper, how many ounces does the alloy weigh?

(A) $19\frac{1}{2}$ (B) 26 (C) $58\frac{1}{2}$ (D) 65 (E) 104

Solution:

Method 1: *Using a Ratio*

Since there are 15 ounces of silver to 25 ounces of copper, the ratio of the weights of silver and copper is $\frac{15}{25}$, or $\frac{3}{5}$.

Let $3x$ and $5x$ represent, respectively, the weights of the silver and copper in ounces. Since there are 39 ounces of silver, $3x = 39$ and $x = 13$.

The weight of the alloy $= 3x + 5x = 8x$.

Since $x = 13$, then $8x = 104$. Hence, the alloy weighs 104 ounces.

Method 2: *Using a Proportion* (See Reviewing Proportions, page 197.)

Let $x =$ the number of ounces of alloy containing 39 ounces of silver.

x ounces of alloy contain 39 ounces of silver.

40 ounces of alloy contain 15 ounces of silver.

Proportion: $\frac{x}{40} = \frac{39}{15}$

$$x = 40\left(\frac{39}{15}\right)$$

$$= 40\left(\frac{13}{5}\right) = 104$$

Ans. (E)

After completing each test, match your answers with the numbers in the boxes.

Matching Test A

1. The perimeter of a quadrilateral is 65. If one side is 9 and the other three sides are in the ratio 1:2:4, find the length of the shortest side.

2. A board 44 inches in length is divided into three parts in the ratio 3:4:5. Find the number of inches in the shortest piece.

3. The perimeter of a triangle is 36 inches. If the sides are in the ratio 3:4:5, how many inches are there in the length of the longest side?

4. Find the number of inches in the length of the shorter part of a line segment of 45 inches if it is divided into two parts in the ratio 2:3.

5. Find the number of inches in the length of the longest part of a line segment of 55 inches if it is divided into three parts in the ratio 2:3:5.

a. 11	*b.* 15	*c.* 8	*d.* $27\frac{1}{2}$	*e.* 18

Matching Test B

1. Orange juice is mixed with grapefruit juice in the ratio 3:4. In 35 quarts of the mixture, how many quarts are grapefruit juice?
2. In an alloy of 13 pounds, the ratio of the weights of the 3 components is 16:9:7. Find the number of pounds of the heaviest component.
3. In a mixture of 15 quarts, the volumes of the three chemicals being mixed are in the ratio 2:2:5. Find the number of quarts in the largest of the three volumes.
4. The weights of nickel, silver, and zinc in an alloy are in the ratio 3:5:7. Find the number of pounds of the entire alloy if the weight of the zinc is 42 pounds.
5. Harry and Tom received $75 for a job on which Harry worked 6 days and Tom worked 9 days. If the money is divided between them in accordance with the ratio of the numbers of days spent on the work, how many more dollars will Tom get than Harry?

a. 90 *b.* $8\frac{1}{3}$ *c.* 15 *d.* 20 *e.* $6\frac{1}{2}$

Matching Test C

1. If a profit of $2400 is divided in the ratio 3:5:8, find the number of dollars in the smallest share.
2. Profits totaling $2800 are to be shared in the ratio 2:3:5. Find the number of dollars in the largest share.
3. A profit of $450 is divided by three partners in the ratio 2:3:5. Find the number of dollars by which the largest share exceeds the smallest.
4. An estate of $4488 is to be divided among the heirs in the ratio 2:4:5. Find the number of dollars in the smallest share.
5. An estate of $1500 was bequeathed to two children and one grandchild. Find the number of dollars that the grandchild should get if his share is one-half that of each of the two children.

a. 135 *b.* 816 *c.* 300 *d.* 1400 *e.* 450

Matching Test D

1. A cake recipe calls for $1\frac{1}{4}$ teaspoons of yeast for every 6 ounces of flour. How many ounces of flour should be used with $2\frac{1}{2}$ teaspoons of yeast?
2. A recipe requires 2 parts of sugar to 9 parts of flour. If 5 ounces of sugar are to be used, how many ounces of flour are needed?
3. In a juice blend, 3 cups of grapefruit juice are used with 5 cups of orange juice. If 16 quarts of the blend are made, how many quarts of orange juice were used?

4. A salmon salad uses $\frac{1}{2}$ cup of celery to 3 cups of salmon. If 4 ounces of celery were used to make the salad, how many ounces does the salad weigh?

5. A formula requires 8 ounces of milk to 15 ounces of water. If 11 ounces of milk are to be used, how many ounces of water are needed to the nearest ounce?

a. $22\frac{1}{2}$	*b.* 10	*c.* 21	*d.* 12	*e.* 28

1.4 PROBLEMS INVOLVING SCALES IN THE FORM OF RATIOS

Model Problem

A blueprint is drawn to a scale of 1:120. How many feet is the actual length of a boat if the length of the boat on the blueprint is 5.5 inches?

(A) 77 (B) 72 (C) 55 (D) 44 (E) 33

Solution:

If the scale is 1:120, the actual length of the boat is 120 times as much as the blueprint length.

Hence, the actual length of the boat $= 120 \times 5.5$ inches

$$= \frac{120 \times 5.5}{12} \text{ feet} \quad \text{(Convert inches to feet.)}$$

$= 55$ feet *Ans.* (C)

Matching Test

After completing this test, match your answers with the numbers in the box.

1. A scale of 1:120 is used on a blueprint. How many feet is the actual length of a boat if the length of the boat on the blueprint is 8.2 inches?

2. A scale of 1:60 is used in a scale model of a plane. How many feet is the width of the plane if the width of the model is 8.8 inches?

3. In a magazine illustration of an appliance, a scale of 1:8 is used. How many feet long is the appliance if its length in the illustration is 4.8 inches?

4. An architectural drawing uses a scale of 1:12. How many inches should a line in the drawing be to represent an actual length of 8.5 feet?

5. In a book illustration, a scale of 1:24 is used to represent an object whose actual length is 11 feet. How many inches long is the representation of this length in the illustration?

a. 44	*b.* 3.2	*c.* 8.5	*d.* 5.5	*e.* 82

1.5 GEOMETRIC PROBLEMS INVOLVING SCALES ON A GRAPH CHART

Model Problem

Find the area of the rectangle on the right if each unit along the x-axis is 4 and each unit along the y-axis is 3.

(A) 30 (B) 120 (C) 270 (D) 360 (E) 480

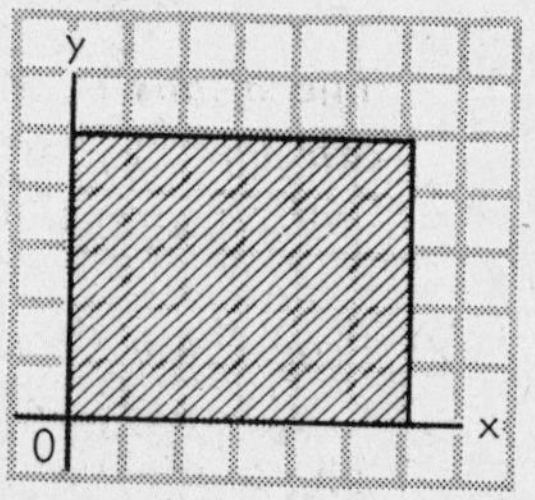

Solution:

Since each x-unit = 4, the base = $6 \cdot 4 = 24$.

Since each y-unit = 3, the altitude = $5 \cdot 3 = 15$.

The area of a rectangle is the product of its base and altitude.

Hence, the required area = bh

$= 24 \cdot 15$

$= 360$ *Ans.* (D)

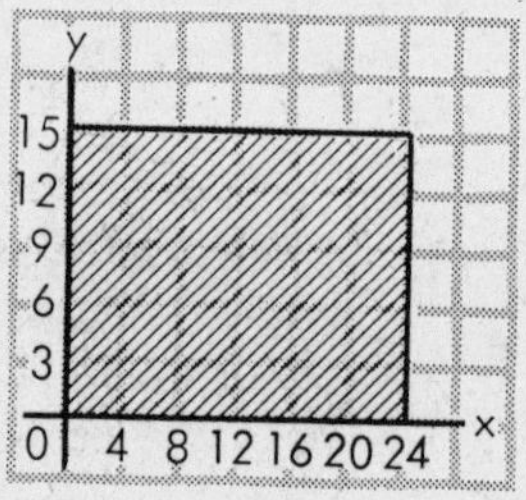

Matching Test

After completing this test, match your answers with the numbers in the box.

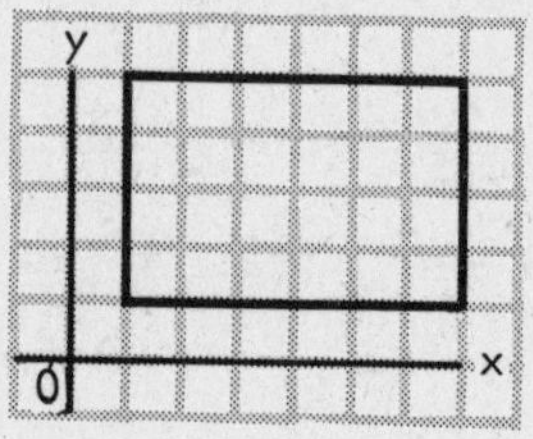

1. Each x-unit = 2.
Each y-unit = 3.

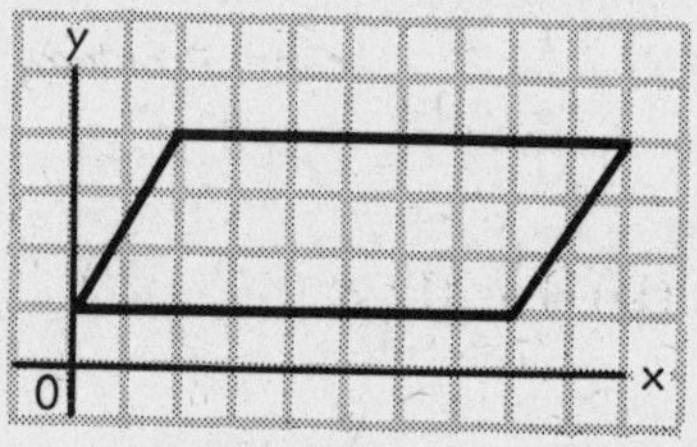

2. Each x-unit = 2.
Each y-unit = 2.

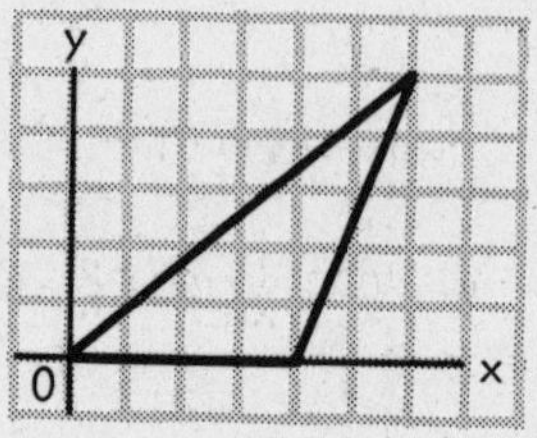

3. Each x-unit = 4.
Each y-unit = 3.

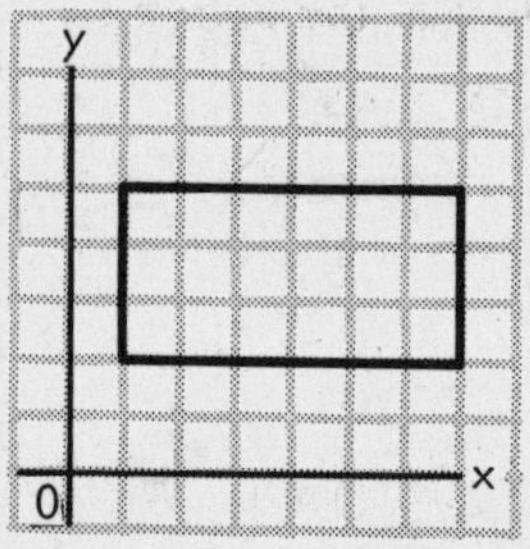

4. Each x-unit equals each y-unit.

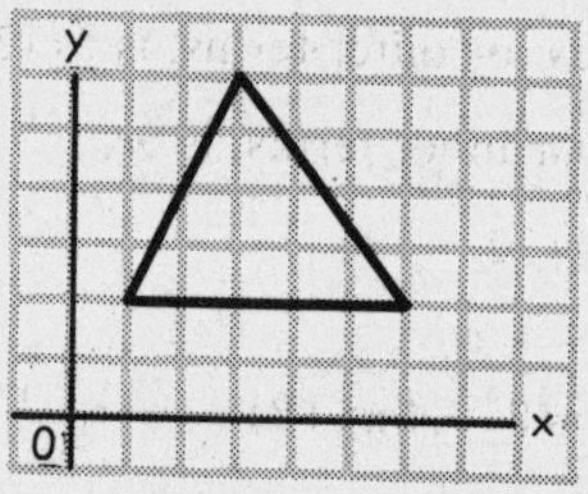

5. Each x-unit = 2.

1. Find the area of the rectangle.
2. Find the area of the parallelogram.
3. Find the area of the triangle.
4. If the area of the rectangle is 72, find the value of each x-unit and each y-unit.
5. If the area of the triangle is 60, find the value of each y-unit.

a. 120
b. 2
c. 3
d. 96
e. 144

2. REVIEWING PROPORTIONS

A ***proportion*** is an equality of two ratios. For example, 2:3 = 20:30 and $\frac{a}{b} = \frac{c}{d}$ are proportions.

The ***means*** of a proportion are the middle terms, that is, the second and the third terms. The ***extremes*** of a proportion are the end terms, that is, the first and the fourth terms. For example, in the proportion $a:b = c:d$, the means are b and c, the extremes are a and d.

Proportion Principle. If $a:b = c:d$, then $ad = bc$.

The proportion principle may be stated in either of two ways:

1. In a proportion, the product of the means equals the product of the extremes. For example, if $x:5 = 4:3$, then $3x = 20$. Hence, $x = 6\frac{2}{3}$.
2. When the proportion is in fraction form, such as $\frac{x}{3} = \frac{5}{2}$, the cross-products are equal. For example, if $\frac{x}{3} = \frac{5}{2}$, then $2x = 15$. Hence, $x = 7\frac{1}{2}$.

2.1 SOLVING A PROPORTION IN COLON FORM

Model Problem

Solve for r: $6:r = 9:12$

(A) $\frac{1}{8}$ (B) 8 (C) $\frac{2}{9}$ (D) $4\frac{1}{2}$ (E) 9

Solution:

Solve the proportion $6:r = 9:12$ by applying the following proportion principle:

In a proportion, the product of the means equals the product of the extremes.

The product of the extremes, or outer terms, is 6(12), or 72.

The product of the means, or inner terms, is $9r$.

Hence, $9r = 72$

Divide each side by 9.

$r = 8$ *Ans.* (B)

Alternate Method

Change $6:r = 9:12$ into the fraction form:	$\frac{6}{r} = \frac{9}{12}$
Multiply each side by $12r$ or cross-multiply:	$72 = 9r$
Divide each side by 9:	$8 = r$

Matching Test

Solve each proportion for r. After completing this test, match your answers with the numbers in the box.

1. $r:12 = 9:6$ **2.** $r:12 = 6:9$ **3.** $r:8 = 9:12$ **4.** $r:8 = 9:6$ **5.** $r:6 = 8:12$

a. 4	*b.* 8	*c.* 6	*d.* 18	*e.* 12

2.2 SOLVING A PROPORTION IN FRACTION FORM

Model Problem

Solve for x: $\frac{2x}{3} = \frac{x+1}{2}$

(A) $\frac{1}{2}$ (B) $\frac{1}{3}$ (C) 2 (D) 3 (E) -3

Solution:

To solve the given equation for x, first clear the equation of fractions by cross-multiplying.

$$\frac{2x}{3} = \frac{x+1}{2}$$

$$2(2x) = 3(x+1)$$

$$4x = 3x + 3$$

Subtract $3x$ from each side: $\quad -3x = -3x$

$$x = 3 \quad \textit{Ans.}\ (D)$$

Matching Test

Solve each proportion for x. After completing this test, match your answers with the numbers in the box.

1. $\frac{x+3}{4} = \frac{2x}{5}$ **2.** $\frac{x+4}{3} = \frac{2x}{5}$ **3.** $\frac{x+3}{2} = \frac{5x}{4}$ **4.** $\frac{x+5}{2} = \frac{3x}{4}$ **5.** $\frac{x+5}{4} = \frac{3x}{2}$

a. 10	*b.* 2	*c.* 1	*d.* 20	*e.* 5

2.3 EVALUATING IN A PROPORTION

Model Problem

In the proportion $\frac{a}{b} = \frac{c+1}{d}$, find c if $a = 7$, $b = 8$, and $d = 4$.

(A) $3\frac{3}{8}$ (B) $2\frac{1}{2}$ (C) 27 (D) $\frac{3}{4}$ (E) $1\frac{5}{8}$

Solution:

Since $a = 7$, $b = 8$, and $d = 4$, substitute these values in the given proportion.

$$\frac{a}{b} = \frac{c+1}{d}$$

$$\frac{7}{8} = \frac{c+1}{4}$$

Clear the proportion of fractions by cross-multiplying.

$$8c + 8 = 28$$

Subtract 8 from each side: $\quad -8 = -8$

$$8c = 20$$

Divide each side by 8: $\quad \frac{8c}{8} = \frac{20}{8}$

$$c = \frac{5}{2} \text{ or } 2\frac{1}{2} \qquad \textit{Ans. } \text{(B)}$$

> *Alternate Method*
>
> If each side is multiplied by 8, then $\frac{7}{8} = \frac{c+1}{4}$ becomes $7 = 2c + 2$. By subtracting 2 from each side and then dividing each side by 2, $c = 2\frac{1}{2}$.

Matching Test

Substitute the given values in the proportion $\frac{a+2}{b} = \frac{c+1}{2d}$. Find the missing values indicated by the question marks. After completing this test, match your answers with the numbers in the box.

1. $a = 2,\ b = 8,\ c = 6,\ d = ?$
2. $a = 2,\ b = 8,\ d = 6,\ c = ?$
3. $a = 6,\ b = 2,\ d = 8,\ c = ?$
4. $b = 8,\ c = 6,\ d = 2,\ a = ?$
5. $a = 8,\ c = 2,\ d = 6,\ b = ?$

a. 5
b. 40
c. 7
d. 12
e. 63

2.4 USING PROPORTIONS TO SOLVE SCALE PROBLEMS (MAPS, BLUEPRINTS, SCALE DRAWINGS, ETC.)

Model Problem

The scale on a map is $\frac{3}{4}$ inch = 6 miles. What is the actual distance between two towns if the towns are $1\frac{1}{4}$ inches apart on the map?

(A) $7\frac{1}{2}$ mi. (B) 8 mi. (C) 9 mi. (D) $10\frac{1}{2}$ mi. (E) 10 mi.

Solution:

Method 1: *Using a Proportion*

Let $d =$ the distance in miles between the two towns.

Form a proportion in which the ratio of the map lengths equals the ratio of the actual distances:

$1\frac{1}{4}$ inches represents d miles.

$\frac{3}{4}$ inch represents 6 miles.

$$\text{Proportion:} \quad \frac{\frac{5}{4}}{\frac{3}{4}} = \frac{d}{6}$$

$$\frac{5}{3} = \frac{d}{6}$$

$$6\left(\frac{5}{3}\right) = d, \quad d = 10$$

Method 2: *Using a Unit Measure*

Since a length of $\frac{3}{4}$ inch represents 6 miles, then a length of 1 inch will represent a length of $6 \div \frac{3}{4}$ or 8 miles.

Since the scale is 1 inch to 8 miles, a line that is $1\frac{1}{4}$ inches long will represent a distance of $8 \times 1\frac{1}{4}$ or 10 miles.

Ans. (E)

Matching Test

After completing this test, match your answers with the numbers in the box.

1. How many feet long is an actual length if it is represented by a $1\frac{1}{2}$ inch line on a blueprint having a scale of $\frac{3}{4}$ inch to 6 feet?

2. How many miles is the actual distance between two cities if the distance is represented by a 1.2 inch line on a map having a scale of $\frac{1}{2}$ inch to 10 miles?

3. How many miles is a distance represented by a 3.2 inch line on a map whose scale is 25 miles per inch?

4. How many miles per inch is the scale of a map in which a line 1.8 inches long represents a distance of 27 miles?

5. Using the scale $\frac{1}{2}$ inch = 5 yards, how many inches long should a line on a diagram be in order to represent the 100-yard length of a football field?

a. 10	*b.* 24	*c.* 15	*d.* 80	*e.* 12

2.5 USING PROPORTIONS TO SOLVE COST PROBLEMS

Model Problems

(1) If n articles cost c cents, write an expression that represents the cost, in cents, of 40 articles.

(A) $\frac{40c}{n}$ (B) $\frac{40n}{c}$ (C) $\frac{nc}{40}$ (D) $\frac{c}{40n}$ (E) $\frac{40}{nc}$

Solution:

Method 1: Using a Proportion

Let $x =$ the cost of the 40 articles in cents.

Form a proportion in which the ratio of the numbers of articles equals the ratio of the costs:

$x¢$ is the cost of 40 articles.

$c¢$ is the cost of n articles.

$$\text{Proportion: } \frac{x}{c} = \frac{40}{n}$$

$$\text{Hence, } x = \frac{40c}{n}$$

Method 2: Using a Unit Value

If n articles cost c cents, then 1 article costs $\frac{c}{n}$ cents.

For example, if 10 pencils cost 30 cents, then 1 pencil costs $\frac{30}{10}$ or 3 cents.

Since 1 article costs $\frac{c}{n}$ cents, 40 articles cost $40\left(\frac{c}{n}\right)$ or $\frac{40c}{n}$ cents.

Ans. (A)

(2) If 4 oranges cost c cents, how many cents will 3 dozen oranges cost?

(A) $\frac{3}{4}c$ (B) $9c$ (C) $\frac{4}{3}c$ (D) $\frac{c}{9}$ (E) $\frac{9}{c}$

Solution:

Method 1: Using a Proportion

Let $x =$ the cost in cents of 3 dozen, or 36, oranges.

Form a proportion in which the ratio of the numbers of oranges equals the ratio of the costs:

$x¢$ is the cost of 36 oranges.

$c¢$ is the cost of 4 oranges.

$$\text{Proportion: } \frac{x}{c} = \frac{36}{4}$$

$$\text{Hence, } x = 9c$$

Method 2: Using a Unit Value

If 4 oranges cost c¢, then 1 orange costs $\frac{1}{4}c$¢.

Since 1 orange costs $\frac{1}{4}c$¢, 3 dozen, or 36, oranges cost $36\left(\frac{1}{4}c\right)$ or $9c$¢.

Ans. (B)

After completing each test, match your answers with the answers in the boxes.

Matching Test A

1. How many dollars will 10 books cost if 5 books of the same kind cost n dollars?
2. How many books can be bought for n dollars if \$10 is the cost of 5 books of the same kind?
3. How many dollars will n books cost if \$2 is the cost of p books of the same kind?
4. How many books can be bought for \$2 if n dollars is the cost of p books of the same kind?
5. How many dollars will n books cost if p dollars is the cost of q books?

a. $\frac{np}{q}$	*b.* $2n$	*c.* $\frac{2n}{p}$	*d.* $\frac{n}{2}$	*e.* $\frac{2p}{n}$

Matching Test B

1. If n oranges sell for 25¢, how much will a half-dozen cost?
2. If 5 oranges sell for n¢, how many can be bought for 120¢?
3. If 4 oranges sell for 15¢, how much will n oranges cost?
4. If 5 apples sell for 19¢, how much will n apples cost?
5. If n apples sell for 30¢, how many can be bought for 70¢?

a. $\frac{600}{n}$
b. $\frac{19n}{5}$¢
c. $\frac{15n}{4}$¢
d. $\frac{7n}{3}$
e. $\frac{150}{n}$¢

Matching Test C

1. The price of a train ticket is directly proportional to the distance traveled. If the cost of going 24 miles is $1.56, how many cents must be paid for a train ticket to travel 32 miles?

2. For every $25 in sales, a salesman gets a commission of 80¢. If his commission is $4, how many dollars worth of goods did he sell?

3. At a rate of 90¢ for $1\frac{1}{2}$ gallons of a solvent, what is the cost in cents of $2\frac{1}{2}$ gallons?

4. How many cents must be paid for $2\frac{1}{4}$ dozen apples at the rate of 3 for 14¢?

5. If $1.20 is paid for 100 grams of a certain metal, at that rate, how many cents will 175 grams of the same metal cost?

a. 210	*b.* 208	*c.* 125	*d.* 150	*e.* 126

2.6 USING PROPORTIONS TO SOLVE WORK, RATE, AND CONSUMPTION PROBLEMS

Model Problems

(1) A furnace uses $1\frac{1}{2}$ tons of coal in 30 days. At this rate, how many days will a supply of 6 tons of coal last?

(A) 90 (B) 120 (C) 135 (D) 270 (E) 360

Solution:

Method 1: *Using a Proportion*

Let $x =$ the number of days that 6 tons of coal will last.

Form a proportion in which the ratio of the numbers of days that the coal will last equals the ratio of the weights of the coal. For example, twice as much coal will last twice as long.

In x days, 6 tons are used.

In 30 days, $1\frac{1}{2}$ tons are used.

$$\text{Proportion: } \frac{x}{30} = \frac{6}{1\frac{1}{2}}$$

$$\text{Hence, } x = 30(4) = 120$$

Method 2: *Using a Unit Value*

If $1\frac{1}{2}$ tons of coal last 30 days, then 1 ton of coal will last $30 \div 1\frac{1}{2}$ or 20 days.

Since 1 ton of coal lasts 20 days, 6 tons will last 6×20 or 120 days.

Ans. (B)

(2) If $\frac{3}{5}$ of a job is completed in 6 days, how many days will it take to do the rest of the job at this rate?

(A) 4 (B) 6 (C) 10 (D) 12 (E) 16

Solution:

Method 1: *Using a Proportion*

Let $x =$ the number of days needed to finish the remaining $\frac{2}{5}$ of the job.

Form a proportion in which the ratio of the numbers of days worked equals the ratio of the amounts of work done. For example, in twice as many days, twice as much work can be done.

x days are needed for $\frac{2}{5}$ of the job. $\left(\frac{3}{5}\text{ is already done}\right)$

6 days are needed for $\frac{3}{5}$ of the job.

$$\frac{x}{6} = \frac{\frac{2}{5}}{\frac{3}{5}}$$

$$\frac{x}{6} = \frac{2}{3}$$

$$x = 6\left(\frac{2}{3}\right) = 4$$

Method 2: *Using a Unit Value*

If $\frac{3}{5}$ of the job is done in 6 days, then the time required to do the entire job is $6 \div \frac{3}{5}$ or 10 days.

Thus, 1 job takes 10 days. Since $\frac{2}{5}$ of the job remains to be done, the number of days required to do $\frac{2}{5}$ of the job $= \frac{2}{5}$ of 10, or 4 days.

Ans. (A)

Matching Test

After completing this test, match your answers with one of the numbers in the box.

1. If $\frac{2}{5}$ of a job is completed in 6 days, how many days will it take to do the rest of the job at this rate?

2. If $\frac{4}{7}$ of a pool is filled in 10 minutes, how many minutes will it take to fill the rest of the pool at this rate?

3. A car owner used 25 tankfuls of gas during the first 8 months of the year. At the same rate of consumption, how many tankfuls will he need for the rest of the year?

4. A furnace uses $2\frac{1}{4}$ tons of coal in $3\frac{1}{2}$ weeks. At this rate, how many weeks will a supply of 9 tons of coal last?

5. A furnace uses $2\frac{1}{2}$ tons of coal in 11 days. At this rate, how many tons of coal are needed for 44 days?

a. $7\frac{1}{2}$	*b.* $12\frac{1}{2}$	*c.* 14	*d.* 10	*e.* 9

3. REVIEWING VARIATION

Direct Variation

If the side of a square is made three times as long, then the perimeter of the new square will also be three times as long. For example, if the original square has a side of 2 and a perimeter of 8, the new square will have a side of 6 and a perimeter of 24. Thus, the ratio of the perimeter to the side is $\frac{8}{2}$ for the first square and $\frac{24}{6}$ for the second square. Note that these ratios are equal; that is, $\frac{24}{6} = \frac{8}{2} = 4$.

The relation between the perimeter, p, of a square and its side, s, is expressed in the formula $p = 4s$, or $\frac{p}{s} = 4$. The perimeter and the side of the square are two quantities that are said to *vary directly as each other.*

Principles of Direct Variation

1. If x and y vary directly as each other and a value of either x or y is multiplied (or divided) by a number, then the corresponding value of the other is multiplied (or divided) by the same number. (Note the use of the *same operation* upon both variables.)

   For example, if the side of a square is doubled, then the perimeter of the square is also doubled.

2. If x and y vary directly as each other, their relation may be expressed in either of the following forms, using k as a constant:

   $$(a)\ y = kx \qquad (b)\ \frac{y}{x} = k$$

3. If x and y vary directly as each other and if x varies from x_1 to x_2 while y varies from y_1 to y_2, this relation may be expressed in the following proportion:

   $$\frac{x_2}{x_1} = \frac{y_2}{y_1}$$

Inverse Variation

If the base of a rectangle is made three times as long and the area does not change, then the altitude of the new rectangle will be only one-third as long. For example, if the area is to remain fixed at 48 and the base is changed from 8 to 24, then the altitude will change from 6 to 2. Thus, the ratio of the first base to the second is $\frac{8}{24}$, or $\frac{1}{3}$, and the ratio of the first altitude to the second is $\frac{6}{2}$, or $\frac{3}{1}$. Note that the two ratios are reciprocals of each other. Hence, $\frac{24}{8} = \frac{6}{2}$.

The relation between the base, b, and the altitude, h, of a rectangle having a fixed area is expressed in the formula $bh = k$, where k is a constant. For a fixed area, the base and altitude of a rectangle are said to *vary inversely as each other.*

Principles of Inverse Variation

1. If x and y vary inversely as each other and a value of x or y is multiplied by a number, then the corresponding value of the other variable is divided by the same number. Similarly, if a value of x or y is divided by a number, then the corresponding value of the other variable is multiplied by the same number. (Note the use of *inverse operations* upon both variables.)

   For example, if the area of a rectangle is fixed and the base is doubled, then the altitude is halved; that is, one dimension is multiplied by 2 and the other is divided by 2.

2. If x and y vary inversely as each other, their relation may be expressed in the following form, using k as a constant:

   $$xy = k$$

   For example, for a fixed distance, rate and time vary inversely; that is, $RT = k$.

3. If x and y vary inversely as each other and if x varies from x_1 to x_2 while y varies from y_1 to y_2, this relation may be expressed in the following proportion:

   $$\frac{x_2}{x_1} = \frac{y_1}{y_2}$$

   Note the inverting of the ratio $\frac{y_2}{y_1}$ into the ratio $\frac{y_1}{y_2}$.

   For example, if a man covers a distance in 3 hours at 40 mph, then at 60 mph he will cover the same distance in 2 hours. The rate in miles per hour has changed from 40 to 60, while the time in hours has changed from 3 to 2. The inverse proportion is $\frac{60}{40} = \frac{3}{2}$.

   In set 3.3, note three important applications of inverse variation: (1) meshed gears, (2) belted pulleys, and (3) a balanced lever.

Power Variation

If the radius of a circle is made three times as long, then the area of the new circle will be nine times as great. The area of a circle is said to *vary directly as the square of the radius.*

If the radius of a sphere is made three times as long, then the new volume will be 27 times as great. The volume of a sphere is said to *vary directly as the cube of the radius.*

Principles of Power Variation

1. If x varies directly as the square of y and if x varies from x_1 to x_2 while y varies from y_1 to y_2, then this relation may be expressed in the following proportion:

$$\frac{x_2}{x_1} = \left(\frac{y_2}{y_1}\right)^2 \text{ or } \frac{x_2}{x_1} = \frac{y_2^2}{y_1^2}$$

For example, a circle having twice the radius of another circle will have 4 times the area of the smaller circle.

2. If x varies directly as the cube of y and if x varies from x_1 to x_2 while y varies from y_1 to y_2, then this relation may be expressed in the following proportion:

$$\frac{x_2}{x_1} = \left(\frac{y_2}{y_1}\right)^3 \text{ or } \frac{x_2}{x_1} = \frac{y_2^3}{y_1^3}$$

For example, a solid sphere having twice the radius of another sphere will weigh 8 times as much, assuming that the two spheres are the same in all respects except size.

3.1 APPLYING DIRECT-VARIATION PRINCIPLES

Model Problems

(1) If x varies directly as y and if $x = 12$ when $y = 8$, find x when $y = 10$.

(A) 9.6 (B) 10 (C) 12 (D) 14 (E) 15

Solution:

Form a proportion in which the ratio of the x-values equals the ratio of the corresponding y-values:

Table of Values

	x	y
Find x when $y = 10$ ⟶	x	10
if $x = 12$ when $y = 8$. ⟶	12	8

Proportion: $\frac{x}{12} = \frac{10}{8}$

$$x = 12\left(\frac{10}{8}\right) = 15 \quad \textit{Ans.} \text{ (E)}$$

(2) Which one of the following statements is false?

(A) If $y = 5x$ and x is tripled, then y is tripled.
(B) If $C = \pi D$ and D is quadrupled, then C is quadrupled.
(C) If $p = 4s$ and s is divided by 3, then p is divided by 3.
(D) If $A = LW$ and L is 10, then, when A is divided by 5, W is divided by 5.
(E) If $A = LW$ and A is 120, then, when L is multiplied by 4, W is multiplied by 4.

Solution:

The equations in the first four statements are all of the form $y = kx$, where k is a constant. When the relation between two variables can be expressed in the form $y = kx$, the variables are said to vary directly as each other. In direct variation, if either variable is multiplied (or divided) by a number, the other variable is also multiplied (or divided) by that number. Hence, the first four statements are true.

In statement (E), the equation is of the form $xy = k$. This can be readily seen if A is replaced with its given value, 120. The equation then becomes $LW = 120$. When the relation between two variables can be expressed in the form $xy = k$, the variables are said to vary inversely as each other. Thus, if L is multiplied by 4, then W is *divided* by 4. Hence, statement (E) is false. *Ans.* (E)

Matching Test

After completing this test, match your answers with the numbers in the box.

1. If x varies directly as y and if $x = 6$ when $y = 10$, find x when $y = 15$.

2. If x varies directly as y and if $x = 9$ when $y = 15$, find x when $y = 10$.

3. Given: $\frac{y}{x} = k$, and k is a constant. If $x = 6$ when $y = 9$, find y when $x = 10$.

4. Given: $y = 10x$, and x is multiplied by 5. By what is y multiplied?

5. The price of a train ticket varies directly as the distance of the trip. A ticket for a 30-mile trip costs \$1.65. How many miles can one travel for a ticket costing 99¢?

a. 6	*b.* 5	*c.* 9	*d.* 18	*e.* 15

3.2 APPLYING INVERSE-VARIATION PRINCIPLES

Model Problems

(1) If x varies inversely as y and if $x = 12$ when $y = 8$, find x when $y = 10$.

(A) 9.6 (B) 10 (C) 12 (D) 14 (E) 15

Solution:

Method 1: *Using Inverse Ratios in a Proportion*

Form a proportion in which the ratio of the x-values and the ratio of the corresponding y-values are *inversely equal*; that is, keep one ratio unchanged and invert the other:

Table of Values

	x	y
Find x when $y = 10$ ⟶	x	10
if $x = 12$ when $y = 8$ ⟶	12	8

Proportion: $\frac{x}{12} = \frac{8}{10}$ (Note that the ratio $\frac{10}{8}$ is inverted.)

$$x = 12\left(\frac{8}{10}\right) = 9.6$$

Method 2: Using a Constant

Find x by applying the following principle:

If x varies inversely as y, then the product of any value of x and the corresponding value of y is a constant.

Hence, $xy = k$

Since $x = 12$ when $y = 8$, substitute 12 for x and 8 for y.

$(12)(8) = k, \ 96 = k$

To find x when $y = 10$, substitute 10 for y and 96 for k.

$$xy = k$$
$$x(10) = 10x = 96$$
$$x = 9.6$$

Ans. (A)

(2) If x varies inversely as y, then when x is

(A) increased by 2, y is decreased by 2 (B) increased by 2, y is increased by 2
(C) multiplied by 2, y is multiplied by 2 (D) multiplied by 2, y is divided by 2
(E) divided by 2, y is divided by 2

Solution:

Make the correct choice by applying the following rule:

If x varies inversely as y, then if x is multiplied by a number, y is divided by the same number.

Hence, if x varies inversely as y, and x is multiplied by 2, then y is divided by 2.

Ans. (D)

Wrong Choices:

(C) and (E) If x varies directly as y, then if x is multiplied by a number, y is multiplied by the same number. Similarly, if x is divided by a number, then y is also divided by that number.

(A) and (B) involve addition and subtraction. These choices are not cases in which variables vary directly or inversely.

Matching Test

After completing this test, match your answers with the numbers in the box.

1. If x varies inversely as y and if $x = 6$ when $y = 10$, find x when $y = 15$.
2. If x varies inversely as y and if $x = 9$ when $y = 10$, find x when $y = 15$.
3. Given: $xy = k$, and k is a constant. If $x = 15$ when $y = 6$, find y when $x = 10$.
4. If $xy = 20$ and x is multiplied by 5, by what is y divided?
5. For a fixed area, the length of a rectangle varies inversely as the width. If the length of a rectangle is doubled and the area remains the same, by what is the width multiplied?

a. 9	*b.* 6	*c.* 5	*d.* $\frac{1}{2}$	*e.* 4

3.3 USING INVERSE-VARIATION PRINCIPLES TO SOLVE PROBLEMS INVOLVING GEARS, PULLEYS, AND BALANCED LEVERS

Model Problems

(1) A gear having 30 teeth and turning at 360 revolutions per minute (rpm) is meshed with another gear having 72 teeth. At what speed in rpm is the second gear turning?

(A) 120 (B) 150 (C) 180 (D) 240 (E) 864

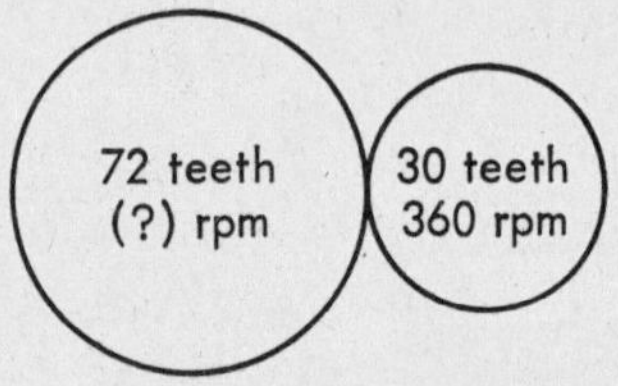

Meshed Gears

Solution:

Let $n =$ the number of rpm of the second gear.

Form a proportion in which the ratio of gear teeth is *inversely* equal to the ratio of the speeds:

Gear turning at n rpm has 72 teeth.

Gear turning at 360 rpm has 30 teeth.

Inverse proportion: $\frac{n}{360} = \frac{30}{72}$ $\left(\text{Note that the ratio } \frac{72}{30} \text{ has been inverted.}\right)$

$$n = \frac{30}{72}(360) = 150 \quad \textit{Ans.} \text{ (B)}$$

> *Note.* In the case of meshed gears, the product of the number of teeth and the rate of rotation is constant.
>
> Hence, in the given problem,
>
> $$72n = (30)(360)$$
>
> $$n = 150$$

(2) A pulley having a diameter of 8 inches and turning at 270 rpm is belted to another pulley turning at 360 rpm. How many inches is the diameter of the second pulley?

(A) 12 (B) $10\frac{2}{3}$ (C) 9 (D) 6 (E) $5\frac{1}{3}$

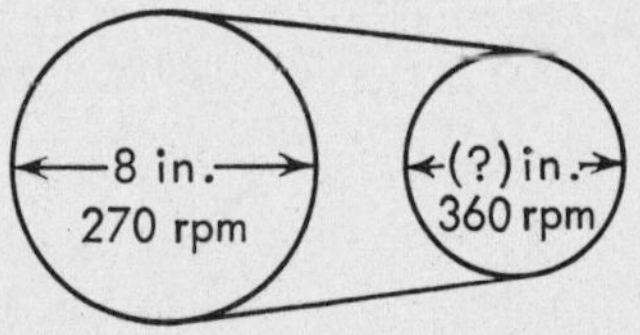

Belted Pulleys

Solution:

Let $n =$ the number of inches in the diameter of the second pulley.

Form a proportion in which the ratio of the speeds is *inversely equal* to the ratio of the diameters:

n-inch pulley turns at 360 rpm.
8-inch pulley turns at 270 rpm.

Inverse proportion: $\frac{n}{8} = \frac{270}{360}$ (Note that the ratio $\frac{360}{270}$ has been inverted.)

$$n = 8\left(\frac{270}{360}\right) = 6 \quad \textit{Ans.}\ \text{(D)}$$

> *Note.* In the case of belted pulleys, the product of the length of the diameter and the rate of rotation is constant.
>
> Hence, in the given problem,
>
> $$360n = (8)(270)$$
> $$n = 6$$

(3) A boy weighing 99 pounds and a man are balanced at opposite ends of a seesaw. How many pounds is the weight of the man if he is 3 feet from the fulcrum, or pivot, while the boy is 4 feet from it?

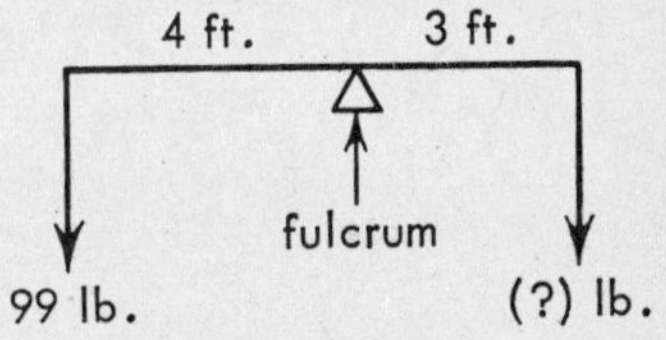

Balanced Lever

(A) $74\frac{1}{4}$ (B) 120 (C) 132 (D) 144 (E) 165

Solution:

Let $n =$ the weight in pounds of the man.

Form a proportion in which the ratio of the weights is *inversely equal* to the ratio of the distances from the fulcrum:

Weight of n lb. is 3 ft. from fulcrum.
Weight of 99 lb. is 4 ft. from fulcrum.

Inverse proportion: $\frac{n}{99} = \frac{4}{3}$ (Note that the ratio $\frac{3}{4}$ has been inverted.)

$$n = 99\left(\frac{4}{3}\right) = 132 \quad \textit{Ans.}\ \text{(C)}$$

> *Note.* In the case of a balanced lever, the product of the weight of an object and its distance from the fulcrum is constant.
>
> Hence, in the given problem,
>
> $$3n = (99)(4)$$
> $$n = 132$$

Matching Test

After completing this test, match your answers with the numbers in the box.

1. How many teeth does a gear have if it turns at 360 rpm and is in mesh with another gear of 30 teeth turning at 240 rpm?
2. How many revolutions per minute does a gear of 40 teeth make if it is in mesh with another gear having 30 teeth and turning at 240 rpm?
3. How many inches is the diameter of a pulley that is turning at 270 rpm if it is belted to another pulley having a diameter of 9 inches and turning at 360 rpm?
4. How many revolutions per minute does a pulley make if its diameter is 8 inches and it drives a pulley having a diameter of 9 inches and turning at 360 rpm?
5. Two boys weighing 90 pounds and 120 pounds are balanced at opposite ends of a seesaw. How many feet from the fulcrum is the lighter boy if the heavier one is 6 feet from it?

a. 12	*b.* 8	*c.* 180	*d.* 20	*e.* 405

3.4 USING INVERSE-VARIATION PRINCIPLES TO SOLVE PROBLEMS INVOLVING WORK AND CONSUMPTION

Model Problem

A farm can be plowed by 4 machines in 12 hours. If 2 more machines of the same kind are added, how many hours will it now take to plow the field?

(A) 4 (B) 8 (C) 12 (D) 16 (E) 24

Solution:

Method 1: Using a Proportion

Let n = the number of hours that it will take the 6 machines to plow the field.

Form a proportion in which the ratio of the times of plowing is *inversely equal* to the numbers of machines used:

The field can be plowed in n hours by 6 machines
and in 12 hours by 4 machines.

Inverse proportion: $\frac{n}{12} = \frac{4}{6}$ (Note that the ratio $\frac{6}{4}$ has been inverted.)

$$n = 12\left(\frac{4}{6}\right) = 8$$

Method 2: Using a Constant

Since the same field is being plowed, *the number of machine-hours is a constant.*

Since each of the 4 machines works 12 hours, the number of machine-hours required for the job is $4 \cdot 12$, or 48.

If 6 machines are to perform 48 machine-hours of work, then each machine will require $\frac{48}{6}$ or 8 hours.

Ans. (B)

Matching Test

After completing this test, match your answers with the numbers in the box.

1. If 6 men can do a certain job in 10 days, at the same rate of work, how many men are needed to do the same job in 4 days?

2. If 10 tractors are needed to plow a certain field in 4 hours, how many tractors of the same kind are needed to plow the field in 5 hours?

3. It is found that 12 trucks of the same capacity will require 6 trips each to carry away a load of earth, but 4 of these trucks cannot be used because of mechanical defects. How many *more* trips will each truck now have to make to carry away the same load?

4. Food for 30 days is on hand for a group of 9 men. If 6 more men join the group and are put on the same rations, how many fewer days will the food supply last?

5. A supply of gasoline used at the rate of 15 gallons per hour lasts 24 days. If it is used at the rate of 18 gallons per hour, how many days will it last?

a. 8	*b.* 20	*c.* 15	*d.* 3	*e.* 12

3.5 APPLYING POWER-VARIATION PRINCIPLES

Model Problems

(1) A pipe having a diameter of 3 inches has a rate of flow of 9 gallons per second. Under the same conditions, how many gallons per second is the rate of flow of a pipe having a 5-inch diameter?

(A) $3\frac{6}{25}$ (B) $5\frac{2}{5}$ (C) $13\frac{1}{2}$ (D) 15 (E) 25

Solution:

Let $n =$ the number of gallons per second through the 5-inch pipe.

Form a proportion in which *the ratio of the rates of flow equals the square of the ratio of the diameters:*

n gal. per sec. flows through the 5-inch pipe.

9 gal. per sec. flows through the 3-inch pipe.

$$\frac{n}{9} = \left(\frac{5}{3}\right)^2 \quad \text{(See Note, below.)}$$

$$n = \left(\frac{25}{9}\right)9 = 25 \quad \textit{Ans.} \text{ (E)}$$

> *Note.* The ratio of the rates of flow equals the ratio of the areas of the circular cross-sections of the pipes. For any two circles, the ratio of the areas equals the square of the ratio of the diameters.

(2) A metal spherical ball having a diameter of 2 inches weighs 5 pounds. How many pounds will a spherical ball of the same metal composition weigh if its diameter is 4 inches?

(A) 10 (B) 20 (C) 40 (D) 60 (E) 80

Solution:

Let $n =$ the number of pounds in the weight of the 4-inch sphere.

Form a proportion in which *the ratio of the weights equals the cube of the ratio of the diameters*:

n lb. is the weight of a 4-inch sphere.
5 lb. is the weight of a 2-inch sphere.

$$\frac{n}{5} = \left(\frac{4}{2}\right)^3 \quad \text{(See Note, below.)}$$

$$n = 5(2^3) = 5(8) = 40 \quad \textit{Ans.}\ \text{(C)}$$

> *Note.* The ratio of the weights equals the ratio of the volumes for the same metal composition. For any two spheres, the ratio of the volumes equals the cube of the ratio of the diameters.

Matching Test

After completing this test, match your answers with the numbers in the box.

1. The price of a glass sphere having a diameter of 3 inches is $4. At this rate, how many dollars should be paid for a sphere of the same composition if the diameter is 6 inches?

2. Two cubes weigh 108 pounds and 4 pounds. If they are made of the same material and the side of the smaller cube is 3 inches, how many inches is the side of the larger cube?

3. How many spheres having a radius of 2 inches have a total volume equal to that of a sphere having a 6-inch radius?

4. If a pipe having a diameter of 4 feet has a rate of flow of 8 gallons per second, how many gallons per second is the rate of flow of a pipe having a diameter of 6 feet and operating under the same conditions?

5. How many circles having a 3-inch radius will have a total area equal to that of a circle with a 12-inch radius?

a. 18	*b.* 16	*c.* 9	*d.* 27	*e.* 32

4. PRACTICE PROBLEMS

1. Express the ratio 80%:60%:45% in lowest terms.

(A) 16:12:9 (B) 4:3:2 (C) 7:4:3 (D) $8:6:4\frac{1}{2}$ (E) 80:60:45 **[1.1]**

2. Of 22 students taking mathematics and 25 taking science, it was found that 10 take both subjects. What is the ratio of the number of students taking only mathematics to the number of those taking only science? (*Hint.* If a student takes both mathematics and science, he will be found among the 22 who take mathematics and also among the 25 who take science.)

(A) 2:3 (B) 3:2 (C) 4:5 (D) 5:6 (E) 3:4 **[1.1]**

3. In an acid solution, the ratio of acid to water is 3:11. What part of the solution is the acid?

(A) $\frac{3}{14}$ (B) $\frac{3}{11}$ (C) $\frac{8}{14}$ (D) $\frac{8}{11}$ (E) $\frac{3}{8}$ **[1.1]**

4. What is the ratio of the number of games won to the number of games played if a team lost three-sevenths as many games as it won?

(A) $\frac{7}{3}$ (B) $\frac{7}{10}$ (C) $\frac{3}{10}$ (D) $\frac{3}{7}$ (E) $\frac{10}{7}$ **[1.1]**

5. The sides of triangle ABC have been doubled. What is the ratio of the unshaded part to the shaded part?

(A) $\frac{1}{2}$ (B) $\frac{3}{2}$ (C) $\frac{2}{1}$ (D) $\frac{3}{1}$ (E) $\frac{3}{4}$ **[1.2]**

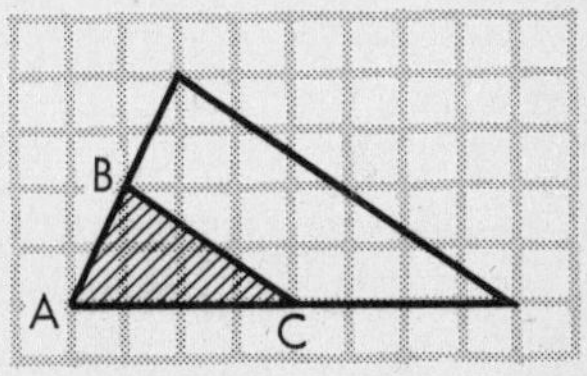

6. One of the angles of a right triangle is 25°. What is the ratio of the largest angle of the triangle to the difference between the acute angles?

(A) 9:4 (B) 4:9 (C) 13:8 (D) 9:5 (E) 5:9 **[1.2]**

7. A board 27 inches long is cut into two parts in the ratio 4:5. Find the number of inches in the length of the longer part.

(A) 3 (B) 10 (C) 12 (D) 15 (E) 18 **[1.3]**

8. To earn a total of $24, Mary worked 3 hours while Joan worked 5 hours. If their earnings are shared in the ratio of the numbers of hours they worked, how many dollars did Mary get?

(A) 3 (B) 9 (C) 10 (D) 14 (E) 15 **[1.3]**

9. In a school, the ratio of seniors to all other students is 1:7. If a total of 1120 students are enrolled in the school, how many of them are not seniors?

(A) 140 (B) 160 (C) 740 (D) 960 (E) 980 **[1.3]**

10. A candy recipe calls for 6 parts of milk, 5 parts of syrup, 4 parts of cocoa, 2 parts of sugar, and 1 part of butter. How many ounces of candy are obtained if 10 ounces of milk are used?

(A) 20 (B) 22 (C) 24 (D) 28 (E) 30 **[1.3]**

11. What is the scale in ratio form that is needed to represent an actual length of 8.3 feet by a line that is 8.3 inches?

(A) 1:1 (B) 1:12 (C) 1:10 (D) 10:1 (E) 12:1 **[1.4]**

12. In the figure at the right, the area of the trapezoid is 176. If each x-unit equals each y-unit, find the value of each x-unit and each y-unit.

(A) 3 (B) 1 (C) 2 (D) 4 (E) 6 **[1.5]**

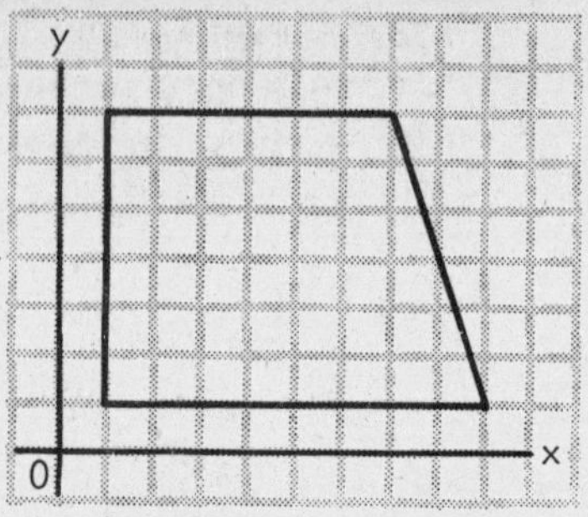

13. If $6:r = 8:12$, find r.

(A) 7.2 (B) 7.5 (C) 8.4 (D) 9 (E) 10 **[2.1]**

14. Find x: $\frac{x+5}{4} = \frac{2x}{3}$

(A) $2\frac{1}{2}$ (B) 3 (C) 4 (D) 5 (E) 8 **[2.2]**

15. Find x in the proportion $\frac{a+2}{x} = \frac{c+1}{2d}$ if $a = 8$, $c = 2$, and $d = 6$.

(A) 14 (B) 20 (C) $26\frac{2}{3}$ (D) 30 (E) 40 **[2.3]**

16. Using the scale "$\frac{3}{4}$ inch = 24 miles," how many inches in length is a line representing a distance of 192 miles?

(A) 4 (B) 6 (C) 8 (D) 10 (E) $10\frac{2}{3}$ **[2.4]**

17. How many books can be bought for q dollars if n dollars is the cost of p books?

(A) $\frac{pn}{q}$ (B) $\frac{nq}{p}$ (C) $\frac{n}{pq}$ (D) npq (E) $\frac{pq}{n}$ **[2.5]**

18. If 3 apples sell for n¢, how many cents will 20 apples cost?

(A) $\frac{3n}{20}$ (B) $\frac{20n}{3}$ (C) $60n$ (D) $\frac{3}{20n}$ (E) $\frac{20}{3n}$ **[2.5]**

19. A merchant paid \$48 for $3\frac{1}{2}$ dozen articles. He now wishes to buy $12\frac{1}{4}$ dozen of the same articles. How many dollars must he pay if the price per article is the same?

(A) 84 (B) 96 (C) 160 (D) 168 (E) 192 **[2.5]**

20. Henry found that it took him 9 minutes to walk $\frac{3}{8}$ of a mile. At this rate, how many minutes will it take him to walk the rest of the mile?

(A) 12 (B) 15 (C) 18 (D) 21 (E) 24 **[2.6]**

21. If the speed does not change, time varies directly as distance. If a man moving at a uniform speed travels a distance of 14 miles in 4 hours, how many miles farther will he travel in 6 hours?

(A) $3\frac{1}{2}$ (B) 4 (C) 6 (D) 7 (E) 21 **[3.1]**

22. For a fixed distance, rate (or speed) varies inversely as time. If Charles covers a certain distance in 10 hours at the rate of 24 mph, how many hours will he take to travel the same distance at 30 mph?

(A) 6 (B) 8 (C) 10 (D) 12 (E) $12\frac{1}{2}$ **[3.2]**

23. A man and his son sit on opposite ends of a seesaw. The man weighs 150 pounds and sits at a distance of 4 feet from the fulcrum in order to balance the weight of his son, whose distance from the fulcrum is 10 feet. How many pounds does the son weigh?
(A) 40 (B) 48 (C) 54 (D) 60 (E) 80 **[3.3]**

24. A work crew of 4 men takes 5 days to complete one-half of a job. If 6 men are then added to the crew and the men continue to work at the same rate, how many days will it take the enlarged crew to do the rest of the job?
(A) 2 (B) 3 (C) $3\frac{1}{3}$ (D) 4 (E) $4\frac{4}{5}$ **[3.4]**

25. The contents of a full cylindrical container of juice weighs 4 ounces. If the dimensions of the container are tripled, how many ounces will the contents weigh?
(A) 12 (B) 36 (C) 104 (D) 108 (E) 144 **[3.5]**

VI. REVIEWING PLANE GEOMETRY: LINES AND ANGLES

Reviewing the Basic Terms of Plane Geometry: Point, Line, Surface

A ***point*** has position only. It has no length, no width, and no thickness. A point is represented by a dot and designated by a capital letter next to the dot. For example, points A and B are shown at the right.

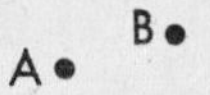

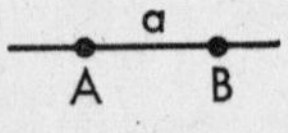

A ***line*** has length but no width and no thickness. A line may be straight, curved, or both straight and curved.

A ***straight line*** is generated by a point moving in a fixed or unchanging direction and is designated by the capital letters of any two of its points or by a small letter. The straight line shown is designated as either AB or a.

A ***curved line*** is generated by a point moving in a continuously changing direction. The curved line shown may be designated as arc AB or $\widehat{AB}$.

A ***broken line*** is a combination of straight lines.

A straight line may be extended in either direction since its length is unlimited. However, a ***straight line segment*** is the part of a straight line contained between two of its points. The straight line segment is designated by its two endpoints or by a small letter. The distance between two points is the straight line segment between the points. Two nonparallel straight lines intersect in a point.

A ***surface*** has length and width but no thickness. Some common examples of a surface are the side of a box, the outside of a sphere, or a blackboard.

A ***plane*** is a surface that entirely contains any straight line joining two of its points. *Plane Geometry* deals only with those figures that can be drawn on a plane.

Reviewing Circles

Parts of Circles

The symbol for a circle is $\odot$. The circle whose center is point O is symbolized as $\odot O$.

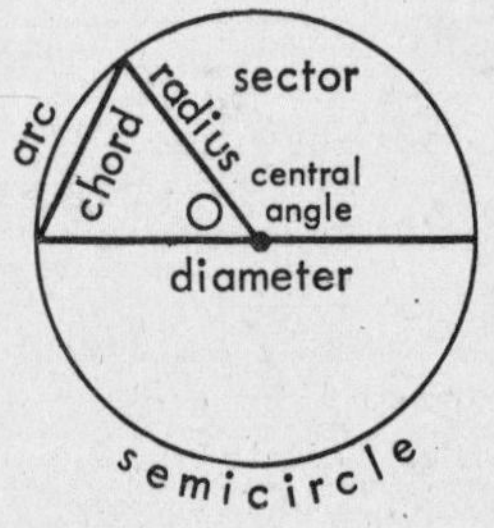

The ***circumference*** of a circle is a closed curved line such that each of its points is the same distance from the center. An ***arc*** is part of the circumference of the circle. The symbol for arc is $\frown$. The arc whose endpoints are A and B is symbolized as $\widehat{AB}$. A ***semicircle*** is an arc that is one-half the circumference.

A ***radius*** is the distance of a point on the circumference from the center. A ***chord*** is any line joining two points of the circumference. A ***diameter*** is a chord that passes through the center of the circle.

A ***central angle*** is an angle formed by two radii. A ***sector*** is a part of the circle bounded by two radii and an intercepted arc.

Properties of Circles

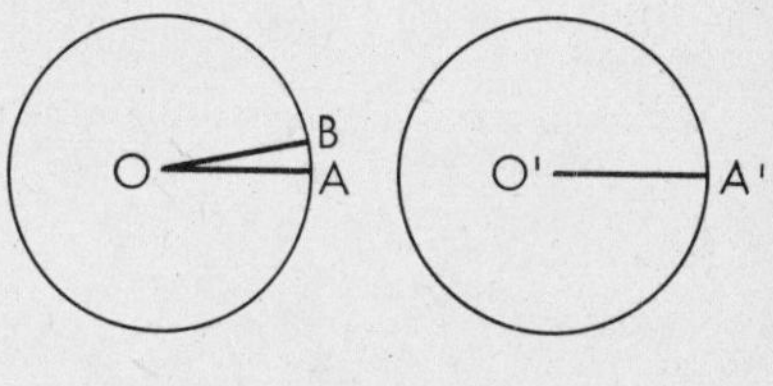

The circumference of a circle contains 360°. Hence, an arc of 1° is $\frac{1}{360}$ of the circumference. A central angle of 1° intercepts an arc of 1°. For example, if central angle *BOA* is an angle of 1°, then arc *AB* is an arc of 1°.

Equal circles have equal radii. Conversely, if circles have equal radii, then the circles are equal. For example, if $OA = O'A'$, then circle O = circle O'.

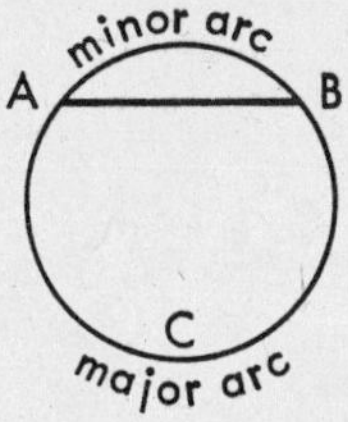

A diameter divides a circumference into two semicircles. Any chord other than a diameter divides the circumference into two arcs, a ***minor arc*** and a ***major arc***. In the diagram, note the minor and major arcs of chord *AB*. To symbolize the major arc of chord *AB*, use a third letter, for example, $\widehat{ACB}$. The major arc is greater than a semicircle; hence, it contains more than 180°. The minor arc is less than a semicircle; hence, it contains less than 180°.

Radii of a circle are equal. Diameters of a circle are equal.

Reviewing Angles

An ***angle*** is the figure formed by two straight lines meeting in a point. The ***sides*** of the angle are the two straight lines; the ***vertex*** of the angle is the point.

The symbol for angle is ∠; that for angles is ∡.

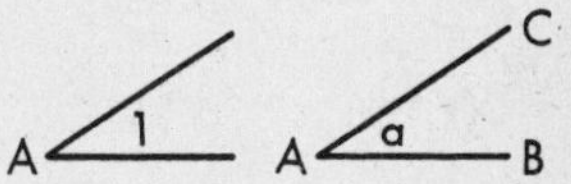

An angle may be designated in three ways: (1) by using a capital letter at the vertex, (2) by using a small letter or number placed between the sides and near the vertex, or (3) by using three capital letters of which the middle one is the letter at the vertex. For example, the angle shown may be written as $\angle A$, $\angle 1$, $\angle a$, $\angle BAC$, or $\angle CAB$.

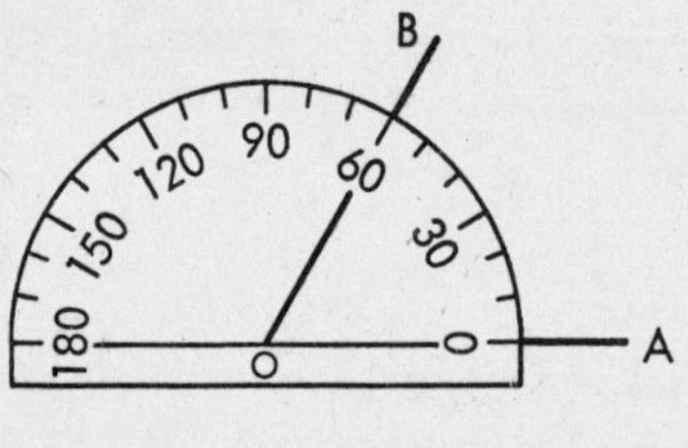

The size of an angle depends on the amount of rotation about the vertex needed to turn one side so that it may be made to coincide with the other side. For example, in the protractor at the left, $\angle AOB = 60°$, since side *OA* must be turned about *O* through 60° to make it coincide with side *OB*.

The *measure of an angle* is the number of degrees in the angle. For example, if $\angle AOB = 60°$, the number 60 is the measure of $\angle AOB$. Because the measure of an angle is a number, it is unnecessary to add the word "degrees" or to use the symbol for degrees, °, when giving the measure of an angle.

The size of an angle does not depend on the length of its sides.

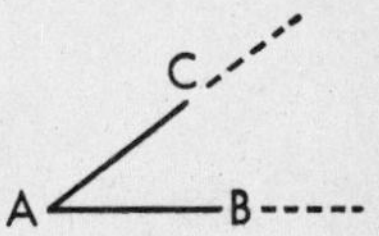

Note that extending the lengths of the sides of $\angle CAB$ does not change the size of the angle.

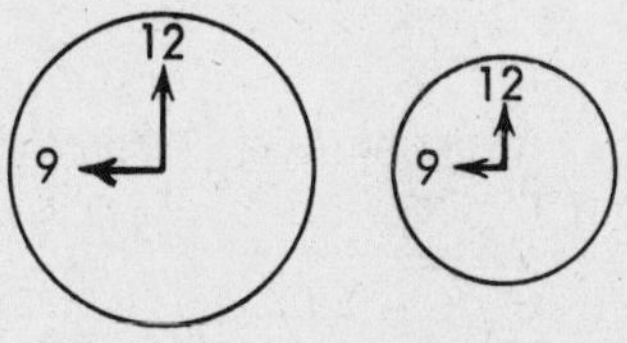

Watch A *Watch B*

Note also that although the hands of watch *A* are longer than those of watch *B*, the angle formed by the hands of watch *A* at 9 o'clock is equal to the angle formed by the hands of watch *B* at the same time.

The compass on the right indicates the eight major directions at 45° intervals. When an airplane turns clockwise through successive 45° intervals beginning at N (North), the airplane will face NE (Northeast) after a turn of 45°, E (East) after a turn of 90°, SE (Southeast) after a turn of 135°, and so on. A pilot or navigator of a plane must be thoroughly familiar with the compass intervals.

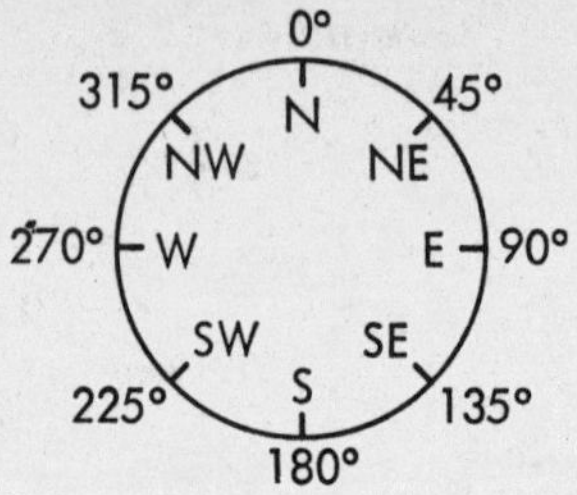

Kinds of Angles

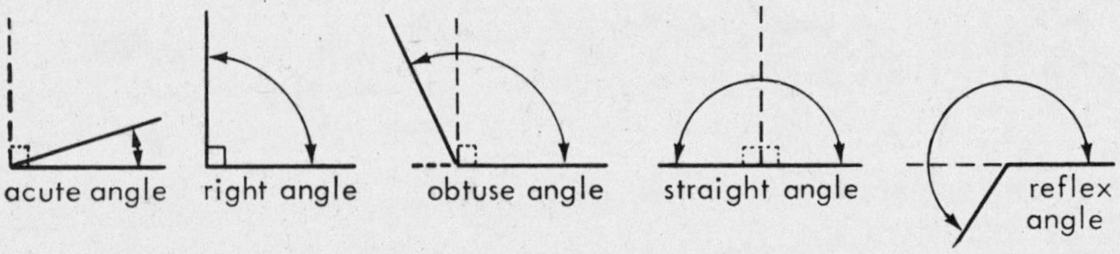

A ***right angle*** equals 90°.

A ***straight angle*** equals 180°.

An ***acute angle*** is an angle that is less than 90°; hence, an acute angle is smaller than a right angle.

An ***obtuse angle*** is more than 90° and less than 180°; hence, it is larger than a right angle and smaller than a straight angle.

A ***reflex angle*** is more than 180° and less than 360°.

Pairs of Angles

Adjacent angles. Line AD in Fig. 1 cuts the angle of $c°$ into two adjacent angles of $a°$ and $b°$. Hence, $a + b = c$. Angles are adjacent if they have a common vertex and a common side between them.

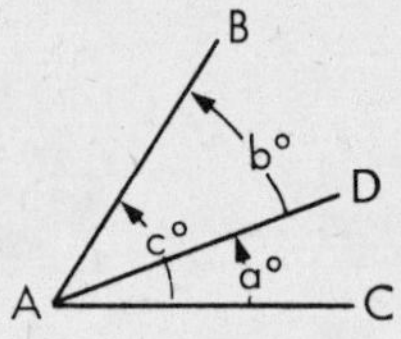

Fig. 1

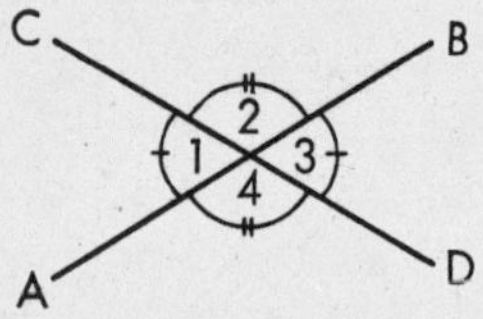

Fig. 2

Vertical angles. Lines AB and CD in Fig. 2 form two pairs of vertical angles: $\angle 1$ and $\angle 3$, also $\angle 2$ and $\angle 4$. Vertical angles are two nonadjacent angles formed by two intersecting straight lines. Vertical angles are equal angles. Hence, $\angle 1 = \angle 3$ and $\angle 2 = \angle 4$.

Complementary angles. If the sum of two angles is 90°, the angles are complementary. Complementary angles may be adjacent, as in Fig. 3, or nonadjacent, as in Fig. 4. In either case, $a + b = 90$.

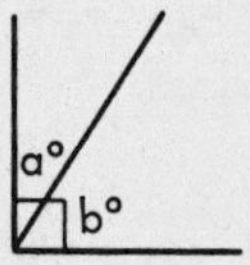

Fig. 3

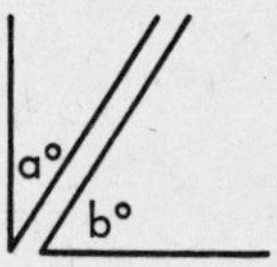

Fig. 4

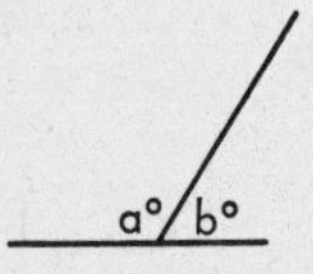

Fig. 5

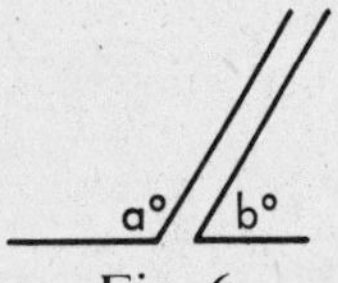

Fig. 6

Supplementary angles. If the sum of two angles is 180°, the angles are supplementary. Supplementary angles may be adjacent, as in Fig. 5, or nonadjacent, as in Fig. 6. In either case, $a + b = 180$.

Reviewing Polygons

A ***polygon*** is a closed plane figure bounded by straight lines. The following table indicates the most important kinds of polygons classified according to the number of sides:

No. of Sides	*Polygon*	*No. of Sides*	*Polygon*
3	triangle	8	octagon
4	quadrilateral	10	decagon
5	pentagon	12	dodecagon
6	hexagon	*n*	*n*-gon

If a polygon has equal sides, it is *equilateral*; if it has equal angles, it is *equiangular*. A polygon that is both equilateral and equiangular is a ***regular polygon***. In the diagram, if $ABCDE$ is a regular pentagon, then $AB = BC = CD = DE = EA$ and $\angle A = \angle B = \angle C = \angle D = \angle E$.

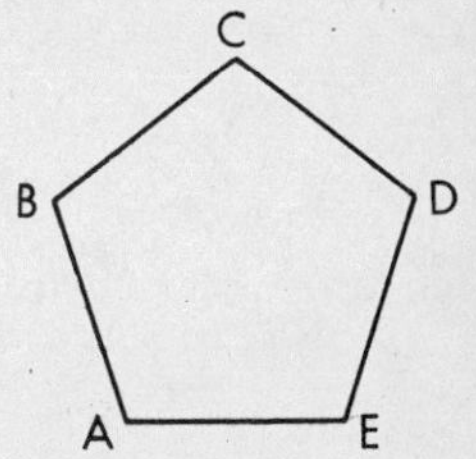

Regular Pentagon

Reviewing Triangles

A ***triangle*** is a polygon of three sides. The symbol for a triangle is $\triangle$; that for triangles is ⚠. A triangle may be designated by the capital letters at its vertices or by a Roman numeral placed inside it. Thus, the triangle on the right may be called $\triangle ABC$ or $\triangle$I.

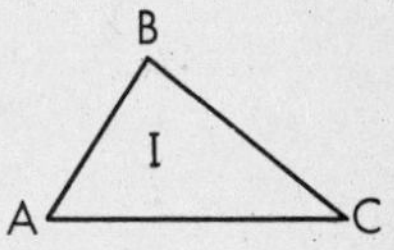

CLASSIFYING TRIANGLES ACCORDING TO THE EQUALITY OF THEIR SIDES

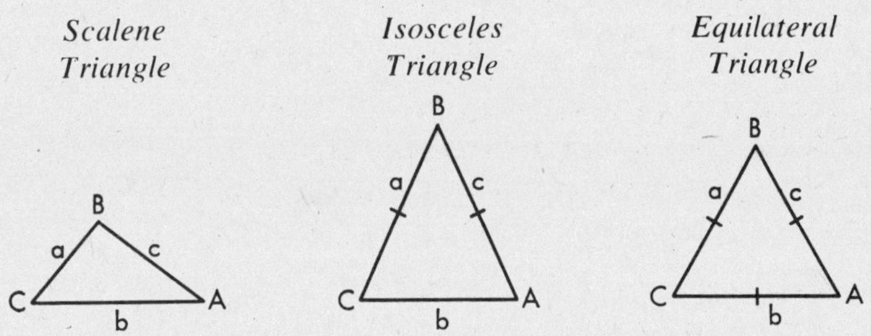

Fig. 1 Fig. 2 Fig. 3

A ***scalene triangle*** is a triangle having no equal sides. In Fig. 1, $a \neq b \neq c$.

An ***isosceles triangle*** is a triangle having at least two equal sides. In Fig. 2, $a = c$. The equal sides of an isosceles triangle are its ***arms***, or ***legs***; the remaining side is its ***base***. The angle opposite the base is the ***vertex angle***. If $a = c$, then b is the base and $\angle B$ is the vertex angle of isosceles triangle ABC.

An ***equilateral triangle*** is a triangle having all equal sides. In Fig. 3, $a = b = c$.

Angle Properties of Scalene, Isosceles, and Equilateral Triangles

1. A scalene triangle has no equal angles.
2. The base angles of an isosceles triangle are equal. In an isosceles triangle, the equal angles are opposite the equal sides.
3. An equilateral triangle is equiangular. Hence, each angle of an equilateral triangle is equal to 60°.

Classifying Triangles According to Their Kinds of Angles

Right Triangle

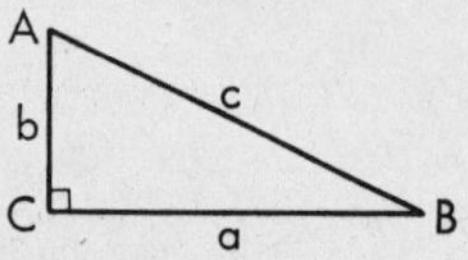

Fig. 4

Obtuse Triangle

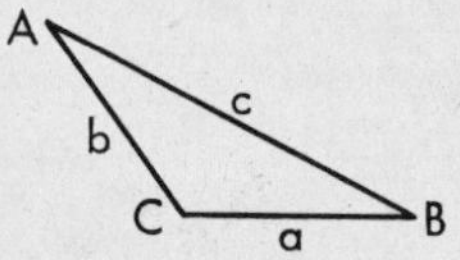

Fig. 5

Acute Triangle

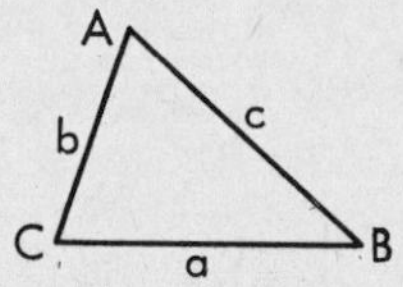

Fig. 6

A ***right triangle*** is a triangle having a right angle. In Fig. 4, $\angle C$ is a right angle. The sides of the right angle are the ***legs***, or ***arms***, of the right triangle; the side opposite the right angle is the ***hypotenuse***. Hence, a and b are legs, or arms, of $\triangle ABC$ and c is the hypotenuse.

An ***obtuse triangle*** is a triangle having an obtuse angle. In Fig. 5, $\angle C$ of obtuse $\triangle ABC$ is an obtuse angle.

An ***acute triangle*** is a triangle having all acute angles. The triangle in Fig. 6 is an acute triangle, since $\angle A$, $\angle B$, and $\angle C$ are each acute angles.

The Law of Pythagoras Applied to the Sides of Right Triangles, Obtuse Triangles, and Acute Triangles

1. In a right triangle, the square of the hypotenuse equals the sum of the squares of the legs. In Fig. 4, $c^2 = a^2 + b^2$.
2. In an obtuse triangle, the square of the side opposite the obtuse angle is greater than the sum of the squares of the other two sides. In Fig. 5, $c^2 > a^2 + b^2$.
3. In an acute triangle, the square of the side opposite the largest angle is smaller than the sum of the squares of the other two sides. In Fig. 6, if $\angle C$ is the largest angle, then $c^2 < a^2 + b^2$.

Special Lines in a Triangle

The special lines in a triangle are angle bisectors, medians, perpendicular bisectors, altitudes, and lines that join the midpoints of the sides.

Angle Bisectors

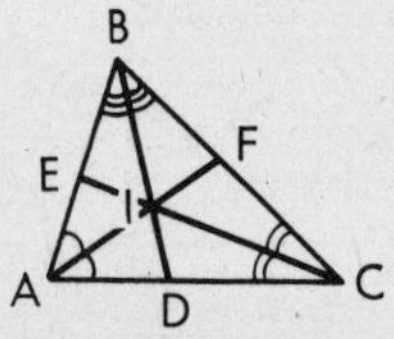

Fig. 7

Medians

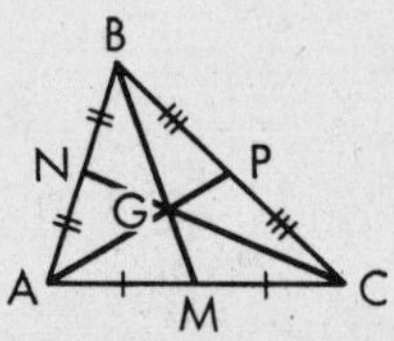

Fig. 8

Perpendicular Bisectors

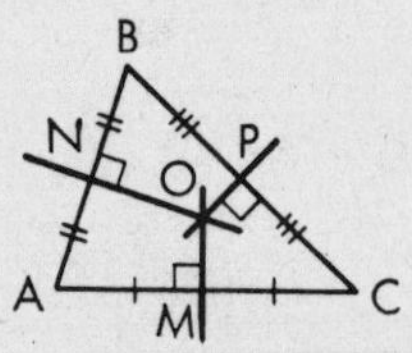

Fig. 9

Fig. 7. The ***angle bisectors*** of a triangle bisect the angles of the triangle. Their point of concurrency, I, is the *incenter* of the triangle. Point I is the center of the inscribed circle of the triangle.

Fig. 8. The ***medians*** of a triangle bisect the sides of the triangle. Each median is drawn from the vertex opposite the bisected side. Their point of concurrency is G, the *centroid* of the triangle. The centroid of a triangle divides each median into two segments so that one segment is twice the other. (The centroid of a triangle is also its center of gravity.)

Fig. 9. The ***perpendicular bisector*** of a side of a triangle is, as the name indicates, perpendicular to the side and bisects the side. The point of concurrency of the three perpendicular bisectors is O, the *circumcenter* of the triangle. Point O is the center of the circumscribed circle of the triangle.

Altitudes

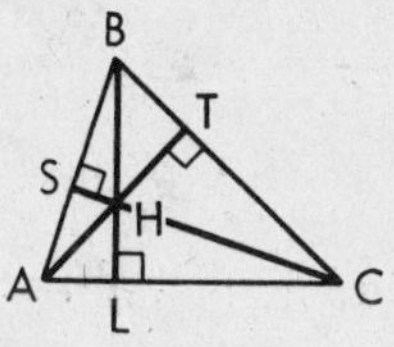

Fig. 10*a*

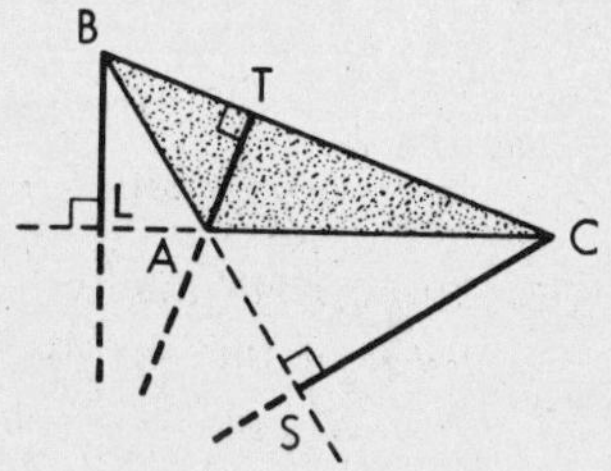

Fig. 10*b*

Line Joining the Midpoints of Two Sides

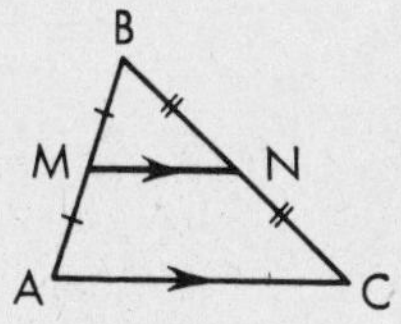

Fig. 11

Fig. 10*a*. The ***altitudes*** of a triangle are perpendicular to the sides of the triangle. Each altitude is drawn from the vertex opposite the side to which it is perpendicular. Their point of concurrency is H, the *orthocenter* of the triangle.

Fig. 10*b*. If the triangle is obtuse, two of its altitudes fall outside the triangle. Hence, the point of concurrency lies outside the obtuse triangle.

Fig. 11. A ***line joining the midpoints of two sides*** of a triangle is parallel to the third side and equal to one-half of the third side. (Arrows are used to represent parallel lines.)

Hence, in $\triangle ABC$, if M and N are the respective midpoints of AB and BC, then $MN \parallel AC$ and $MN = \frac{1}{2}AC$.

Reviewing Quadrilaterals, Trapezoids, and Parallelograms

A ***quadrilateral*** is a four-sided polygon. The trapezoid and the parallelogram are the two most important types of quadrilaterals. In the trapezoid, two and only two sides are parallel, while in the parallelogram, two pairs of opposite sides are parallel.

TRAPEZOIDS

Trapezoid

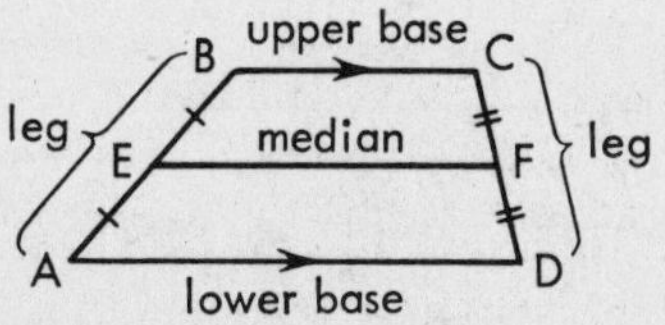

Fig. 1

Isosceles Trapezoid

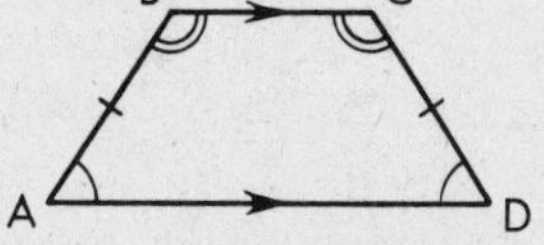

Fig. 2

A ***trapezoid*** is a quadrilateral having two and only two parallel sides. The parallel sides are the ***bases***, the nonparallel sides are the ***legs***, and the line joining the midpoints of the legs is the ***median*** of the trapezoid. In Fig. 1, AD is the lower base, BC is the upper base, AB and CD are the legs, and EF is the median of trapezoid $ABCD$. The median of a trapezoid equals one-half the sum of the bases. Hence, the median is the average of the bases.

An ***isosceles trapezoid*** is a trapezoid whose legs are equal. In an isosceles trapezoid, the base angles are equal. In Fig. 2, the equal parts of isosceles trapezoid $ABCD$ are: $AB = CD$, $\angle A = \angle D$, and $\angle B = \angle C$. Note also that each lower-base angle is the supplement of each upper-base angle.

PARALLELOGRAMS AND SPECIAL PARALLELOGRAMS

Fig. 3. A ***parallelogram*** is a quadrilateral whose opposite sides are parallel.

Fig. 4. A ***rectangle*** is an equiangular parallelogram.

Fig. 5. A ***rhombus*** is an equilateral parallelogram.

Fig. 6. A ***square*** is an equilateral and an equiangular parallelogram.

Parallelogram

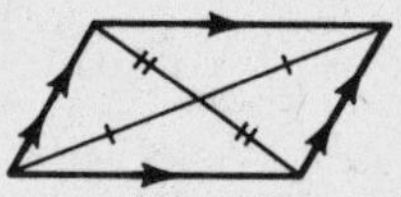

Fig. 3

Rectangle

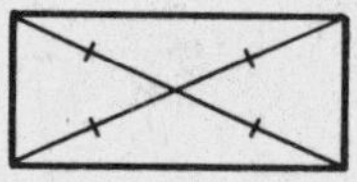

Fig. 4

Rhombus

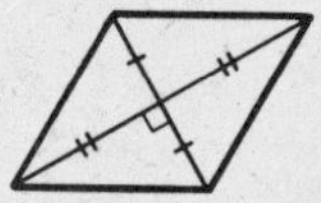

Fig. 5

Square

Fig. 6

Since their opposite sides are parallel to each other, the rectangle, the square, and the rhombus are special types of parallelograms. Hence, they have all the general properties of the parallelogram:

1. The opposite sides are equal.
2. The opposite angles are equal.
3. A diagonal divides the parallelogram into two congruent triangles.
4. Consecutive angles are supplementary.
5. The diagonals bisect each other.

PROPERTIES OF DIAGONALS IN PARALLELOGRAMS, RECTANGLES, RHOMBUSES, AND SQUARES

Property	*Parallelogram*	*Rectangle*	*Rhombus*	*Square*
Diagonals bisect each other.	X	X	X	X
Diagonals are equal.		X		X
Diagonals are perpendicular.			X	X
Diagonals bisect vertex angles.			X	X
Diagonals form congruent triangles.	X (2 pairs)	X (2 pairs)	X (all 4)	X (all 4)

The square is both a rectangle and a rhombus. Hence, the square has the properties of both the rectangle and the rhombus. Note this in the preceding table.

Principles Involving Lines and Angles

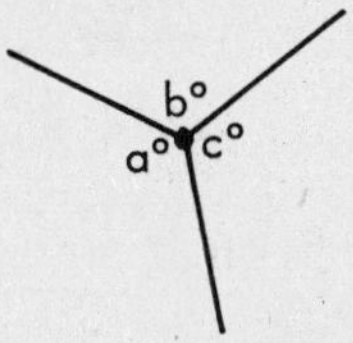

Fig. 1

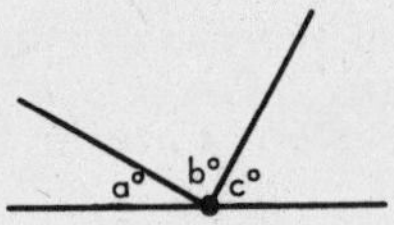

Fig. 2

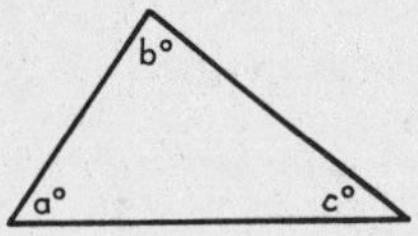

Fig. 3

1. The sum of the angles about a point equals 360°.

   In Fig. 1, $a + b + c = 360$.

2. The sum of the angles about a point on one side of a straight line equals 180°.

   In Fig. 2, $a + b + c = 180$.

3. The sum of the angles of a triangle equals 180°.

   In Fig. 3, $a + b + c = 180$.

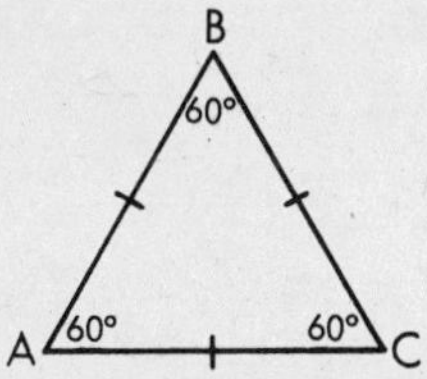

Fig. 4

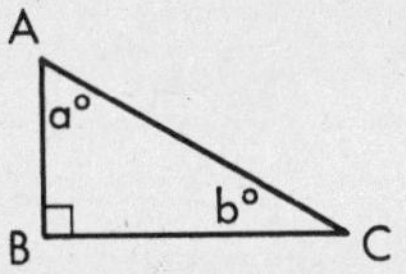

Fig. 5

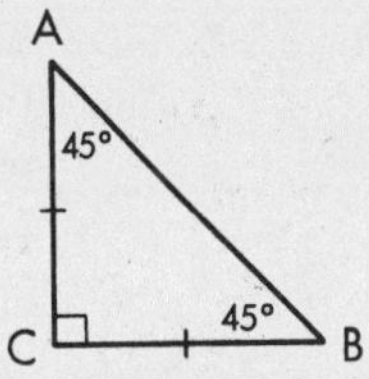

Fig. 6

4. Each angle of an equilateral triangle equals 60°.

   In Fig. 4, $\triangle ABC$ is an equilateral triangle; hence, $\angle A = \angle B = \angle C = 60°$.

5. The acute angles of a right triangle are complementary.

   In Fig. 5, $\angle B$ is a right angle: hence, $a + b = 90$.

6. Each acute angle of an isosceles right triangle equals 45°.

   In Fig. 6, $\triangle ABC$ is an isosceles right triangle; hence, $\angle A = \angle B = 45°$.

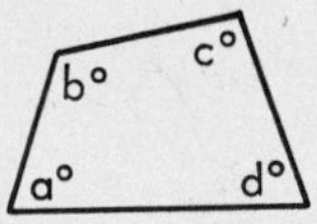

Fig. 7

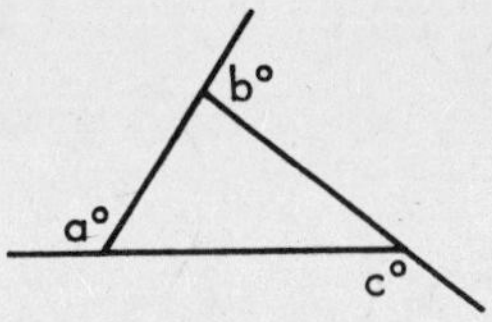

Fig. 8

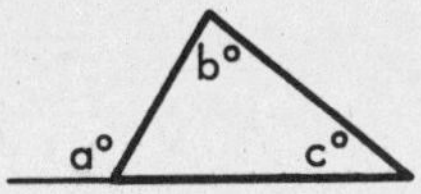

Fig. 9

7. The sum of the angles of a quadrilateral equals 360°.

   In Fig. 7, $a + b + c + d = 360$.

8. The sum of the exterior angles of a triangle equals 360°.

   In Fig. 8, $a + b + c = 360$.

9. An exterior angle of a triangle is equal to the sum of the two nonadjacent interior angles.

   In Fig. 9, $a = b + c$.

10*a*. The sum of the exterior angles of any polygon equals 360°.

In Fig. 10, $a + b + c + d + e + f = 360$.

10*b*. The sum of the interior angles of a polygon of n sides equals $(n - 2)$ straight angles, or $(n - 2)180°$.

For the hexagon in Fig. 10,

$$u + v + w + x + y + z = (6 - 2)180 = 4(180) = 720$$

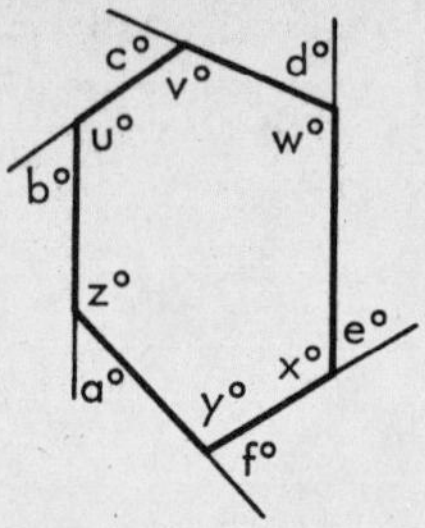

Fig. 10

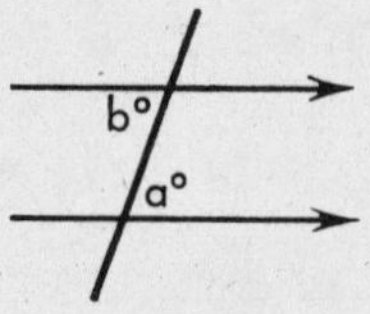

Fig. 11

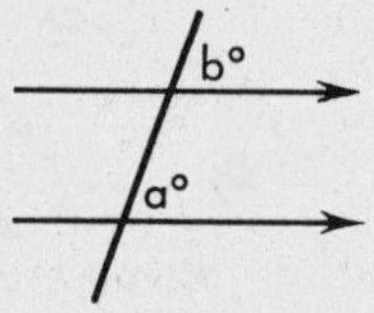

Fig. 12

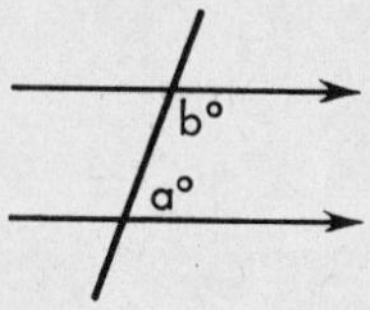

Fig. 13

11. If two parallel lines are cut by a transversal, the alternate interior angles are equal.

    In Fig. 11, $a = b$. (Parallel lines are indicated by arrowheads.)

12. If two parallel lines are cut by a transversal, the corresponding angles are equal.

    In Fig. 12, $a = b$.

13. If two parallel lines are cut by a transversal, the interior angles on the same side of the transversal are supplementary.

    In Fig. 13, $a + b = 180$.

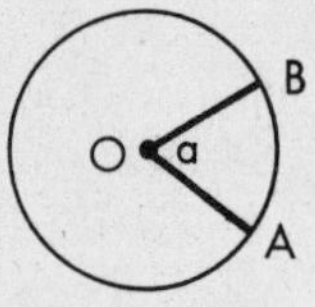

Fig. 14

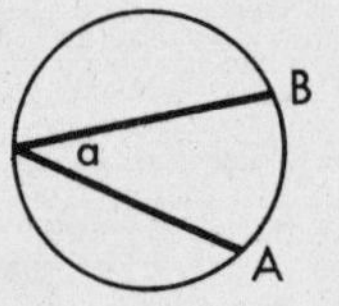

Fig. 15

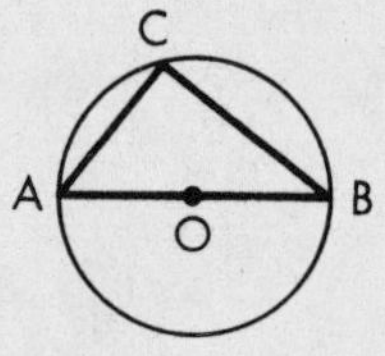

Fig. 16

14. A central angle is measured by its intercepted arc.

    In Fig. 14, $\angle a \stackrel{\circ}{=} \widehat{AB}$. (The symbol $\stackrel{\circ}{=}$ means "is measured by.")

15. An inscribed angle is measured by one-half its intercepted arc.

    In Fig. 15, $\angle a \stackrel{\circ}{=} \frac{1}{2}\widehat{AB}$.

16. An angle inscribed in a semicircle is a right angle.

    In Fig. 16, AB is a diameter; hence, $\angle C$ is a right angle.

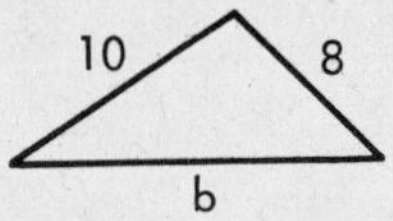

Fig. 17

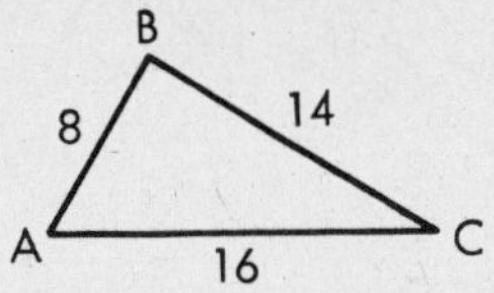

Fig. 18

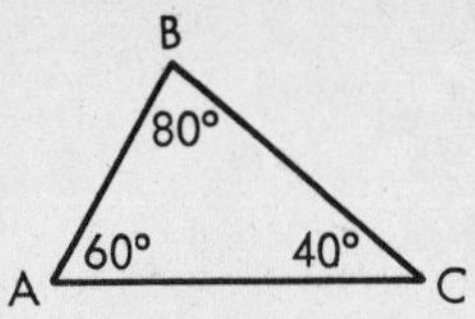

Fig. 19

17. The length of a side of a triangle must be less than the sum of the other two sides and greater than their difference.

In Fig. 17, side b is less than 18 and greater than 2. If b is an integer, then b may have any integral value from 3 to 17.

18. If two sides of a triangle are unequal, the angles opposite these sides are unequal, and the greater of the two angles lies opposite the greater side. Hence, the greatest angle of a triangle lies opposite its greatest side.

In Fig. 18, the greatest angle of $\triangle ABC$ is $\angle B$, while the smallest is $\angle C$.

19. If two angles of a triangle are unequal, the sides opposite these angles are unequal, and the greater of the two sides lies opposite the greater angle. Hence, the greatest side of a triangle lies opposite its greatest angle.

In Fig. 19, the greatest side of $\triangle ABC$ is AC, while the smallest is AB.

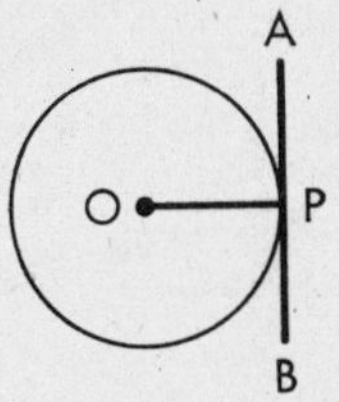

Fig. 20

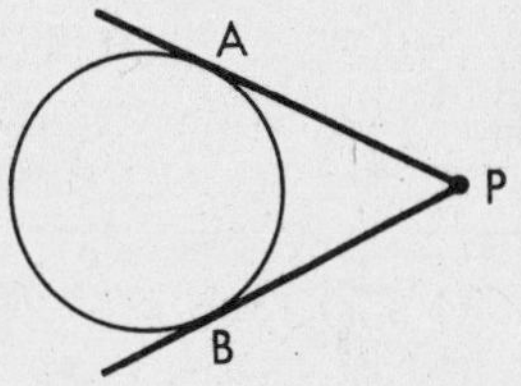

Fig. 21

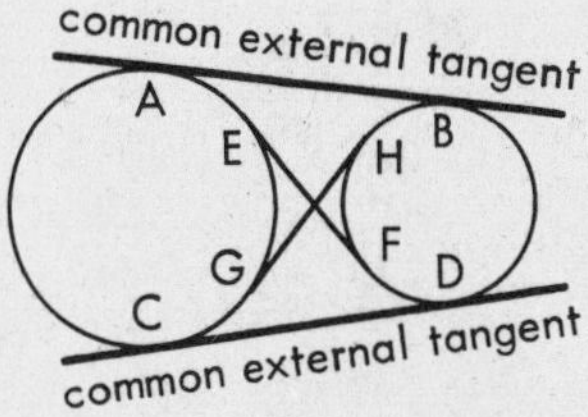

Fig. 22

20. A tangent to a circle is perpendicular to the radius drawn to the point of contact.

(A tangent touches a circle at one and only one point.)

In Fig. 20, AB is a tangent and OP is a radius to the point of contact; hence $AB \perp OP$.

21. Tangents to a circle from an outside point are equal.

In Fig. 21, AP and BP are tangents; hence, $AP = BP$.

22. If two circles are outside each other, the number of common tangents to both circles is 4; two of them are common external tangents and the other two are common internal tangents.

In Fig. 22, AB and CD are the common external tangents, while EF and GH are the common internal tangents.

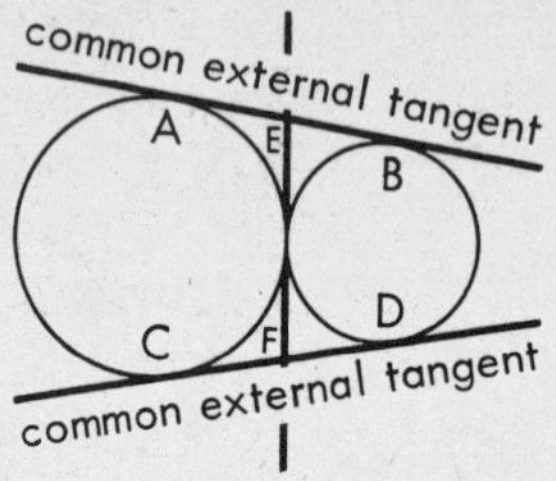

Fig. 23

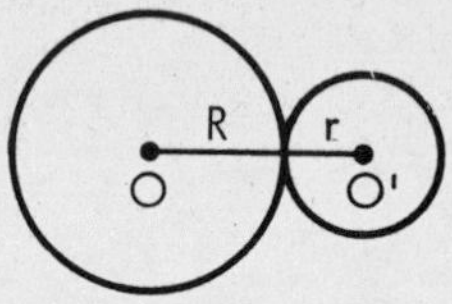

Fig. 24

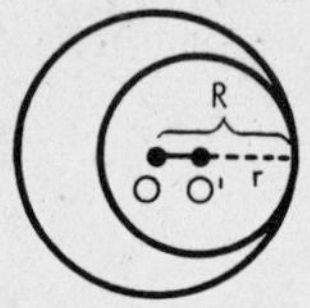

Fig. 25

23. If two circles are tangent externally, the number of common tangents to both circles is 3; two of them are common external tangents and the third is a common internal tangent.

    In Fig. 23, AB and CD are the common external tangents, and EF is the common internal tangent.

24. If two circles are tangent externally, the length of the line of centers is the sum of the radii.

    In Fig. 24, OO' is the line of centers; hence, $OO' = R + r$.

25. If two circles are tangent internally, the length of the line of centers is the difference of the radii.

    In Fig. 25, the line of centers $OO' = R - r$.

1. ANGLES

1.1 PROBLEMS INVOLVING THE SUM OF THE ANGLES ABOUT A POINT: SUPPLEMENTARY AND COMPLEMENTARY ANGLES

Model Problems

(1) In the accompanying figure, $\angle ACB$ is a straight angle and DC is perpendicular to CE. If the number of degrees in angle ACD is represented by x, the number of degrees in angle BCE is represented by

(A) $90 - x$ (B) $x - 90$ (C) $90 + x$ (D) $180 - x$ (E) $x - 180$

Solution:

Since a straight angle equals 180°,

$$y + x + 90 = 180$$

Subtract $x + 90$ from each side.

$$y = 90 - x$$

Ans. (A)

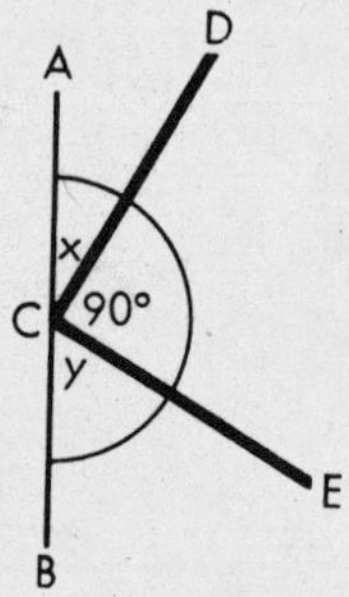

(2) The difference between the supplement and the complement of an acute angle is

(A) an acute angle (B) a right angle (C) an obtuse angle (D) a straight angle (E) a reflex angle

Solution:

The difference between the supplement and the complement of the same angle can be shown to be a right angle as follows:

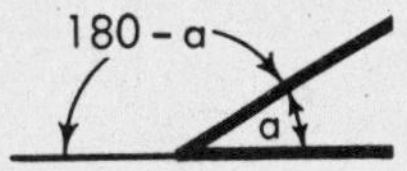

Fig. 1

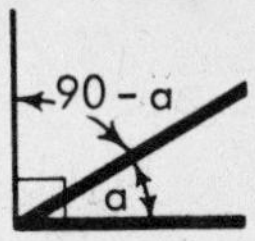

Fig. 2

$$\begin{aligned} \text{Supplement of } \angle a &= 180 - a \text{ (Fig. 1)} \\ \text{Complement of } \angle a &= \underline{\ 90 - a} \text{ (Fig. 2)} \\ \text{By subtraction, difference} &= \ 90 \end{aligned}$$

Ans. (B)

> *Note.* Figure 3 at the right shows that the difference between the supplement and the complement of an acute angle is a right angle.

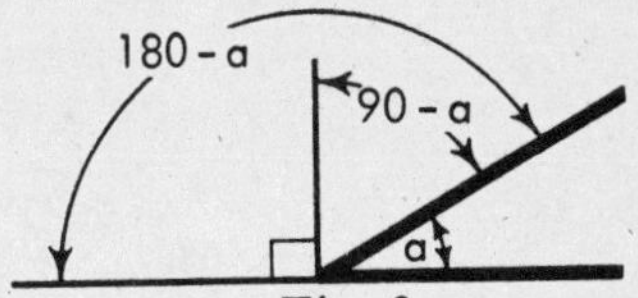

Fig. 3

(3) The difference between two supplementary angles is 40°. Find the number of degrees in the larger of the two angles.

(A) 65 (B) 90 (C) 100 (D) 110 (E) 120

Solution:

Find x by using two equations as follows:

Since the sum of two supplementary angles equals 180°,

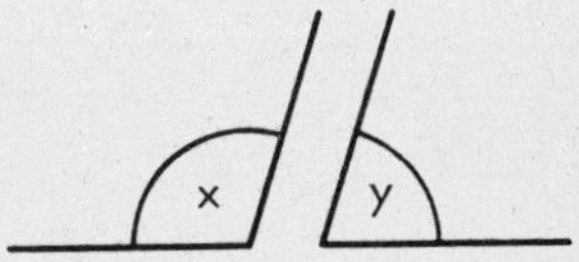

$$\begin{aligned} x + y &= 180 \\ \text{It is given that } \quad x - y &= \ \ 40 \\ \text{By adding: } \quad 2x \quad &= 220 \\ x \quad &= 110 \quad \textit{Ans.} \text{ (D)} \end{aligned}$$

(4) The number of degrees in angle A is equal to $\frac{1}{5}$ the number of degrees in its complement. Find the number of degrees in angle A.

(A) 15 (B) 30 (C) 60 (D) 75 (E) 150

Solution:

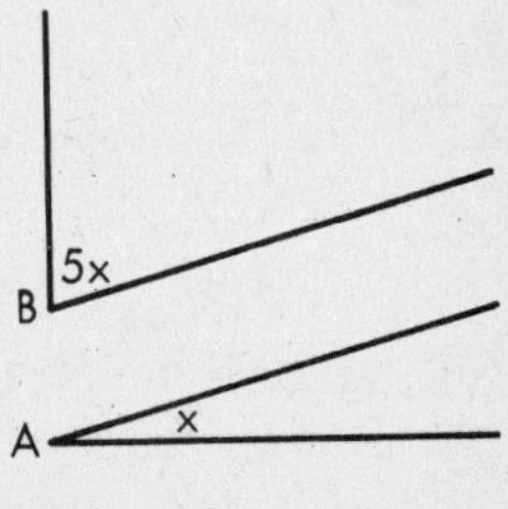

If x = the number of degrees in $\angle A$, then

$5x$ = the number of degrees in $\angle B$.

Find x by applying the following definition:

Complementary angles are two angles whose sum equals 90°.

Hence, $x + 5x = 90$

$6x = 90$

$x = 15$ *Ans.* (A)

After completing each test, match your answers with the numbers in the boxes.

Matching Test A

In each figure, find the number of degrees in $\angle x$.

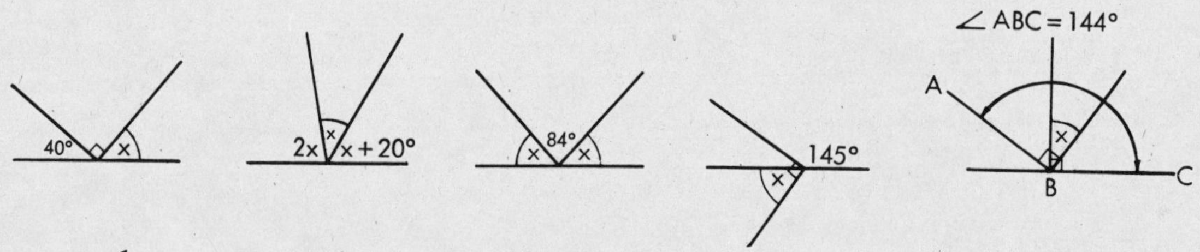

1. **2.** **3.** **4.** **5.**

a. 40	*b.* 36	*c.* 50	*d.* 48	*e.* 55

Matching Test B

Find the number of degrees in the larger of the two angles.

1. The difference between two supplementary angles is 50°.
2. The difference between two complementary angles is 50°.
3. The difference between an angle and twice its complement is 51°.
4. An angle is equal to four times its supplement.
5. An angle is equal to four times its complement.

a. 72
b. 115
c. 144
d. 77
e. 70

1.2 SUM OF THE ANGLES OF A TRIANGLE

Model Problems

(1) The angles of a triangle are in the ratio 1:4:5. Find the number of degrees in the smallest angle of the triangle.

(A) 15 (B) 18 (C) $22\frac{1}{2}$ (D) 30 (E) 36

Solution:

Since the angles are in the ratio 1:4:5, let x, $4x$, and $5x$ represent the number of degrees in the angles of the triangle.

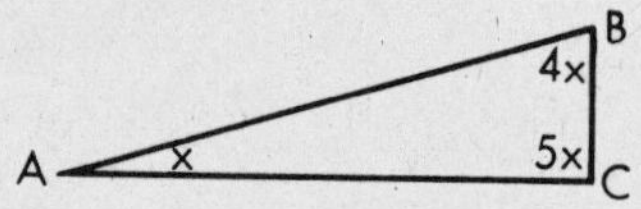

Find x as follows:

The sum of the angles of a triangle equals 180°.

Hence, $x + 4x + 5x = 180$

$10x = 180$

Divide each side by 10.

$x = 18$ *Ans.* (B)

(2) The angles of a triangle are in the ratio 2:3:7. Find the number of degrees in the largest angle.

(A) 30 (B) 54 (C) 90 (D) 105 (E) 120

Solution:

Since the angles of the triangle are in the ratio of 2:3:7, let $2x$, $3x$, and $7x$ represent the number of degrees in the angles of the triangle. Find $7x$ as follows:

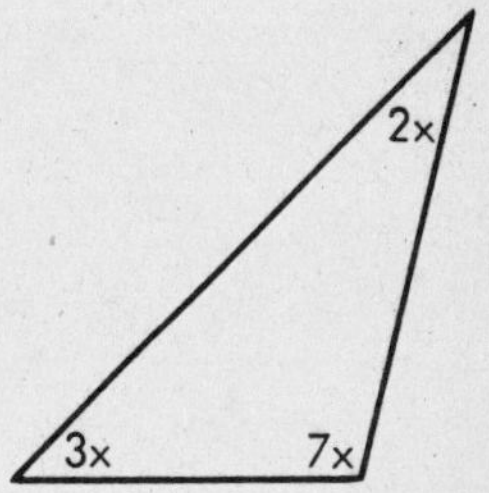

The sum of the angles of a triangle equals 180°.

Hence, $2x + 3x + 7x = 180$
$12x = 180$

Divide each side by 12.

$x = 15$
$7x = 7(15) = 105$ *Ans.* (D)

(3) The number of degrees in one acute angle of a right triangle is four times the number of degrees in the other acute angle. Find the number of degrees in the smaller acute angle.

(A) 18 (B) $22\frac{1}{2}$ (C) 30 (D) $67\frac{1}{2}$ (E) 72

Solution:

Since the larger acute angle is four times the smaller, let x = number of degrees in smaller acute angle A and let $4x$ = number of degrees in larger acute angle B.

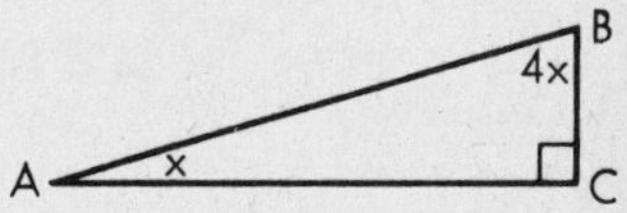

Find x as follows:

The sum of the acute angles of a right triangle equals 90°.

Hence, $x + 4x = 90$
$5x = 90$

Divide each side by 5.

$x = 18$ *Ans.* (A)

(4) The number of degrees in a base angle of an isosceles triangle is twice the number of degrees in the vertex angle. Find the number of degrees in the vertex angle.

(A) 20 (B) 30 (C) 36 (D) 60 (E) 72

Solution:

Since each of the equal base angles is twice the vertex angle, let x = number of degrees in the vertex angle and $2x$ = number of degrees in each base angle.

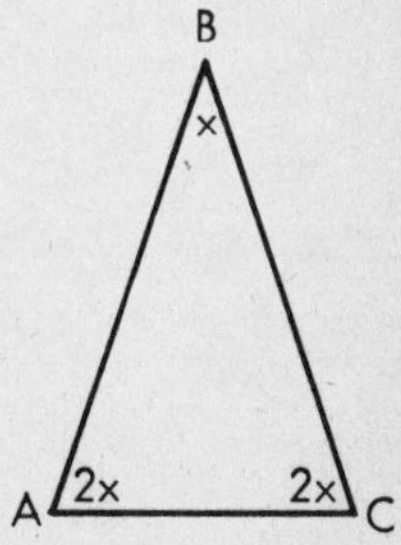

Find x as follows:

The sum of the angles of a triangle equals 180°.

$x + 2x + 2x = 180$
$5x = 180$

Divide each side by 5.

$x = 36$ *Ans.* (C)

(5) The sum of two angles of a triangle is 75° and their difference is 9°. Find the number of degrees in each of the *three* angles of the triangle. Check your answer.

(A) 42, 51, 87 (B) 33, 42, 95 (C) 33, 42, 105 (D) 42, 51, 107 (E) 37, 46, 97

Solution:

Let $x =$ the number of degrees in angle A of the triangle.

Let $y =$ the number of degrees in angle B of the triangle.

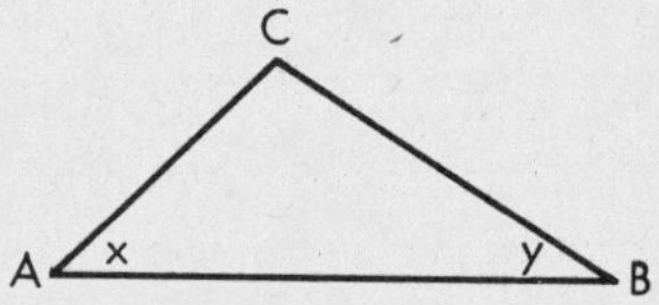

The sum of the angles is 75°. Hence, $x + y = 75$
The difference of the angles is 9°. Hence, $x - y = 9$
Combine both equations by addition: $2x = 84$

Divide each side by 2.

$$x = 42$$

Substitute 42 for x in the equation: $x + y = 75$

$$42 + y = 75$$

Subtract 42 from each side.

$$y = 33$$

Since the sum of the three angles of a triangle equals 180°, the remaining angle at C must equal 180° minus the sum of the other two angles.

Hence, angle $C = 180° - 75° = 105°$.

Hence, the three angles of the triangle are 42°, 33°, and 105°. *Ans.* (C)

Check in the original problem.

The sum of two angles of a triangle is 75°. $42° + 33° \stackrel{?}{=} 75°$

$$75° = 75°$$

The difference of the two angles is 9°. $42° - 33° \stackrel{?}{=} 9°$

$$9° = 9°$$

The sum of the three angles is 180°. $105° + 42° + 33° \stackrel{?}{=} 180°$

$$180° = 180°$$

(6) In triangle ABC, angle B is twice as large as angle A. The exterior angle at C is 111°. Find the number of degrees in angle A.

(A) 23 (B) 35 (C) 37 (D) 46 (E) 74

Solution:

Let $x =$ the number of degrees in $\angle A$.

Since $\angle B$ is twice $\angle A$, then $2x$ is the number of degrees in $\angle B$.

Find x as follows:

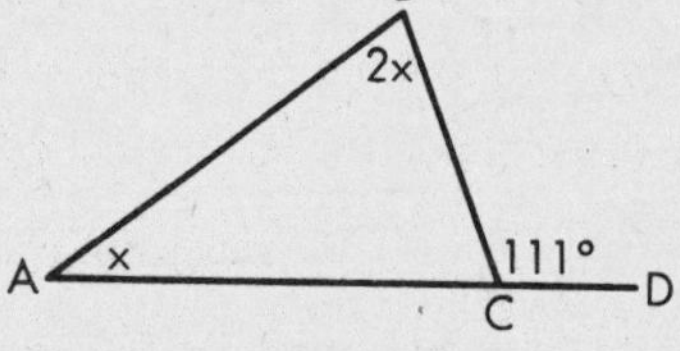

An exterior angle of a triangle equals the sum of the two nonadjacent interior angles.

Hence, $x + 2x = 111$

$$3x = 111$$

Divide each side by 3.

$$x = 37 \quad \textit{Ans.}\ \text{(C)}$$

(7) In triangle ABC, $AC = BC$. The number of degrees in an exterior angle at vertex C is represented by $5x + 10$. If angle $A = 40°$, find the value of x.

(A) 80 (B) 20 (C) 18 (D) 14 (E) 6

Solution:

Since the angles opposite the equal sides of a triangle are equal, $\angle B = \angle A = 40°$.

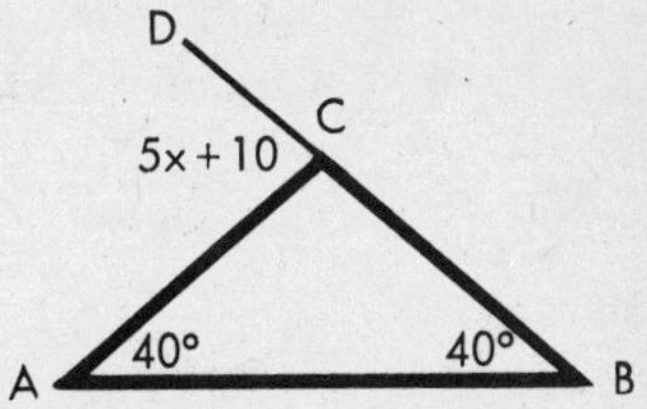

Find x as follows:

An exterior angle of a triangle equals the sum of the two nonadjacent interior angles.

Hence, $5x + 10 = 40 + 40$

Subtract 10 from each side.

$$5x = 70$$

Divide each side by 5.

$$x = 14 \quad \textit{Ans.}\ \text{(D)}$$

(8) The number of degrees in the angles of a triangle is represented by y, $y + 20$, and $2y$, respectively. The triangle is

(A) acute (B) right (C) obtuse (D) equiangular (E) isosceles

Solution:

It is given that y, $y + 20$, and $2y$ are the number of degrees in the angles of the triangle.

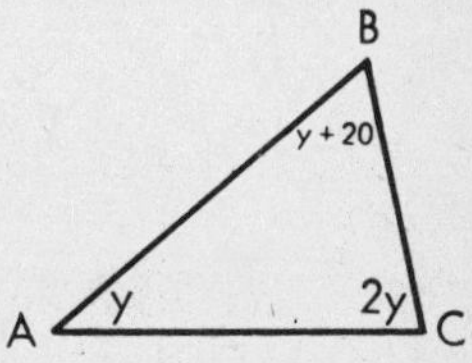

Find the angles as follows:

The sum of the angles of a triangle equals 180°.

Hence, $y + y + 20 + 2y = 180$
$4y + 20 = 180$

Subtract 20 from each side.

$$4y = 160$$
$$y = 40$$
$$y + 20 = 60$$
$$2y = 80$$

Thus, the angles are found to be 40°, 60°, and 80°.

Since each of the angles is an acute angle, the triangle must be an acute triangle. *Ans.* (A)

(9) Side AC of triangle ABC is extended through C to D. Angle $BCD = 108°$ and the number of degrees in angle A is twice the number of degrees in angle B. Triangle ABC is

(A) right (B) obtuse (C) isosceles (D) scalene (E) equilateral

Solution:

Since $\angle A$ is twice $\angle B$, let $2x$ and x represent the number of degrees in $\angle A$ and $\angle B$, respectively.

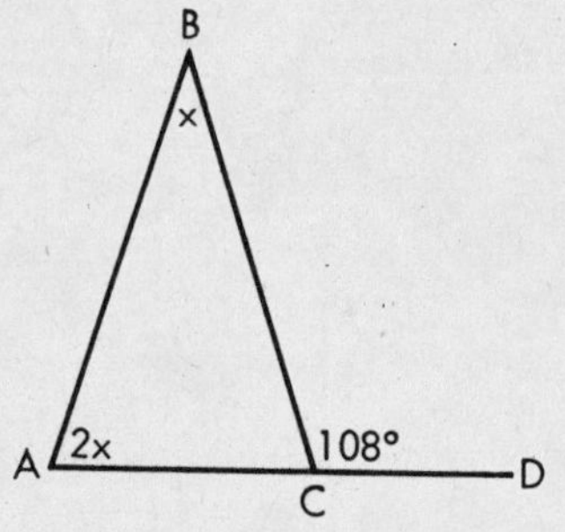

Find x by applying the following principle:

An exterior angle of a triangle equals the sum of the two nonadjacent interior angles.

Hence, $2x + x = 108$
$3x = 108$
$x = 36$ and $2x = 72$.

The remaining angle of $\triangle ABC$ is the supplement of 108°.

Therefore, $\angle BCA = 180° - 108° = 72°$.

Since two of the angles are equal, $\triangle ABC$ is an isosceles triangle. *Ans.* (C)

Wrong Choices:

(A) Since $\triangle ABC$ does not have a right angle, it cannot be a right triangle.

(B) Since $\triangle ABC$ does not have an obtuse angle, it cannot be an obtuse triangle.

(D) To be scalene, $\triangle ABC$ must not have any equal sides. Since two of the angles are equal, the triangle has two equal sides.

(E) Since $\angle A$ is not equal to $\angle B$, $\triangle ABC$ cannot be equilateral.

After completing each test, match your answers with the answers in the boxes.

Matching Test A

Find the number of degrees in each angle of the given triangle.

1. The sum of two angles is 75° and their difference is 15°.
2. The sum of two angles is 75° and one is twice the other.
3. In a right triangle, one of the acute angles is twice the other.
4. In an isosceles triangle, the vertex angle is twice as much as either base angle.
5. In an isosceles triangle, either base angle is twice the vertex angle.

a. 36, 72, 72
b. 45, 45, 90
c. 45, 30, 105
d. 30, 60, 90
e. 50, 25, 105

Matching Test B

In each figure, find the number of degrees in $\angle x$.

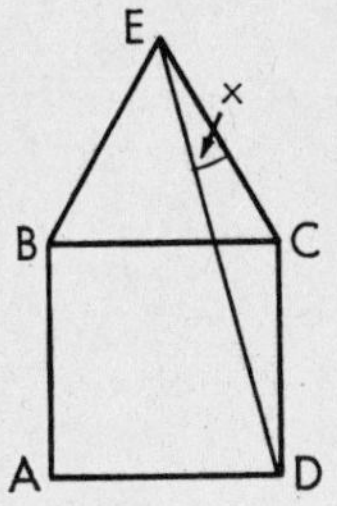

1.

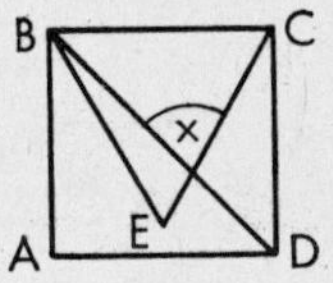

2.

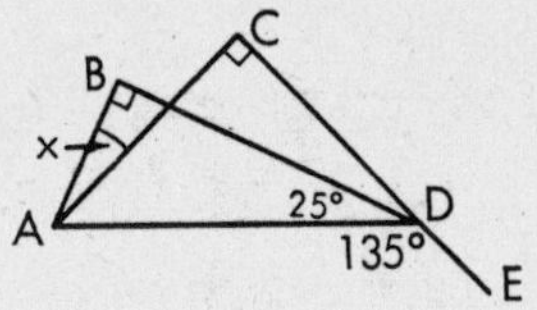

3.

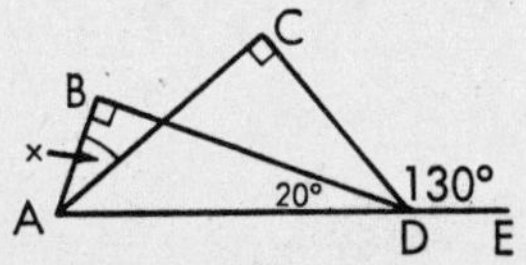

4.

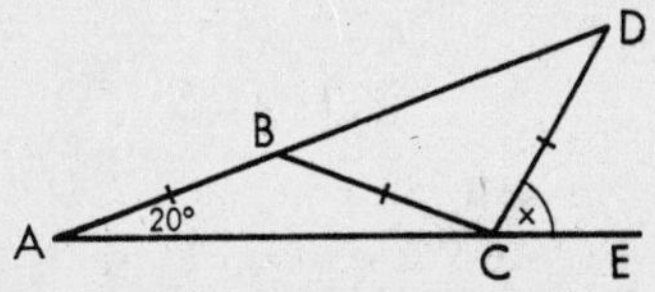

5.

1. Equilateral triangle BEC surmounts square $ABCD$.
2. Equilateral triangle BEC is inside square $ABCD$.
3. Triangles ABD and ACD are right triangles.
4. Triangles ABD and ACD are right triangles.
5. In isosceles triangle ABC, $AB = BC$.
   In isosceles triangle BCD, $BC = CD$.

a. 60
b. 20
c. 30
d. 15
e. 75

Matching Test C

A triangle classified according to its angles may be (1) an obtuse triangle, (2) a right triangle, (3) an equiangular triangle, (4) an acute triangle having no equal angles, or (5) an acute triangle having only two equal angles. In each of the following, determine the largest angle of the triangle; then determine the classification of the triangle according to angles.

1. The angles of the triangle are in the ratio 2:3:4.
2. The angles of the triangle are in the ratio 2:3:5.
3. Two angles of the triangle are in the ratio 2:5 and the third angle is 75°.
4. The angles of the triangle are in the ratio 1:3:5.
5. The angles contain $(y + 40)°$, $(2y + 20)°$, and $(4y - 20)°$.

a. 100°, obtuse △	*b*. 60°, equiangular △	*c*. 90°, right △
d. 80°, acute △, no equal angles	*e*. 75°, acute △, two equal angles	

Matching Test D

In the following figures, $\triangle ABC$ is an isosceles triangle having equal sides AB and BC. In each figure, find x in terms of a.

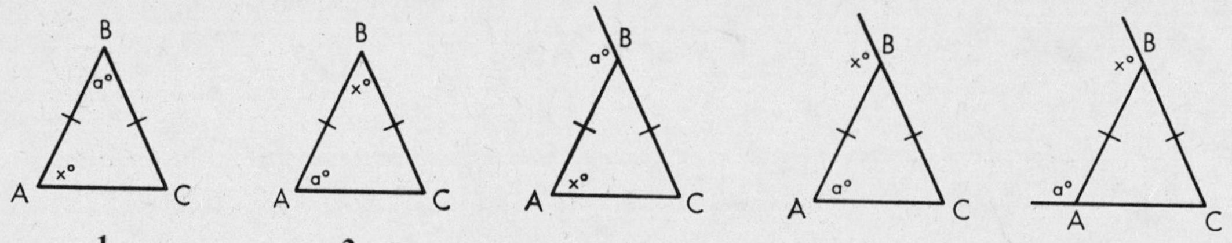

1. **2.** **3.** **4.** **5.**

a. $180 - 2a$	*b.* $2(180 - a)$	*c.* $\dfrac{180 - a}{2}$	*d.* $\dfrac{a}{2}$	*e.* $2a$

Matching Test E

In the figures for **2** to **5**, a beam of light directed at a mirror is reflected so that the original direction and the new direction of the light ray make equal angles with the mirror.

In each figure, express x in terms of a.

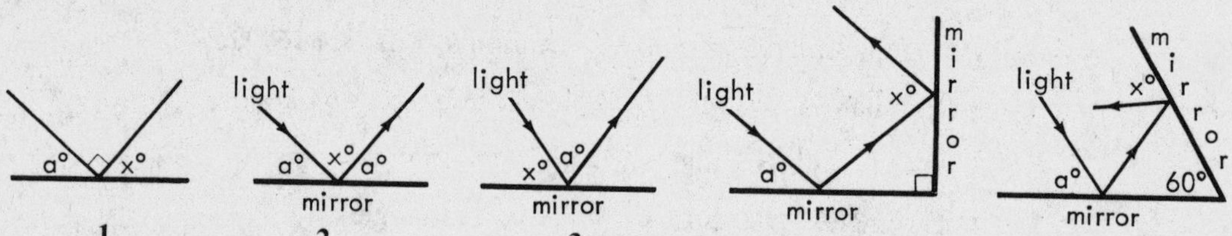

1. **2.** **3.** **4.** **5.**

a. $2a$	*b.* $\dfrac{180 - a}{2}$	*c.* $120 - a$	*d.* $90 - a$	*e.* $180 - 2a$

1.3 ANGLES FORMED BY PARALLEL LINES

Model Problems

(1) Parallel lines AB and CD are intersected by transversal EF. The number of degrees in a pair of alternate interior angles is represented by $2x - 60$ and $x + 40$, respectively. Find the value of x.

(A) 40 (B) 80 (C) 100 (D) 120 (E) 140

Solution:

Find x by applying the following principle:

If two parallel lines are cut by a transversal, the alternate interior angles are equal.

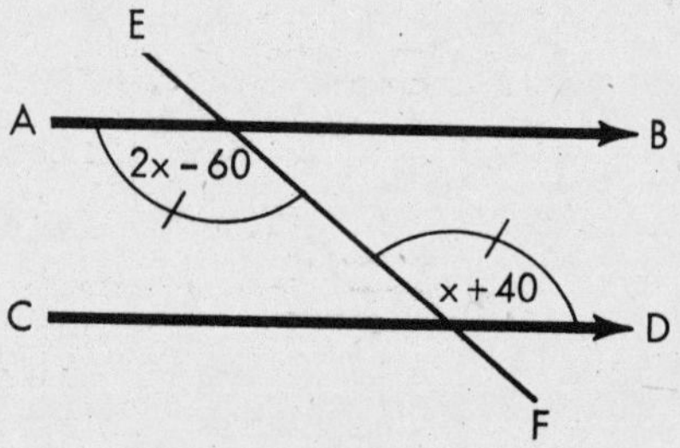

Hence, $2x - 60 = x + 40$

Subtract x from each side.

$$x - 60 = 40$$

Add 60 to each side.

$$x = 100 \quad \textit{Ans. (C)}$$

(2) Two parallel lines are cut by a transversal. If two interior angles on the same side of the transversal contain x degrees and $(3x - 40)$ degrees, find the number of degrees in the smaller angle.

(A) 20 (B) $32\frac{1}{2}$ (C) 40 (D) 55 (E) 125

Solution:

Find x by applying the following principle:

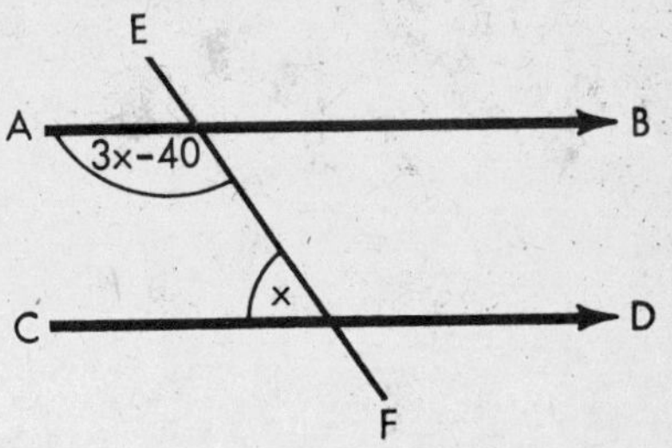

If two parallel lines are cut by a transversal, the interior angles on the same side of the transversal are supplementary.

Hence, $x + 3x - 40 = 180$

$4x - 40 = 180$

Add 40 to each side.

$4x = 220$

Divide each side by 4.

$x = 55$ *Ans.* (D)

(3) In the diagram at the right, if $AB \parallel CD$, what is the relation between x, a, and b? Find x if $a = 44$ and $b = 40$.

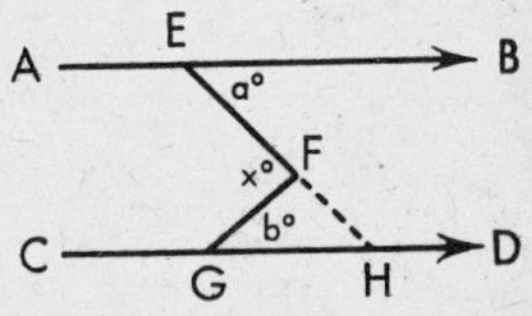

(A) $x = a - b$, $x = 4$ (B) $x = a$, $x = 44$ (C) $x = b$, $x = 40$

(D) $x = a + b$, $x = 84$ (E) $x + a + b = 180$, $x = 96$

Solution:

By extending EF to intersect CD at H, triangle FGH is formed. Since alternate interior angles of parallel lines are equal, $\angle FHG = a$.

Angle x is an exterior angle of $\triangle FGH$. Since the exterior angle of a triangle is the sum of the two nonadjacent angles, $x = a + b$.

Hence, if $a = 44$ and $b = 40$, $x = 44 + 40 = 84$. *Ans.* (D)

After completing each test, match your answers with the numbers in the boxes. (The arrowheads in the diagrams indicate that the lines are parallel.)

Matching Test A

In each figure, find the value of x.

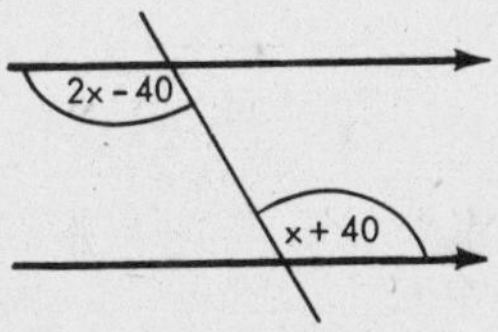

1.

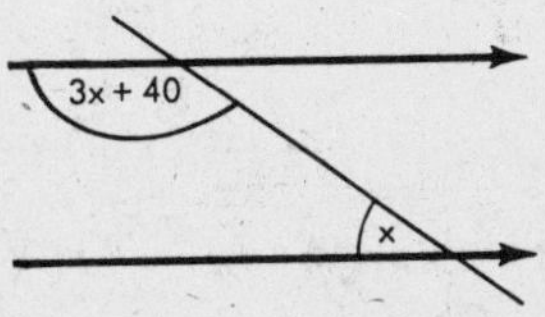

2.

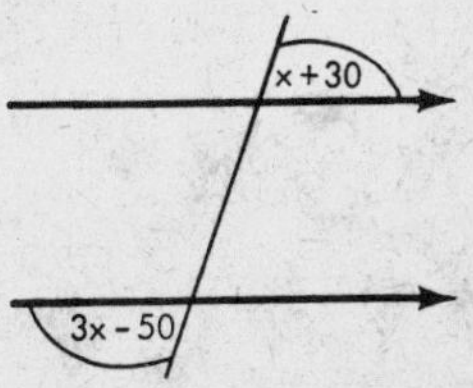

3.

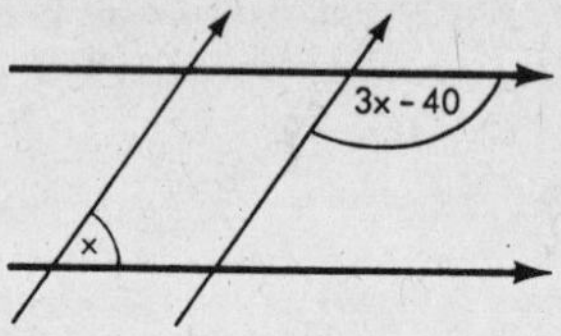

4.

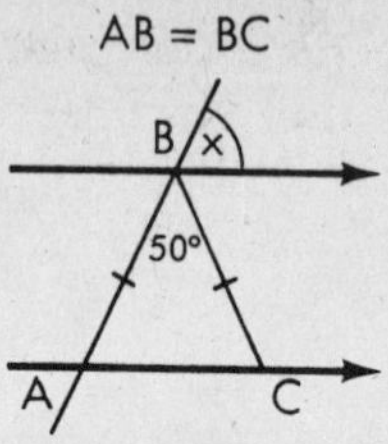

5.

a. 40 *b.* 55 *c.* 35 *d.* 80 *e.* 65

Matching Test B

In each figure, find the value of x.

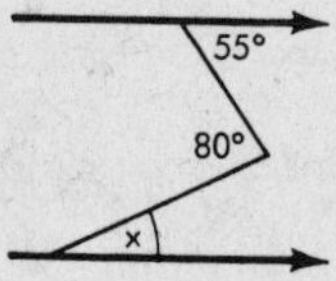

1.

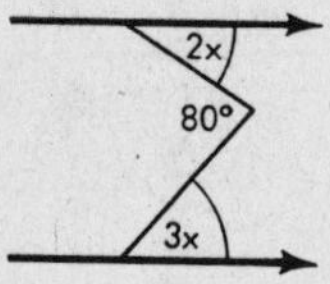

2.

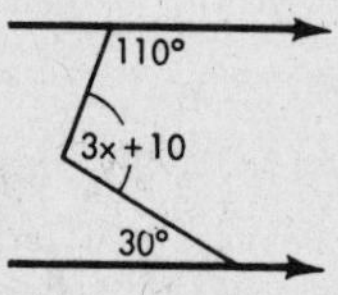

3.

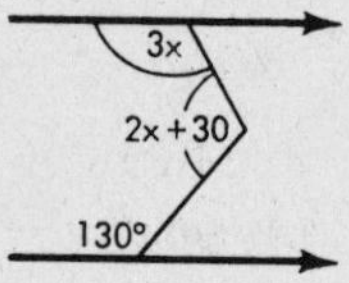

4.

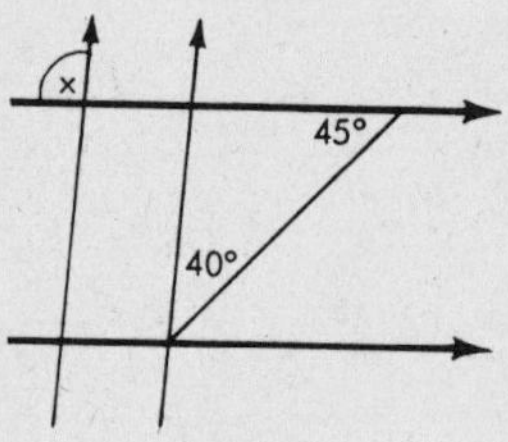

5.

a. 40 *b.* 30 *c.* 25 *d.* 16 *e.* 95

1.4 MEASUREMENT OF ANGLES (CENTRAL AND INSCRIBED ANGLES)

Model Problems

(1) The number of degrees in angle ABC inscribed in circle O is 70. Find the number of degrees in central angle AOC.

(A) 220 (B) 140 (C) 110 (D) 70 (E) 35

Solution:

Let $x =$ the number of degrees in $\angle AOC$.

Since $\angle B = 70°$ and an inscribed angle is measured by one-half its intercepted arc, $\widehat{AC} = 140°$.

Since a central angle is measured by its intercepted arc, $x = 140$. *Ans.* (B)

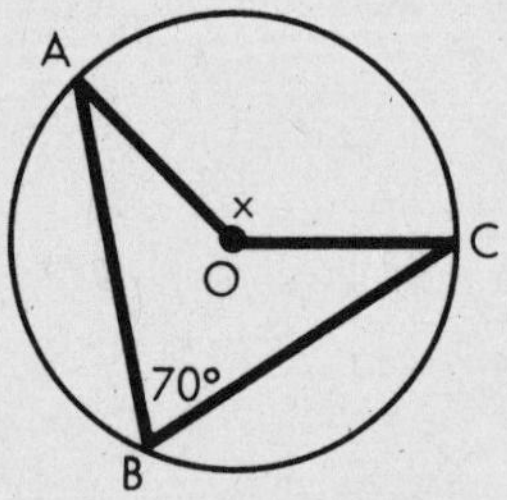

(2) In circle O, central angle AOB is equal to 50°. If point C is on minor arc AB, what is the number of degrees in angle ACB?

(A) 65 (B) $67\frac{1}{2}$ (C) 105 (D) 130 (E) 155

Solution:

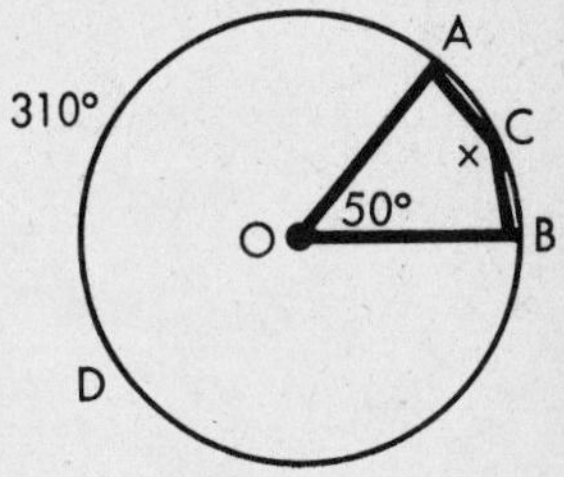

Since $\angle O = 50°$ and a central angle is measured by its intercepted arc, minor arc $AB = 50°$.

Since the circle contains 360° and minor arc $AB = 50°$, major arc $ADB = 360° - 50° = 310°$.

Find x, the number of degrees in $\angle ACB$, by applying the following principle:

An inscribed angle is measured by one-half its intercepted arc.

Hence, $x = \frac{310}{2} = 155$. *Ans.* (E)

(3) The angle formed by two tangents to a circle from the same point is 70°. Find the number of degrees in the smaller intercepted arc.

(A) 70 (B) 110 (C) 140 (D) 220 (E) 290

Solution:

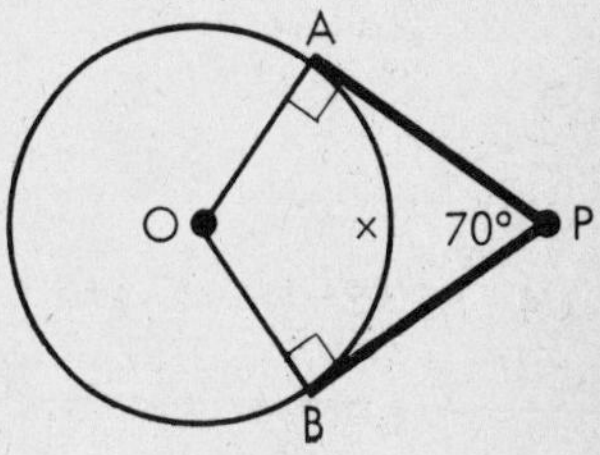

In the diagram at the right, PA and PB are tangents to circle O from point P.

Draw radii OA and OB to tangents PA and PB.

Since a radius drawn to the point of contact of a tangent is perpendicular to the tangent, then $\angle A$ and $\angle B$ are right angles.

The sum of the angles of a quadrilateral is 360°.

Hence, central $\angle O = 360° - (70° + 90° + 90°)$

$= 360° - 250° = 110°$

Find x, the number of degrees in $\widehat{AB}$, as follows:

A central angle is measured by its intercepted arc.

Hence, since $\angle O = 110°$, then $x = 110$. *Ans.* (B)

> *Note.* The sum of the number of degrees in the angle formed by two intersecting tangents and the minor intercepted arc is 180°. Using this principle, $x = 180 - 70 = 110$.

After completing each test, match your answers with the numbers in the boxes.

Matching Test A

In each figure, find x.

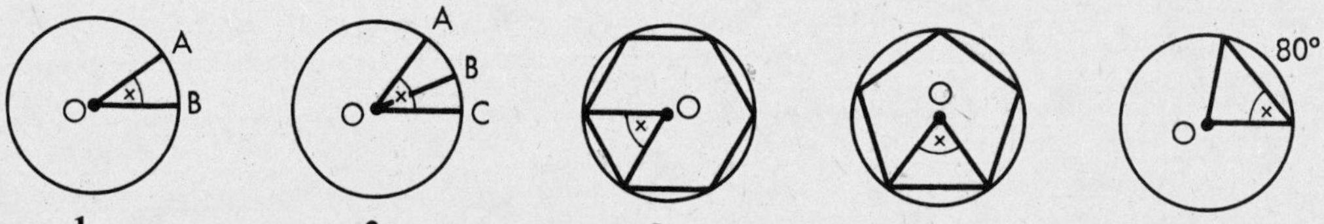

1. **2.** **3.** **4.** **5.**

1. $\widehat{AB}$ is one-tenth of the circumference.	*a.* 50
2. $\widehat{AB}$ is $\frac{1}{12}$ and $\widehat{BC}$ is $\frac{1}{15}$ of the circumference.	*b.* 36
3. The polygon is an inscribed regular hexagon.	*c.* 72
4. The polygon is an inscribed regular pentagon.	*d.* 60
5. (The figure contains all necessary information.)	*e.* 54

Matching Test B

In each figure, find x.

AB and BC are tangents

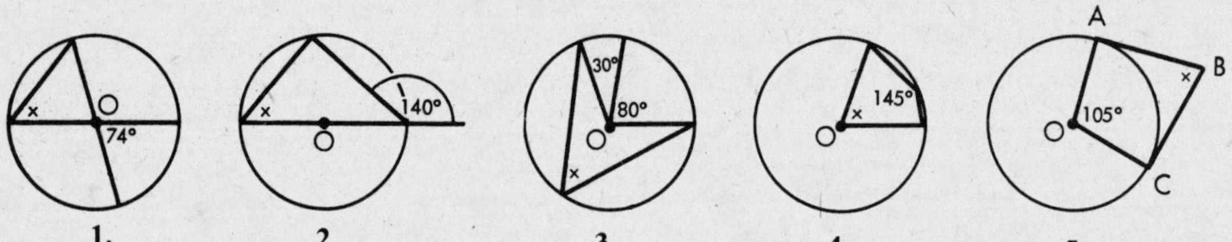

1. **2.** **3.** **4.** **5.**

a. 50	*b.* 70	*c.* 53	*d.* 55	*e.* 75

2. LINES

2.1 PROBLEMS INVOLVING COMMON TANGENTS AND THE LINE OF CENTERS

Model Problems

(1) Two circles with radii 4 and 8, respectively, are internally tangent to each other. The distance between their centers is

(A) 0 (B) 2 (C) 12 (D) 6 (E) 4

Solution:

Find the distance between the centers of the circles, OO', by applying the following principle:

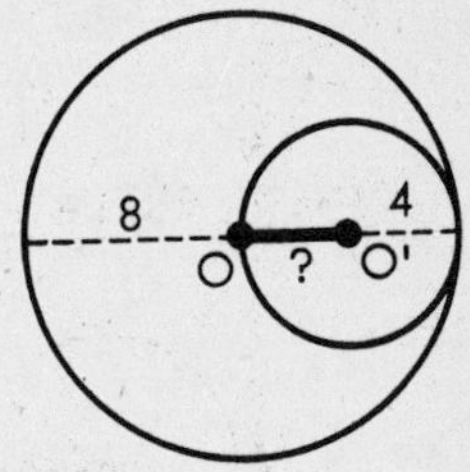

The distance between the centers of two circles that are internally tangent is the difference between their radii.

Hence, if the given radii are 8 and 4, then $OO' = 8 - 4 = 4$. *Ans.* (E)

(2) If two circles are externally tangent to each other, what is the greatest number of common tangents that can be drawn to both circles?

(A) 0 (B) 1 (C) 2 (D) 3 (E) 4

Solution:

As shown in the diagram, there are three common tangents that can be drawn to two externally tangent circles.

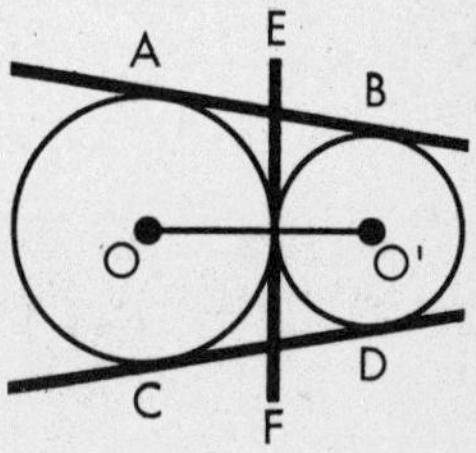

Tangents AB and CD are common external tangents, while tangent EF is a common internal tangent. *Ans.* (D)

Matching Test

After completing this test, match your answers with the numbers in the box.

1.

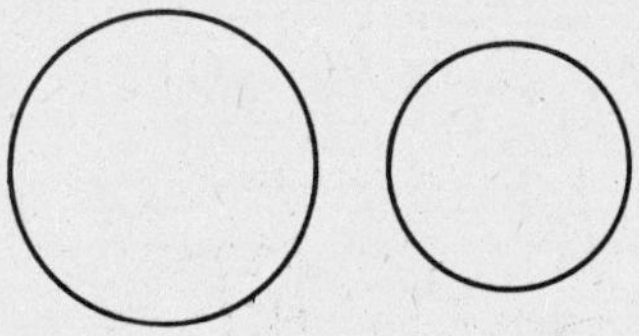

2.

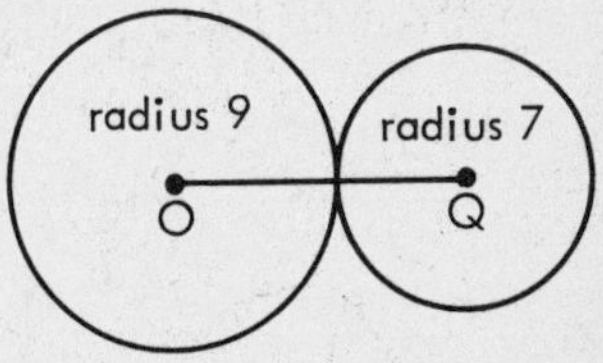

3.

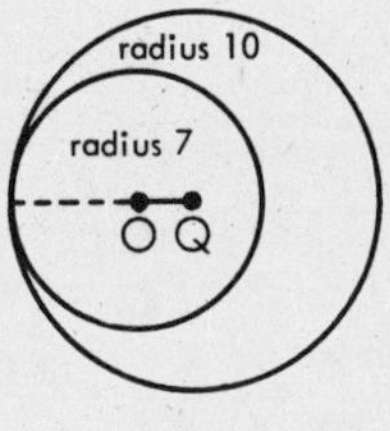

4.

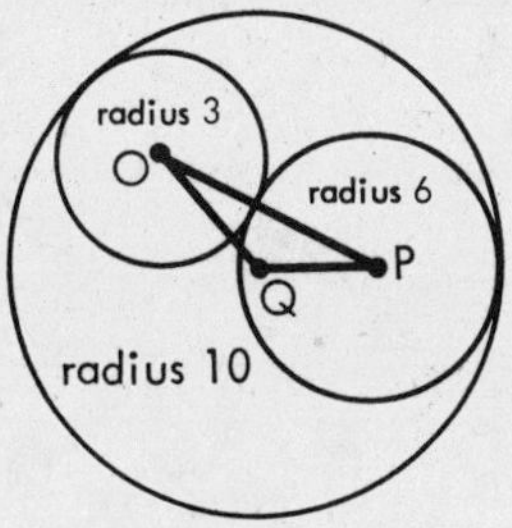

5.

1. How many common tangents may be drawn?
2. How many common tangents may be drawn?
3. Find the length of OQ, the line of centers.
4. Find the length of OQ, the line of centers.
5. Find the semiperimeter of $\triangle OPQ$, whose vertices are centers of the circles tangent to each other.

a. 16	*b.* 10	*c.* 2	*d.* 3	*e.* 4

2.2 LENGTHS OF THE THREE SIDES OF A TRIANGLE

Model Problems

(1) If two sides of a triangle are 5 and 7, the third side may be

(A) 1 (B) 2 (C) 11 (D) 12 (E) 13

Solution:

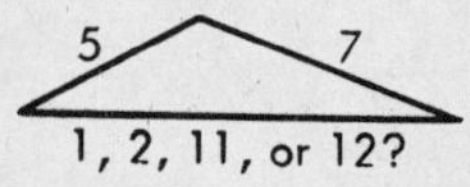

A possible length for a side of a triangle may be determined by applying the following principle:

The length of a side of a triangle must be less than the sum of the other two sides and greater than their difference.

Since the two given sides of the triangle are 5 and 7, their sum is 12 and their difference is 2. Hence, the remaining side must be less than 12 and more than 2. This leaves 11 as the only possible value for the third side among the choices. *Ans.* (C)

(2) Two sides of a triangle are 1 and 2. The third side may be

(A) 1 (B) 2 (C) 3 (D) 4 (E) none of these

Solution:

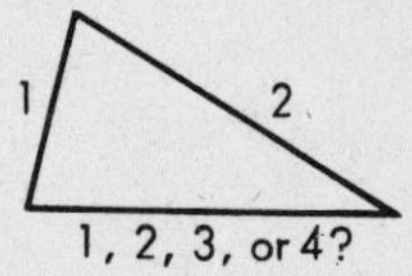

A possible length for a side of a triangle may be determined by applying the following principle:

The length of a side of a triangle must be less than the sum of the other two sides and greater than the difference.

Since the two given sides of the triangle are 1 and 2, their sum is 3 and their difference is 1. Hence, the remaining side must be less than 3 and more than 1. Thus, the only value the third side may have among the choices is 2. *Ans.* (B)

Matching Test

In each of the following, find the integral values that the third side of a triangle may have if the other two sides are as indicated. After completing this test, match your answers with the sets of numbers in the box.

1. 2 and 2 **2.** 3 and 4 **3.** 3 and 3 **4.** 2 and 3 **5.** 2 and 5

a. 2, 3, 4	*b.* 1, 2, 3	*c.* 4, 5, 6	*d.* 2, 3, 4, 5, 6	*e.* 1, 2, 3, 4, 5

2.3 UNEQUAL SIDES IN A TRIANGLE

Model Problems

(1) In $\triangle ABC$, if angle $A = 60°$ and angle $B > 60°$, which statement concerning the sides of the triangle is true?

(A) AB is the shortest. (B) AB is the longest. (C) BC is the shortest.

(D) BC is the longest. (E) AC is the shortest.

Solution:

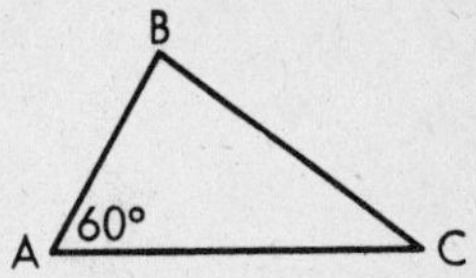

It is given that $\angle A = 60°$ and that $\angle B$ is greater than 60°. Since the sum of the angles of a triangle is 180°, $\angle C$ must be less than 60°.

The statement in choice (A) can be shown to be true by applying the following principle:

The shortest side of a triangle lies opposite the smallest angle.

Hence, AB must be the shortest side since it lies opposite the smallest angle, $\angle C$. *Ans.* (A)

(2) In $\triangle ABC$, angle A is greater than angle B and the bisectors of angle A and angle B meet in D. Then

(A) $DB > DA$ (B) $DB = DA$ (C) $DB < DA$

(D) DB is the longest side of $\triangle ADB$ (E) DA is the longest side of $\triangle ADB$

Solution:

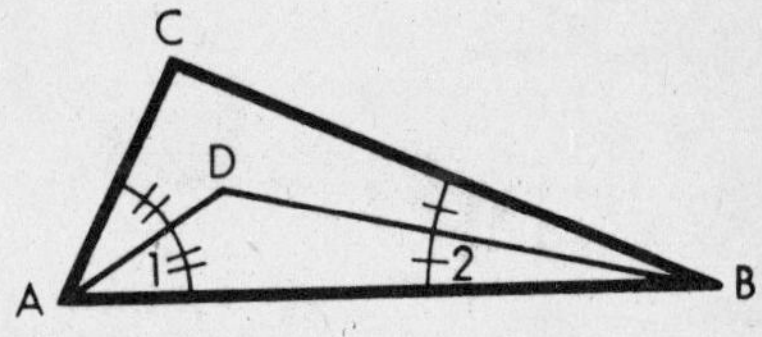

Determine that DB is greater than DA by applying the following principle:

In a triangle, the greater side lies opposite the greater angle.

In $\triangle DAB$, DB lies opposite $\angle 1$ and DA lies opposite $\angle 2$. Since $\angle A$ is greater than $\angle B$, it follows that $\angle 1$ is greater than $\angle 2$, because halves of unequals are unequal in the same order. Hence, DB, which lies opposite the greater angle, must be the greater side.

Also, since $\angle A + \angle B$ is less than 180°, it follows that $\angle 1 + \angle 2$ is less than 90°. Hence, $\angle D$ is an obtuse angle and AB is the longest side of $\triangle ADB$. This rules out choices (D) and (E). *Ans.* (A)

Matching Test

After completing this test, match your answers with the sides in the box. In Fig. 1, $\angle B = 60°$, $\angle C > 60°$, BE and ED are angle bisectors. In Fig. 2, $PQ \perp QS$, $\angle P = 46°$, $\angle R = 100°$.

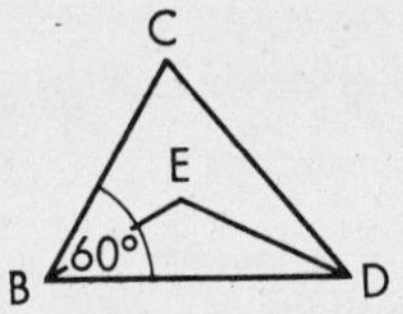

Fig. 1

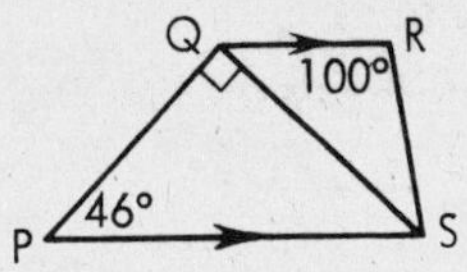

Fig. 2

1. (Fig. 1) Which is longer, CB or CD?	*a.* ED
2. (Fig. 1) Which is longer, CD or BD?	*b.* PQ
3. (Fig. 1) Which is longer, BE or ED?	*c.* QR
4. (Fig. 2) Which is shorter, PQ or QS?	*d.* CD
5. (Fig. 2) Which is shorter, QR or RS?	*e.* BD

3. PRACTICE PROBLEMS

1. Express x in terms of a.
(A) $90 - a$ (B) $180 - a$ (C) $270 - a$ (D) $90 + a$ (E) $180 + a$ **[1.1]**

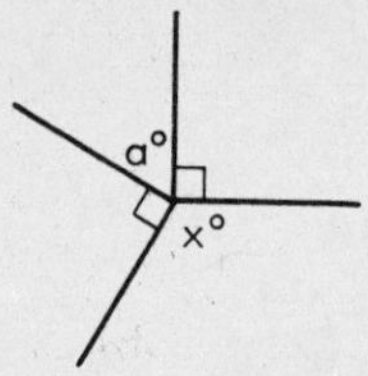

2. The supplement of an angle is four times its complement. The number of degrees in the angle is
(A) 20 (B) 30 (C) 45 (D) 60 (E) 75 **[1.1]**

3. In the figure at the right, ACE is a straight line. Find x.
(A) 51 (B) 54 (C) 65 (D) 81 (E) 88 **[1.2]**

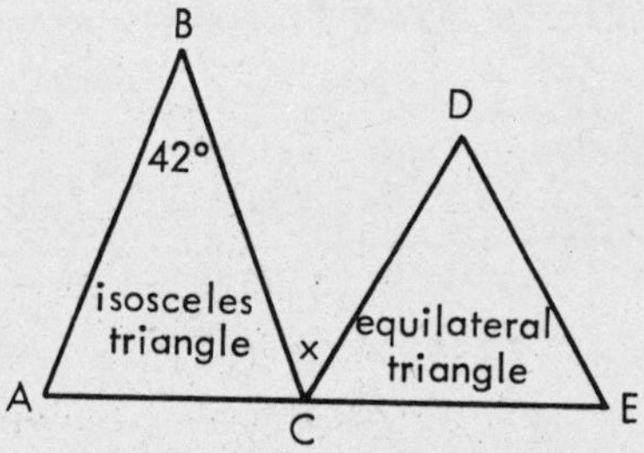

4. In an isosceles triangle, either exterior angle at the base is twice the exterior angle at the vertex. Find the number of degrees in each base angle of the triangle.
(A) 30 (B) 36 (C) 54 (D) 60 (E) 72 **[1.2]**

5. Find x, the number of degrees in the angle formed by secants AD and AE.
(A) 18 (B) 24 (C) 27 (D) 30 (E) 36 **[1.2]**

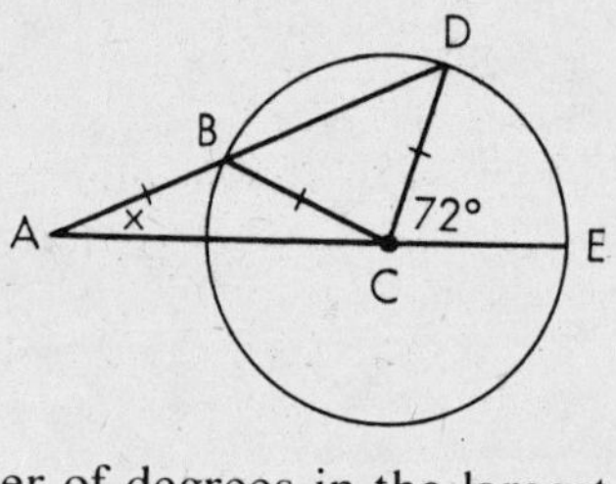

6. The angles of a triangle are in the ratio 2:5:8. Find the number of degrees in the largest angle.
(A) 12 (B) 30 (C) 60 (D) 84 (E) 96 **[1.2]**

7. Triangle ABC is an isosceles triangle having equal sides AB and BC. Find x in terms of a.
(A) $\frac{a}{2}$ (B) $90 - \frac{a}{2}$ (C) $\frac{90 - a}{2}$ (D) $\frac{360 - a}{2}$ (E) $180 + \frac{a}{2}$ **[1.2]**

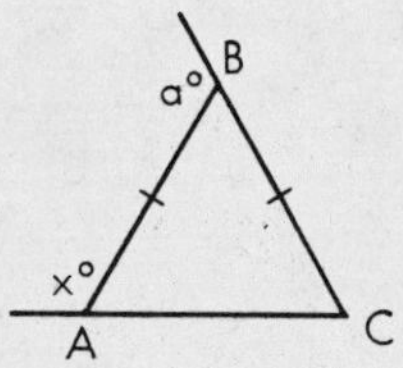

8. DE is parallel to FG. BC and AC are angle bisectors. Find x.
(A) 18 (B) 24 (C) 36 (D) 48 (E) 54 **[1.3]**

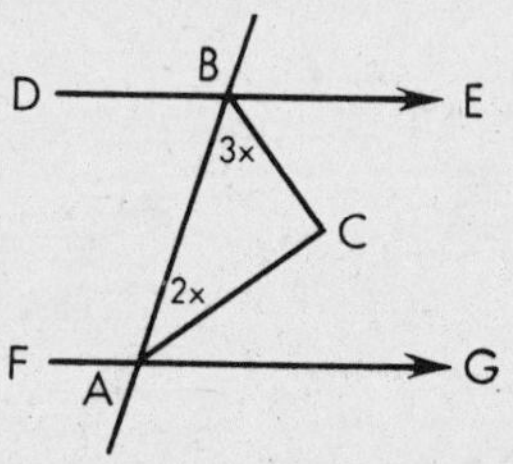

9. AB is parallel to CD. AB bisects $\angle EAD$. Find x.
(A) 36 (B) 44 (C) 48 (D) 90 (E) 92
[1.3]

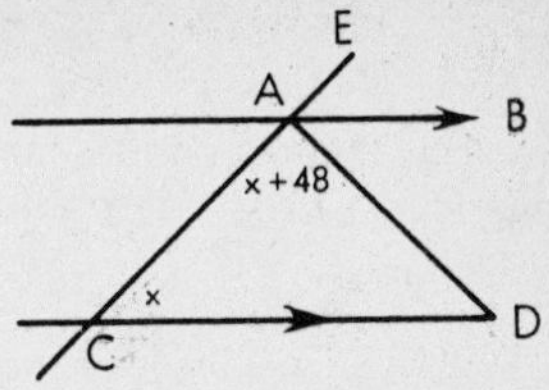

10. Find the number of degrees in arc ACB.
(A) 88 (B) 92 (C) 184 (D) 268 (E) 272
[1.4]

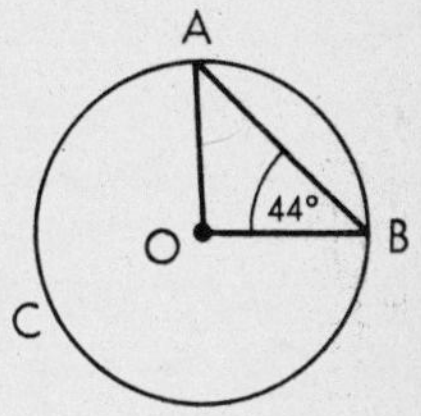

11. AB and BC are tangents. Find x, the number of degrees in $\angle O$.
(A) 15 (B) 75 (C) 95 (D) 105 (E) 85
[1.4]

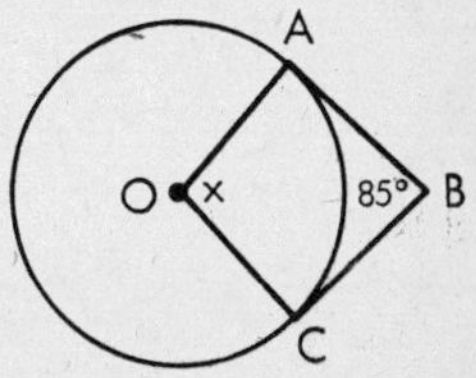

12. The circles shown at the right are externally tangent to each other. Find the semiperimeter of $\triangle OPQ$, whose vertices are the centers of the three circles.
(A) 18 (B) 24 (C) 27 (D) 32 (E) 36
[2.1]

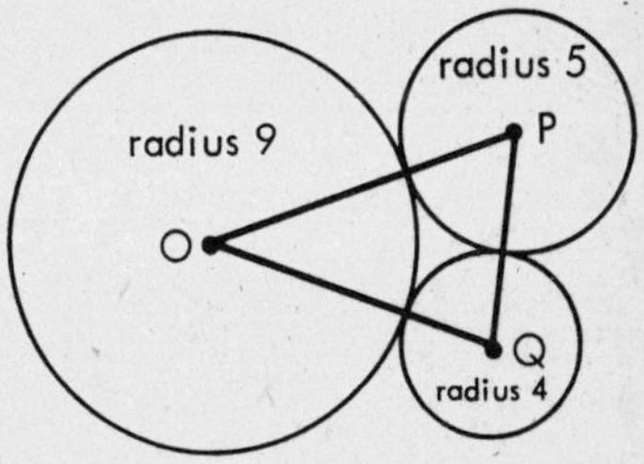

13. If two sides of a triangle are 3 and 7, what integral values may the third side have?
(A) 3, 4, 5 (B) 4, 5, 6 (C) 8, 9, 10 (D) 6, 7, 8, 9, 10 (E) 5, 6, 7, 8, 9
[2.2]

14. In $\triangle BCD$, angle $B = 60°$ and angle C is greater then 60°. BE and ED are angle bisectors of $\triangle BCD$. Find the longest side of $\triangle BED$.
(A) ED (B) BD (C) BE (D) BC (E) CD
[2.3]

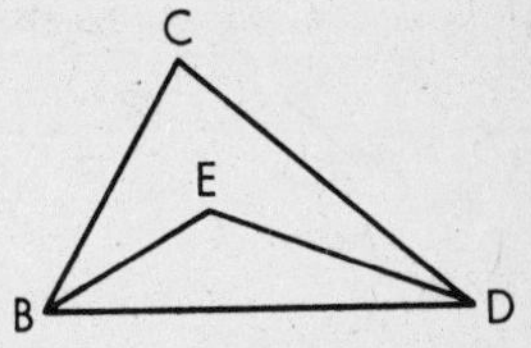

VII. REVIEWING SIMILAR POLYGONS, PROPORTIONAL LINES, AND THE LAW OF PYTHAGORAS

By definition, similar polygons are polygons whose corresponding sides are in proportion and whose corresponding angles are equal.

Principles Involving Similar Polygons and Proportional Lines

1. If a line intersecting two sides of a triangle is parallel to the third side, the two intersected sides are divided so that (*a*) their corresponding segments are in proportion and (*b*) the two sides and either pair of corresponding segments are in proportion.

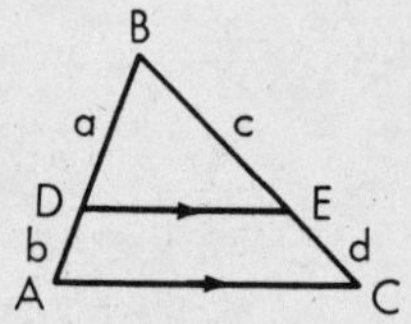

Fig. 1

For example, in Fig. 1, if $DE \parallel AC$, then (*a*) $\frac{a}{b} = \frac{c}{d}$

and (*b*) $\frac{a}{AB} = \frac{c}{BC}$ or $\frac{b}{AB} = \frac{d}{BC}$

The proportion $\frac{a}{b} = \frac{c}{d}$ can be used to make new proportions by changing the direction in which the terms are arranged to form the two ratios. For example, if the terms are rearranged in a leftward direction, that is, $\frac{a \leftarrow c}{b \leftarrow d}$, the proportion becomes $\frac{c}{a} = \frac{d}{b}$.

In the following diagrams of Fig. 2, note how each direction in which the line segments are chosen results in a different proportion:

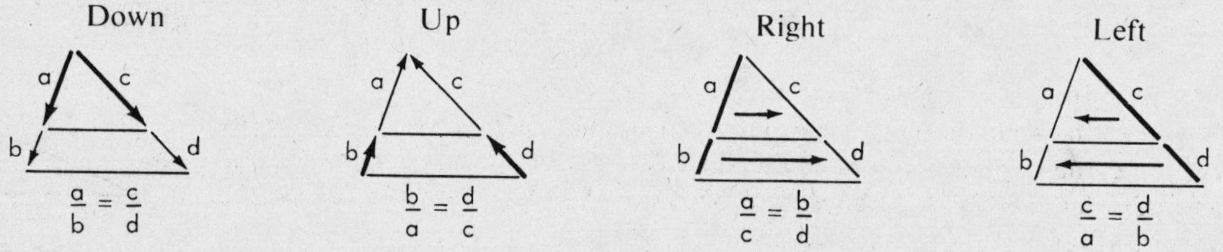

Fig. 2

2. If a line intersecting two sides of a triangle is parallel to the third side, the triangle cut off is similar to the given triangle.

   For example, in Fig. 3, if $DE \parallel AC$, then $\triangle DBE \sim \triangle ABC$. In the similar triangles, proportions such as $\frac{a}{AB} = \frac{e}{AC}$ or $\frac{c}{BC} = \frac{e}{AC}$ may be obtained by applying the definition of similar polygons.

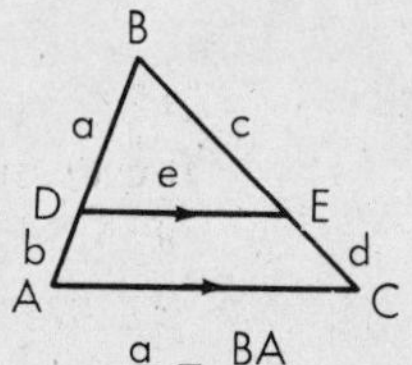

Fig. 3

3. Corresponding sides of similar triangles are in proportion.

4. Corresponding lines of similar triangles are in proportion.

   Compare principles 3 and 4 by noting that principle 4 is more general, since "lines" not only includes sides, but also broken lines such as perimeters and curved lines such as circumferences of inscribed or circumscribed circles.

   Hence, corresponding altitudes of similar triangles have the same ratio as corresponding sides, or corresponding medians, or corresponding angle bisectors. In Fig. 4, for example, if $\triangle ABC \sim \triangle A'B'C'$, then $\frac{s}{s'} = \frac{h}{h'} = \frac{m}{m'}$.

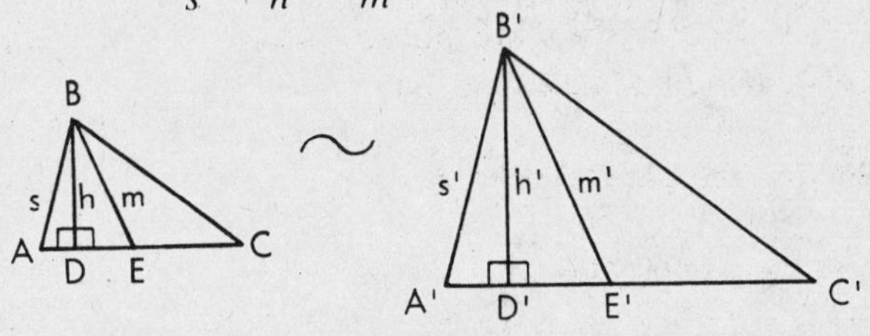

Fig. 4

5. Corresponding lines of similar polygons are in proportion.

   Compare principles 4 and 5 by noting that principle 5 is more general, since "polygons" not only includes triangles, but also quadrilaterals, pentagons, hexagons, etc.

   Hence, in Fig. 5, if quadrilateral $ABCD \sim$ quadrilateral $A'B'C'D'$, then $\frac{s}{s'} = \frac{d}{d'} = \frac{p}{p'}$, where s and s' are corresponding sides, d and d' are corresponding diagonals, and p and p' are the respective perimeters.

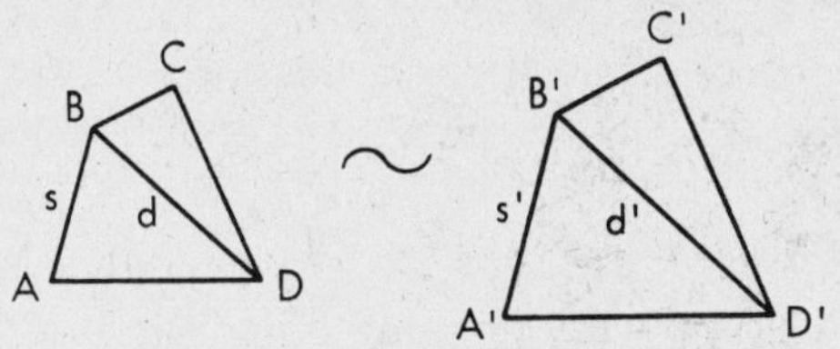

Fig. 5

Principles Involving the Law of Pythagoras

LAW OF PYTHAGORAS

6. In a right triangle, the square of the hypotenuse equals the sum of the squares of the legs. Hence, in right $\triangle ABC$, $c^2 = a^2 + b^2$. Note the correspondence of the small letter used for a side and the large letter used for the vertex opposite the side. Generally, C is the vertex of the right angle.

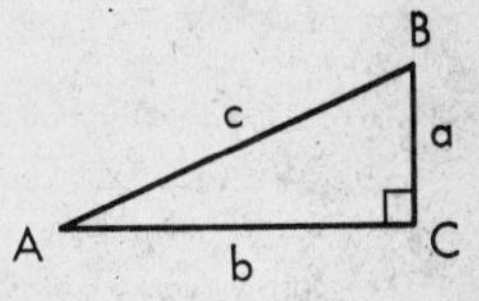

Fig. 6

CONVERSE OF THE LAW OF PYTHAGORAS

7. If $c^2 = a^2 + b^2$ applies to a triangle whose sides are a, b, and c, then the triangle is a right triangle. For example, a triangle having sides of 3, 4, and 5 is a right triangle because $3^2 + 4^2 = 5^2$; that is, $9 + 16 = 25$.

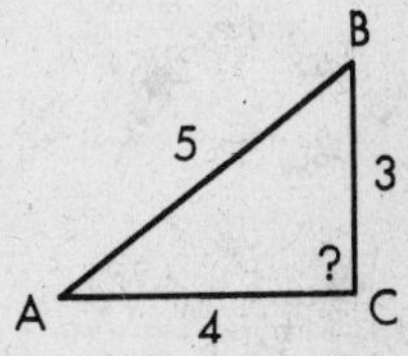

Fig. 7

Principles of the 30°–60°–90° Triangle

8. The leg opposite the 30° angle of a right triangle is one-half the hypotenuse. In Fig. 8, note that 30°–60°–90° triangle ABC is one-half of equilateral triangle ABD. Hence, b, which lies opposite a 30° angle, is one-half of hypotenuse c; that is, $b = \frac{c}{2}$.

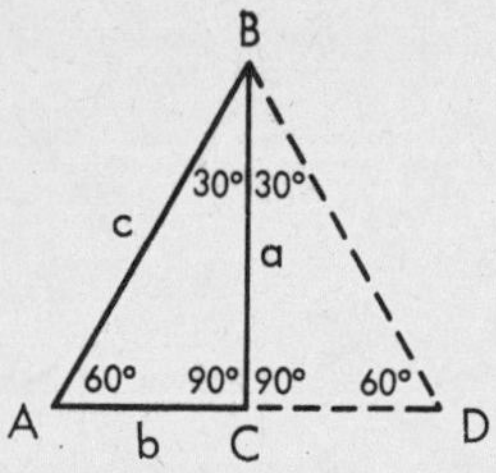

Fig. 8

9. The leg opposite 60° is the product of one-half the hypotenuse and $\sqrt{3}$. Hence, a, which lies opposite a 60° angle, is one-half of hypotenuse c times $\sqrt{3}$; that is, $a = \frac{c}{2}\sqrt{3}$.

10. The leg opposite 60° equals the leg opposite 30° times $\sqrt{3}$. Hence, $a = b\sqrt{3}$.

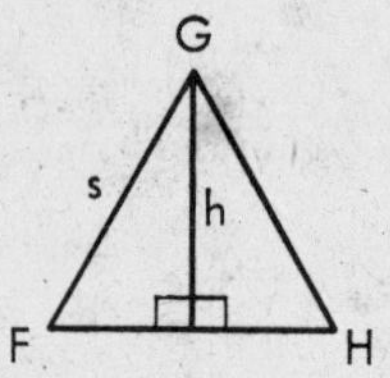

Fig. 9

11. In an equilateral triangle, the altitude equals the product of one-half a side and $\sqrt{3}$. Hence, if $\triangle FGH$ is equilateral, then $h = \frac{s}{2}\sqrt{3}$. (Since principle 11 follows directly from principle 9, 11 is a corollary of 9.)

Principles of the 45°–45°–90° Triangle

12. The leg opposite a 45° angle of a right triangle is the product of one-half the hypotenuse and $\sqrt{2}$. Hence, in right $\triangle ABC$, $a = b = \frac{c}{2}\sqrt{2}$. (In Fig. 10, note that a 45°–45°–90° triangle is one-half a square.)

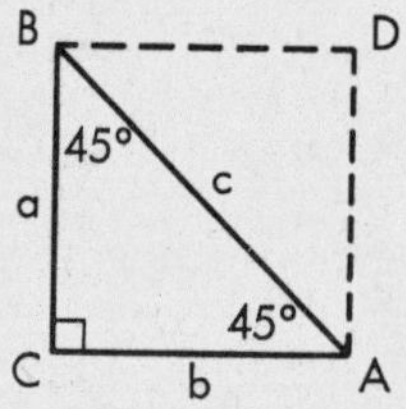

Fig. 10

13. In an isosceles right triangle (45°–45°–90° triangle), the hypotenuse is the product of either leg and $\sqrt{2}$. Hence, in Fig. 10, $c = a\sqrt{2} = b\sqrt{2}$.

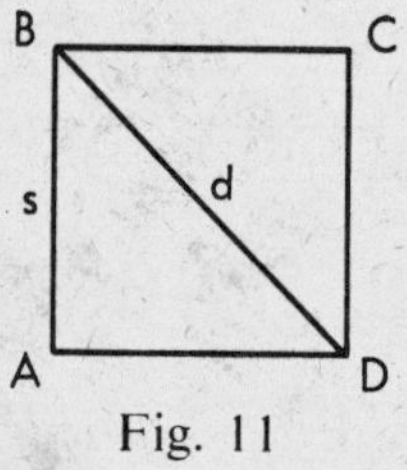

Fig. 11

14. In a square, a diagonal is the product of a side and $\sqrt{2}$. Hence, if $ABCD$ is a square, then $d = s\sqrt{2}$. (Principle 14 is a corollary of principle 13.)

1. SIMILAR POLYGONS

1.1 PROBLEMS INVOLVING PROPORTIONS OF SIDES OF SIMILAR TRIANGLES

Model Problem

The two triangles at the right are similar and angle A = angle D, angle B = angle E, angle C = angle F. Find the value of x.

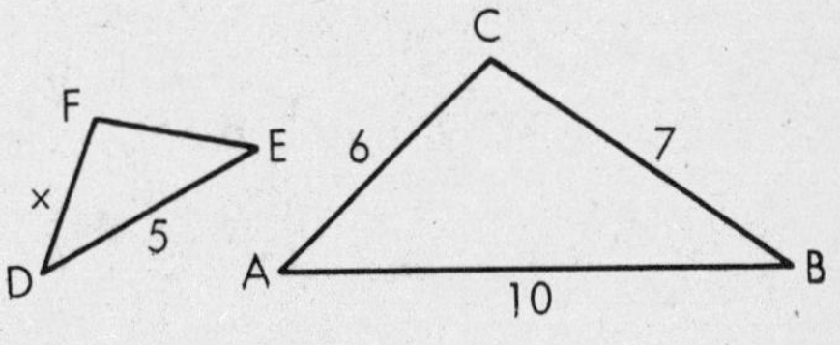

(A) 3 (B) $3\frac{1}{2}$ (C) 4 (D) $5\frac{5}{6}$ (E) $8\frac{1}{3}$

Solution:

Find x by applying the following principle:

If two triangles are similar, the ratio of any pair of corresponding sides equals the ratio of any other pair of corresponding sides.

$$\text{Hence, } \frac{x}{6} = \frac{5}{10}$$

$$\frac{x}{6} = \frac{1}{2}$$

Multiply each side by 6.

$$x = \left(\frac{1}{2}\right) 6 = 3 \quad \textit{Ans.} \text{ (A)}$$

Matching Test

In similar triangles ABC and $A'B'C'$, $\angle A = \angle A'$, $\angle B = \angle B'$, and $\angle C = \angle C'$ (refer to the accompanying figure). Use the value given in each problem to find the missing values indicated by the question marks. After completing this test, match your answers with the numbers in the box.

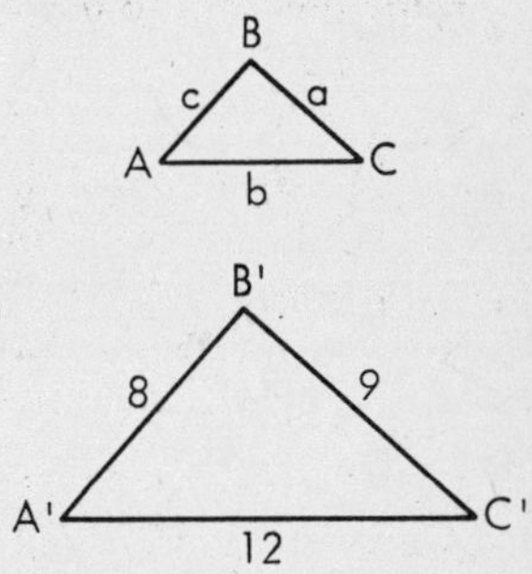

1. $b = 6$, $c = ?$
2. $c = 6$, $b = ?$
3. $b = 6$, $a = ?$
4. $a = 6$, $b = ?$
5. $a = 6$, $c = ?$

a. 8
b. $4\frac{1}{2}$
c. $5\frac{1}{3}$
d. 9
e. 4

1.2 PROPORTIONS OF LINES IN SIMILAR POLYGONS

Model Problems

(1) The ratio of the corresponding sides of two similar polygons is 3:2. What is the ratio of the perimeters of the polygons?

(A) $\frac{2}{3}$ (B) $\frac{4}{9}$ (C) $\frac{9}{4}$ (D) $\frac{3}{2}$ (E) $\frac{3}{4}$

Solution:

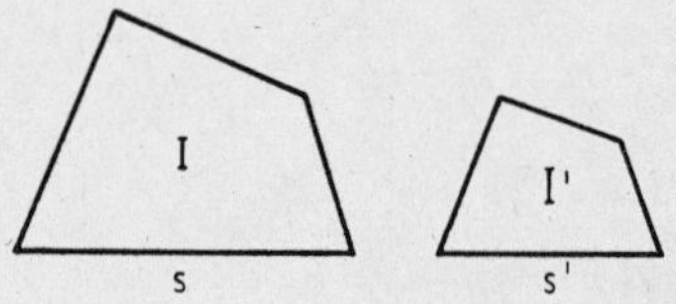

Find the ratio of the perimeters by the following method:

In similar polygons, the ratio of the perimeters equals the ratio of any two corresponding sides.

Hence, $\dfrac{\text{perimeter of I}}{\text{perimeter of I}'} = \dfrac{s}{s'}$

Substitute $\frac{3}{2}$ for $\frac{s}{s'}$.

$\dfrac{\text{perimeter of I}}{\text{perimeter of I}'} = \dfrac{3}{2}$ *Ans.* (D)

(2) Corresponding altitudes of two similar triangles are 6 and 4. If the perimeter of the larger triangle is 18, what is the perimeter of the smaller triangle?

(A) 8 (B) 12 (C) 16 (D) 27 (E) $40\frac{1}{2}$

Solution:

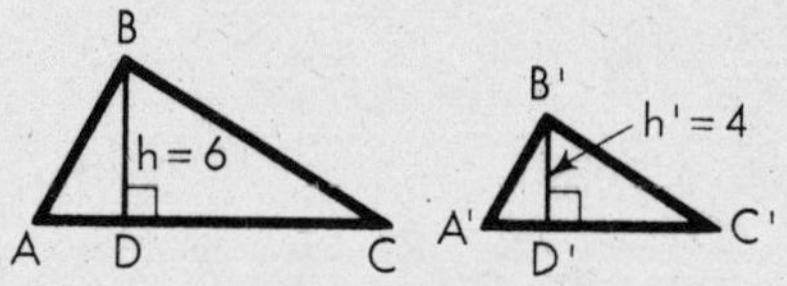

Find the perimeter of the smaller triangle by means of the following principle:

If two triangles are similar, the ratio of their perimeters equals the ratio of any two corresponding altitudes.

Hence, $\dfrac{\text{perimeter of } \triangle A'B'C'}{\text{perimeter of } \triangle ABC} = \dfrac{h'}{h}$

Substitute 6 for h, 4 for h', and 18 for the perimeter of $\triangle ABC$. Let p' = the perimeter of $\triangle A'B'C'$.

$$\frac{p'}{18} = \frac{4}{6}$$

Multiply each side by 18.

$$p' = 18\left(\frac{4}{6}\right) = 12 \quad \textit{Ans. } (B)$$

(3) The ratio of the perimeters of two similar pentagons is 4:1. The ratio of two corresponding sides is

(A) 5:1 (B) 2:1 (C) 16:1 (D) 4:1 (E) 1:4

Solution:

Find the ratio of any two corresponding sides by applying the following principle:

If two polygons are similar, the ratio of any two corresponding sides equals the ratio of their perimeters.

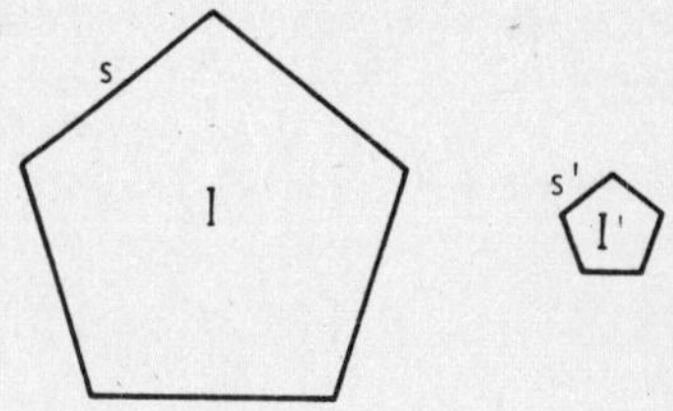

Hence, $\frac{s}{s'} = \frac{\text{perimeter of I}}{\text{perimeter of I}'}$

Substitute the given ratio, $\frac{4}{1}$, for $\frac{\text{perimeter of I}}{\text{perimeter of I}'}$.

$\frac{s}{s'} = \frac{4}{1}$ *Ans.* (D)

After completing each test, match your answers with the answers in the boxes.

Matching Test A

Complete the following statements by filling in the blanks.

1. If the ratio of the sides of two similar pentagons is 7:5, then the ratio of their perimeters is . . . .
2. If the ratio of the perimeters of two similar triangles is 11:17, then the ratio of two corresponding medians is . . . .
3. If the perimeter of one equilateral triangle is 20% greater than the perimeter of another equilateral triangle, then the altitude of the first is . . . greater than the corresponding altitude of the second.
4. Tripling the side of an equilateral triangle . . . the perimeter.
5. If the side of an isosceles right triangle is multiplied by 5, the altitude to the hypotenuse is multiplied by . . . .

a. 11:17	*b.* 20%	*c.* 7:5	*d.* 5	*e.* triples

Matching Test B

In each of the following, triangle I is similar to triangle I′ ($\triangle I \sim \triangle I'$).

1. In $\triangle I$, a side is 5 and the perimeter is 12. If the corresponding side of $\triangle I'$ is 15, what is the perimeter of $\triangle I'$?
2. In $\triangle I$, the perimeter is 12 and a median is 6. If the perimeter of $\triangle I'$ is 20, what is the corresponding median of $\triangle I'$?
3. In $\triangle I$, a median is 6 and an altitude is 9. If the corresponding median of $\triangle I'$ is 20, what is the corresponding altitude of $\triangle I'$?
4. In $\triangle I$, an altitude is 9 and an angle bisector is 15. If the corresponding altitude of $\triangle I'$ is 21, what is the corresponding angle bisector of $\triangle I'$?
5. In $\triangle I$, an altitude is 24 and a median is 15. If the corresponding median of $\triangle I'$ is 10, what is the corresponding altitude of $\triangle I'$?

a. 10	*b.* 16	*c.* 30	*d.* 35	*e.* 36

1.3 SIMILAR TRIANGLES FORMED BY HEIGHTS AND SHADOWS

Model Problem

A man six feet tall casts a shadow 5 feet long at the same time that a nearby building casts a shadow 60 feet long. How many feet high is the building?

(A) 48 (B) 50 (C) 72 (D) 75 (E) 80

Solution:

Find the height of the building by applying the following principle:

At the same time of the day in localities near each other, the ratio of the heights of vertical objects equals the ratio of their horizontal shadows.

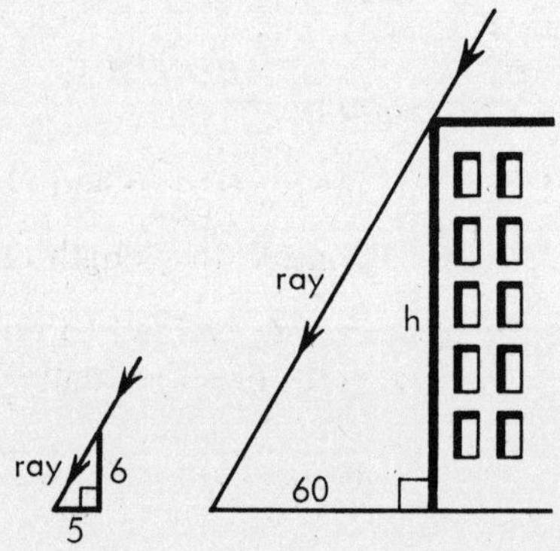

Hence, if h = the height of the building in feet,

$$\frac{h}{6} = \frac{60}{5}$$

$$\frac{h}{6} = 12$$

Multiply each side by 6.

$$h = 6(12) = 72 \quad \textit{Ans. } (C)$$

> *Note.* The triangles are similar because of two pairs of equal angles: the right angles and the angles between the parallel rays of the sun and the horizontal ground.

Matching Test

In the figure, h and h' represent the heights of two nearby objects; s and s' represent their shadows cast at the same time. Find the missing number of feet indicated by the question marks in the following table. After completing this test, match your answers with the numbers in the box.

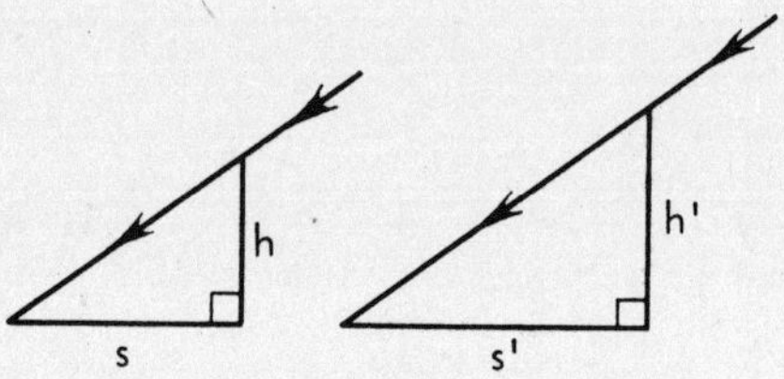

	h	s	h'	s'
1.	6 ft.	5 ft.	54 ft.	?
2.	5 ft.	4 ft.	30 ft.	?
3.	3 ft. 6 in.	2 ft. 4 in.	75 ft.	?
4.	5 ft. 8 in.	?	34 ft.	48 ft.
5.	9 ft.	?	72 ft.	56 ft.

a. 8
b. 50
c. 7
d. 45
e. 24

2. PROPORTIONAL SEGMENTS AND SIDES IN A TRIANGLE

2.1 PROBLEMS INVOLVING PROPORTIONAL SEGMENTS CUT OFF BY A LINE PARALLEL TO ONE SIDE OF A TRIANGLE

Model Problem

In triangle ABC, a line parallel to base AC cuts AB at D and BC at E. If $BD = 6$, $DA = 2$, and $BC = 12$, find BE.

(A) 3 (B) 4 (C) 8 (D) 9 (E) 10

Solution:

Since $BD = 6$ and $DA = 2$, then $AB = 8$.

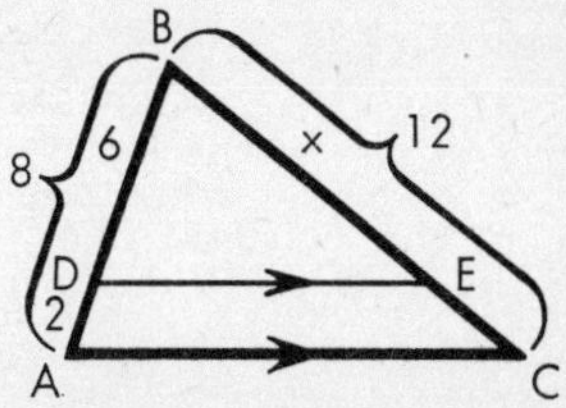

Find x, the length of BE, by applying the following principle:

A line parallel to one side of a triangle divides the other two sides proportionally.

Hence, $\frac{x}{12} = \frac{6}{8}$

Multiply each side by 12.

$$x = \frac{6}{8}(12) = 9 \quad \textit{Ans.}\ \text{(D)}$$

Matching Test

In $\triangle ABC$, $DE \parallel AC$. Use the values given in each problem to find the missing values indicated by the question marks. After completing this test, match your answers with the numbers in the box.

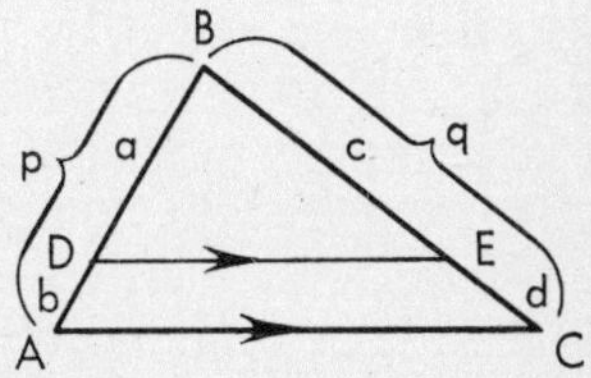

1. $a = 9,\ b = 3,\ c = 12,\ d = ?$	*a.* 8
2. $a = 9,\ b = 3,\ d = 12,\ c = ?$	*b.* 36
3. $a = 9,\ b = 3,\ q = 12,\ c = ?$	*c.* 18
4. $p = 9,\ b = 3,\ c = 12,\ q = ?$	*d.* 4
5. $p = 9,\ a = 3,\ q = 12,\ d = ?$	*e.* 9

2.2 PROPORTIONAL SIDES FORMED BY A LINE PARALLEL TO ONE SIDE OF A TRIANGLE

Model Problem

In $\triangle ABC$, a line parallel to AC intersects AB at D and CB at E. If $DB = 10$, $AD = 5$, and $AC = 6$, find the length of DE.

(A) $1\frac{1}{2}$ (B) 2 (C) $2\frac{1}{2}$ (D) 3 (E) 4

Solution:

Since a line parallel to one side of a triangle cuts off a triangle similar to the given triangle,

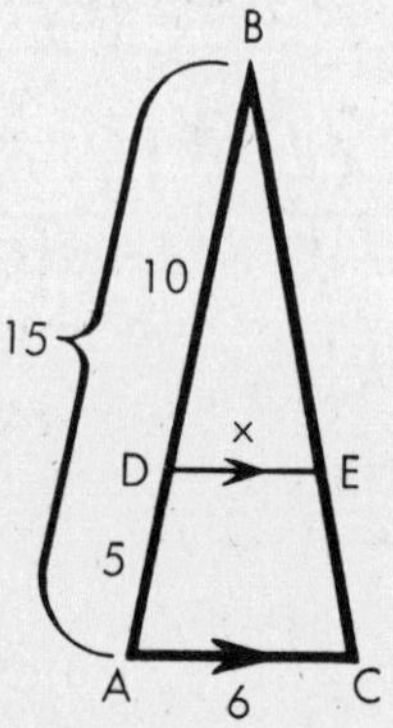

$$\triangle DBE \sim \triangle ABC$$

Also, $AB = 10 + 5 = 15$.

Find x, the length of DE, by applying the following principle:

Corresponding sides of similar triangles are in proportion.

Hence, $\frac{x}{6} = \frac{DB}{AB}$

Substitute 10 for DB and 15 for AB.

$$\frac{x}{6} = \frac{10}{15}$$

Multiply each side by 6.

$$x = 6\left(\frac{10}{15}\right) = 4 \quad \textit{Ans.} \text{ (E)}$$

Matching Test

In $\triangle ABC$, $DE \parallel AC$. Use the values given in each problem to find the missing lengths indicated by the question marks. After completing this test, match your answers with the numbers in the box.

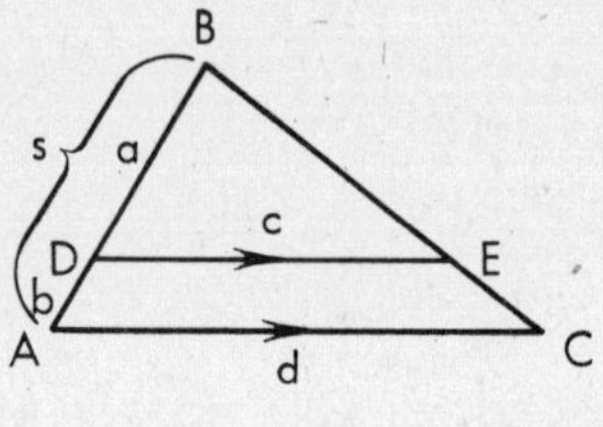

1. $a = 8$, $b = 4$, $c = 12$, $d = ?$
2. $a = 8$, $b = 4$, $d = 12$, $c = ?$
3. $a = 4$, $b = 8$, $c = 12$, $d = ?$
4. $a = 4$, $b = 8$, $d = 12$, $c = ?$
5. $c = 4$, $d = 8$, $a = 12$, $b = ?$

a. 18
b. 12
c. 8
d. 36
e. 4

3. LAW OF PYTHAGORAS

3.1 PROBLEMS INVOLVING THE LAW OF PYTHAGORAS

Model Problems

(1) Find the length of the diagonal of a rectangle whose sides are 2 and 1.

(A) $1\frac{1}{2}$ (B) 3 (C) $\sqrt{3}$ (D) $\sqrt{5}$ (E) $2\sqrt{3}$

Solution:

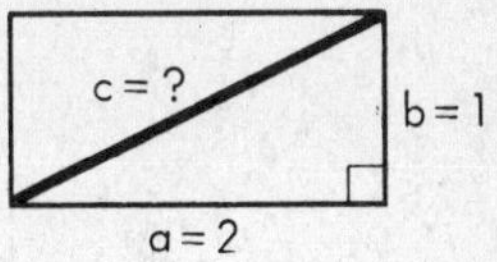

Since the diagonal of the rectangle is the hypotenuse of a right triangle, use the Law of Pythagoras.

In a right triangle, the square of the hypotenuse equals the sum of the squares of the legs.

Hence, $(\text{hypotenuse})^2 = (\text{leg})^2 + (\text{leg})^2$

Substitute 2 and 1 for the legs.

$$c^2 = 2^2 + 1^2$$
$$= 4 + 1 = 5$$

Take the square root of each side and keep in mind that c is positive.

$$c = \sqrt{5} \quad \textit{Ans. } \text{(D)}$$

(2) The hypotenuse of a right triangle is 9 and one side is 3. Find the length of the other side.

(A) $3\sqrt{2}$ (B) $3\sqrt{10}$ (C) 6 (D) $6\sqrt{2}$ (E) $6\sqrt{3}$

Solution:

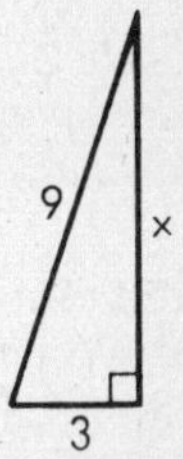

Find x, the length of the other leg, by applying the Law of Pythagoras.

In a right triangle, the square of the hypotenuse equals the sum of the squares of the legs.

Hence, $(\text{leg})^2 + (\text{leg})^2 = (\text{hypotenuse})^2$

$$x^2 + 3^2 = 9^2$$
$$x^2 + 9 = 81$$

Subtract 9 from each side.

$$x^2 = 72$$

Take the square root of each side and keep in mind that x is positive.

$$x = \sqrt{72} = 6\sqrt{2} \quad \textit{Ans. } \text{(D)}$$

(3) If the hypotenuse of a right triangle is 10 and one leg is 8, find the other leg.

(A) $2\sqrt{82}$ (B) 9 (C) $6\sqrt{3}$ (D) $6\sqrt{2}$ (E) 6

Solution:

Find the other leg by using the Law of Pythagoras.

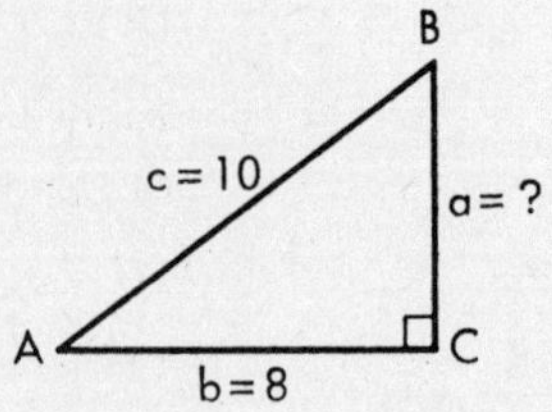

In a right triangle, the square of the hypotenuse equals the sum of the squares of the legs.

Hence, $c^2 = a^2 + b^2$

Substitute 10 for c and 8 for b.

$$10^2 = a^2 + 8^2$$

$$100 = a^2 + 64$$

Subtract 64 from each side of the equation.

$$36 = a^2$$

Take the square root of each side and keep in mind that a is positive.

$6 = a$ *Ans.* (E)

> *Note.* The three sides, 6, 8, and 10, of the right triangle are in the ratio 3:4:5. This type of triangle is referred to as a 3-4-5 right triangle.

(4) The formula $b = \sqrt{c^2 - a^2}$ is used in finding one leg b of a right triangle in which a is the other leg and c is the hypotenuse. In a right triangle, one leg is 16 and the hypotenuse is 20. Find the other leg.

(A) $8\sqrt{82}$ (B) 9 (C) 12 (D) 15 (E) 18

Solution:

Find b in the formula $b = \sqrt{c^2 - a^2}$ by substituting 20 for c and 16 for a.

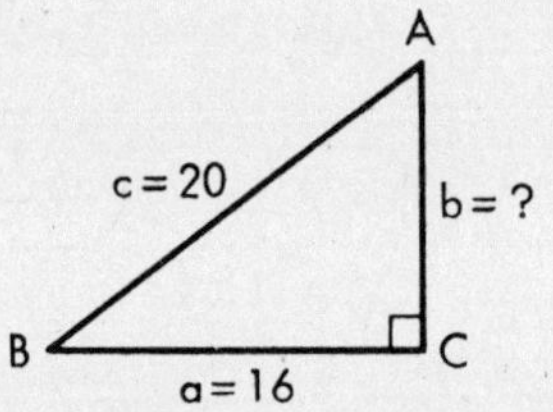

$$b = \sqrt{20^2 - 16^2}$$

$$= \sqrt{400 - 256}$$

$$= \sqrt{144}$$

$= 12$ *Ans.* (C)

> *Note.* Since the three sides, 12, 16, and 20, of $\triangle ABC$ are in the ratio 3:4:5, the triangle is a 3-4-5 right triangle.

(5) In rectangle $ABCD$, diagonal $AC = 12$ and $BC = 7$. Find the length of AB.

(A) $\sqrt{193}$ (B) $\sqrt{105}$ (C) 9.5 (D) $\sqrt{95}$ (E) $\sqrt{85}$

Solution:

Find AB, or x, as follows:

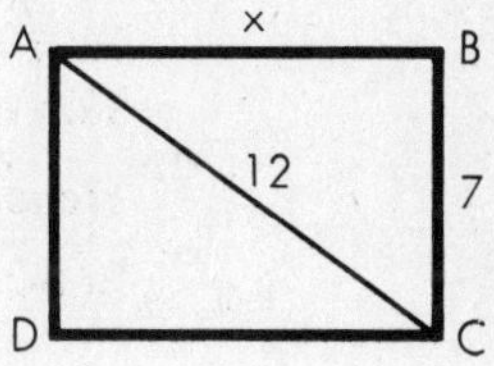

In a right triangle, the square of the hypotenuse equals the sum of the squares of the legs.

Hence, $(\text{leg})^2 + (\text{leg})^2 = (\text{hypotenuse})^2$

In rt. $\triangle ABC$, $7^2 + x^2 = 12^2$

$$49 + x^2 = 144$$

Subtract 49 from each side.

$$x^2 = 95$$

Take the square root of each side and keep in mind that x is positive.

$x = \sqrt{95}$ *Ans.* (D)

After completing each test, match your answers with the numbers in the boxes.

Matching Test A

Each row in the following table gives the lengths of two sides of a right triangle. In each case, find the length of the third side.

	Leg	*Leg*	*Hypotenuse*
1.	3	5	?
2.	2	?	7
3.	?	6	8
4.	$\sqrt{10}$	$\sqrt{15}$	?
5.	$4\sqrt{3}$	?	8

a. $\sqrt{45}$ or $3\sqrt{5}$
b. 5
c. 4
d. $\sqrt{28}$ or $2\sqrt{7}$
e. $\sqrt{34}$

Matching Test B

1. Find the hypotenuse of a right triangle whose legs are 4 and 5..
2. Find the remaining leg of a right triangle whose hypotenuse is 8 and whose shorter leg is 4.
3. Find the diagonal of a square whose perimeter is 16.
4. Find the side of a rhombus whose diagonals are 4 and 6.
5. A spotlight is 5 feet from each of the walls of a stage. If these walls are at right angles to each other, how many feet is the spotlight from the intersection of the walls?

a. $\sqrt{50}$ or $5\sqrt{2}$ *b.* $\sqrt{48}$ or $4\sqrt{3}$ *c.* $\sqrt{13}$ *d.* $\sqrt{41}$ *e.* $\sqrt{32}$ or $4\sqrt{2}$

3.2 CONVERSE OF THE LAW OF PYTHAGORAS

Model Problem

Which set of numbers may represent the lengths of the sides of a right triangle?

(A) 20, 30, 40 (B) 30, 40, 50 (C) 30, 40, 60 (D) 40, 50, 90 (E) none of these

Solution:

Make the correct choice by applying the converse of the Law of Pythagoras:

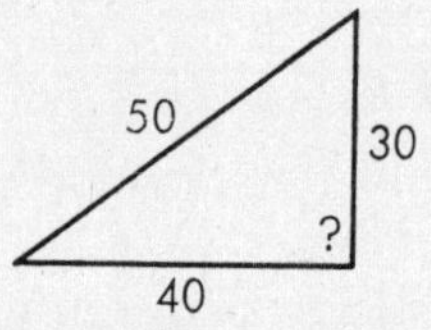

If the square of one side of a triangle equals the sum of the squares of the other two sides, then the triangle is a right triangle.

The only set of numbers that satisfies the converse of the Law of Pythagoras is the set in choice (B), 30, 40, 50, since $50^2 = 30^2 + 40^2$; that is, $2500 = 900 + 1600$. *Ans.* (B)

Wrong Choices:

(A) Since $20^2 = 400$, $30^2 = 900$, and $40^2 = 1600$, note that the square of the largest side is *larger* than the sum of the squares of the other two sides. In this case, the triangle is an obtuse triangle.

(C) Since $30^2 = 900$, $40^2 = 1600$, and $60^2 = 3600$, note that the square of the largest side is *larger* than the sum of the squares of the other two sides. In this case, the triangle is an obtuse triangle.

(D) Since $90 = 40 + 50$, it is impossible to have a triangle with such lengths of sides. The largest side of a triangle must be smaller than the sum of the other two sides.

Selection Test

Of the following sets of numbers, which set *cannot* be used as the lengths of the sides of a right triangle?

a. 3, 4, 6 *b.* 2, 2, $2\sqrt{2}$ *c.* 2, $2\sqrt{3}$, 4 *d.* 7, 24, 25 *e.* 1, $\frac{4}{3}$, $\frac{5}{3}$

a. 3, 4, 6

3.3 LAW OF PYTHAGORAS IN A 3-4-5 TRIANGLE

If the sides of a triangle are in the ratio 3:4:5, the triangle is a right triangle and is referred to as a 3-4-5 triangle. For example, a triangle whose sides are 6, 8, and 10 is a 3-4-5 triangle, since the ratio 6:8:10 is equivalent to 3:4:5.

Short-cut method for finding a side of a 3-4-5 triangle. If the legs of a right triangle are in the ratio 3:4, or if the ratio of a leg to the hypotenuse is either 3:5 or 4:5, the triangle is a 3-4-5 triangle. Hence, its sides can be represented by $3x$, $4x$, and $5x$. In the following model problem, note how this can greatly simplify problems involving 3-4-5 triangles:

Model Problem

Find the diagonal of a rectangle whose sides are 9 and 12.

(A) 6 (B) 15 (C) $3\sqrt{7}$ (D) $6\sqrt{7}$ (E) 16

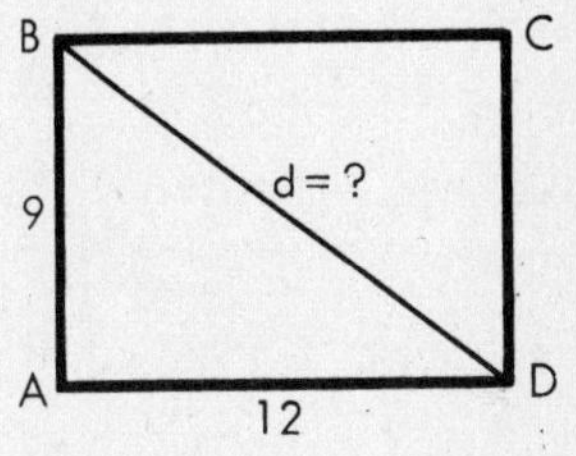

Solution:

Find diagonal d as follows:

If the legs of a right triangle are in the ratio 3:4, the triangle is a 3-4-5 triangle.

The legs of right $\triangle ABD$ are 9 and 12. Since $\frac{9}{12} = \frac{3}{4}$, the ratio of the legs is 3:4. Hence, $\triangle ABD$ is a 3-4-5 triangle.

Represent the sides by $3x$, $4x$, and $5x$. Since the shorter leg is 9, $3x = 9$ and $x = 3$. Hence, hypotenuse BD, or $d = 5x = 5 \cdot 3 = 15$. *Ans.* (B)

> *Note.* Diagonal d can also be found by the direct application of the Law of Pythagoras.
>
> $$(\text{hypotenuse})^2 = (\text{leg})^2 + (\text{leg})^2$$
>
> In right $\triangle ABD$, $d^2 = 9^2 + 12^2$
>
> $$= 81 + 144 = 225$$
>
> Hence, $d = 15$

After completing these tests, match your answers with the answers in the boxes.

Matching Test A

In each of the following, the triangle is a 3-4-5 triangle and one of the sides is given. Find the two remaining sides.

	Shorter leg, 3x	*Longer leg, 4x*	*Hypotenuse, 5x*
1.	12	?	?
2.	?	12	?
3.	?	?	55
4.	?	?	12.5
5.	?	200	?

a. 16, 20
b. 7.5, 10
c. 9, 15
d. 150, 250
e. 33, 44

Matching Test B

The problems in this test can be solved by using the short-cut method for finding a side of a 3-4-5 triangle.

1. A fireman's ladder, 100 feet long, reaches a point on a building that is 80 feet above the ground. If the ground is horizontal, how many feet from the foot of the building is the foot of the ladder?

2. Two boats leave the same dock at the same time, one traveling due north at the rate of 3 mph, the other due east at the rate of 4 mph. How many miles apart are the boats at the end of 5 hours?

3. The diagonal of a rectangle is 40 and its base is 32. Find the altitude of the rectangle.

4. A plane flies 240 miles due north, then 320 miles due west. How many miles must it fly to return to its starting point by the shortest route?

5. Two legs of a right triangle are in the ratio 3:4. Find the shorter of the legs if the hypotenuse is 75.

a. 24 *b.* 60 *c.* 45 *d.* 400 *e.* 25

Selection Test

Of the following sets of numbers, which set *cannot* be used as the lengths of the sides of a 3-4-5 triangle?

a. $1\frac{1}{2}, 2, 2\frac{1}{2}$ *b.* $\frac{3}{5}, \frac{4}{5}, 1$ *c.* 555, 333, 444 *d.* 8.8, 6.6, 11 *e.* 75, 150, 125

e. 75, 150, 125

3.4 LAW OF PYTHAGORAS IN A 5-12-13 TRIANGLE

If the sides of a triangle are in the ratio 5:12:13, the triangle is a right triangle and is referred to as a 5-12-13 triangle. For example, a triangle whose sides are 10, 24, and 26 is a 5-12-13 triangle, since the ratio 10:24:26 is equivalent to 5:12:13.

Short-cut method for finding a side of a 5-12-13 triangle. If the legs of a right triangle are in the ratio 5:12, or if the ratio of a leg to the hypotenuse is either 5:13 or 12:13, the triangle is a 5-12-13 triangle. Hence, its sides can be represented by $5x$, $12x$, and $13x$.

For example, if the legs of a right triangle are 10 and 24, note that $\frac{10}{24} = \frac{5}{12}$; that is, the legs are in the ratio 5:12. Hence, the triangle is a 5-12-13 triangle. Let $5x =$ the shorter leg and $13x =$ the hypotenuse. Since the shorter leg = 10, then $5x = 10$ and $x = 2$. Hence, the hypotenuse, $13x = 13 \cdot 2 = 26$.

The following model problem shows a simple application of this method, since the sides of right $\triangle ABD$ are 5 and 13. Note that no computation is needed in this case.

Model Problem

In triangle ABC, $AB = AC = 13$ inches and $BC = 10$ inches. Find the number of inches in the altitude drawn from A.

(A) $\sqrt{69}$ (B) 12.5 (C) 12 (D) 10 (E) 9

Solution:

Since the altitude to the base of an isosceles triangle bisects the base, $BD = DC = 5$.

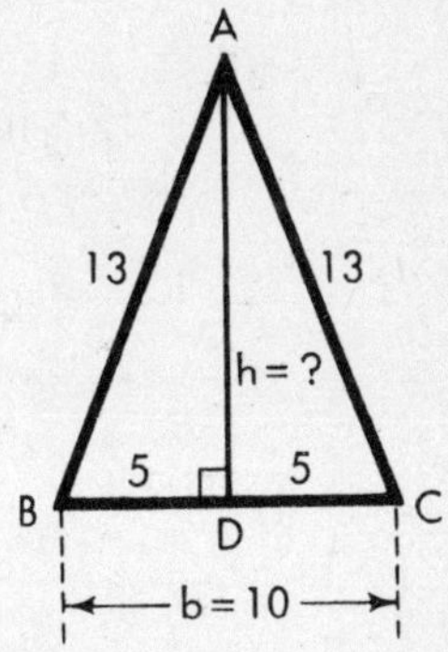

Find altitude h as follows:

If a leg and the hypotenuse of a right triangle are in the ratio 5:13, the triangle is a 5-12-13 triangle.

In right $\triangle ABD$, leg $BD = 5$ and hypotenuse $AB = 13$. Hence, $\triangle ABD$ is a 5-12-13 triangle and AD, or $h = 12$. *Ans.* (C)

> *Note.* Altitude h can also be found by the direct application of the Law of Pythagoras.
>
> $$(\text{leg})^2 + (\text{leg})^2 = (\text{hypotenuse})^2$$
> $$h^2 + 5^2 = 13^2$$
> $$h^2 = 13^2 - 5^2$$
> $$= 169 - 25 = 144$$
>
> Hence, $h = 12$

After completing each test, match your answers with the answers in the boxes.

Matching Test A

In each of the following, the triangle is a 5-12-13 triangle; one of the sides is given. Find the two remaining sides.

	Shorter leg, 5x	*Longer leg, 12x*	*Hypotenuse, 13x*
1.	10	?	?
2.	?	36	?
3.	?	?	65
4.	$2\frac{1}{2}$	?	?
5.	?	1200	?

a. 15, 39
b. 500, 1300
c. 24, 26
d. 25, 60
e. 6, $6\frac{1}{2}$

Matching Test B

The problems in this test can be solved by using the short-cut method for finding a side of a 5-12-13 triangle.

1. The sides of a rectangular field are 50 feet and 120 feet. How many seconds are saved if a man walking at the rate of 4 feet per second takes the diagonal shortcut across the field instead of walking along the two sides?

2. A ladder 52 feet long stands against a tree so that the foot of the ladder is 20 feet from the foot of the tree. How many feet up the tree does the ladder extend?

3. A right triangle is inscribed in a circle whose radius is 13. One of the sides of the triangle is 10. Find the length of the other side.

4. Dr. Smith's home is 5 miles due north of the town hospital, while Dr. Adams lives 12 miles due west of the same hospital. On a map having a scale of 2 miles per inch, how many inches apart are their homes?

5. The diagonals of a rhombus are 20 and 48. Find the perimeter of the rhombus.

a. 6.5 *b.* 24 *c.* 104 *d.* 48 *e.* 10

Selection Test

Of the following sets of numbers, which set *cannot* be used as the lengths of the sides of a 5-12-13 triangle?

a. $2\frac{1}{2}, 6, 6\frac{1}{2}$ *b.* 40, 96, 104 *c.* $\frac{5}{7}, \frac{12}{7}, \frac{13}{7}$ *d.* $\frac{1}{2}, \frac{6}{5}, \frac{13}{10}$ *e.* 65, 132, 169

e. 65, 132, 169

3.5 LAW OF PYTHAGORAS IN AN 8-15-17 TRIANGLE

If the sides of a triangle are in the ratio 8:15:17, the triangle is a right triangle and is referred to as an 8-15-17 triangle. For example, a triangle whose sides are 40, 75, and 85 is an 8-15-17 triangle, since the ratio 40:75:85 is equivalent to 8:15:17.

Short-cut method for finding a side of an 8-15-17 triangle. If the legs of a right triangle are in the ratio 8:15, or if the ratio of a leg to the hypotenuse is either 8:17 or 15:17, the triangle is an 8-15-17 triangle. Hence, its sides can be represented by $8x$, $15x$, and $17x$.

For example, if one leg of a right triangle is 75 and the hypotenuse is 85, note that $\frac{75}{85} = \frac{15}{17}$. Hence, the triangle is an 8-15-17 triangle. Let $8x =$ the shorter leg and $15x =$ the given leg. Since the given leg $= 75$, then $15x = 75$ and $x = 5$. Hence, the shorter leg, $8x = 8(5) = 40$.

Model Problems

(1) What is the length of the hypotenuse of a right triangle whose legs are 8 and 15?

(A) 11.5 (B) 16.5 (C) 17 (D) 20 (E) $\sqrt{161}$

Solution:

Find hypotenuse AB, or c, as follows:

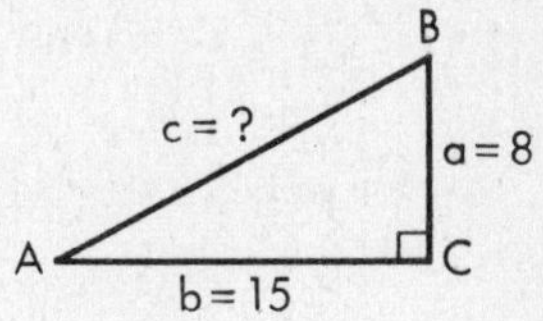

Since $a = 8$ and $b = 15$, the legs of the triangle are in the ratio 8:15. Hence, $\triangle ABC$ is an 8-15-17 triangle and $c = 17$.

Ans. (C)

> *Note.* The hypotenuse can also be found by the direct application of the Law of Pythagoras.
>
> $$(\text{hypotenuse})^2 = (\text{leg})^2 + (\text{leg})^2$$
> $$c^2 = 8^2 + 15^2$$
> $$= 64 + 225 = 289$$
>
> Hence, $c = 17$

(2) A diagonal of a rectangle is 17 and the length is 15. Find the width of the rectangle.

(A) 8 (B) 10 (C) 13 (D) 16 (E) $\sqrt{514}$

Solution:

Find width w as follows:

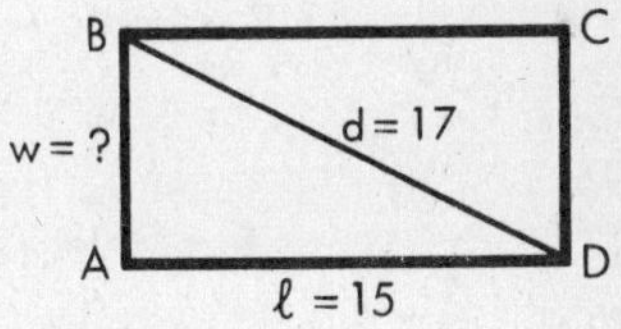

In right $\triangle ABD$, leg $l = 15$ and hypotenuse $d = 17$. Hence, $\triangle ABD$ is an 8-15-17 triangle and shorter leg $w = 8$. *Ans.* (A)

> *Note.* The Law of Pythagoras can also be applied directly as follows:
>
> $$(\text{leg})^2 + (\text{leg})^2 = (\text{hypotenuse})^2$$
> $$w^2 + 15^2 = 17^2$$
> $$w^2 = 17^2 - 15^2$$
> $$= 289 - 225 = 64$$
>
> Hence, $w = 8$

(3) In circle O with radius 17, chord CD is 30. Find the distance of chord CD from the center of the circle.

(A) 6.5 (B) 8 (C) 10 (D) 13 (E) 16

Solution:

To find the distance of chord CD from O, it is necessary to find the length of OA, the perpendicular from O to CD.

Since $OA \perp CD$, then $AD = \frac{1}{2}(30) = 15$.

To find OA, or x, note that right $\triangle OAD$ is an 8-15-17 triangle, since leg AD and hypotenuse OD are in the ratio 15:17. Hence, $x = 8$. *Ans.* (B)

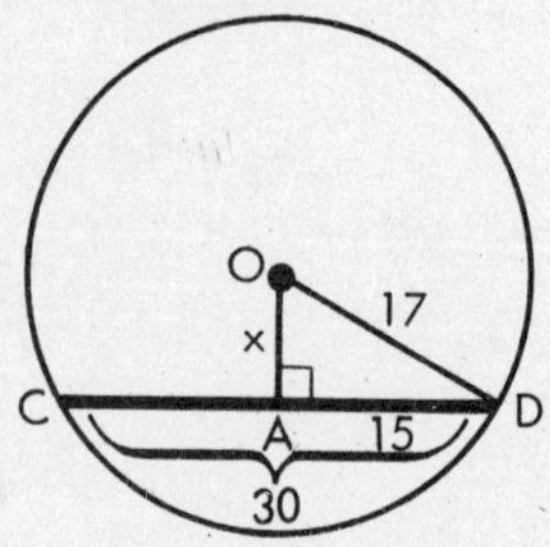

Note. The length of OA can also be found by applying the Law of Pythagoras directly as follows:

$$(\text{leg})^2 + (\text{leg})^2 = (\text{hypotenuse})^2$$
$$x^2 + 15^2 = 17^2$$
$$x^2 = 17^2 - 15^2$$
$$= 289 - 225 = 64$$

Hence, $x = 8$

Matching Test

The figures in this test contain 8-15-17 triangles. In each figure, find the length represented by x. After completing this test, match your answers with the numbers in the box.

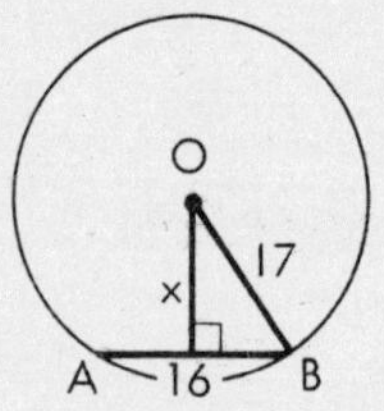

1.

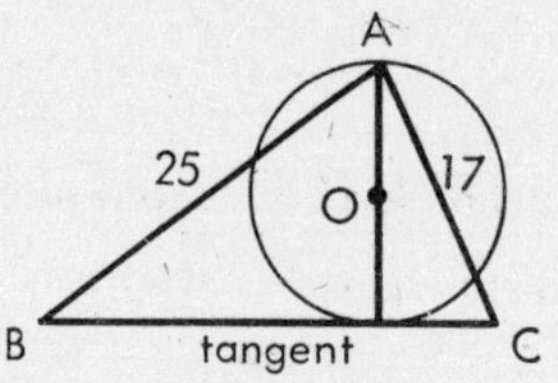

2.

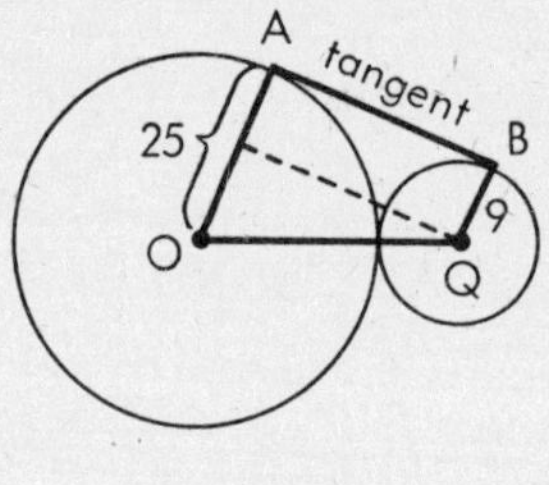

3.

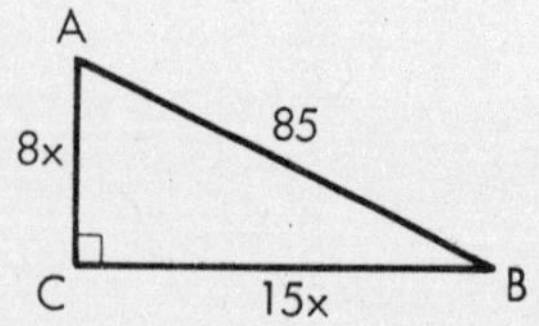

4.

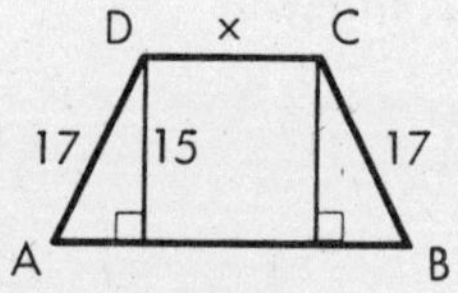

5.

1. $x =$ distance from O to chord AB.

2. Diameter = 15. $x =$ length of BC.

3. $x =$ length of common tangent AB.

4. $x =$ common factor of AC and BC.

5. $ABCD$ is an isosceles trapezoid. $AB = 30$.

a. 30
b. 5
c. 14
d. 28
e. 15

3.6 SPECIAL RIGHT TRIANGLES: 30°-60°-90° AND 45°-45°-90° TRIANGLES

Model Problems

(1) In triangle ABC, angle $A = 30°$, angle $B = 60°$, and $BC = 4$. Find the length of AC.

(A) 4 (B) $4\sqrt{2}$ (C) $4\sqrt{3}$ (D) 8 (E) $8\sqrt{3}$

Solution:

Since $\angle A = 30°$ and $\angle B = 60°$, then $\angle C = 180° - 30° - 60° = 90°$. Leg AC, or b, may be found as follows:

Method 1	*Method* 2
In a right triangle, the leg opposite the 60° angle equals the leg opposite the 30° angle multiplied by $\sqrt{3}$. Hence, $b = a\sqrt{3}$ Substitute 4 for a. $b = 4\sqrt{3}$	In a right triangle, the leg opposite the 30° angle equals one-half the hypotenuse. Since $a = 4$, then $c = 8$. In a right triangle, the leg opposite the 60° angle equals one-half the hypotenuse multiplied by $\sqrt{3}$. Hence, $b = \frac{c}{2}\sqrt{3}$ Substitute 8 for c. $b = \frac{8}{2}\sqrt{3} = 4\sqrt{3}$

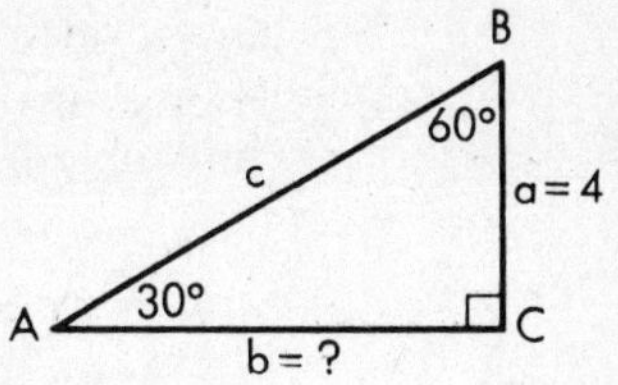

Ans. (C)

(2) If the hypotenuse of a right triangle is $6\sqrt{2}$ and one leg is 6, find the length of the other leg.

(A) 4 (B) $4\sqrt{2}$ (C) $4\sqrt{3}$ (D) 8 (E) 6

Solution:

Leg x may be found as follows:

Method 1	*Method* 2
In a right triangle, $(\text{leg})^2 + (\text{leg})^2 = (\text{hyp.})^2$ $6^2 + x^2 = (6\sqrt{2})^2$ $36 + x^2 = 72$ $x^2 = 36$ $x = 6$	In a right triangle, if the hypotenuse equals a leg multiplied by $\sqrt{2}$, the triangle is isosceles. Hence, leg x = leg $AC = 6$

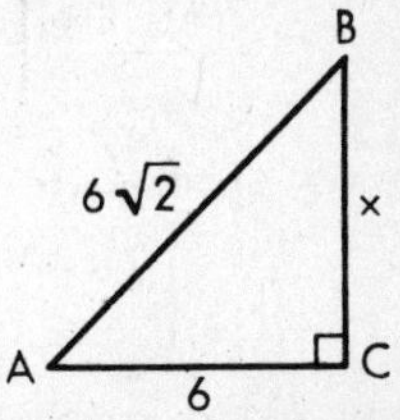

Ans. (E)

(3) An altitude of an equilateral triangle is $3\sqrt{3}$. Find the length of a side of the triangle.

(A) 6 (B) 12 (C) $4\sqrt{3}$ (D) $6\sqrt{3}$ (E) 9

Solution:

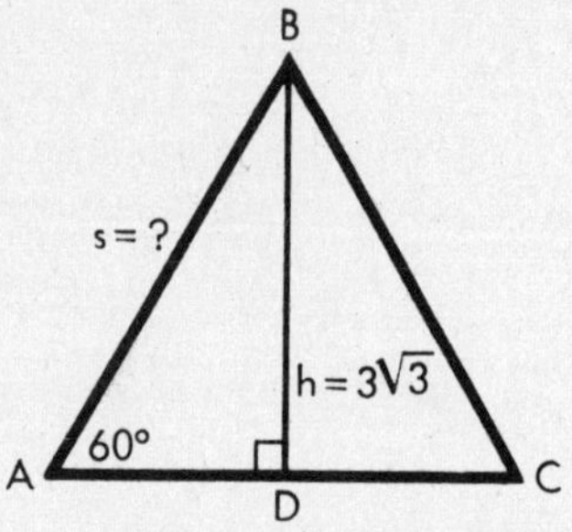

Each angle of an equilateral triangle is 60°. Hence, in equilateral $\triangle ABC$, $\angle A = 60°$. Find side s as follows:

In the 30°-60°-90° $\triangle ABD$, h is the leg opposite 60°.

In a right triangle, the leg opposite the 60° angle equals one-half the hypotenuse multiplied by $\sqrt{3}$.

Hence, $h = \frac{s}{2}\sqrt{3}$

Substitute $3\sqrt{3}$ for h.

$$3\sqrt{3} = \frac{s}{2}\sqrt{3}$$

Divide each side by $\sqrt{3}$.

$$3 = \frac{s}{2}, \ s = 6 \quad \textit{Ans.}\ \text{(A)}$$

(4) Find the length of an altitude of an equilateral triangle whose side is 2.

(A) 1 (B) $\sqrt{3}$ (C) $\frac{\sqrt{3}}{2}$ (D) $\frac{2\sqrt{3}}{3}$ (E) $\frac{\sqrt{3}}{3}$

Solution:

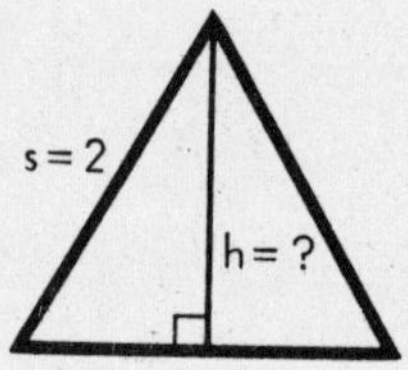

Find an altitude, h, of the equilateral triangle by applying the following principle:

In an equilateral triangle, an altitude equals one-half a side multiplied by $\sqrt{3}$.

Hence, $h = \frac{s}{2}\sqrt{3}$

Substitute 2 for s.

$$h = \frac{2}{2}\sqrt{3}$$
$$= \sqrt{3} \quad \textit{Ans.}\ \text{(B)}$$

(5) Find a diagonal of a square whose side is 6.

(A) $6\sqrt{2}$ (B) $6\sqrt{3}$ (C) 8.4 (D) 9 (E) 12

Solution:

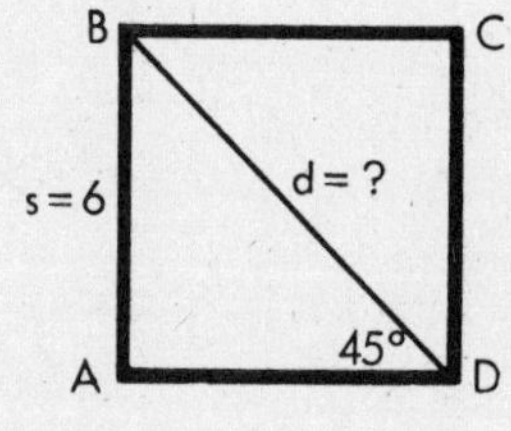

A diagonal of a square bisects each vertex angle. Hence, in square $ABCD$, angle $BDA = 45°$.

Find diagonal d as follows:

In an isosceles right triangle, the hypotenuse equals a leg opposite a 45° angle multiplied by $\sqrt{2}$.

Hence, if a side $s = 6$, then $d = 6\sqrt{2}$. *Ans.* (A)

(6) Find the length of a side of a square whose diagonal is $3\sqrt{2}$.

(A) 1.5 (B) $2\sqrt{2}$ (C) 3 (D) $\frac{3}{\sqrt{2}}$ (E) $\frac{2}{\sqrt{3}}$

Solution:

A diagonal of a square bisects each vertex angle. Hence, in square *ABCD*, angle $BAC = 45°$ and $\triangle ABC$ is a 45°-45°-90° right triangle.

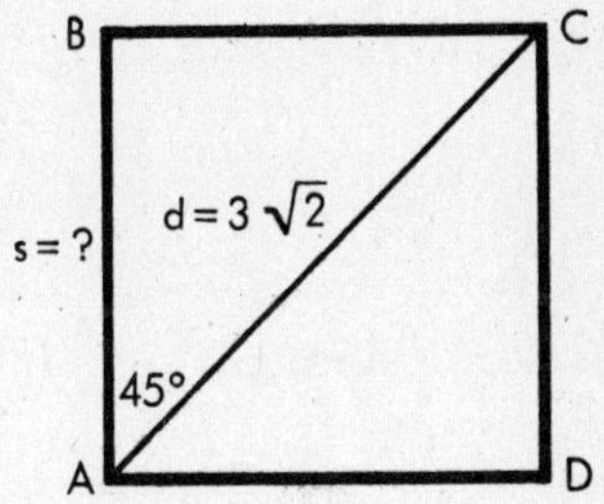

Find side *s* as follows:

In a 45°-45°-90° right triangle, the hypotenuse equals a leg multiplied by $\sqrt{2}$.

Hence, $d = s\sqrt{2}$

Substitute $3\sqrt{2}$ for *d*.

$3\sqrt{2} = s\sqrt{2}$

Divide each side by $\sqrt{2}$.

$3 = s$ *Ans.* (C)

(7) If the apothem of a regular hexagon is $6\sqrt{3}$, find a side of the hexagon.

(A) 6 (B) 9 (C) 18 (D) $12\sqrt{3}$ (E) 12

Solution:

An apothem of a regular polygon is a line drawn from its center perpendicular to one of its sides. Thus, in the hexagon at the right, $OH \perp AB$.

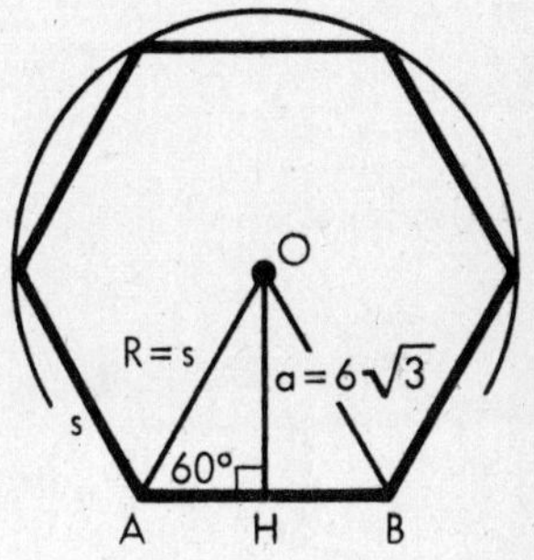

Since a side of a regular hexagon equals a radius of its circumscribed circle, $\triangle OAB$ is an equilateral triangle and $R = s$.

Find *s* by applying the following rule:

In a right triangle, a leg opposite the 60° angle is one-half the hypotenuse multiplied by $\sqrt{3}$.

Hence, $a = \frac{s}{2}\sqrt{3}$

Substitute $6\sqrt{3}$ for *a*.

$6\sqrt{3} = \frac{s}{2}\sqrt{3}$

Divide each side by $\sqrt{3}$.

$6 = \frac{s}{2}$, $s = 12$ *Ans.* (E)

After completing each test, match your answers with the numbers in the boxes.

Matching Test A

1. If the perimeter of a square is 24, find a diagonal.	*a*. 12
2. If a diagonal of a square is 6, find its perimeter.	*b*. $12\sqrt{2}$
3. If the perimeter of an equilateral triangle is 24, find the altitude.	*c*. $12\sqrt{3}$
4. If an altitude of an equilateral triangle is $6\sqrt{3}$, find a side.	*d*. $4\sqrt{3}$
5. If an altitude of an equilateral triangle is 6, find its perimeter.	*e*. $6\sqrt{2}$

Matching Test B

In each figure, find the length represented by x.

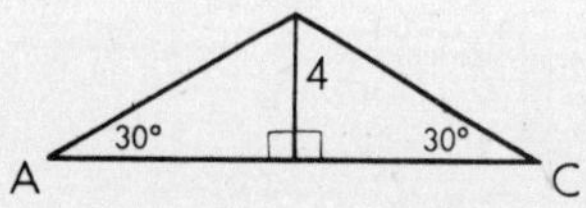

1.

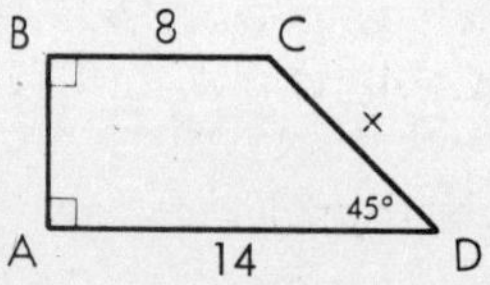

2.

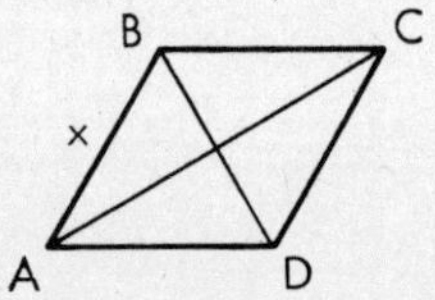

3.

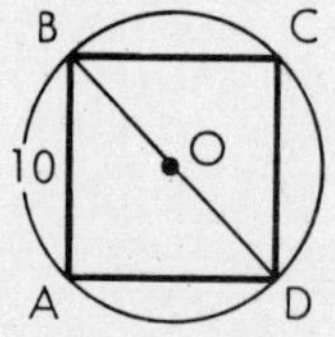

4.

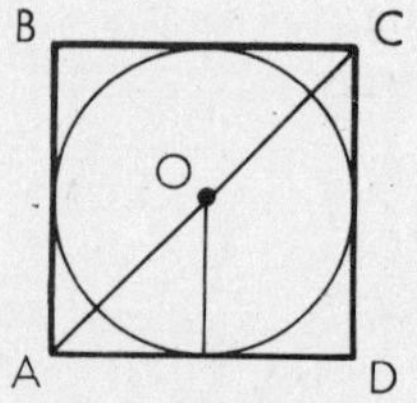

5.

1. $x =$ length of AC.	*a*. $4\sqrt{2}$
2. (The figure contains all necessary information.)	*b*. 12
3. $ABCD$ is a rhombus. $AC = 12\sqrt{3}$. $BD = 12$.	*c*. $5\sqrt{2}$
4. $ABCD$ is a square. $x =$ radius of the circumscribed circle.	*d*. $6\sqrt{2}$
5. $ABCD$ is a square. $AC = 16$. $x =$ radius of the inscribed circle.	*e*. $8\sqrt{3}$

4. PRACTICE PROBLEMS

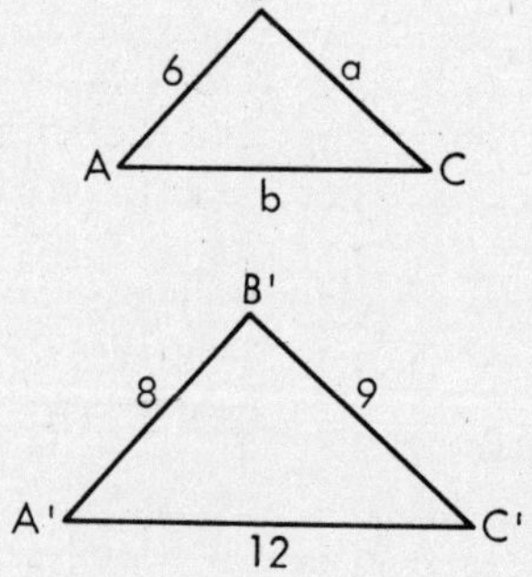

1. In the figure at the right, $\angle A = \angle A'$, $\angle B = \angle B'$, $\angle C = \angle C'$. Find a.

(A) $6\frac{3}{4}$ (B) 4 (C) $4\frac{1}{2}$ (D) $5\frac{1}{3}$ (E) 8 **[1.1]**

2. If the circumference of a circle is to be made four times as large, then the diameter must be made ______ times as large.

(A) 2 (B) 4 (C) $\frac{4}{\pi}$ (D) 2π (E) 8 **[1.2]**

3. Corresponding medians of two similar triangles are 24 and 30. If an altitude of the first triangle is 20, find the corresponding altitude in the other triangle.
(A) 18 (B) 21 (C) 24 (D) 25 (E) 27 **[1.2]**

4. A tree 44 feet tall casts a shadow 33 feet long at the same time that a nearby upright pole casts a shadow of 3 feet 9 inches. Find the number of feet in the height of the pole.
(A) $2\frac{13}{16}$ (B) 4 (C) $4\frac{1}{2}$ (D) 5 (E) 60 **[1.3]**

5. In $\triangle ABC$, DE is parallel to AC and cuts AB at D and BC at E. If $AB = 12$, $AD = 3$, and $EC = 9$, find BE.
(A) 36 (B) 30 (C) 27 (D) 24 (E) 18 **[2.1]**

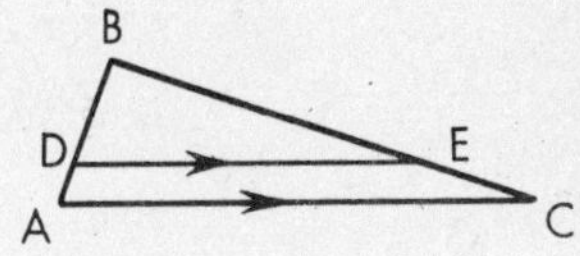

6. In $\triangle ABC$, DE is parallel to AC and cuts AB at D and BC at E. If $DE = 4$, $AC = 8$, and $AB = 6$, find DB.
(A) 3 (B) 2 (C) 4 (D) $4\frac{1}{2}$ (E) 6 **[2.1]**

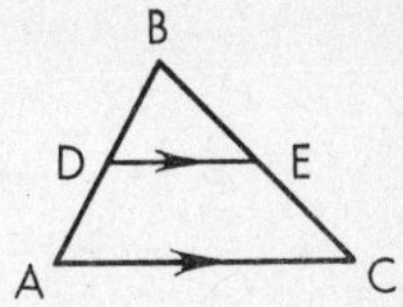

7. The legs of a right triangle are each $5\sqrt{2}$. Find the hypotenuse.
(A) 20 (B) 10 (C) $\frac{5}{\sqrt{2}}$ (D) $\frac{10}{\sqrt{2}}$ (E) $10\sqrt{2}$ **[3.1]**

8. A chord 8 inches long is 3 inches from the center of the circle. How many inches long is the radius of the circle?
(A) $2\frac{1}{2}$ (B) $\sqrt{7}$ (C) 5 (D) $\sqrt{73}$ (E) 10 **[3.1]**

9. Which of the following sets of numbers cannot be used as the lengths of the sides of a right triangle?
(A) 3, 5, 6 (B) 3, 3, $3\sqrt{2}$ (C) 3, $3\sqrt{3}$, 6 (D) 5, 12, 13 (E) $\frac{3}{5}$, $\frac{4}{5}$, 1 **[3.2]**

10. A leg of a right triangle is 24 inches. If the ratio of the other leg to the hypotenuse is 4:5, find the number of inches in the hypotenuse.
(A) 25 (B) 30 (C) 40 (D) 45 (E) 60 **[3.3]**

11. A rectangular lot has sides of 33 feet and 44 feet. How many feet are saved by walking diagonally across the lot instead of along two adjacent sides?
(A) 11 (B) 16.5 (C) 22 (D) 55 (E) 77 **[3.3]**

12. Which of the following sets of numbers cannot be used as the lengths of the sides of a right triangle whose legs are in the ratio 3:4?
(A) $\frac{3}{4}$, 1, $\frac{5}{4}$ (B) 6, 8, 12 (C) 55, 33, 44 (D) 150, 200, 250 (E) .3, .4, .5 **[3.3]**

13. The hypotenuse of a right triangle is 78. If the ratio of the legs of the triangle is 5:12, find the length of the shorter leg.
(A) 6 (B) 15 (C) 30 (D) 45 (E) 60 **[3.4]**

14. In acute $\triangle ABC$, BD is the altitude to AC. If $AB = 15$, $BC = 13$, and $BD = 12$, how long is AC?
(A) 5 (B) 9 (C) 13 (D) 14 (E) 15 **[3.4]**

15. Which of the following sets of numbers cannot be used as the lengths of the sides of a 5-12-13 triangle?

(A) $\frac{5}{11}, \frac{12}{11}, \frac{13}{11}$ (B) .5, 1.2, 1.3 (C) 10, 24, 26 (D) 52, 48, 20 (E) 25, 60, 75

[3.4]

16. Find x, the length of AB.

(A) 6 (B) 9 (C) 12 (D) 13 (E) 15 **[3.5]**

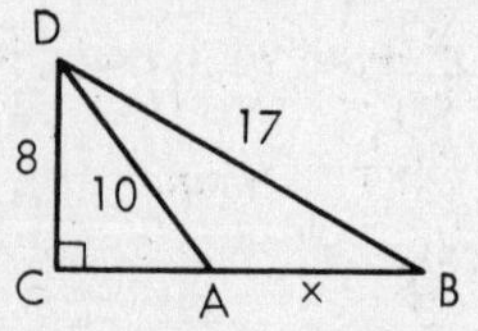

17. A side of a square is 8. Find the diagonal to the nearest integer.

(A) 9 (B) 10 (C) 11 (D) 12 (E) 15

[3.6]

18. Find the radius of the circumscribed circle and the perimeter of regular hexagon $ABCDEF$.

(A) 4, 24 (B) 12, 48 (C) 8, 32 (D) 8, 48 (E) $8\sqrt{3}$, $48\sqrt{3}$

[3.6]

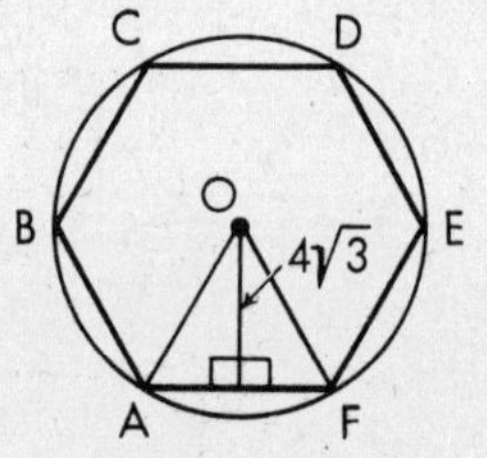

VIII. REVIEWING AREAS AND VOLUMES

Reviewing Areas of Plane Figures

A ***square unit*** is the surface of a square whose side is 1 unit. Hence, a square inch is a square whose side is 1 inch.

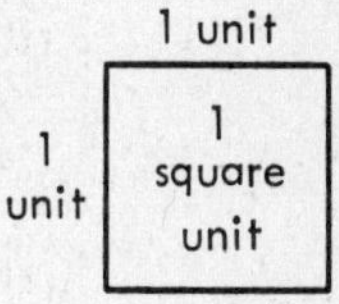

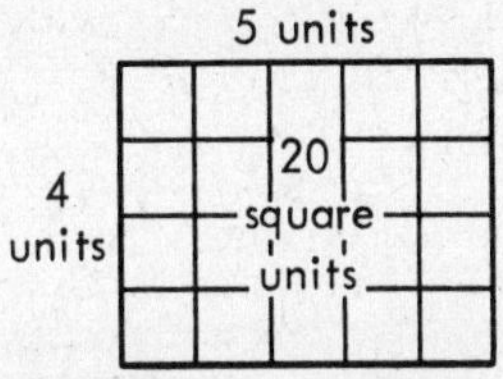

The ***area of a polygon*** is the number of square units contained in its surface.

For example, the area of a rectangle 5 inches long and 4 inches wide is 20 square inches, since its area can be shown to contain 20 squares each of which is a square inch.

Principles of Area

1. The *area of a rectangle* equals the product of its base and altitude.
   Formula: $K = bh$

   For example, if $b = 5$ and $h = 4$, then $K = bh = 5(4) = 20$.

Rectangle

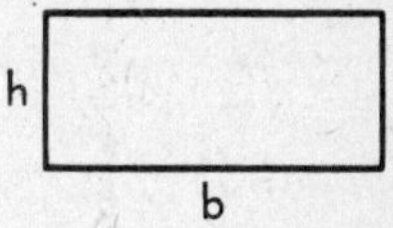

$K = bh$

Square

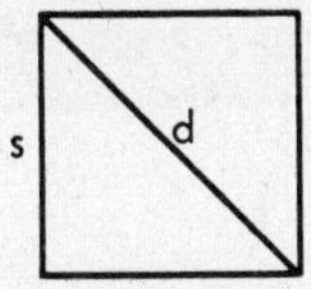

(1) $K = s^2$

(2) $K = \frac{1}{2}d^2$

2. The *area of a square* equals:
   (1) the square of a side. Formula: $K = s^2$
   For example, if $s = 5$, then $K = s^2 = 25$.

   (2) one-half the square of a diagonal. Formula: $K = \frac{1}{2}d^2$

   For example, if $d = 5$, then $K = \frac{1}{2}d^2 = \frac{1}{2}(25) = 12\frac{1}{2}$.

3. The *area of a parallelogram* equals:
   (1) the product of a side and the altitude to that side.
   Formula: $K = bh$
   For example, if $b = 10$ and $h = 3.4$,
   then $K = bh = 10(3.4) = 34$.

   (2) the product of two adjacent sides and the sine of the included angle. Formula: $K = ab \sin C$
   For example, if $a = 4$, $b = 5$, and $C = 30°$,

   then $K = ab \sin C = 4(5)\left(\frac{1}{2}\right) = 10$.

Parallelogram

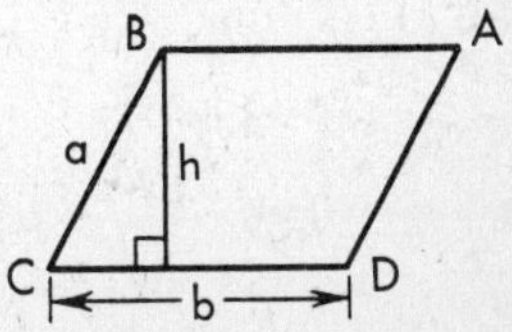

(1) $K = bh$
(2) $K = ab \sin C$

4. The *area of a triangle* equals:

   (1) one-half the product of one side and the altitude to that side.

   Formula: $K = \frac{1}{2}bh$

   For example, if $b = 10$ and $h = 3.4$,
   then $K = \frac{1}{2}bh = \frac{1}{2}(10)(3.4) = 17$.

   (2) one-half the product of any two adjacent sides and the sine of the included angle. Formula: $K = \frac{1}{2}ab \sin C$

   For example, if $a = 4$, $b = 5$, and $C = 30°$,
   then $K = \frac{1}{2}ab \sin C = \frac{1}{2}(4)(5)\left(\frac{1}{2}\right) = 5$.

Triangle

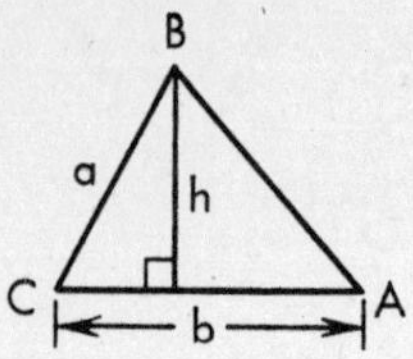

(1) $K = \frac{1}{2}bh$

(2) $K = \frac{1}{2}ab \sin C$

5. The *area of an equilateral triangle* equals:

   (1) one-fourth the square of a side times $\sqrt{3}$. Formula: $K = \frac{s^2}{4}\sqrt{3}$

   For example, if $s = 8$,
   then $K = \frac{s^2}{4}\sqrt{3} = \frac{64}{4}\sqrt{3} = 16\sqrt{3}$.

   (2) one-third the square of an altitude times $\sqrt{3}$.

   Formula: $K = \frac{h^2}{3}\sqrt{3}$

   For example, if $h = 6$, then $K = \frac{h^2}{3}\sqrt{3} = \frac{36}{3}\sqrt{3} = 12\sqrt{3}$.

Equilateral Triangle

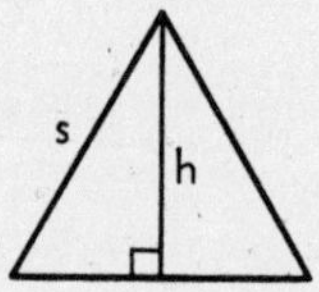

(1) $K = \frac{1}{4}s^2\sqrt{3}$

(2) $K = \frac{1}{3}h^2\sqrt{3}$

Rhombus

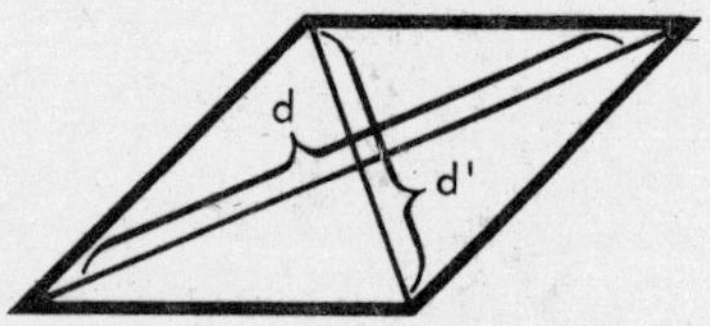

$K = \frac{1}{2}dd'$

6. The *area of a rhombus* equals one-half the product of the diagonals. Formula: $K = \frac{1}{2}dd'$

   For example, if $d = 7$ and $d' = 4$,
   then $K = \frac{1}{2}dd' = \frac{1}{2}(7)(4) = 14$.

7. The *area of a trapezoid* equals:

   (1) one-half the product of its altitude and the sum of the bases.

   Formula: $K = \frac{1}{2}h(b + b')$

   For example, if $h = 8$, $b = 13$, and $b' = 7$,
   then $K = \frac{1}{2}h(b + b') = \frac{1}{2}(8)(13 + 7) = \frac{1}{2}(8)(20) = 80$.

   (2) the product of its altitude and median. Formula: $K = hm$

Trapezoid

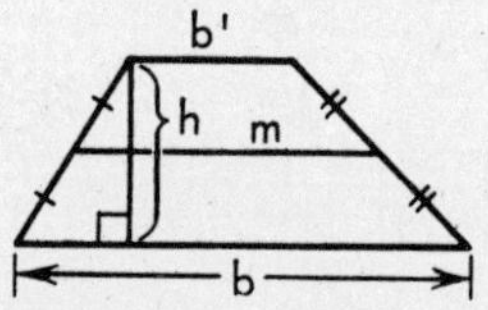

(1) $K = \frac{1}{2}h(b + b')$

(2) $K = hm$

> *Note.* The median of a trapezoid is one-half the sum of the bases.

For example, if $b = 13$ and $b' = 7$, then the median, $m = \frac{1}{2}(13 + 7) = \frac{1}{2}(20) = 10$. Using $m = 10$ and $h = 8$, $K = hm = (8)(10) = 80$.

8. The *area of a regular polygon* equals:

   (1) one-half the product of the perimeter and apothem.

   Formula: $K = \frac{1}{2}pr$

   (2) one-half the product of the number of sides, the length of a side, and the apothem. Formula: $K = \frac{1}{2}nsr$

   For example, if $n = 4$, $s = 6$, and $r = 3$, then $K = \frac{1}{2}nsr = \frac{1}{2}(4)(6)(3) = 36$.

   Here, the perimeter, $p = ns = 4(6) = 24$. Using $K = \frac{1}{2}pr$, $K = \frac{1}{2}(24)(3) = 36$.

Regular Polygon

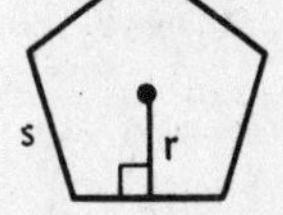

(1) $K = \frac{1}{2}pr$

(2) $K = \frac{1}{2}nsr$

9. The *areas of similar polygons* are to each other as the squares of any two corresponding lines.

   Hence, if $\triangle ABC \sim \triangle A'B'C'$, then

$$\frac{K}{K'} = \frac{s^2}{s'^2} = \frac{h^2}{h'^2} = \frac{m^2}{m'^2} = \frac{p^2}{p'^2}$$

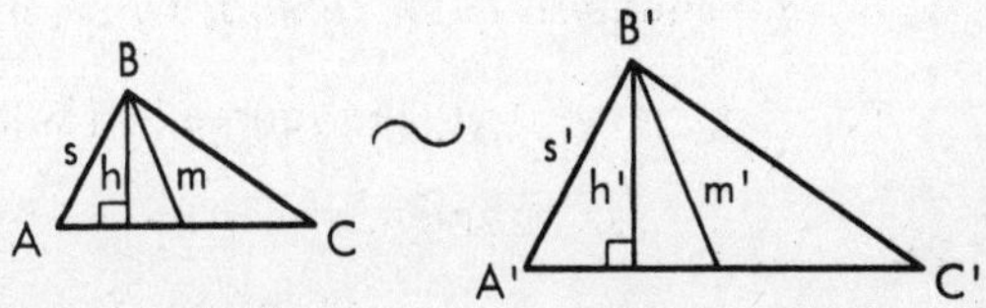

   For example, if the area of $\triangle A'B'C'$ is 49 times the area of $\triangle ABC$ and the two triangles are similar, then the corresponding sides, medians, altitudes, and perimeters are in the ratio 7:1.

   Principle 9 may be restated as follows:

   The *ratio of the areas of two similar polygons* equals the ratio of the squares of any two corresponding lines.

Area Principles Involving Equivalent Polygons

Equivalent polygons, or ***equal polygons,*** are polygons having the same area, or size. For example, a rectangle of base 8 and altitude 2 is equivalent, or equal, to a square of side 4.

Equivalent Polygons

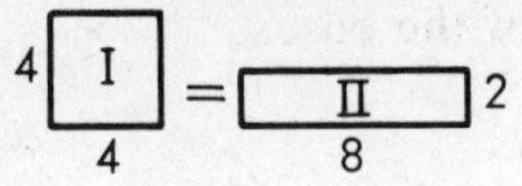

I = II

The polygons have the same size.

Similar Polygons

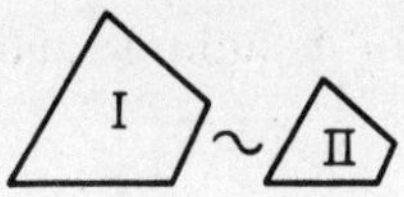

I ∼ II

The polygons have the same shape.

Congruent Polygons

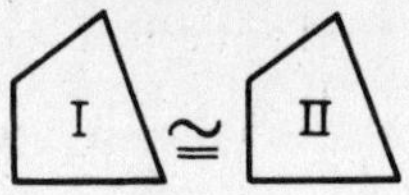

I ≅ II

The polygons have both the same size and the same shape.

10. Parallelograms are equal in area if they have equal bases and equal altitudes.

11. Triangles are equal in area if they have equal bases and equal altitudes.

12. A median divides a triangle into two equal triangles. Thus, a median bisects a triangle.

    In $\triangle ABC$, if BM is a median, then $\triangle I = \triangle II$. Note that $\triangle I$ and $\triangle II$ have equal bases and the same altitude.

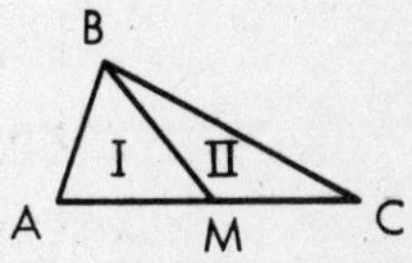

Reviewing Volumes of Rectangular Solids and Cubes

A ***cubic unit*** is a cube having an edge of 1 unit. Hence, a cubic inch is a cube whose edge is 1 inch. Note that a cube is bounded by six faces and that each face is a square. A cube has twelve edges: four on top, four on bottom, and four along the vertical sides.

1 Cubic Inch

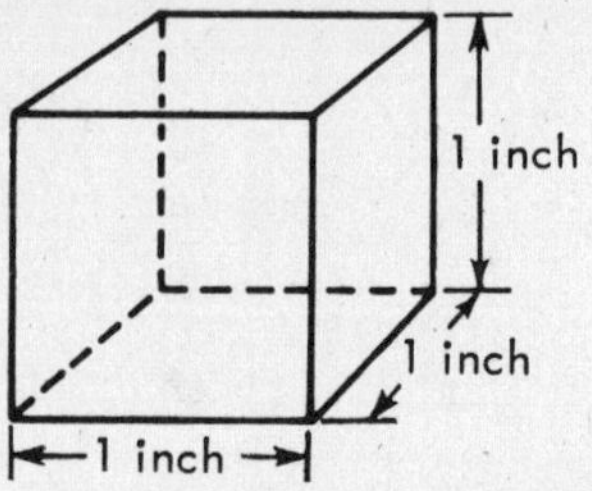

24 Cubic Units

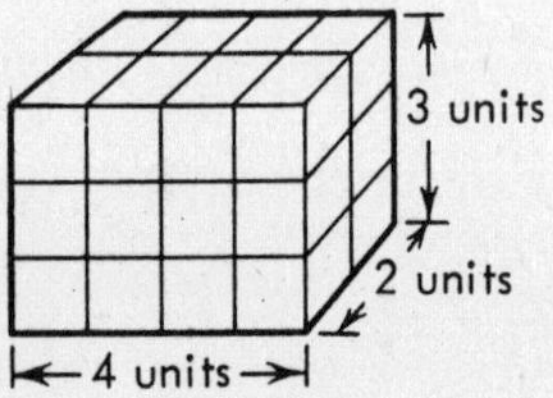

The ***volume of a solid*** is the number of cubic units that the solid can contain. For example, a box 4 units long, 2 units wide, and 3 units high has a volume of 24 cubic units; that is, its volume is large enough to contain 24 cubic units, or 24 cubes, each having an edge of 1 unit.

Principles of Volume

13. The *volume of a rectangular solid* equals the product of the length, the width, and the height. Formula: $V = lwh$

Rectangular Solid

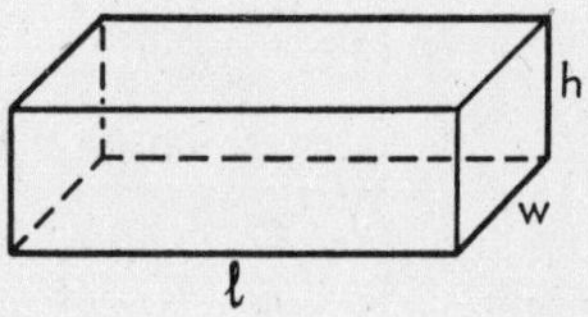

Cube

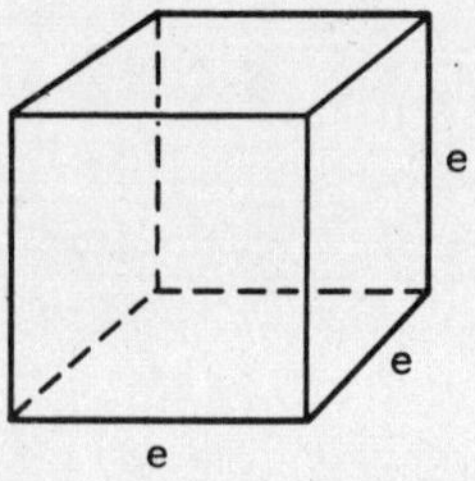

14. The *volume of a cube* equals the cube of an edge. Formula: $V = e^3$

1. AREAS OF RECTANGLES AND SQUARES

1.1 AREA OF A SQUARE

Model Problems

(1) If the length of a diagonal of a square is 8 inches, find the number of square inches in the area of the square.

(A) 64 (B) 32 (C) $16\sqrt{2}$ (D) 16 (E) $8\sqrt{2}$

Solution:

The area of a square equals the square of a side or one-half the square of a diagonal. Let $K =$ the number of square inches in the area of the square.

Method 1	*Method* 2
In rt. $\triangle ABD$, $s^2 + s^2 = d^2$ Substitute 8 for d. $2s^2 = 8^2 = 64$ $s^2 = 32$ Since $K = s^2$, $K = 32$	$K = \frac{d^2}{2}$ Substitute 8 for d. $K = \frac{8^2}{2}$ $= \frac{64}{2} = 32$

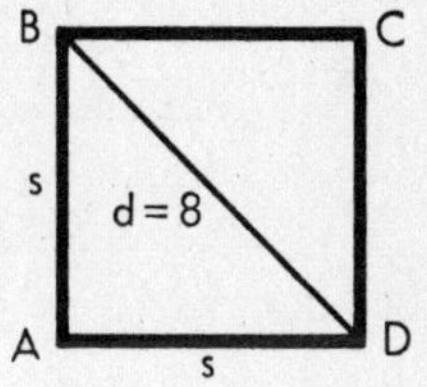

Ans. (B)

(2) Square $ABCD$ is inscribed in a circle as shown in the diagram. If the radius of the circle is 6, find the area of the square.

(A) 36 (B) 72 (C) $72\sqrt{2}$ (D) 144 (E) $144\sqrt{2}$

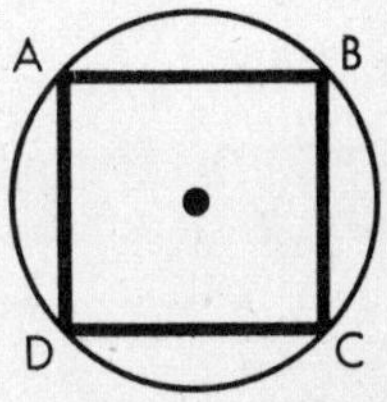

Solution:

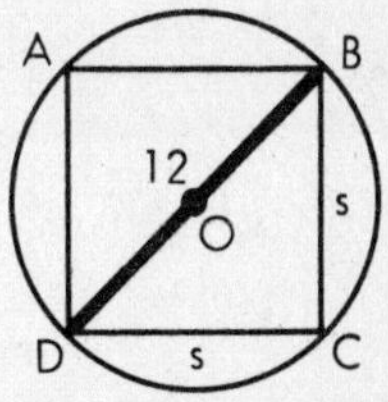

The area of a square equals the square of a side. Thus, if s is a side of the square, find s^2 as follows:

The diagonal of a square inscribed in a circle is a diameter of the circle.

Hence, if the radius of the circle is 6, then the diagonal is 12.

Apply the Law of Pythagoras:

In a right triangle, the square of the hypotenuse equals the sum of the squares of the legs.

$$(\text{leg})^2 + (\text{leg})^2 = (\text{hypotenuse})^2$$

$$s^2 + s^2 = 12^2$$

$$2s^2 = 144$$

Divide each side by 2: $$\frac{2s^2}{2} = \frac{144}{2}$$

$$s^2 = 72$$

Hence, the area of square $ABCD = s^2 = 72$. *Ans.* (B)

Matching Test

Refer to the figures at the right to find the missing lengths and areas indicated by the question marks in the following tables. After completing this test, match your answers with the pairs of numbers in the box.

	Side	*Diagonal*	*Area*
1.	8	?	?
2.	?	8	?

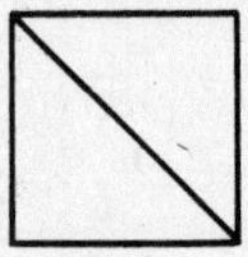

Square

	Side of the square	*Radius of the circle*	*Area of the square*
3.	8	?	?
4.	?	8	?

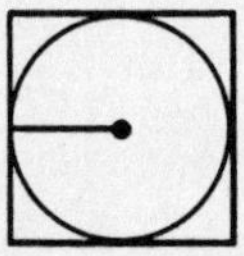

Circumscribed Square

	Side of the square	*Radius of the circle*	*Area of the square*
5.	8	?	?

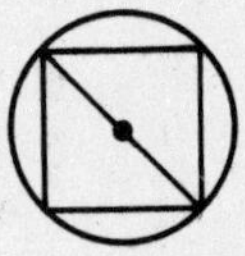

Inscribed Square

a. 16, 256	*b.* 4, 64	*c.* $4\sqrt{2}$, 64	*d.* $8\sqrt{2}$, 64	*e.* $4\sqrt{2}$, 32

1.2 RATIOS OF AREAS OF RECTANGLES (TILES, BLOCKS, ETC.)

Model Problem

How many square tiles 3 inches on a side are needed to cover a floor 6 feet by 8 feet?

(A) 16 (B) 64 (C) 192 (D) 768 (E) 2304

Solution:

Since 1 square foot = 144 square inches, the area of the rectangular floor in square inches $= 6 \cdot 8 \cdot 144$.

The area of each square tile in square inches $= 3 \cdot 3$.

Divide to find n, the number of tiles: $n = \dfrac{6 \cdot 8 \cdot 144}{3 \cdot 3}$

$= 768$ *Ans.* (D)

Matching Test

After completing this test, match your answers with the numbers in the box.

1. How many square tiles 4 inches on a side are needed to cover a floor 6 feet by 8 feet?
2. How many rectangular tiles 3 inches by 4 inches are needed to cover a floor 4 feet by 8 feet?
3. In order to cover a floor, 512 tiles are used. If each tile is a square 3 inches on a side, find the floor area covered in square feet.
4. If a floor is covered by 1152 square tiles, each $2\frac{1}{2}$ inches on a side, find the area of the floor in square feet.
5. A square floor 12 feet on a side is covered by square tiles, 8 inches on a side. How many tiles are there in the floor?

a. 32	*b*. 50	*c*. 432	*d*. 324	*e*. 384

1.3 AREAS OF RECTANGLES IN COMBINATION FIGURES

Model Problem

A rectangle is formed by 5 equal squares placed side by side in a line. If the perimeter of each square is 20, find the perimeter and the area of the rectangle.

(A) 60, 100 (B) 30, 100 (C) 100, 125 (D) 30, 125 (E) 60, 125

Solution:

Since the perimeter of each square is 20, each side is $\frac{1}{4}(20)$, or 5.

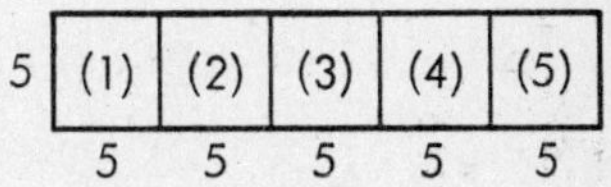

Since the length and width of the rectangle are 25 and 5, respectively,

(1) the perimeter $= 2 \cdot 25 + 2 \cdot 5 = 60$

(2) the area $= 25 \cdot 5 = 125$ *Ans.* (E)

Matching Test

After completing this test, match your answers with the numbers in the box.

1. (Fig. 1) The area of each of the five equal squares is 16. Find the perimeter of the entire figure.
2. (Fig. 1) The perimeter of each of the equal squares is 24. Find the area of the entire figure.
3. (Fig. 1) The area of the entire figure consisting of five equal squares is 125. Find its perimeter.

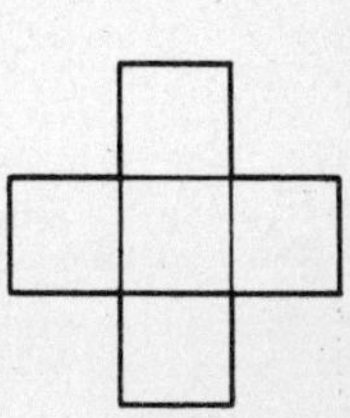

Fig. 1

4. (Fig. 2) The entire figure, a square, contains two squares, A and B, whose areas are 64 and 25. Find the area of the shaded part.

5. (Fig. 2) The area of the entire square is 121. If the area of square A is 64, find the perimeter of square B.

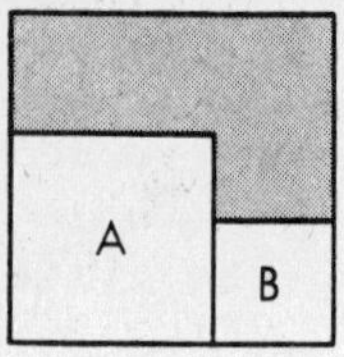

Fig. 2

a. 60	*b.* 80	*c.* 180	*d.* 48	*e.* 12

1.4 DIFFERENCES OF AREAS OF RECTANGLES (WALKS, FRAMES, ETC.)

Model Problem

A rectangular flower bed 8 yards by 5 yards is surrounded by a walk whose width is 1 yard. Find the number of square yards in the area of the walk.

(A) 54 (B) 32 (C) 30 (D) 18 (E) 14

Solution:

Since the walk, 1 yard wide, lies on each side of the flower bed, the dimensions of the larger rectangle formed by the outer edges of the walk are 10 yards by 7 yards. Hence, the area of the larger rectangle in square yards $= 10 \cdot 7 = 70$.

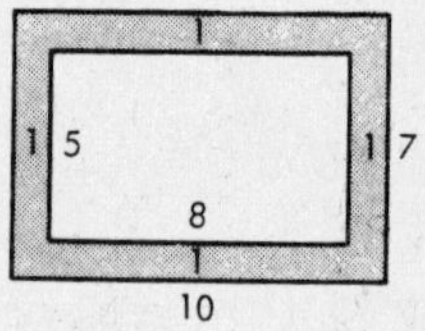

The area of the flower bed in square yards $= 8 \cdot 5 = 40$.

By subtracting, the area of the walk in square yards $= 70 - 40 = 30$. *Ans.* (C)

Matching Test

After completing this test, match your answers with the numbers in the box.

1. A rectangular flower bed 8 yards by 5 yards is surrounded by a walk whose width is 2 yards. Find the number of square yards in the area of the walk.
2. A picture 24 inches by 15 inches has a frame with a width of 3 inches. Find the number of square inches in the area of the frame.
3. A picture 24 inches by 15 inches has a frame with a width of $1\frac{1}{2}$ inches. Find the total area of the picture and the frame in square inches.
4. A swimming pool 24 feet by 15 feet is bordered by a walk that is $2\frac{1}{2}$ feet wide. Find the total area of the pool and the walk in square feet.
5. A square $3\frac{1}{2}$ inches on a side is cut out of a square 5 inches on a side. How many square inches remain?

a. 486	*b.* 580	*c.* 68	*d.* $12\frac{3}{4}$	*e.* 270

1.5 EXPRESSING THE RELATION BETWEEN THE AREAS OF A RECTANGLE AND A SQUARE

Model Problem

The area of a square of side a is 58 square units less than the area of a rectangle whose length is b and whose width is 9. An equation that expresses this relation is

(A) $a^2 - 58 = 9b$ (B) $a^2 + 58 = 9b$ (C) $4a + 58 = 2(9 + b)$

(D) $4a - 58 = 2(9 + b)$ (E) $4a + 58 = 2(9b)$

Solution:

The area of a square is the square of its side. Hence, if the side is a, the area of the square is a^2.

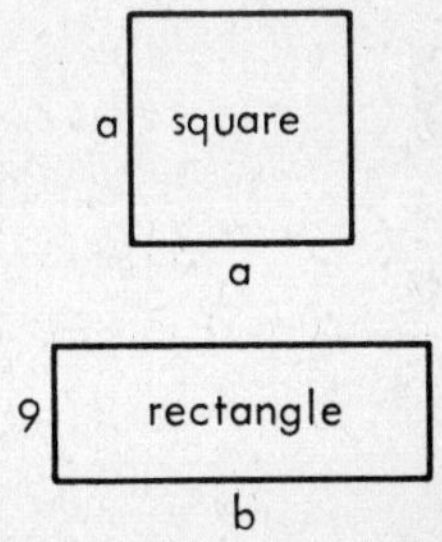

The area of a rectangle is the product of its length and width. Hence, if the length is b and the width is 9, the area of the rectangle is $9b$.

The relationship "The area of the square is 58 less than the area of the rectangle" is expressed by the equation

$$a^2 = 9b - 58$$

By adding 58 to each side or by transposing -58, we obtain

$$a^2 + 58 = 9b \quad \textit{Ans.}\ (B)$$

Matching Test

Each statement describes a relation between a square of side a and a rectangle whose dimensions are b and 9. For each statement, write an equation expressing the relation between a and b. (Refer to the diagram in the preceding model problem.) After completing this test, match your answers with the expressions in the box.

1. The areas of the rectangle and the square are equal.

2. The sum of the areas of the rectangle and the square is 50.

3. The area of the square is 50 more than twice the area of the rectangle.

4. The area of the rectangle is 50 less than twice the area of the square.

5. The sum of twice the area of the square and one-third the area of the rectangle is 50.

a. $a^2 = 18b + 50$	*b.* $2a^2 = 9b + 50$	*c.* $a^2 = 9b$	*d.* $a^2 + 9b = 50$	*e.* $2a^2 + 3b = 50$

2. AREAS OF PARALLELOGRAMS AND TRIANGLES

2.1 PROBLEMS INVOLVING AREAS OF PARALLELOGRAMS AND TRIANGLES

Model Problems

(1) The lengths of two sides of a triangle are 12 inches and 15 inches, and the included angle contains 30 degrees. Find the number of square inches in the area of the triangle.

(A) 45 (B) $45\sqrt{3}$ (C) 80 (D) 90 (E) 180

Solution:

Method 1	*Method* 2
Find the area of $\triangle ABC$ by applying the following rule: The area of a triangle is equal to one-half the product of any two sides and the sine of the angle included by these sides. Hence, in the figure at the right, if K = area of $\triangle ABC$, then $K = \frac{1}{2}(12)(15) \sin 30°$ Substitute .5 for sin 30°. $K = \frac{1}{2}(12)(15)(.5)$ $= 45$	Draw $BD \perp AC$ as shown. In right $\triangle ABD$, BD is a leg opposite a 30° angle. Hence, $BD = \frac{1}{2}(12) = 6$. Find the area of $\triangle ABC$ by applying the following rule: The area of a triangle is equal to one-half the product of the base and altitude. Hence, $K = \frac{1}{2}bh$ Substitute 15 for b and 6 for h. $K = \frac{1}{2}(15)(6) = 45$

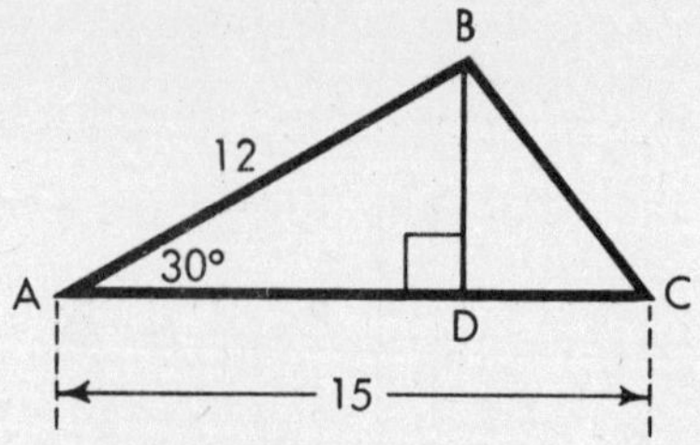

Ans. (A)

(2) In parallelogram $ABCD$, $AB = 8$ inches, $AD = 6$ inches, and angle $A = 30°$. Find the number of square inches in the area of the parallelogram.

(A) 48 (B) 12 (C) 28 (D) $24\sqrt{3}$ (E) 24

Solution:

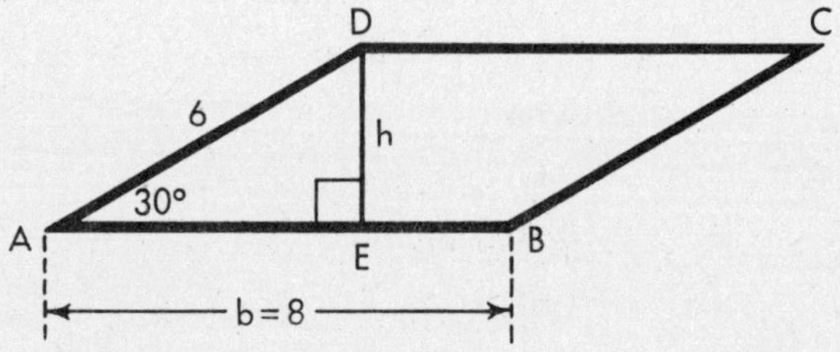

In right $\triangle ADE$, leg DE, or h, is opposite a 30° angle. Since a leg opposite 30° equals one-half the hypotenuse,

$$h = \frac{6}{2} = 3$$

The area of a parallelogram equals the product of the base and the altitude.

Area of $\square ABCD = bh$

Substitute 8 for b and 3 for h.

Area of $\square ABCD = 8(3) = 24$

Ans. (E)

Matching Test

After completing this test, match your answers with the numbers in the box.

1. Find the area of a parallelogram whose sides are 12 and 6 and include an angle of 30°.

2. Find the area of a triangle whose sides are 12 and 6 and include an angle of 30°.

3. Find the area of a rhombus having a side of 12 if one of its angles is 30°.

4. Find the area of a parallelogram whose sides are 12 and 8 and include an angle of 150°.

5. Find the base of a parallelogram whose area is 72 and whose base is twice its altitude.

a. 72	*b.* 48	*c.* 36	*d.* 12	*e.* 18

2.2 AREA OF A RIGHT TRIANGLE

Model Problems

(1) Find the area of a right triangle whose legs are 4 and 5.

(A) 5 (B) 9 (C) 10 (D) 20 (E) 40

Solution:

In a right triangle, either leg may be regarded as the base and the other leg as the altitude.

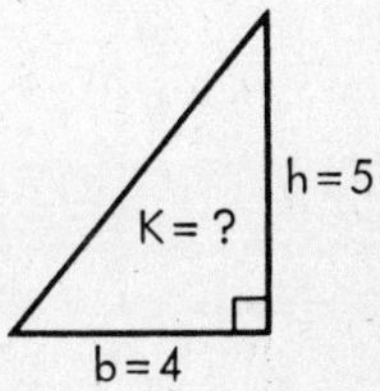

Hence, in the triangle at the right, leg $b = 4$ and altitude $h = 5$.

The area of a triangle equals one-half the product of the base and altitude.

If K = area of triangle,

$$K = \frac{1}{2}bh$$

$$= \frac{1}{2}(4)(5) = 10 \quad \textit{Ans.}\ (C)$$

(2) The area of a triangle is equal to one-half the product of two of its sides if the triangle is

(A) a right triangle (B) an obtuse triangle (C) an acute triangle (D) equilateral (E) equiangular

Solution:

The correct choice may be made by applying the following principle:

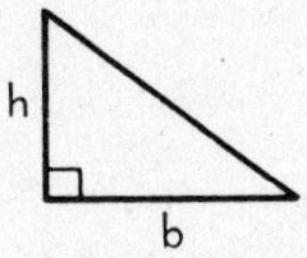

Fig. 1

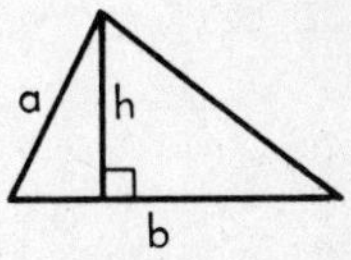

Fig. 2

The area of a triangle equals one-half the product of any side and the altitude drawn to that side.

Hence, the area of a triangle is equal to one-half the product of two of its sides *only* if the triangle is a right triangle and the two sides are its legs (Fig. 1).

If the triangle is not a right triangle, the area can never be equal to one-half the product of two of its sides (Fig. 2).

Ans. (A)

(3) In an isosceles right triangle the hypotenuse is 18. Find the area of the triangle.

(A) $40\frac{1}{2}$ (B) 81 (C) $81\sqrt{2}$ (D) 162 (E) 324

Solution:

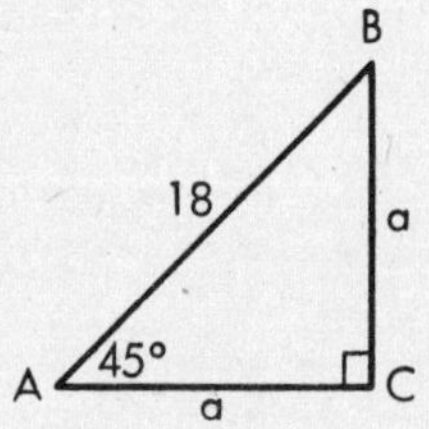

Triangle ABC, at the right, is an isosceles right triangle; hence, either acute angle equals 45°.

Either leg a may be found as follows:

In a right triangle, a leg opposite 45° equals one-half the hypotenuse multiplied by $\sqrt{2}$.

Hence, either leg $a = \frac{1}{2}(18)\sqrt{2} = 9\sqrt{2}$.

The area of a triangle equals one-half the product of a side and the altitude drawn to that side.

Hence, the area of $\triangle ABC = \frac{1}{2}a(a)$

$= \frac{1}{2}(9\sqrt{2})(9\sqrt{2})$

$= 81$ *Ans.* (B)

Matching Test

After completing this test, match your answers with the numbers in the box.

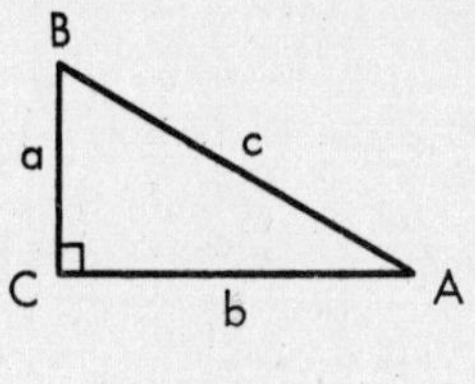

Fig. 1

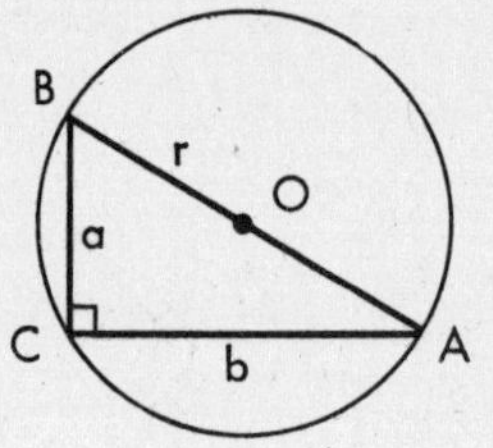

Fig. 2

Find the area of right $\triangle ABC$ if

1. (Fig. 1) $a = 6$ and $b = 10$.	*a.* 50
2. (Fig. 1) $a = 6$ and $c = 10$.	*b.* 96
3. (Fig. 1) $\triangle ABC$ is isosceles and $a = 10$.	*c.* 25
4. (Fig. 1) $\triangle ABC$ is isosceles and $c = 10$.	*d.* 30
5. (Fig. 2) $\triangle ABC$ is inscribed, $r = 10$, and $a = 16$.	*e.* 24

2.3 AREA OF AN EQUILATERAL TRIANGLE

Model Problems

(1) Find the area of an equilateral triangle whose side is 8.

(A) 16 (B) $16\sqrt{3}$ (C) 32 (D) 64 (E) $64\sqrt{3}$

Solution:

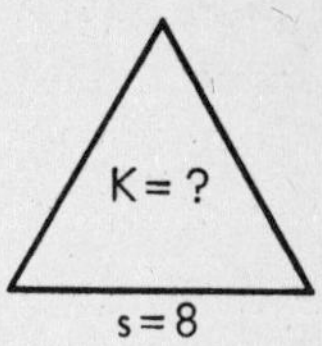

The area of an equilateral triangle equals the product of the square of a side and $\sqrt{3}$ divided by 4.

If K = area of equilateral triangle,

$$K = \frac{s^2\sqrt{3}}{4}$$

$$= \frac{8^2\sqrt{3}}{4}$$

$$= \frac{64\sqrt{3}}{4} = 16\sqrt{3} \quad \textit{Ans. } (B)$$

(2) The perimeter of an equilateral triangle is 12. Find its area.

(A) $4\sqrt{3}$ (B) $12\sqrt{3}$ (C) 36 (D) $36\sqrt{3}$ (E) 72

Solution:

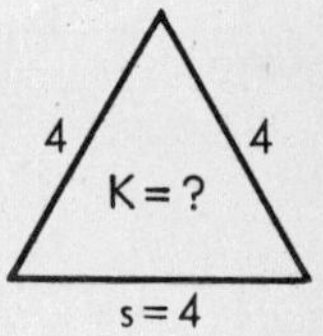

The area of an equilateral triangle equals the product of the square of a side and $\sqrt{3}$ divided by 4.

If K = area of equilateral triangle,

$$K = \frac{s^2\sqrt{3}}{4}$$

Since the perimeter = 12, each of the three sides = 4.

Substitute 4 for s.

$$K = \frac{4^2\sqrt{3}}{4}$$

$$= \frac{16\sqrt{3}}{4} = 4\sqrt{3} \quad \textit{Ans. } (A)$$

(3) One side of a rhombus is 12 and one of the angles contains 60°. Find the area of the rhombus.

(A) $12\sqrt{3}$ (B) 36 (C) $36\sqrt{3}$ (D) 72 (E) $72\sqrt{3}$

Solution:

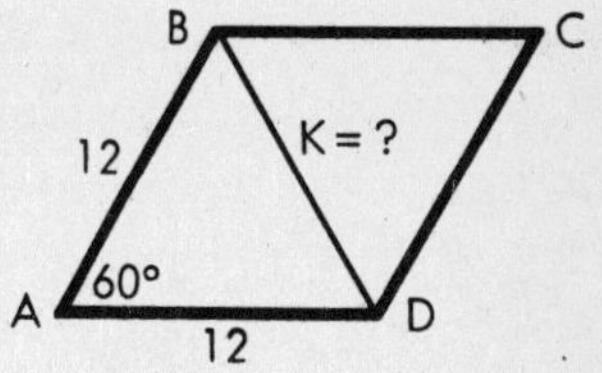

An isosceles triangle that has a 60° angle is equilateral; hence, in the figure at the right, $\triangle ABD$ is equilateral.

Find the area of $\triangle ABD$ by using the formula for the area of an equilateral triangle in terms of its side, s:

$$\text{Area of } \triangle ABD = \frac{s^2}{4}\sqrt{3}$$

Substitute 12 for s.

$$= \frac{144}{4}\sqrt{3}$$

$$= 36\sqrt{3}$$

Since a diagonal of a parallelogram divides it into two congruent triangles,

area of rhombus $ABCD$ = twice the area of $\triangle ABD$

$$= 2(36\sqrt{3}) = 72\sqrt{3} \quad \textit{Ans. } (E)$$

(4) Find the area of a regular hexagon whose side is 6.

(A) $18\sqrt{3}$ (B) $36\sqrt{3}$ (C) 54 (D) $54\sqrt{3}$ (E) $108\sqrt{3}$

Solution:

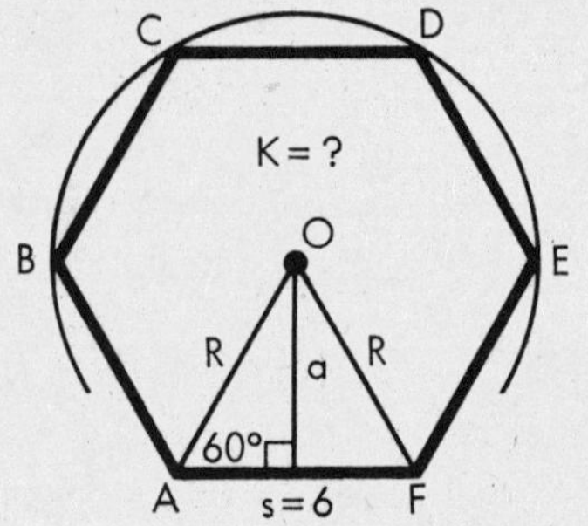

In the regular hexagon at the right, a is the apothem and $\triangle AOF$ is equilateral.

In $\triangle AOF$, R, the radius of the circumscribed circle, equals s, the side of the regular hexagon. If $s = 6$, then $R = 6$.

In the 30°-60°-90° right triangle, a is opposite the 60° angle.

Hence, $a = \frac{6}{2}\sqrt{3} = 3\sqrt{3}$.

Method 1	*Method* 2
The area of the regular hexagon equals 6 times the area of equilateral $\triangle AOF$. If K = area of regular hexagon, $K = 6\left(\frac{s^2\sqrt{3}}{4}\right)$ $= 6\left(\frac{36\sqrt{3}}{4}\right)$ $= 54\sqrt{3}$	The area of a regular polygon equals one-half the product of the perimeter and apothem. $K = \frac{1}{2}pa$ Substitute 36 for p and $3\sqrt{3}$ for a. $K = \frac{1}{2}(36)(3\sqrt{3})$ $= 54\sqrt{3}$

Ans. (D)

Matching Test

Find the areas of the following polygons. After completing this test, match your answers with the numbers in the box.

1. The side of an equilateral triangle is 6.
2. The perimeter of an equilateral triangle is 6.
3. The side of a rhombus having a 60° angle is 6.
4. The side of a regular hexagon is 6.
5. In a right triangle, the leg opposite a 30° angle is 4.

a. $\sqrt{3}$
b. $8\sqrt{3}$
c. $9\sqrt{3}$
d. $54\sqrt{3}$
e. $18\sqrt{3}$

2.4 AREAS OF POLYGONS CONTAINING SPECIAL RIGHT TRIANGLES

Model Problem

In the diagram at the right, $AB = 17$, $BD = 10$, and $BC = 8$. Find the area of $\triangle ABD$.

(A) 36 (B) 20 (C) 28 (D) 35 (E) 84

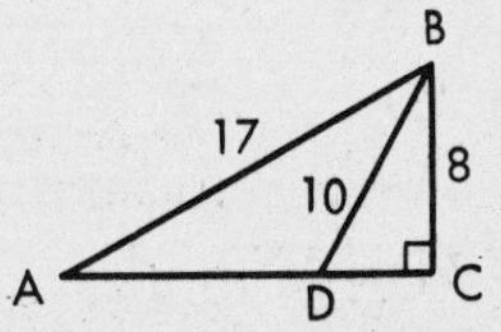

Solution:

Since the hypotenuse is 10 and one leg is 8, $\triangle DBC$ is a 3-4-5 triangle. Since the sides of $\triangle DBC$ have a common factor of 2, leg $DC = 6$.

Note that $\triangle ABC$ is an 8-15-17 triangle. Hence, leg $AC = 15$.

In $\triangle ABD$, base $AD = AC - DC$. Hence, base $AD = 15 - 6 = 9$.

The area of $\triangle ABD = \frac{1}{2}bh$. Since $b = 9$ and $h = 8$, the area of $\triangle ABD = \frac{1}{2} \cdot 9 \cdot 8 = 36$.

Ans. (A)

Matching Test

In each figure, find the area of the entire polygon by using the given data. After completing this test, match your answers with the numbers in the box.

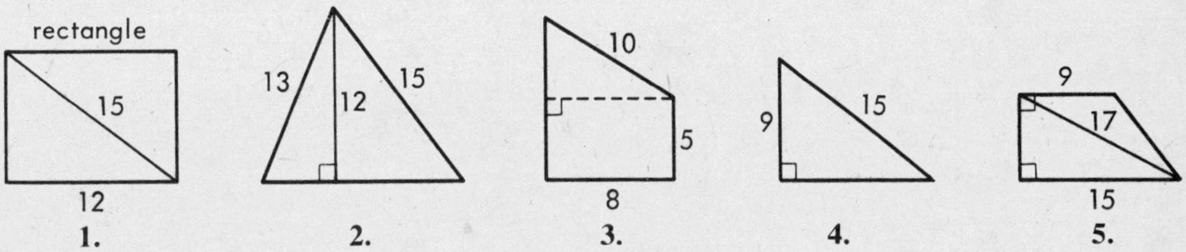

a. 54	*b*. 84	*c*. 108	*d*. 64	*e*. 96

2.5 AREA OF A RHOMBUS

Model Problems

(1) Find the area of a rhombus whose diagonals are 4 and 9.

(A) 9 (B) 18 (C) 24 (D) 30 (E) 36

Solution:

The area of a rhombus equals one-half the product of the diagonals.

If K = area of rhombus,

$$K = \frac{1}{2}dd'$$

Substitute 4 for d and 9 for d'.

$$K = \frac{1}{2}(4)(9) = 18 \quad \textit{Ans.} \text{ (B)}$$

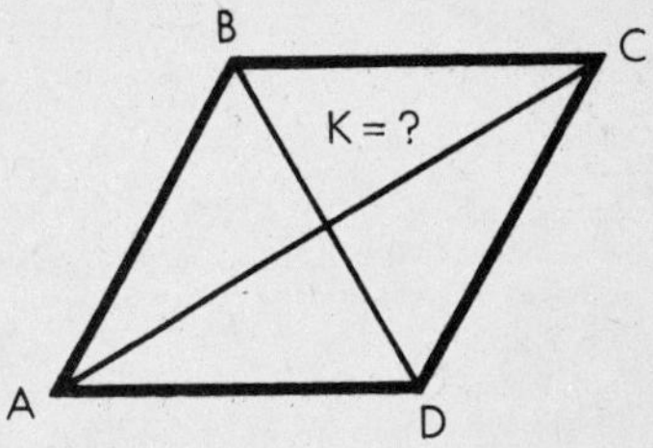

(2) The area of a rhombus is 20. The length of one diagonal is 10. Find the length of the other diagonal.

(A) 4 (B) 6 (C) 8 (D) 10 (E) 12

Solution:

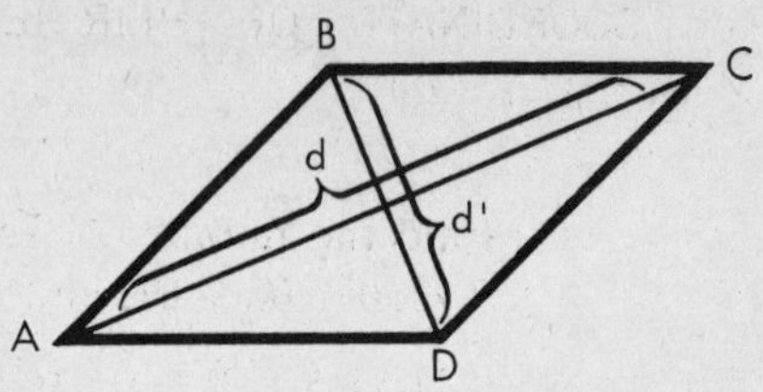

The area of a rhombus equals one-half the product of the diagonals.

If K = area of rhombus,

$$K = \frac{1}{2}dd'$$

Substitute 20 for K and 10 for d.

$$20 = \frac{1}{2}(10)d'$$

$$20 = 5d', \ d' = 4 \quad \textit{Ans.} \ \text{(A)}$$

(3) The area of a rhombus is 49 and the length of one diagonal is twice that of the other. Find the length of the shorter diagonal.

(A) $3\frac{1}{2}$ (B) 7 (C) 14 (D) $7\sqrt{2}$ (E) $14\sqrt{2}$

Solution:

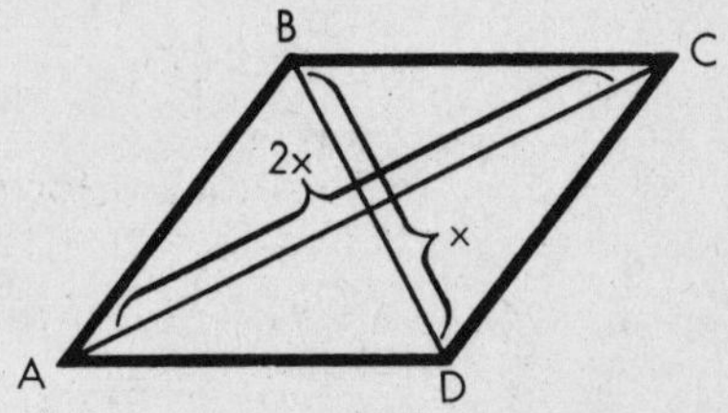

Since one diagonal is twice the other, let x = the shorter diagonal and $2x$ = the longer diagonal.

The area of a rhombus equals one-half the product of the diagonals.

If K = area of rhombus,

$$K = \frac{1}{2}dd'$$

Substitute 49 for K, x for d, and $2x$ for d'.

$$49 = \frac{1}{2}(x)(2x)$$

$$49 = x^2, \ x = 7 \quad \textit{Ans.} \ \text{(B)}$$

Matching Test

After completing this test, match your answers with the numbers in the box.

1. Find the area of a rhombus whose diagonals are 9 and 12.

2. Find the area of a rhombus having a side of 13 and a short diagonal of 10.

3. Find the area of a rhombus having a perimeter of 40 and a long diagonal of 16.

4. Find the long diagonal of a rhombus having an area of 16 if the long diagonal is twice the short diagonal.

5. Find the short diagonal of a rhombus whose area is 12 and whose long diagonal is two more than the short diagonal.

a. 120	*b.* 4	*c.* 8	*d.* 96	*e.* 54

2.6 FINDING AREAS OF TRIANGLES AND QUADRILATERALS GIVEN THE COORDINATES OF THEIR VERTICES ON A GRAPH

Model Problems

(1) Find the number of square units in the area of the triangle whose vertices are the points $A(2,0)$, $B(6,0)$, and $C(6,5)$.

(A) 5 (B) 10 (C) 15 (D) 20 (E) 30

Solution:

If the given points are plotted and then joined, it can be seen that $\triangle ABC$ is a right triangle.

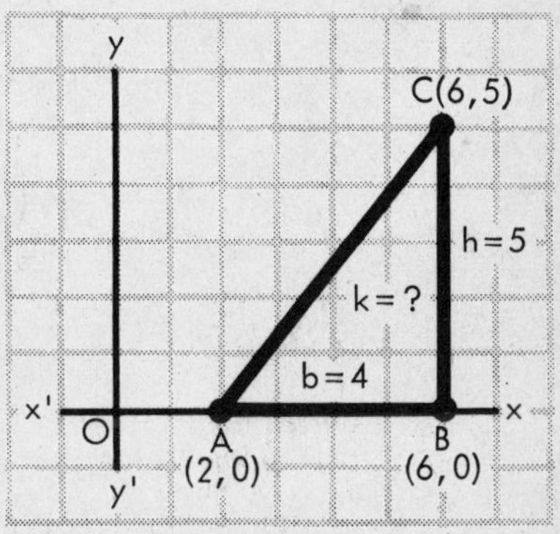

In a right triangle, either leg may be regarded as the base and the other leg as the altitude.

Hence, $b = 4$ and $h = 5$.

The area of a triangle equals one-half the product of the base and the altitude.

If K = area of the triangle,

$$K = \frac{1}{2}bh$$

Substitute 4 for b and 5 for h.

$$K = \frac{1}{2}(4)(5) = 10 \quad \textit{Ans.} \text{ (B)}$$

(2) The coordinates of the vertices of parallelogram $ABCD$ are $A(0,0)$, $B(9,0)$, $C(10,3)$, and $D(1,3)$, respectively. The area of $ABCD$ is

(A) $9\sqrt{10}$ (B) $\frac{9}{2}\sqrt{10}$ (C) $\frac{27}{2}$ (D) 27 (E) $\frac{27}{4}$

Solution:

In ▱$ABCD$, base $b = 9$ and altitude $h = 3$.

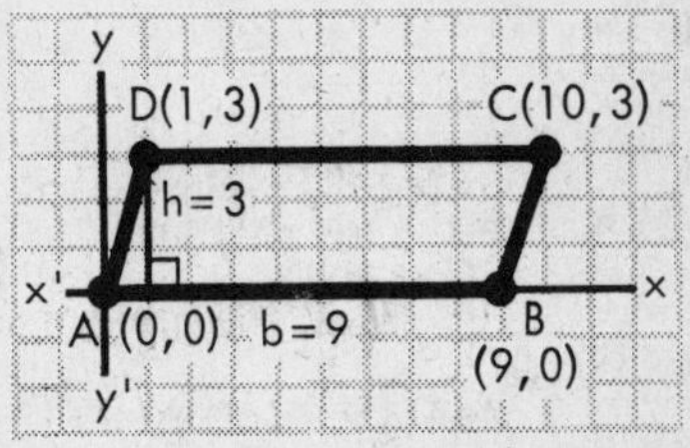

Find the area of ▱$ABCD$ as follows:

The area of a parallelogram equals the product of its base and altitude.

Hence, the area of ▱$ABCD = bh$

Substitute 9 for b and 3 for h.

area of ▱$ABCD = 9(3) = 27$ *Ans.* (D)

(3) The vertices of triangle OAB are the points $(0,0)$, $(a,0)$ and $(0,a)$, respectively. The area of this triangle is

(A) $\frac{a^2}{2}$ (B) $\frac{a^2}{4}$ (C) a^2 (D) $a^2\sqrt{2}$ (E) $\frac{a^2\sqrt{2}}{2}$

Solution:

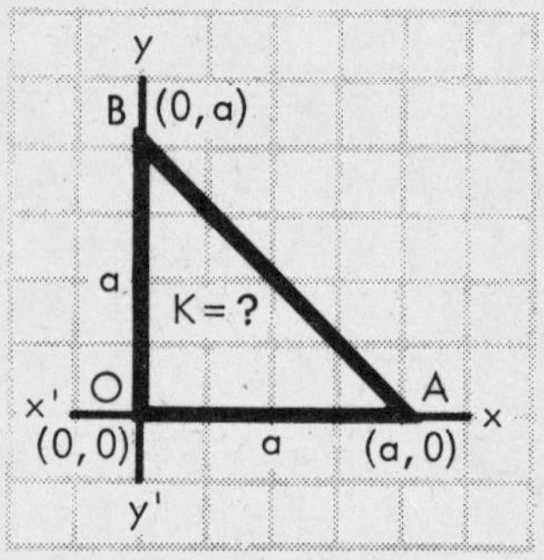

Since the x- and y-axes are perpendicular to each other, triangle OAB is a right triangle.

In a right triangle, either leg may be regarded as the base and the other leg as the altitude. Hence, in rt. $\triangle OAB$, each leg, a, may be the base or the altitude.

The area of a triangle equals one-half the product of the base and altitude.

If K = area of rt. $\triangle OAB$,

$$K = \frac{1}{2}bh$$

Substitute a for b and a for h.

$$K = \frac{1}{2}(a)(a) = \frac{a^2}{2} \quad \textit{Ans.}\ \text{(A)}$$

Matching Test

Find the areas of the following polygons. After completing this test, match your answers with the numbers in the box.

	Polygon	Coordinates of vertices
1.	rectangle $ABCD$	$A(2,0)$, $B(7,0)$, $C(7,6)$, $D(2,6)$
2.	parallelogram $ABCD$	$A(0,0)$, $B(8,0)$, $C(11,7)$, $D(3,7)$
3.	rhombus $ABCD$	$A(2,2)$, $B(12,2)$, $C(18,10)$, $D(8,10)$
4.	right triangle ABC	$A(-2,1)$, $B(6,8)$, $C(6,1)$
5.	triangle ABC	$A(0,0)$, $B\left(3\frac{1}{2},8\right)$, $C(10,0)$

a. 28
b. 56
c. 80
d. 40
e. 30

2.7 EQUIVALENT TRIANGLES, SQUARES, AND RECTANGLES

Model Problems

(1) The base of a triangle is equal to a side of a square and their areas are equal. If a side of the square is 8, find the length of the altitude of the triangle.

(A) 4 (B) 8 (C) 12 (D) 16 (E) 24

Solution:

The area of a square equals the square of a side. ($K_{sq.} = s^2$)

The area of a triangle equals one-half the product of the base and altitude. $\left(K_{\triangle} = \frac{1}{2}bh\right)$

Since the square and triangle have the same area,

$$\frac{1}{2}bh = s^2$$

Substitute 8 for b and 8 for s.

$$\frac{1}{2}(8)h = 8^2$$

$$4h = 64, \quad h = 16 \quad \textit{Ans.}\ \text{(D)}$$

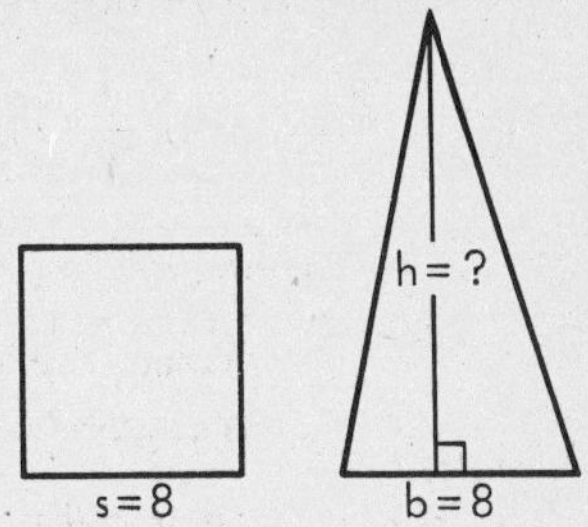

Area of the square = area of the triangle.

(2) If s represents the side of a square and b and h represent the base and altitude of a rectangle, and if the area of the square is equal to the area of the rectangle, then h equals

(A) $\frac{s^2}{b}$ (B) $s^2 - b$ (C) $\frac{b}{s}$ (D) $\frac{b}{s^2}$ (E) $\frac{s}{b}$

Solution:

The area of a square equals the square of a side. ($K_{sq.} = s^2$)

The area of a rectangle equals the product of the base and altitude. ($K_{rect.} = bh$)

Since the areas are equal, $bh = s^2$

Divide by b on each side.

$$h = \frac{s^2}{b} \quad \textit{Ans.}\ \text{(A)}$$

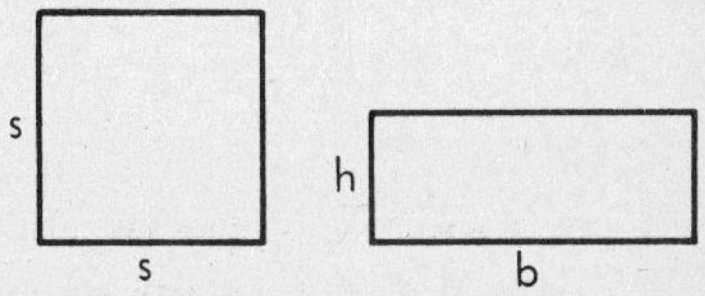

Area of the square = area of the rectangle.

Matching Test

After completing this test, match your answers with the numbers in the box.

1. A square and a rectangle have equal perimeters. If the side of the square is 12 and the base of the rectangle is 18, find the area of the rectangle.

2. A square and a rectangle have equal areas. If the side of the square is 12 and the base of the rectangle is 18, find the perimeter of the rectangle.

3. A square and a triangle have equal areas. If the side of the square and the base of the triangle are each 12, find the altitude of the triangle.

4. A square has twice the area of a triangle. If the side of the square and the base of the triangle are each 12, find the altitude of the triangle.

5. A right triangle has legs of 6 and 8. A rectangle is constructed having the hypotenuse as a base. If the rectangle and the triangle have equal perimeters, find the area of the rectangle.

a. 12	*b.* 108	*c.* 24	*d.* 52	*e.* 20

2.8 RATIOS OF AREAS INVOLVING TRIANGLES AND QUADRILATERALS

Model Problem

In trapezoid $ABCD$ with bases AB and DC, diagonals AC and BD intersect in point E. Name a triangle that is equal in area to $\triangle ABD$.

(A) $\triangle ABC$ (B) $\triangle ADC$ (C) $\triangle BDC$ (D) $\triangle DEC$ (E) $\triangle ABE$

Solution:

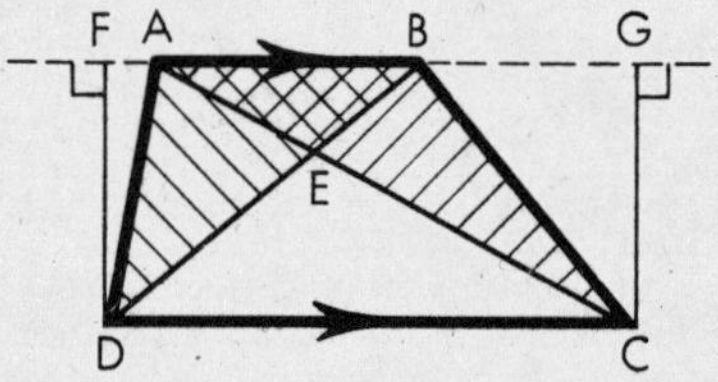

$\triangle ABC$ can be shown to be equal in area to $\triangle ABD$ by applying the following principle:

Triangles having equal bases and equal altitudes are equal in area.

AB is the common base of both $\triangle ABC$ and $\triangle ABD$.

Since parallel lines are everywhere equidistant, the altitudes DF and CG drawn to the common base AB must be equal.

Since $\triangle ABC$ and $\triangle ABD$ have the same base and equal altitudes, the two triangles are equal in area. *Ans.* (A)

Matching Test

In each figure, find the ratio of the area of the shaded part to the area of the entire figure. After completing this test, match your answers with the ratios in the box.

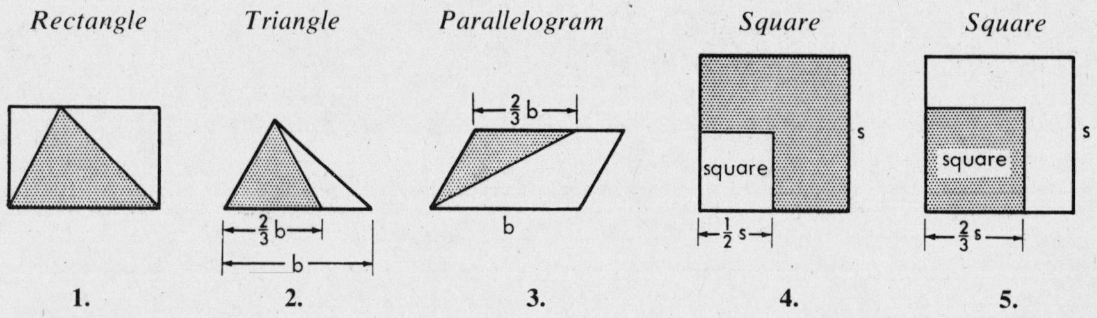

a. $\frac{2}{3}$	*b.* $\frac{3}{4}$	*c.* $\frac{1}{2}$	*d.* $\frac{4}{9}$	*e.* $\frac{1}{3}$

3. AREA OF A TRAPEZOID

3.1 PROBLEMS INVOLVING THE AREA OF A TRAPEZOID

Model Problems

(1) The bases of a trapezoid are 4 and 16 and the altitude is 4. Find the area of this trapezoid.

(A) 20 (B) 32 (C) 40 (D) 64 (E) 80

Solution:

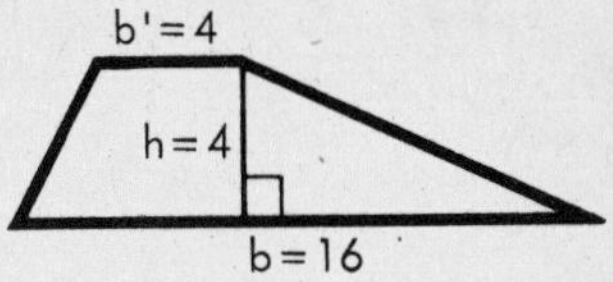

The area of a trapezoid equals one-half the product of the altitude and the sum of the bases.

Area of trapezoid $= \frac{h}{2}(b + b')$

Substitute 4 for h, 16 for b, and 4 for b'.

Area of trapezoid $= \frac{4}{2}(16 + 4)$

$= 2(20) = 40$ *Ans.* (C)

(2) The area of a trapezoid is 42 square inches and the bases are 6 inches and 8 inches. Find the number of inches in the length of the altitude of the trapezoid.

(A) 3 (B) 4 (C) 6 (D) 7 (E) 12

Solution:

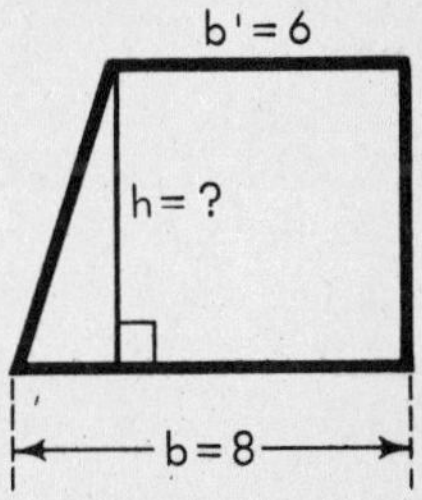

Find altitude h as follows:

The area of a trapezoid equals one-half the product of the altitude and the sum of the bases.

If K = area of the trapezoid,

$K = \frac{h}{2}(b + b')$

Substitute 42 for K, 8 for b, and 6 for b'.

$42 = \frac{h}{2}(8 + 6) = \frac{h}{2}(14)$

$42 = 7h, \quad 6 = h$ *Ans.* (C)

(3) The area of a trapezoid is 21 and the sum of the bases is 14. Find the length of the altitude.

(A) 3 (B) 4 (C) 6 (D) 7 (E) 12

Solution:

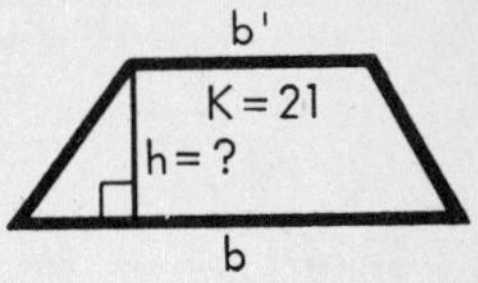

The area of a trapezoid equals one-half the product of the altitude and the sum of the bases.

If K = area of trapezoid,

$K = \frac{h}{2}(b + b')$

Substitute 21 for K and 14 for $b + b'$.

$21 = \frac{h}{2}(14)$

$21 = 7h, \quad h = 3$ *Ans.* (A)

(4) The area of a trapezoid is 48 and its altitude is 6. The longer base is three times the shorter base. Find the length of the shorter base.

(A) 3 (B) 4 (C) 8 (D) 12 (E) 16

Solution:

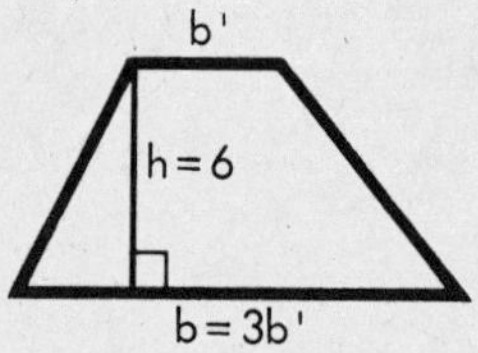

If the longer base, b, is three times the shorter base, b', then $b = 3b'$.

The area of a trapezoid equals one-half the product of the altitude and the sum of the bases.

If K = area of trapezoid,

$$K = \frac{h}{2}(b + b')$$

Substitute 48 for K, 6 for h, and $3b'$ for b.

$$48 = \frac{6}{2}(3b' + b') = 3(4b')$$

$48 = 12b'$, $b' = 4$ *Ans.* (B)

Matching Test

Refer to the trapezoid at the right to find the missing numbers indicated by the question marks in the following table. After completing this test, match your answers with the numbers in the box.

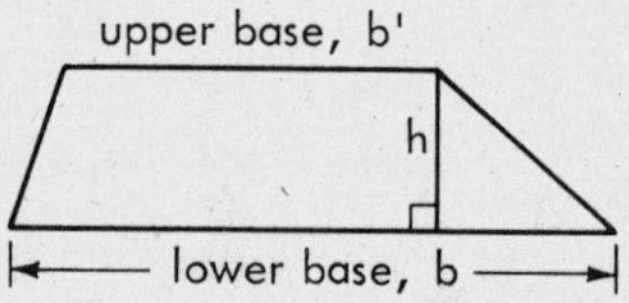

	Bases	*Altitude*	*Area*
1.	$b = 13,\ b' = 5$	4	?
2.	$b = 11,\ b' = 5$	?	32
3.	$b = ?,\ b' = 7$	4	32
4.	$b = 3b',\ b = ?$	4	32
5.	$b' = b - 4,\ b = ?$	4	32

a. 10
b. 4
c. 12
d. 9
e. 36

3.2 AREA OF AN ISOSCELES TRAPEZOID

Model Problems

(1) If the bases of an isosceles trapezoid are 15 feet and 21 feet, respectively, and each leg is 5 feet, find the number of square feet in the area of the trapezoid.

(A) 36 (B) 45 (C) 54 (D) 72 (E) 90

Solution:

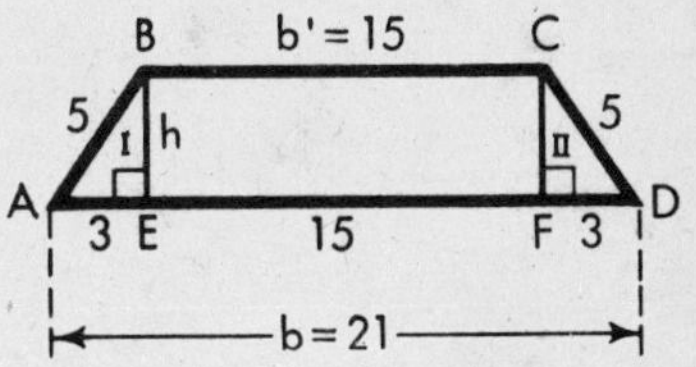

In isosceles trapezoid $ABCD$, if BE and CF are drawn perpendicular to the base, then $BCFE$ is a rectangle. Hence, $EF = BC = 15$.

Since $\triangle \text{I} \cong \triangle \text{II}$, then $AE = FD$. Hence,

$$AE = \frac{1}{2}(21 - 15) = \frac{1}{2}(6) = 3.$$

Using the Law of Pythagoras in $\triangle \text{I}$,

$$h^2 = 5^2 - 3^2$$
$$= 25 - 9 = 16$$
$$h = 4$$

> *Note.* $\triangle \text{I}$ is a 3-4-5 right triangle.

Find the area of the trapezoid as follows:

The area of a trapezoid equals one-half the product of the altitude and the sum of the bases.

If K = area of trapezoid $ABCD$,

$$K = \frac{h}{2}(b + b')$$

Substitute 4 for h, 21 for b, and 15 for b'.

$$K = \frac{4}{2}(21 + 15)$$

$$= \frac{4}{2}(36) = 72 \quad \textit{Ans.}\ \text{(D)}$$

(2) In isosceles trapezoid $ABCD$, the bases are AB and DC, AD is 5 more than DC, and AB is 2 more than twice DC. The perimeter of the trapezoid is 52. Find the area of the trapezoid.

(A) 65 (B) 130 (C) 144 (D) 156 (E) 168

Solution:

Let $x =$ the length of shorter base DC.

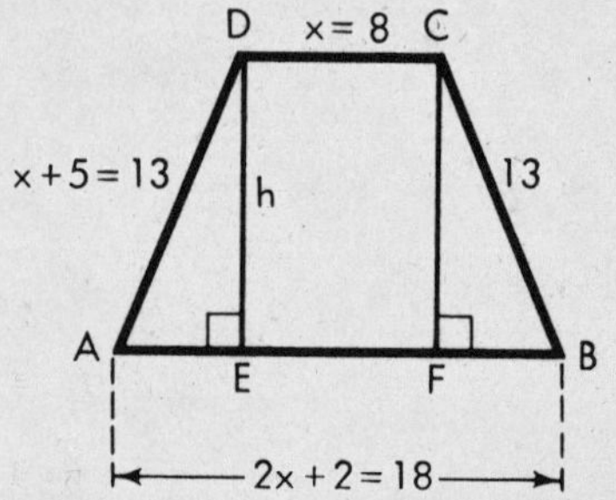

1. Since AD is 5 more than DC, $AD = x + 5$
2. Since AB is 2 more than twice DC, $AB = 2x + 2$
3. The perimeter is 52. Hence,

$$x + (x + 5) + (x + 5) + 2x + 2 = 52$$
$$5x + 12 = 52$$
$$5x = 40, \quad x = 8$$

4. $AB = 2x + 2 = 2(8) + 2 = 18$
5. Draw $DE \perp AB$ and $CF \perp AB$.
6. $EFCD$ is a rectangle.
7. $EF = DC = 8$
8. $\triangle AED \cong \triangle CFB$
9. $AE = FB = \frac{1}{2}(18 - 8) = \frac{1}{2}(10) = 5$
10. $AD = x + 5 = 8 + 5 = 13$
11. Find DE by means of the Law of Pythagoras:

$$\text{If } h = DE, \ h^2 = 13^2 - 5^2$$
$$h^2 = 169 - 25 = 144$$
$$h = 12$$

12. Area of trapezoid $ABCD = \frac{h}{2}(b + b')$

$$= \frac{12}{2}(18 + 8)$$
$$= 6(26) = 156 \quad \textit{Ans. } \text{(D)}$$

Note. $\triangle ADE$ and $\triangle BFC$ are 5-12-13 triangles.

(3) In the accompanying figure, $ABCD$ is an isosceles trapezoid. Bases AB and DC are 30 and 18, respectively, and leg $BC = 10$. AD and BC are extended to meet at E. Find the area of triangle DEC.

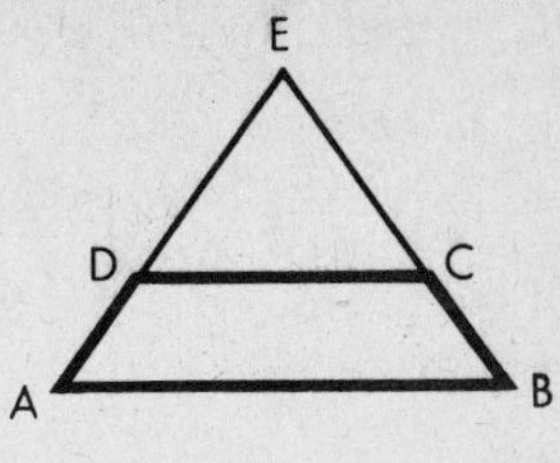

(A) 72 (B) 80 (C) 120 (D) 150 (E) 108

Solution:

1. Draw $EG \perp DC$, $DF \perp AB$, and $CH \perp AB$.
2. $FHCD$ is a rectangle.
3. $FH = DC = 18$.
4. $\triangle AFD \cong \triangle HBC$.
5. $AF = HB = \frac{1}{2}(30 - 18) = \frac{1}{2}(12) = 6$.
6. In rt. $\triangle AFD$, DF may be found as follows, using the Law of Pythagoras:

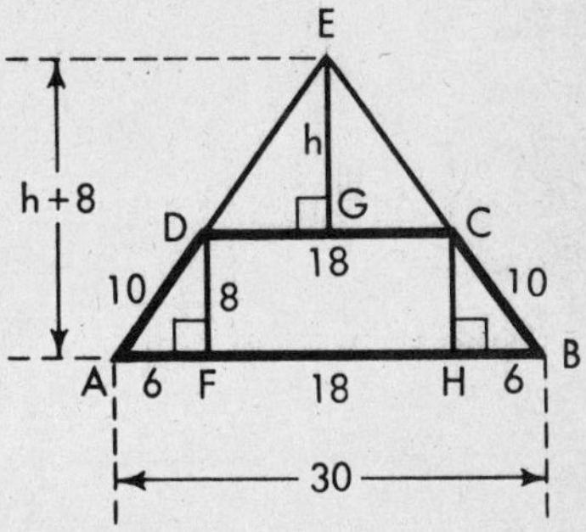

$$\overline{DF}^2 = 10^2 - 6^2$$
$$\overline{DF}^2 = 100 - 36 = 64$$
$$\overline{DF} = 8$$

Note. The sides of $\triangle AFD$ are in the ratio 3:4:5.

7. $DC \parallel AB$.
8. $\triangle DEC \sim \triangle ABE$.
9. Find altitude EG of $\triangle DEC$ by using the following principle:

   The corresponding altitudes of two similar triangles have the same ratio as any two corresponding sides.

   If $h = EG$, $\dfrac{h}{h+8} = \dfrac{18}{30}$

$$\frac{h}{h+8} = \frac{3}{5}$$
$$5h = 3h + 24$$
$$2h = 24, \quad h = 12$$

10. Area of $\triangle DEC = \frac{1}{2}bh$

$$= \frac{1}{2}(18)(12)$$
$$= 108 \quad \textit{Ans.} \text{ (E)}$$

Matching Test

Refer to the figure at the right to find the missing numbers indicated by the question marks in the following table. After completing this test, match your answers with the pairs of numbers in the box.

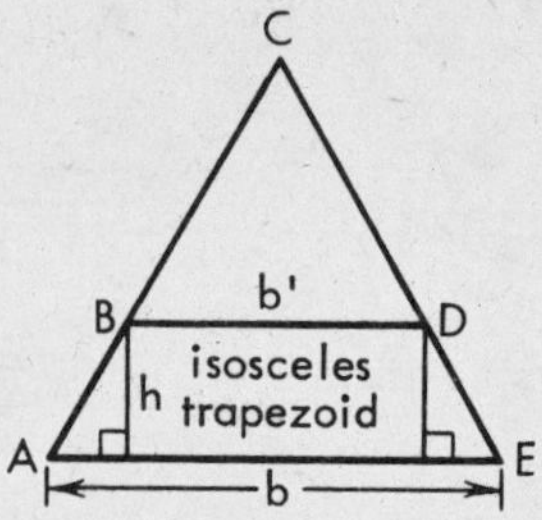

	Bases	*Altitude h*	*Leg AB or DE*	*Area of ABDE*	*Area of △BCD*
1.	$b = 34,\ b' = 10$	5	?	?	
2.	$b = 20,\ b' = 10$		13	?	?
3.	$b = 22,\ b' = 10$	?	10	?	
4.	$b = ?,\ b' = 10$	6	10	?	
5.	$b = 12,\ b' = 8$	9		?	?

a. 13, 110
b. 26, 108
c. 90, 72
d. 8, 128
e. 180, 60

3.3 SUBTRACTING AREAS: TRIANGLES AND TRAPEZOIDS CUT FROM A RECTANGLE

Model Problem

Corner AFE is cut from the rectangle as shown in the figure. The area of the remaining polygon $ABCDE$ in square inches is

(A) 29 (B) 68 (C) 78 (D) 88 (E) 72

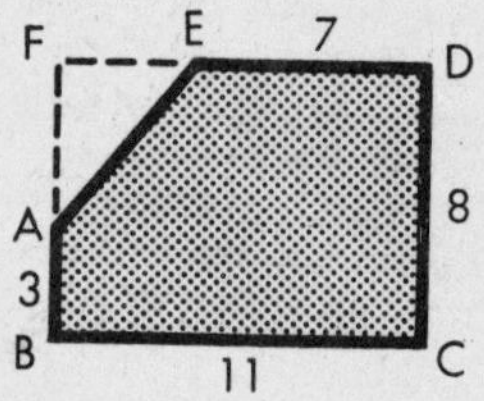

Solution:

To find the area of polygon $ABCDE$, subtract the area of triangle AFE from the area of the rectangle $BFDC$.

1. The area of the rectangle $BFDC$ equals the product of the length and width. Since the length is 11 inches and the width is 8 inches, the area of the rectangle is 88 square inches.

2. The area of triangle AFE equals one-half the product of the base and altitude. The base FE is $11 - 7$, or 4 inches. The altitude AF is $8 - 3$, or 5 inches. Hence, the area of triangle AFE is $\frac{1}{2}(4)(5)$, or 10 square inches.

3. Polygon $ABCDE$ = area of rectangle $BFDC$ − area of triangle AFE

   = 88 square inches − 10 square inches

   = 78 square inches *Ans.* (C)

Matching Test

In each figure, parts of a rectangle have been cut off. The dimensions of the rectangle are 11 by 8. Find the area of the remaining part (shaded). After completing this test, match your answers with the numbers in the box.

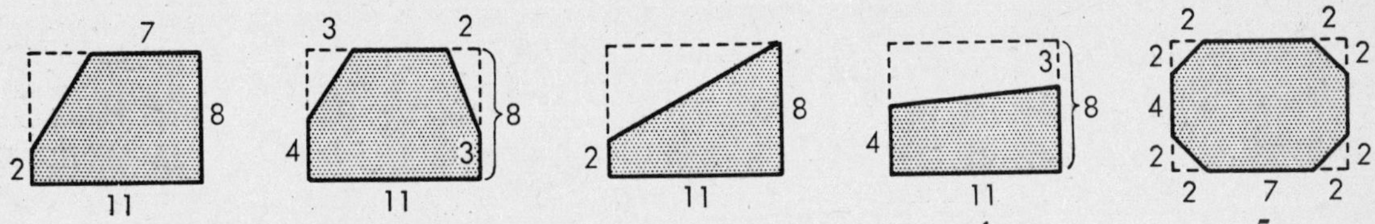

1. **2.** **3.** **4.** **5.**

a. 77	*b.* $49\frac{1}{2}$	*c.* 55	*d.* 80	*e.* 76

4. AREA OF A REGULAR POLYGON

Model Problems

(1) The area of a regular polygon is 81 and its perimeter is 36. Find the length of its apothem.

(A) $2\frac{1}{4}$ (B) $4\frac{1}{2}$ (C) $6\frac{3}{4}$ (D) 9 (E) 18

Solution:

Find the length of the apothem by applying the following principle:

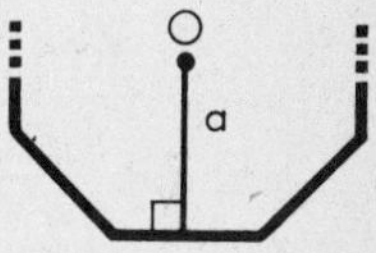

The area of a regular polygon equals one-half the product of the perimeter and the apothem.

Thus, if p is the perimeter and a is the apothem, then

$$\text{area of a regular polygon} = \frac{1}{2}pa$$

Substitute 81 for the area and 36 for p.

$$81 = \frac{1}{2}(36)\,a$$

$$81 = 18a$$

Divide each side by 18.

$$\frac{81}{18} = a, \quad a = 4\frac{1}{2} \quad \textit{Ans.}\ \text{(B)}$$

(2) The area of a regular polygon is 25 square inches and the length of its apothem is 2.5 inches. Find the number of inches in the perimeter of the polygon.

(A) 5 (B) 10 (C) 20 (D) 25 (E) 40

Solution:

The area of a regular polygon equals one-half the product of the perimeter and the apothem.

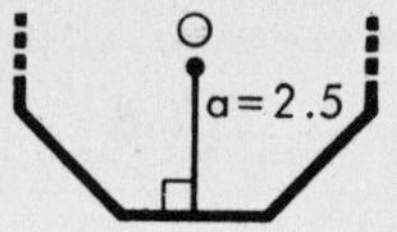

If K = area of regular polygon,

$$K = \frac{1}{2}pa$$

Substitute 25 for K and 2.5 for a.

$$25 = \frac{1}{2}(2.5)\,p$$

Multiply each side by 20.

$500 = 25p, \quad p = 20$ *Ans.* (C)

(3) If an altitude of an equilateral triangle is 12 inches, find the number of inches in the radius of the inscribed circle.

(A) $2\sqrt{3}$ (B) $4\sqrt{3}$ (C) 8 (D) 6 (E) 4

Solution:

Method 1	*Method* 2
Find r by applying the following rule: The radius of a circle inscribed in an equilateral triangle is equal to one-third the altitude of the triangle. Hence, $r = \frac{1}{3}h$ $= \frac{1}{3}(12) = 4$	The center of the inscribed circle is also the center of the circumscribed circle. Hence, OA and OB are radii of the circumscribed circle and $OA = OB$. Since the center of a circle inscribed in a triangle is also the point of concurrency of the angle bisectors of the triangle, OA bisects $\angle A$. Since $\angle A = 60°$, then $\angle OAD = 30°$, and r lies opposite a 30° angle. Hence, $OA = 2r$. Since $OB = OA$, then $OB = 2r$. Therefore, altitude $BD = 3r$. Since $BD = 12$, then $r = 4$.

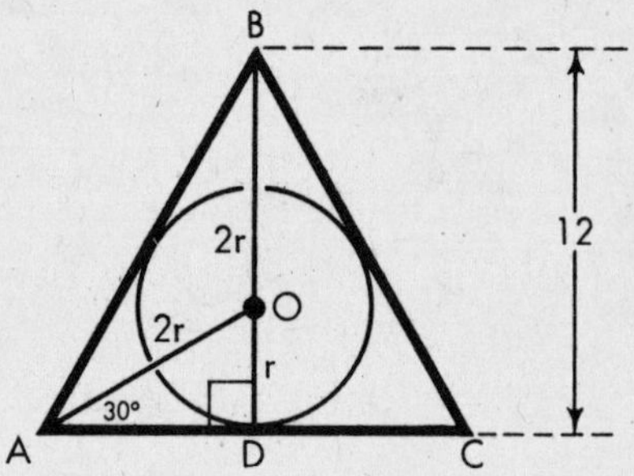

Ans. (E)

(4) The apothem of an inscribed equilateral triangle is $3\sqrt{3}$. Find the area of the triangle.

(A) $72\sqrt{3}$ (B) 27 (C) $27\sqrt{3}$ (D) 81 (E) $81\sqrt{3}$

Solution:

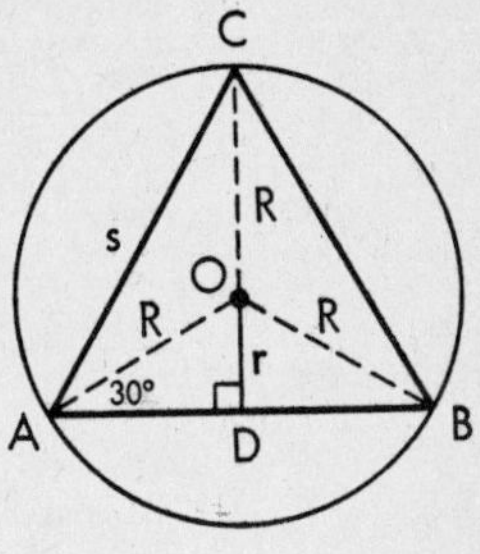

Each angle of equilateral triangle ABC is bisected by R, the radius of the circumscribed circle. Hence, $\angle OAD = 30°$.

In $\triangle ADO$, apothem r is a leg opposite a 30° angle. Therefore, $R = 2r$. Since altitude CD, or h, is the sum of R and r, $h = 3r$.

If $r = 3\sqrt{3}$, then $h = 9\sqrt{3}$.

Since $h = \frac{s}{2}\sqrt{3}$, side $s = 18$.

The area of an equilateral triangle $= \frac{s^2}{4}\sqrt{3}$. Hence,

$$\text{area of } \triangle ABC = \frac{18^2}{4}\sqrt{3} = 81\sqrt{3} \quad \textit{Ans. } \text{(E)}$$

Matching Test

After completing this test, match your answers with the numbers in the box.

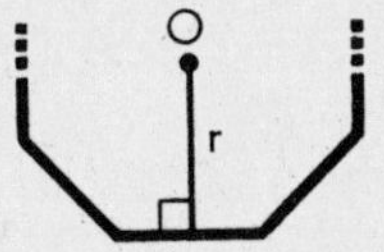

Fig. 1

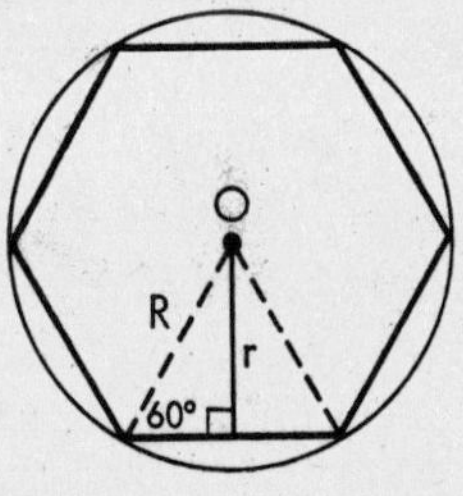

Fig. 2

1. (Fig. 1) The perimeter and apothem of the regular polygon are 32 and 4, respectively. Find the area.
2. (Fig. 1) The area and apothem of the regular polygon are 100 and 5, respectively. Find the perimeter.
3. The apothem of an inscribed equilateral triangle is $2\sqrt{3}$. Find the area.
4. (Fig. 2) The apothem of the inscribed regular hexagon is $2\sqrt{3}$. Find the area.
5. (Fig. 2) The apothem of the inscribed regular hexagon is $3\sqrt{3}$. Find the area.

a. 40	*b.* $24\sqrt{3}$	*c.* 64	*d.* $54\sqrt{3}$	*e.* $36\sqrt{3}$

5. AREAS OF SIMILAR FIGURES

5.1 PROBLEMS INVOLVING AREAS OF SIMILAR TRIANGLES FORMED BY A LINE PARALLEL TO ONE SIDE OF A TRIANGLE

Model Problems

(1) In triangle ABC, D is the midpoint of AB and E is the midpoint of AC. The area of triangle ABC is 24. Find the area of triangle ADE.

(A) 6 (B) 8 (C) 12 (D) 16 (E) 18

Solution:

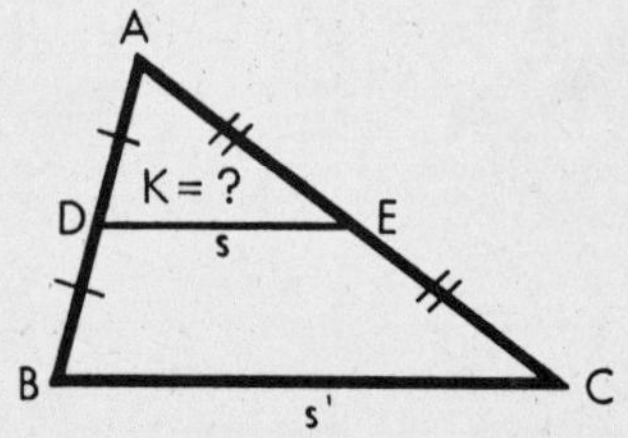

A line joining the midpoints of two sides of a triangle is parallel to the third side and also equal to one-half of it. Hence, the triangle cut off, $\triangle ADE$, is similar to $\triangle ABC$, and the ratio of DE to $BC = 1{:}2$.

Find the area of $\triangle ADE$ by applying the following theorem:

If two triangles are similar, the ratio of their areas equals the square of the ratio of any two corresponding sides.

Hence, $\frac{K}{K'} = \left(\frac{s}{s'}\right)^2$

Substitute 24 for K' and $\frac{1}{2}$ for $\frac{s}{s'}$.

$$\frac{K}{24} = \left(\frac{1}{2}\right)^2 = \frac{1}{4}$$

Cross-multiply to obtain equal products.

$4K = 24, \quad K = 6$ *Ans.* (A)

(2) In triangle ABC, a line parallel to AC intersects AB at D and CB at E. If $DE = 2$ and $AC = 3$, what is the ratio of the area of triangle DBE to the area of triangle ABC?

(A) $\frac{2}{3}$ (B) $\frac{4}{9}$ (C) $\frac{5}{9}$ (D) $\frac{4}{25}$ (E) $\frac{8}{27}$

Solution:

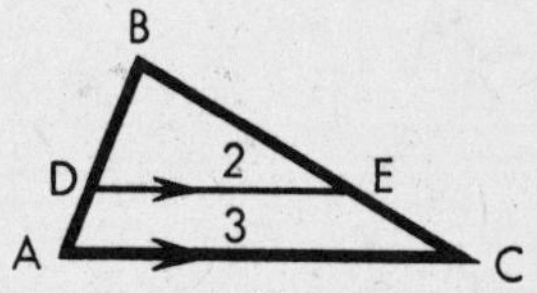

Since a line parallel to one side of a triangle cuts off a triangle similar to the given triangle, $\triangle DBE \sim \triangle ABC$.

Find the ratio of the areas of $\triangle DBE$ and $\triangle ABC$ by applying the following theorem:

If two triangles are similar, the ratio of their areas equals the square of the ratio of any two corresponding sides.

Since the ratio of the corresponding sides DE and AC is equal to $\frac{2}{3}$, the ratio of their areas $= \left(\frac{2}{3}\right)^2 = \frac{4}{9}$. *Ans.* (B)

Matching Test

After completing this test, match your answers with the numbers in the box.

In the accompanying diagram, p is parallel to q.

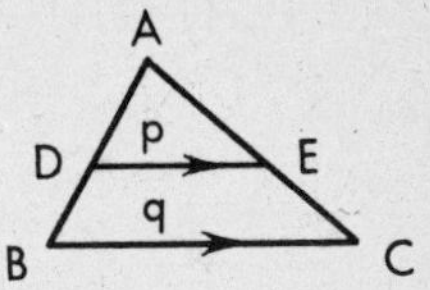

1. If $p = 3$, $q = 6$, and the area of $\triangle ADE = 8$, find the area of $\triangle ABC$.

2. If $p = 4$, $q = 6$, and the area of $\triangle ADE = 8$, find the area of $\triangle ABC$.

3. If $q = 6$, the area of $\triangle ADE = 8$, and the area of $\triangle ABC = 18$, find p.

4. If $p = 6$, the area of $\triangle ADE = 8$, and the area of $\triangle ABC = 32$, find q.

5. If $p = 6$, $q = 10$, and the area of $\triangle ABC = 75$, find the area of $\triangle ADE$.

a. 12	*b.* 4	*c.* 27	*d.* 18	*e.* 32

5.2 AREAS AND SIDES OF SIMILAR POLYGONS

Model Problems

(1) The areas of two similar triangles are in the ratio 4:9. If a side of the smaller triangle is 10, find the length of the corresponding side of the larger triangle.

(A) $4\frac{4}{9}$ (B) $6\frac{2}{3}$ (C) $11\frac{1}{4}$ (D) $12\frac{1}{2}$ (E) 15

Solution:

Find s', the side of the larger triangle, as follows:

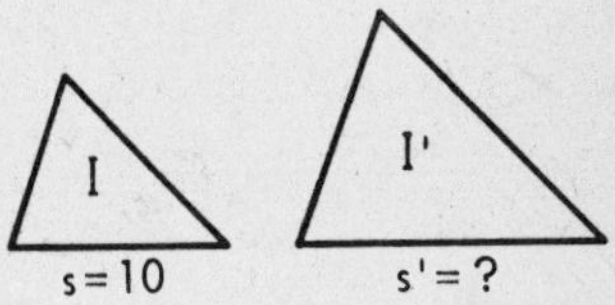

The ratio of the areas of two similar triangles equals the square of the ratio of any two corresponding sides.

Hence, $\dfrac{\text{area of } \triangle \text{I}}{\text{area of } \triangle \text{I}'} = \left(\dfrac{s}{s'}\right)^2$

Substitute $\frac{4}{9}$ for the ratio of the areas and 10 for s.

$$\frac{4}{9} = \left(\frac{10}{s'}\right)^2$$

Take the square root of each side.

$$\frac{2}{3} = \frac{10}{s'}$$

Cross-multiply to obtain equal products.

$$2s' = 30, \quad s' = 15 \quad \textit{Ans.} \text{ (E)}$$

(2) The areas of two similar polygons are in the ratio 1:25. If a side of the smaller polygon is 3, find the length of the corresponding side of the larger polygon.

(A) $8\frac{1}{3}$ (B) $16\frac{2}{3}$ (C) 15 (D) 30 (E) 75

Solution:

Find s, the side of the larger polygon, as follows:

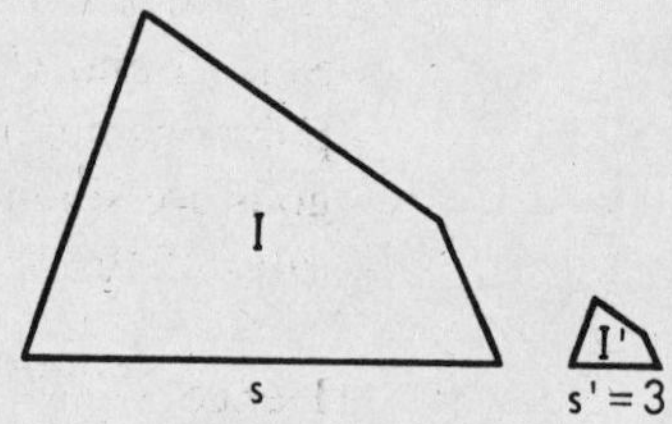

The ratio of the areas of two similar polygons equals the square of the ratio of any two corresponding sides.

Hence, $\frac{\text{area of polygon I}}{\text{area of polygon I}'} = \left(\frac{s}{s'}\right)^2$

Substitute $\frac{25}{1}$ for the ratio of the larger area to the smaller area. Also, substitute 3 for the smaller side, s'.

$$\frac{25}{1} = \left(\frac{s}{3}\right)^2$$

Take the square root of each side.

$$5 = \frac{s}{3}$$

Multiply each side by 3.

$$15 = s \quad \textit{Ans.}\ (C)$$

(3) Corresponding sides of two similar triangles are in the ratio 1:4. If the area of the smaller triangle is 9, find the area of the larger triangle.

(A) 18 (B) 36 (C) 72 (D) 144 (E) 324

Solution:

Find K', the area of the larger triangle, by applying the following theorem:

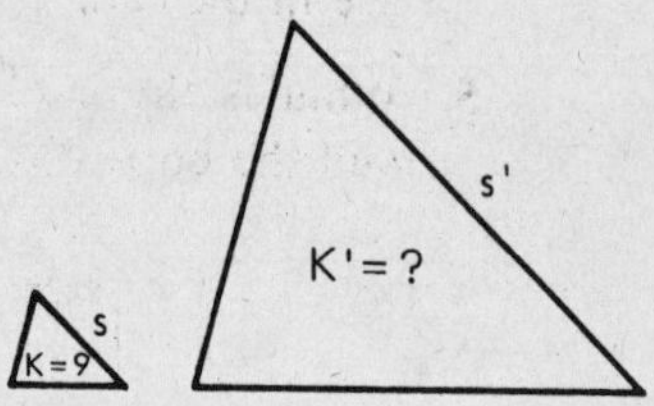

If two triangles are similar, the ratio of their areas equals the square of the ratio of any two corresponding sides.

Hence, $\frac{K}{K'} = \left(\frac{s}{s'}\right)^2$

Substitute $\frac{1}{4}$ for $\frac{s}{s'}$ and 9 for K.

$$\frac{9}{K'} = \left(\frac{1}{4}\right)^2$$

$$\frac{9}{K'} = \frac{1}{16}$$

Cross-multiply to obtain equal products.

$K' = 144$ *Ans.* (D)

(4) The sides of two regular hexagons are, respectively, 1 and 4. Find the ratio of the area of the smaller hexagon to that of the larger.

(A) $\frac{1}{16}$ (B) $\frac{1}{8}$ (C) $\frac{1}{4}$ (D) $\frac{1}{2}$ (E) $\frac{3}{4}$

Solution:

Since I and I′ are regular hexagons, they are regular polygons having the same number of sides. Such polygons are similar, and the ratio of their areas equals the ratio of the squares of any two corresponding sides.

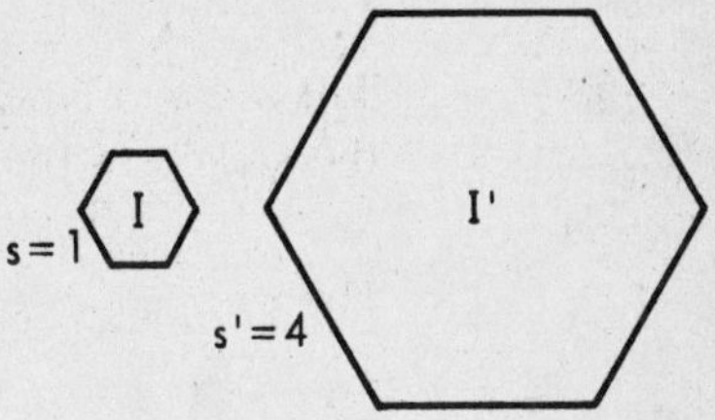

Hence, $\frac{\text{area of I}}{\text{area of I}'} = \frac{s^2}{s'^2}$

Substitute 1 for s and 4 for s'.

$$\frac{\text{area of I}}{\text{area of I}'} = \frac{1^2}{4^2} = \frac{1}{16} \quad \textit{Ans.}\ \text{(A)}$$

Matching Test

After completing this test, match your answers with the numbers or ratios in the box.

1. What is the ratio of the areas of two similar triangles if the ratio of any two corresponding sides is 3:4?

2. What is the area of the smaller of two similar polygons if the ratio of any two corresponding sides is 3:4 and the area of the larger is 64?

3. What is the area of the larger of two similar triangles if two corresponding sides are 18 and 24 and the area of the smaller is 36?

4. What is the ratio of any two corresponding sides of two similar pentagons whose areas are in the ratio 16:9?

5. What is the side of the smaller of two similar triangles if the ratio of their areas is 16:9 and the corresponding side of the larger is 36?

a. 64	*b.* 4:3	*c.* 27	*d.* 36	*e.* 9:16

5.3 AREAS AND PERIMETERS OF SIMILAR POLYGONS; AREAS AND RADII OF CIRCLES

Model Problems

(1) The radius of one circle is 2 and the radius of a second circle is 5. What is the ratio of the area of the first circle to the area of the second circle?

(A) $\frac{2}{5}$ (B) $\frac{3}{5}$ (C) $\frac{4}{9}$ (D) $\frac{4}{25}$ (E) $\frac{21}{25}$

Solution:

Find the ratio of the areas by applying the following theorem:

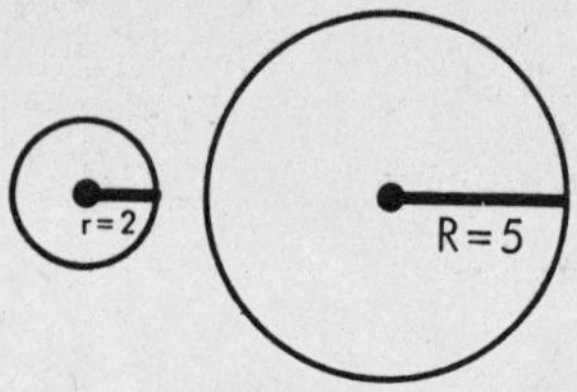

The ratio of the areas of two circles equals the ratio of the squares of their corresponding radii.

Hence, $\dfrac{\text{area of first circle}}{\text{area of second circle}} = \dfrac{r^2}{R^2}$

Substitute 2 for r and 5 for R.

$$\frac{\text{area of first circle}}{\text{area of second circle}} = \frac{2^2}{5^2}$$

$$= \frac{4}{25} \quad \textit{Ans.} \text{ (D)}$$

(2) The areas of two circles are to each other as 1:9. If the radius of the smaller circle is 2, what is the radius of the larger circle?

(A) 6 (B) 9 (C) 12 (D) 16 (E) 18

Solution:

Find the larger radius, R, as follows:

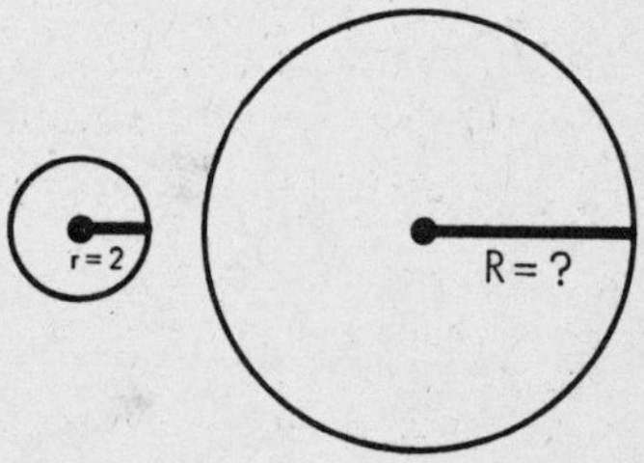

The ratio of the areas of circles (or of similar polygons) is equal to the square of the ratio of any two corresponding lines.

Use the radii as corresponding lines.

$$\frac{\text{area of smaller circle}}{\text{area of larger circle}} = \left(\frac{r}{R}\right)^2$$

Substitute $\frac{1}{9}$ for the ratio of the areas and 2 for r.

$$\frac{1}{9} = \left(\frac{2}{R}\right)^2$$

Take the square root of each side.

$$\frac{1}{3} = \frac{2}{R}$$

Cross-multiply to obtain equal products.

$$R = 6 \quad \textit{Ans.} \text{ (A)}$$

(3) The lengths of the sides of a triangle are 8, 10, and 12. Find the perimeter of a similar triangle whose area is nine times that of the given triangle.

(A) 45 (B) 60 (C) 90 (D) 180 (E) 270

Solution:

The perimeter of $\triangle \text{I}' = 12 + 10 + 8 = 30$.

Find p, the perimeter of $\triangle \text{I}$, as follows:

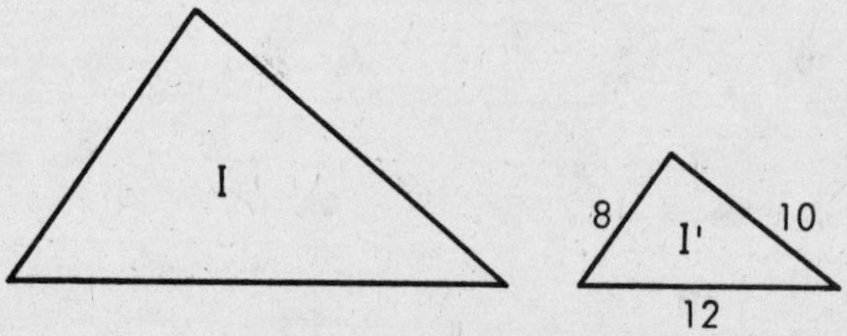

The ratio of the areas of similar triangles equals the square of the ratio of the perimeters. Hence,

$$\frac{\text{area of } \triangle \text{I}}{\text{area of } \triangle \text{I}'} = \left(\frac{p}{p'}\right)^2$$

Since the ratio of the areas is 9:1,

$$\frac{9}{1} = \left(\frac{p}{p'}\right)^2$$

Take the square root of each side, and substitute 30 for p'.

$$\frac{3}{1} = \frac{p}{30}$$

Cross-multiply to obtain equal products.

$$p = 90 \quad \textit{Ans.} \text{ (C)}$$

(4) If the radius of a circle is doubled, then the
- (A) circumference and the area are both doubled
- (B) circumference and the area are both multiplied by 4
- (C) circumference is doubled and the area is multiplied by 4
- (D) circumference is multiplied by 4 and the area is doubled
- (E) circumference is multiplied by 4 and the area is multiplied by 8

Solution:

The correct choice can be made by applying the following two principles:

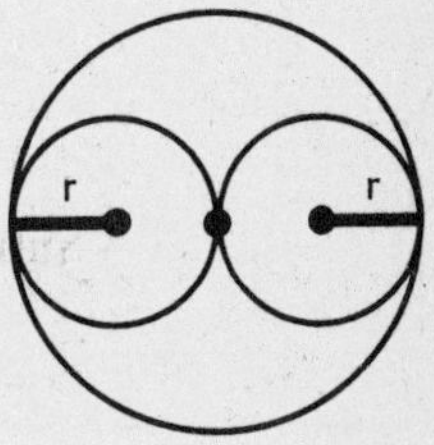

(1) Areas of circles, or of any two similar figures, are to each other as the squares of any two corresponding lines.

(2) Circumferences of circles, or perimeters of any two similar figures, are to each other as the ratio of any two corresponding lines.

Hence, if a radius is doubled, then the circumference is doubled and the area is multiplied by 4.

Ans. (C)

> *Note.* In the diagram, compare the circle having the double radius with the two circles having the single radius. The space left over is equal in area to the two smaller circles.

Matching Test

After completing this test, match your answers with the numbers or ratios in the box.

1. What is the ratio of the areas of two similar polygons if the ratio of their perimeters is 2:3?
2. What is the area of the larger of two similar quadrilaterals if the area of the smaller is 20 and their perimeters are 20 and 30?

3. What is the perimeter of the larger of two similar triangles if the ratio of their areas is 4:9 and the sides of the smaller triangle are 6, 7, and 9?

4. What is the ratio of the areas of two circles if their radii are 15 and 25?

5. What is the ratio of the radii of two circles if their areas are 9π and 25π?

a. 33	*b.* 9:25	*c.* 3:5	*d.* 45	*e.* 4:9

6. VOLUMES OF CUBES AND RECTANGULAR SOLIDS

Model Problem

A tank has the shape of an open cube. Its base and four sides have a total area of 180 square feet. Find the number of cubic feet in the volume of the tank.

(A) 36 (B) 64 (C) 1728 (D) 900 (E) 216

Solution:

A tank in the form of an open cube has five square faces. Since there are five faces, the area of each face $= \frac{180}{5}$ or 36 square feet.

The area of a square is the square of a side. Hence, each edge of the tank $= \sqrt{36}$ or 6 feet.

The volume of a cube is the cube of an edge. Hence, the volume of the cubical tank $= 6^3$ or 216 cubic feet. *Ans.* (E)

Matching Test

After completing this test, match your answers with the numbers in the box.

1. A metal bar in the form of a cube has a total area of 96 square inches for its six sides. Find the number of cubic inches in its volume.

2. The base of a tank is a rectangle 3 feet by 2 feet. The height of the tank is 20 inches. Find the number of cubic feet of water in the tank when it is $\frac{4}{5}$ full.

3. How many cubic yards of air are there in a small room whose dimensions are 7 feet by 15 feet by 9 feet?

4. An open box is made from a rectangular piece of cardboard 14 inches by 8 inches by cutting squares 2 inches on a side from each of the corners and then folding up the remainder. How many cubic inches are there in the volume of the resulting box?

5. A rectangular solid whose dimensions are 11 inches, 3 inches, and 7 inches has a capacity of 1 gallon. How many gallons will it contain if two of the dimensions are tripled and the third dimension is unchanged?

a. 80	*b.* 64	*c.* 9	*d.* 8	*e.* 35

7. PRACTICE PROBLEMS

1. The side of a square inscribed in a circle is 4. Find the radius of the circle and the area of the square.
(A) $2\sqrt{2}$, 8 (B) $2\sqrt{2}$, 16 (C) $4\sqrt{3}$, 16 (D) $4\sqrt{2}$, 32 (E) $4\sqrt{2}$, 16 **[1.1]**

2. Represent the side of a square whose area is $49x^2 - 140x + 100$.
(A) $28x - 40$ (B) $7x - 10$ (C) $7x + 10$ (D) $7x + 25$ (E) $7x - 25$ **[1.1]**

3. A total of 2304 square tiles are used to cover a square floor 12 feet on a side. If the tiles are alike in size, how many inches long is each side of a tile?
(A) 3 (B) 4 (C) 4.8 (D) 6 (E) 12 **[1.2]**

4. The shaded part of the figure at the right consists of five equal rectangles. A and B are equal squares. In each shaded rectangle, the base is twice the altitude. If the perimeter of the entire figure is 56, find the area of each shaded rectangle.
(A) 8 (B) 16 (C) 32 (D) 48 (E) 64 **[1.3]**

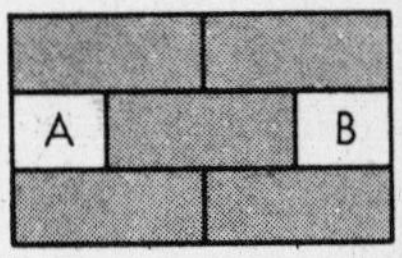

5. A square picture having an area of 36 square inches has a frame of uniform width. Find the number of inches in the width of the frame if the total area of the picture and the frame is 81 square inches.
(A) $1\frac{1}{3}$ (B) $1\frac{1}{2}$ (C) $2\frac{1}{2}$ (D) 3 (E) 9 **[1.4]**

6. The area of a rectangle is 50 more than twice the area of a square. The side of the square is a and the dimensions of the rectangle are b and 9. Write an equation that expresses the relation between a and b.
(A) $9b = 2a^2 - 50$ (B) $b = 2a + 16$ (C) $9b = 2a^2 + 50$ (D) $9b = a^2 + 50$
(E) $9b = 4a + 50$ **[1.5]**

7. Find the base of a parallelogram whose area is 72 and whose base is one more than its altitude.
(A) $4\frac{1}{2}$ (B) 8 (C) 9 (D) 12 (E) 36 **[2.1]**

8. An isosceles right triangle is inscribed in a circle having a radius of 10. Find the area of the triangle.
(A) 50 (B) 100 (C) 125 (D) 100π (E) 200 **[2.2]**

9. Find the area of a right triangle having a $60°$ angle and a hypotenuse of 8.
(A) 8 (B) $8\sqrt{2}$ (C) $8\sqrt{3}$ (D) $16\sqrt{3}$ (E) $64\sqrt{3}$ **[2.3]**

10. Find the area of polygon $ABCD$.
(A) 60 (B) 100 (C) 108 (D) 120 (E) 160 **[2.4]**

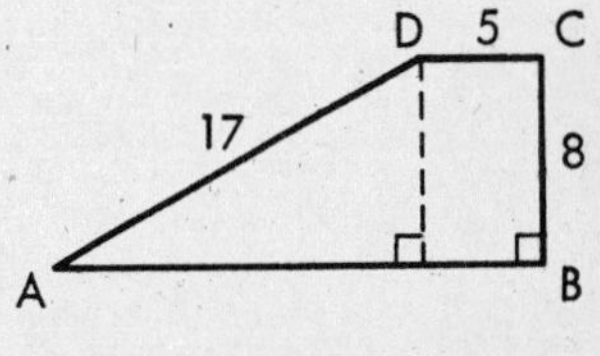

11. Find a diagonal of a square whose area is 18.
(A) 3 (B) $3\sqrt{2}$ (C) 6 (D) 9 (E) 12 **[2.4]**

12. Find the area of trapezoid $ABCD$, whose vertices have the following coordinates:
$A(2, 0)$, $B(3, 6)$, $C(10, 6)$, $D(11, 0)$
(A) 36 (B) 48 (C) 54 (D) 66 (E) 96 **[2.6]**

13. A right triangle has legs of 6 and 8. A rectangle is constructed whose base is 2 more than the hypotenuse of the triangle. If the rectangle and the triangle have equal areas, find the perimeter of the rectangle.

(A) 28 (B) 24.8 (C) 24 (D) 20 (E) 32 **[2.7]**

14. In the rectangle at the right, find the ratio of the area of the shaded part to the area of the entire figure.

(A) $\frac{1}{4}$ (B) $\frac{1}{3}$ (C) $\frac{1}{2}$ (D) $\frac{1}{8}$ (E) $\frac{3}{8}$ **[2.8]**

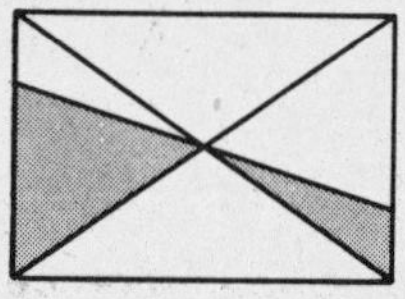

15. The lower base of a trapezoid is twice the sum of the upper base and 5. If its altitude is 4 and its area is 32, find the lower base.

(A) 2 (B) 7 (C) 14 (D) 18 (E) 28 **[3.1]**

16. The bases of an isosceles trapezoid are 10 and 18. If its legs are each 5, find its area.

(A) 70 (B) 140 (C) 42 (D) 56 (E) 84 **[3.2]**

17. The vertices of the inner parallelogram are midpoints of the outer rectangle. The dimensions of the rectangle are 11 by 8. Find the area of the shaded part.

(A) 11 (B) 22 (C) 40 (D) 44 (E) 66 **[3.3]**

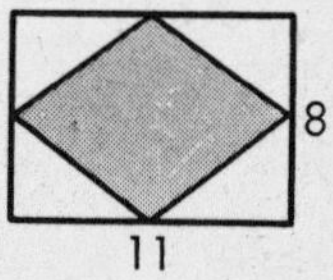

18. The apothem of an inscribed equilateral triangle is 3. Find its area.

(A) $12\sqrt{3}$ (B) $27\sqrt{3}$ (C) 32 (D) 40.5 (E) $\frac{81}{4}\sqrt{3}$ **[4]**

19. In the diagram at the right, FG is parallel to AC. If the area of $\triangle ABC$ is 75, find the area of $\triangle BFG$.

(A) 36 (B) 40 (C) 48 (D) 50 (E) 60 **[5.1]**

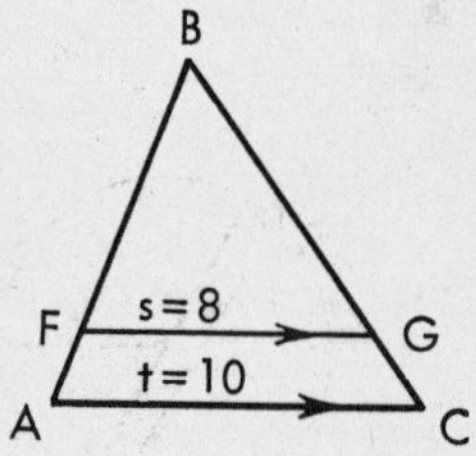

20. The areas of two similar hexagons are 64 and 36. What is the side of the smaller hexagon if the corresponding side of the larger hexagon is 18?

(A) 9 (B) $10\frac{1}{8}$ (C) $13\frac{1}{2}$ (D) 24 (E) 32 **[5.2]**

21. What is the circumference of the larger of two circles if the circumference of the smaller is 18 and the ratio of their areas is 9:25?

(A) 10.8 (B) 30 (C) 30π (D) 50 (E) 50π **[5.3]**

22. A container in the form of a rectangular solid 7 inches by 18 inches by 3 feet is filled with water. If all the water is to be poured into receptacles in the form of cubes 6 inches on a side, how many of these receptacles are needed?

(A) 6 (B) 7 (C) 21 (D) 42 (E) 84 **[6]**

IX. REVIEWING CIRCLES; VOLUME OF A CYLINDER

Circumference and Area of a Circle

The Greek letter π (pi) represents the ratio of the circumference of a circle to its diameter; that is, $\pi = \frac{C}{d}$. Hence, $C = \pi d$.

Some approximate values of π are $\frac{22}{7}$; 3.14, to two decimal places; 3.142, to three decimal places; and 3.1416, to four decimal places.

1. The *circumference of a circle* equals:
   (1) twice the product of π and the radius. Formula: $C = 2\pi r$
   (2) the product of π and the diameter. Formula: $C = \pi d$
   For example, if the diameter of the circle is 10, then $C = \pi d = 10\pi$. Using 3.14 for π, $C = 10(3.14) = 31.4 = 31$, to the nearest integer.

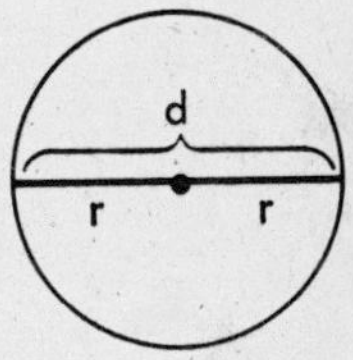

Circumference:
(1) $C = 2\pi r$
(2) $C = \pi d$

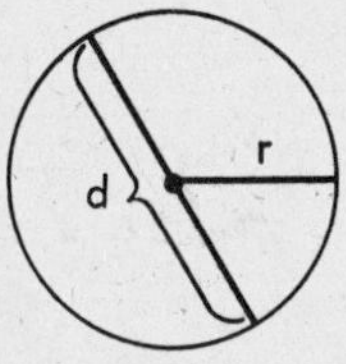

Area:
(1) $K = \pi r^2$
(2) $K = \frac{1}{4}\pi d^2$

2. The *area of a circle* equals:
   (1) the product of π and the square of the radius. Formula: $K = \pi r^2$
   For example, if the radius of a circle is 10, then $K = \pi r^2 = 100\pi$. Using 3.14 for π, $K = (3.14)(100) = 314$.
   (2) one-fourth the product of π and the square of the diameter.
   Formula: $K = \frac{1}{4}\pi d^2$

Concentric Circles

3. The *distance between two concentric circles* equals:
   (1) the difference of their radii, $R - r$, where R and r are the radii of the circles.
   (2) one-half the difference of the diameters, $\frac{1}{2}(D - d)$, where D and d are the diameters of the circles.
   For example, if the outer diameter of a pipe is $3\frac{1}{2}$ inches and the inner diameter is 3 inches, then the thickness of the pipe is $\frac{1}{2}(D - d) = \frac{1}{2}\left(3\frac{1}{2} - 3\right) = \frac{1}{2}\left(\frac{1}{2}\right) = \frac{1}{4}$ inch.

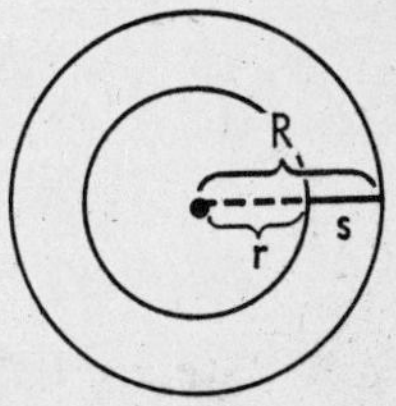

Distance between concentric circles:
(1) $s = R - r$
(2) $s = \frac{1}{2}(D - d)$

4. The *area of a ring* equals:

   (1) the product of π and the difference of the squares of the radii.
   Formula: $K = \pi(R^2 - r^2)$
   For example, if the radii of the concentric circles are 10 and 7, then $K = \pi(R^2 - r^2) = \pi(100 - 49) = 51\pi$.

   (2) one-fourth the product of π and the difference of the squares of the diameters. Formula: $K = \frac{1}{4}\pi(D^2 - d^2)$

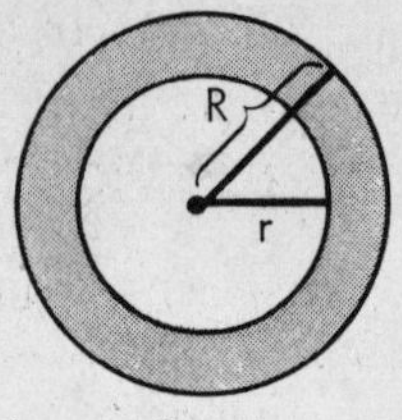

Area of a ring:
(1) $K = \pi(R^2 - r^2)$
(2) $K = \frac{1}{4}\pi(D^2 - d^2)$

Comparing the Circumferences and Areas of Two Circles

Since any two circles have the same shape, circles have the same properties as similar figures. Principles 5 and 6 are corollaries of the more general principles covering any two similar figures.

5. The *circumferences of two circles* are to each other:

   (1) as their radii. Formula: $\frac{C}{C'} = \frac{r}{r'}$

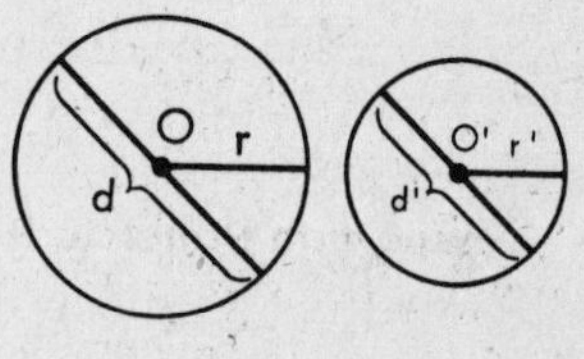

   (2) as their diameters. Formula: $\frac{C}{C'} = \frac{d}{d'}$

   For example, if the radius of the larger circle is twice the radius of the smaller one, then the larger circumference is twice the smaller circumference.

6. The *areas of two circles* are to each other:

   (1) as the squares of their diameters. Formula: $\frac{K}{K'} = \frac{d^2}{d'^2}$

   (2) as the squares of their radii. Formula: $\frac{K}{K'} = \frac{r^2}{r'^2}$

   For example, if the larger radius is 5 times the smaller radius, then the area of the larger circle is 25 times that of the smaller one.

Arcs and Sectors

If the circumference of a circle is divided into 5 equal arcs, the length of each of the 72° arcs is $\frac{1}{5}$ of the circumference; that is, the length of a 72° arc $= \frac{1}{5}(\pi d)$, or $\frac{\pi d}{5}$.

A ***sector*** is the part of a circle that is bounded by two radii and an arc. If the area of the circle is divided into 5 equal sectors, then each 72° sector is $\frac{1}{5}$ of the area of the circle; that is, the area of a 72° sector $= \frac{1}{5}(\pi r^2) = \frac{\pi r^2}{5}$.

7. In a circle of radius r, the *length of an arc of* $n°$ equals:

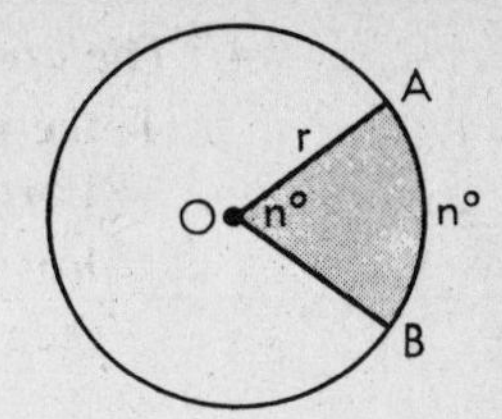

(1) $\frac{n}{360}$ of the circumference of the circle. Formula: $l = \left(\frac{n}{360}\right)C$

(2) $\left(\frac{n}{360}\right)2\pi r$, or $\frac{\pi nr}{180}$. Formula: $l = \frac{\pi nr}{180}$

For example, the length of an arc of 72° in a circle whose radius is 10 equals $\frac{72}{360}(2\pi r) = \frac{1}{5}(20\pi) = 4\pi$.

8. In a circle of radius r, the *area of a sector of* $n°$ equals

(1) $\frac{n}{360}$ of the area of the circle. Formula: $K_{\text{sector}} = \left(\frac{n}{360}\right)K_{\text{circle}}$

(2) $\left(\frac{n}{360}\right)\pi r^2$, or $\frac{\pi nr^2}{360}$. Formula: $K_{\text{sector}} = \frac{\pi nr^2}{360}$

For example, the area of a sector of 72° in a circle whose radius is 10 equals $\frac{72}{360}(\pi r^2) = \frac{1}{5}(100\pi) = 20\pi$.

Proportion Principle Relating Arcs and Sectors Having Equal Central Angles

9. $$\frac{\text{length of an arc of } n°}{\text{circumference}} = \frac{\text{area of a sector of } n°}{\text{area of circle}} = \frac{n}{360}$$

For example, since an arc of 120° is $\frac{1}{3}$ of the circumference and a sector of 120° is $\frac{1}{3}$ of the area of the circle,

$$\frac{\text{arc of } 120°}{\text{circumference}} = \frac{\text{sector of } 120°}{\text{area of circle}} = \frac{1}{3}$$

Segments

A ***minor segment*** of a circle is the part of a circle that is bounded by a chord and its minor arc. In the diagram, note the minor segment bounded by chord AB and its minor arc AB.

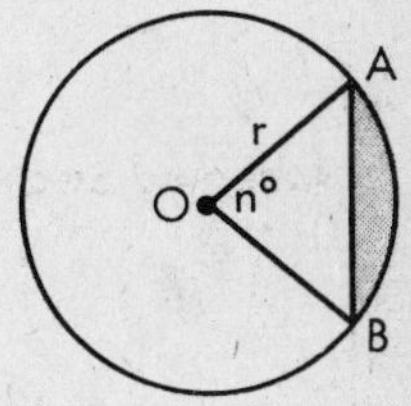

10. The *area of a minor segment* of a circle equals the area of its sector minus the area of the triangle formed by its chord and radii drawn to the endpoints of the chord.

Hence, $K_{\text{segment}} = K_{\text{sector}} - K_{\text{triangle}}$

Volume of a Cylinder

11. The *volume of a cylinder* equals the product of the area of a base and the height. Formula: $V = Bh$, or $V = \pi r^2 h$

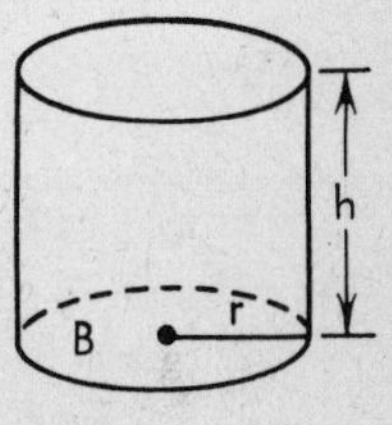

For example, if the circular base of a cylinder has a radius of 3 inches and the height of the cylinder is 7 inches, then the volume of the cylinder $= \pi r^2 h = \pi \cdot 3^2 \cdot 7 = \frac{22}{7} \cdot 9 \cdot 7 = 198$ cubic inches.

1. CIRCUMFERENCE OF A CIRCLE

1.1 PROBLEMS INVOLVING THE CIRCUMFERENCE OF A CIRCLE

Model Problems

(1) Find the circumference of a circle whose diameter is 28.

(A) 14π (B) 28π (C) 84 (D) 56π (E) 56

Solution:

Find the circumference as follows:

The circumference of a circle equals the product of π and the diameter.

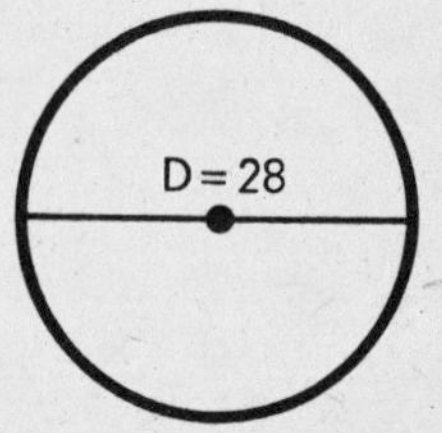

Hence, $C = \pi D$

Substitute 28 for D.

$C = 28\pi$ *Ans.* (B)

(2) The circumference of a circle is 14π. Find the length of the radius.

(A) 7 (B) 14 (C) 7π (D) 14π (E) 21π

Solution:

Find the radius, r, as follows:

The circumference of a circle equals the product of 2π and the radius.

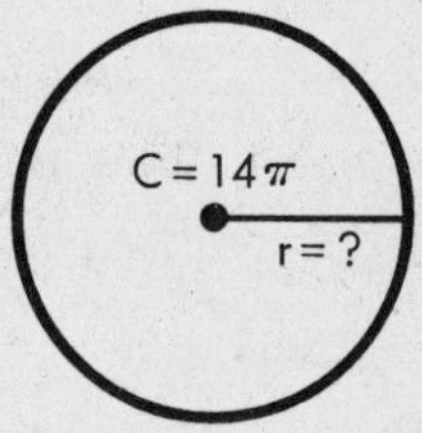

Hence, $C = 2\pi r$

Substitute 14π for C.

$14\pi = 2\pi r$

Divide each side by 2π.

$7 = r$ *Ans.* (A)

(3) The circumference of a circle is 10π. Find the perimeter of a regular hexagon inscribed in this circle.

(A) 15 (B) 20 (C) 30 (D) 60 (E) 120

Solution:

Since the circumference of a circle is $2\pi r$ and the given circumference is 10π,

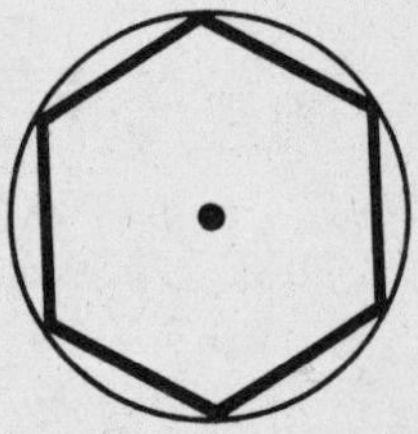

$$10\pi = 2\pi r$$

Divide each side by 2π.

$$5 = r$$

Since a side of an inscribed regular hexagon equals a radius of the circle, then each side $= r = 5$.

Hence, the perimeter of the regular hexagon $= 6(5) = 30$.
Ans. (C)

(4) Find the number of inches in the diagonal of a square inscribed in a circle whose circumference is 80π inches.

(A) $20\sqrt{2}$ (B) $40\sqrt{2}$ (C) 20 (D) 40 (E) 80

Solution:

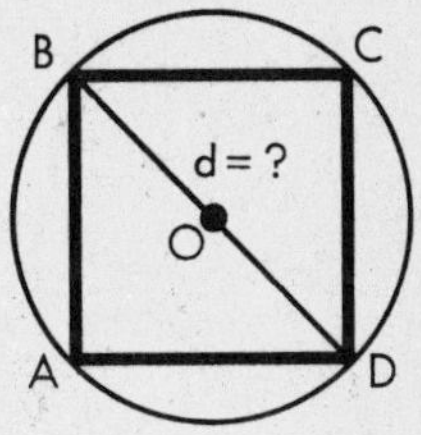

Since a diagonal of an inscribed square is a diameter of the circle, the diagonal d is a diameter.

Find d as follows:

The circumference of a circle equals the product of π and the diameter.

Hence, $C = \pi d$

Substitute 80π for C.

$$80\pi = \pi d$$

Divide each side by π.

$80 = d$ *Ans.* (E)

Matching Test

After completing this test, match your answers with the numbers in the box.

1. Find the circumference of a circle of radius 16.
2. Find the circumference of a circle of area 16π.
3. Find the perimeter of a regular hexagon inscribed in a circle whose circumference is 16π.
4. Find the diagonal of a square inscribed in a circle whose semicircumference is 8π.
5. Find the perimeter of a square circumscribed about a circle whose circumference is 8π.

a. 32	*b.* 48	*c.* 32π	*d.* 16	*e.* 8π

1.2 DISTANCE BETWEEN CONCENTRIC CIRCLES

Model Problem

Find the number of inches in the thickness of a pipe having an outside diameter of $4\frac{1}{4}$ inches and an inside diameter of $3\frac{1}{4}$ inches.

(A) $\frac{1}{2}$ (B) 1 (C) $1\frac{1}{2}$ (D) 2 (E) 3

Solution:

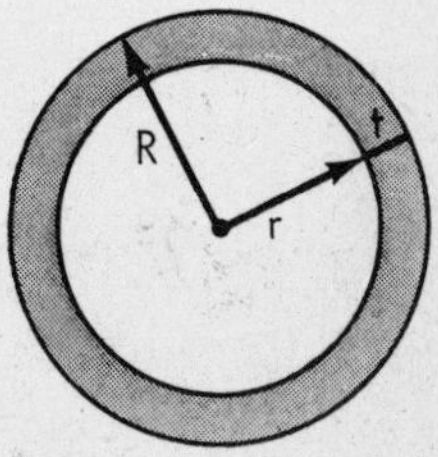

As shown, the thickness, t, of a pipe equals $R - r$, the difference of the radii. Therefore, $t = \frac{1}{2}(D - d)$, where D and d are the diameters. (See principle 3 on page 309.)

Since the thickness is one-half the difference of the diameters,

$$t = \frac{1}{2}\left(4\frac{1}{4} - 3\frac{1}{4}\right) = \frac{1}{2}(1) = \frac{1}{2} \quad \textit{Ans. } (A)$$

Matching Test

After completing this test, match your answers with the numbers in the box.

1. Find the number of inches in the outside diameter of a pipe whose inside diameter is 6 inches and whose thickness is 1 inch.

2. How many inches thick is the rim of a wheel that has an inner edge of diameter 35 inches and an outer edge of diameter 40 inches?

3. Find the number of inches in the inside diameter of a pipe whose thickness is $\frac{3}{4}$ inch and whose outside diameter is $7\frac{1}{2}$ inches.

4. Find the number of inches in the inside diameter of a pipe whose outer circumference is 6π inches and whose thickness is 1 inch.

5. Find the number of inches in the outside diameter of a pipe whose inner circumference is 6π inches and whose thickness is $\frac{1}{2}$ inch.

a. $2\frac{1}{2}$	*b.* 6	*c.* 4	*d.* 8	*e.* 7

1.3 CIRCULAR SPEEDS

Model Problem

A circular saw rotates at a speed of 350 revolutions per minute. If its diameter is 9 inches, what is its cutting speed in feet per minute? $\left(\text{Use } \pi = \frac{22}{7}.\right)$

(A) 825 (B) 1650 (C) 7700 (D) 8250 (E) 16,500

Solution:

Since the speed is to be in *feet* per minute, change the number of inches in the diameter into the equivalent number of feet. Here, a diameter of 9 inches $= \frac{9}{12}$ or $\frac{3}{4}$ feet.

In each complete rotation, a point on the circumference of the saw moves a distance of 1 circumference, or πd, that is, a distance of $\frac{22}{7}\left(\frac{3}{4}\right)$ or $\frac{33}{14}$ feet.

The cutting speed, or rim speed, of a circular saw is the speed of any point on its rim or circumference. Since the saw makes 350 rotations every minute, its cutting speed in feet per minute $= 350\left(\frac{33}{14}\right)$, or 825. *Ans.* (A)

Matching Test

After completing this test, match your answers with the numbers in the box. $\left(\text{Use } \pi = \frac{22}{7}.\right)$

1. A circular saw makes 210 rotations per minute. If its diameter is 3 feet, at what speed in feet per minute will each point on the circumference move?

2. A wheel of diameter 20 inches moves along the ground making 210 revolutions per minute. What is the ground speed of the wheel in feet per minute?

3. If a wheel has a diameter of 28 inches, how many revolutions does it make in covering a distance of 1 mile?

4. What is the diameter in feet of the wheel of a truck if the truck moves a distance of one mile during every 420 revolutions?

5. A gear has a rim speed of 1100 feet per minute if it makes 10 rotations every second. What is its diameter in inches?

a. 1100	*b.* 4	*c.* 1980	*d.* 7	*e.* 720

2. LENGTH OF AN ARC OF A CIRCLE

2.1 PROBLEMS INVOLVING THE LENGTH OF AN ARC OF A CIRCLE

Model Problems

(1) In a circle of diameter 20 inches, find the number of inches in the length of an arc of 72°.

(A) 2π (B) 4π (C) 8π (D) 12π (E) 1.2π

Solution:

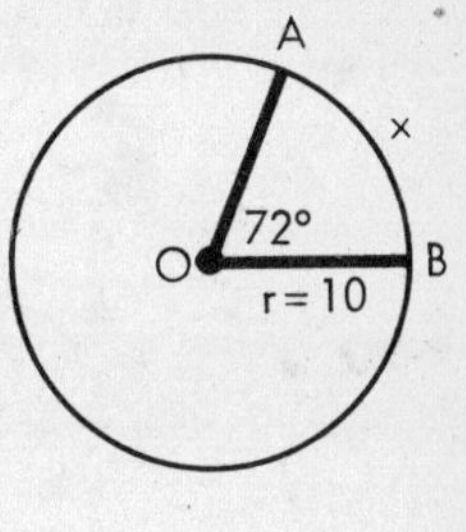

Since the diameter = 20 inches, the radius $= \frac{1}{2}(20)$, or 10 inches.

Find x as follows:

The length of an arc of $n°$ equals $\frac{n}{360}$ of the circumference.

Hence, the length of a 72° arc $= \frac{72}{360}$, or $\frac{1}{5}$ of the circumference.

Therefore, $x = \frac{1}{5}(2\pi r)$

Substitute 10 for r.

$$x = \frac{1}{5}(2\pi 10)$$

$$= 4\pi \quad \textit{Ans.}\ \text{(B)}$$

(2) In a circle whose radius is 6 inches, find the number of degrees in the central angle of an arc whose length is 4π inches.

(A) 60 (B) 72 (C) 120 (D) 150 (E) 240

Solution:

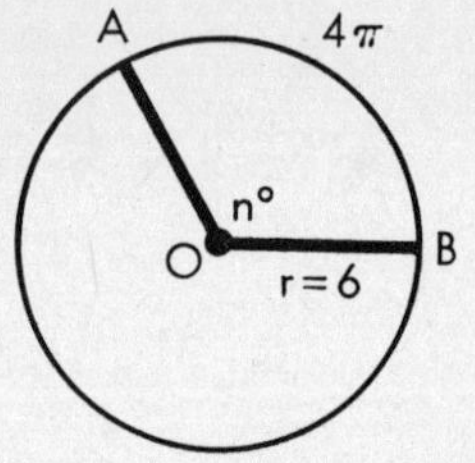

Since the radius, r, is 6, the circumference of the circle $= 2\pi r = 12\pi$.

Find n by applying the following rule:

The number of degrees in a central angle divided by 360 equals the length of its intercepted arc divided by the circumference.

Hence, $\frac{n}{360} = \frac{4\pi}{12\pi}$

$$\frac{n}{360} = \frac{1}{3}$$

$$n = \frac{1}{3}(360) = 120 \quad \textit{Ans. (C)}$$

(3) The circumference of a circle is 24π. If a central angle of 45° intercepts arc AB, find the length of minor arc AB.

(A) 3π (B) 4π (C) 6π (D) 8π (E) 12π

Solution:

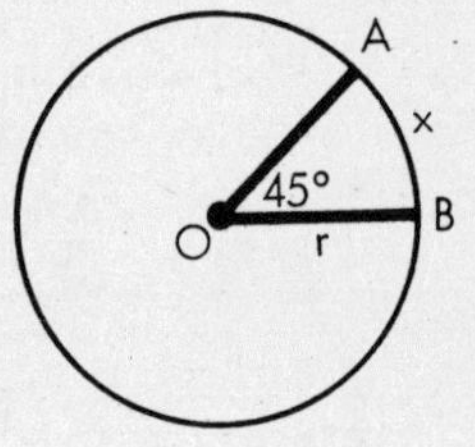

Find x, the length of $\overset{\frown}{AB}$, as follows:

The length of an arc of $n° = \frac{n}{360}$ of the circumference.

Since the central angle $= 45°$, then $\overset{\frown}{AB} = \overset{\frown}{45°}$.

Hence, $x = \frac{45}{360}$, or $\frac{1}{8}$ of the circumference.

Since the circumference is 24π,

$$x = \frac{1}{8}(24\pi) = 3\pi \quad \textit{Ans. (A)}$$

Matching Test

After completing this test, match your answers with the numbers in the box.

1. Find the number of degrees in the central angle of an arc whose length is 4π in a circle of circumference 20π.

2. Find the number of degrees in the central angle of an arc whose length is 4π in a circle of radius 6.

3. Find the length of an arc of 135° in a circle of circumference 16π.

4. Find the length of one of the arcs cut off by the side of an inscribed square in a circle of radius 6.

5. Find the length of arc cut off by the side of an equilateral triangle inscribed in a circle of diameter 15.

a. 120	*b.* 72	*c.* 3π	*d.* 5π	*e.* 6π

2.2 COMBINATION PERIMETERS CONTAINING ARCS

Model Problem

In the diagram at the right, *ABCD* is a square with a side of 6. Find the total length of the arcs and full lines in the figure. (Disregard dotted lines.)

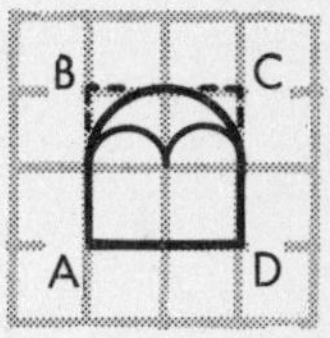

(A) $12 + 3\pi$ (B) $6 + 3\pi$ (C) $6 + 6\pi$ (D) $12 + 6\pi$ (E) $18 + 6\pi$

Solution:

Since the radius of the large semicircle is 3, the length of the semicircumference $= \frac{1}{2}(2\pi r) = \pi r = 3\pi$.

Since the radius of the two smaller semicircles is $1\frac{1}{2}$, the length of each semicircumference $= \frac{1}{2}(2\pi r') = \pi r' = 1\frac{1}{2}\pi$. Hence, the length of the two smaller semicircumferences $= 3\pi$.

The length of the full straight lines $= 6 + 3 + 3 = 12$.

Hence, the total length $= 12 + 3\pi + 3\pi = 12 + 6\pi$. *Ans.* (D)

Matching Test

In each of the following figures, *ABCD* is a square with a side of 6. Find the total length of the arcs and full lines in each figure. (Disregard the dotted lines.) After completing this test, match your answers with the numbers in the box.

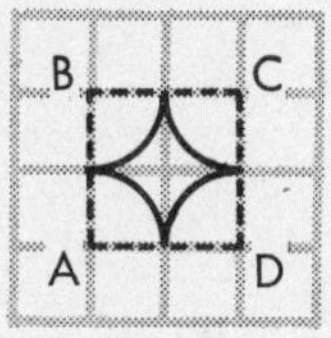

1.

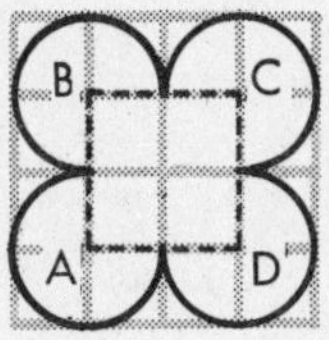

2.

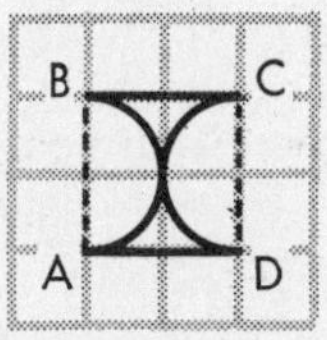

3.

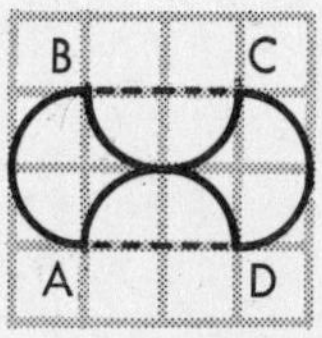

4.

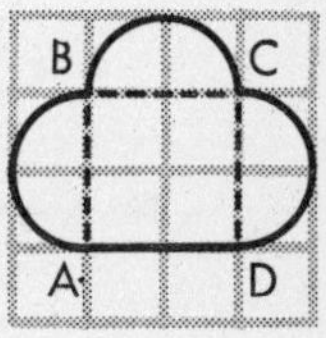

5.

a. $12 + 6\pi$	*b.* 6π	*c.* 12π	*d.* 18π	*e.* $6 + 9\pi$

2.3 ROTATION PROBLEMS

Model Problem

The earth makes a complete rotation of 360° of longitude each day. Through how many degrees does the earth move in 2 hours and 24 minutes?

(A) $62\frac{1}{2}$ (B) 40 (C) $37\frac{1}{2}$ (D) 36 (E) 30

Solution:

Since each day has 24 hours, in 1 hour, the earth will make a rotation of $\frac{360°}{24}$ or 15° of longitude. Hence, in 2 hours and 24 minutes, or in $2\frac{2}{5}$ hours, the earth will rotate $2\frac{2}{5}(15°)$, or 36°. *Ans.* (D)

Matching Test

After completing this test, match your answers with the numbers in the box.

1. Through how many degrees of longitude does the earth rotate in 10 hours?
2. Through how many degrees does the hour hand move in 10 hours?
3. Through how many degrees does the minute hand move in 10 minutes?
4. How many degrees are there in the angle between the hands of the clock at 9:30 P.M.?
5. Through how many degrees does the hour hand move between 3:15 P.M. and 5:45 P.M. of the same day?

a. 60	*b*. 75	*c*. 150	*d*. 300	*e*. 105

3. AREA OF A CIRCLE

3.1 PROBLEMS INVOLVING THE AREA OF A CIRCLE

Model Problems

(1) The area of a circle is 25π. Find the radius of the circle.

(A) 5 (B) $6\frac{1}{4}$ (C) 10 (D) $12\frac{1}{2}$ (E) 25

Solution:

Find the radius, r, as follows:

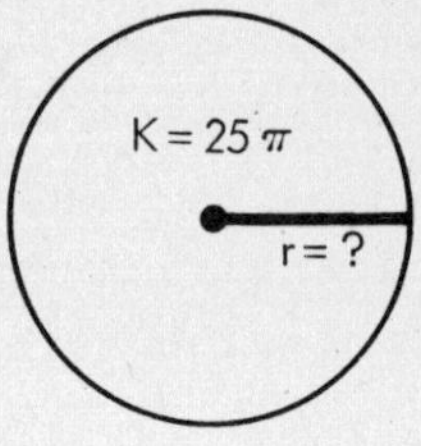

The area of a circle equals the product of π and the square of the radius.

If K = area of circle,

$K = \pi r^2$

Substitute 25π for K.

$25\pi = \pi r^2$

Divide each side by π.

$25 = r^2$

Take the square root of each side.

$5 = r$ *Ans.* (A)

(2) The area of a circle is $\frac{9\pi}{4}\cdot$ Find its circumference.

(A) 3π (B) $4\frac{1}{2}\pi$ (C) 6π (D) 9π (E) 12π

Solution:

1. First find the radius, r, as follows:

   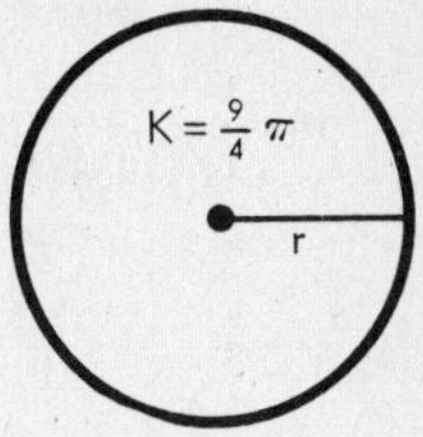

   The area of a circle equals the product of π and the square of the radius.

   If K = area of circle,

   $\pi r^2 = K$

   Since $K = \frac{9}{4}\pi$,

   $\pi r^2 = \frac{9}{4}\pi$

   Divide each side by π.

   $r^2 = \frac{9}{4}$

   Take the square root of each side.

   $r = \frac{3}{2}$

2. Find the circumference, C, as follows:

   The circumference of a circle equals the product of 2π and the radius.

   Hence, $C = 2\pi r$

   Substitute $\frac{3}{2}$ for r.

   $$C = 2\pi\left(\frac{3}{2}\right) = 3\pi \quad \textit{Ans.}\ \text{(A)}$$

(3) The circumference of a circle is 12π. Find the area of the circle.

(A) 12π (B) 72π (C) 48π (D) 36π (E) 24π

Solution:

1. First find the radius, r, as follows:

   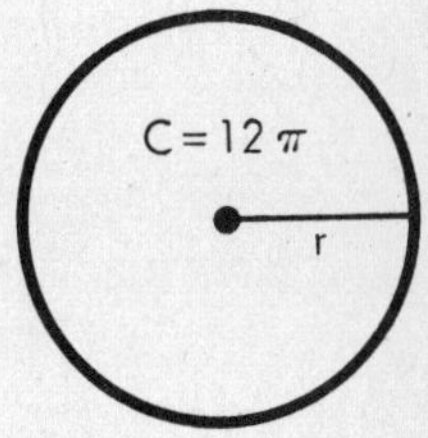

   The circumference of a circle equals the product of 2π and the radius.

   If C = the circumference of the circle, then

   $2\pi r = C$

   Since $C = 12\pi$,

   $2\pi r = 12\pi$

   Divide each side by 2π.

   $r = 6$

2. Find the area, K, as follows:

   The area of a circle equals the product of π and the square of the radius.

   Hence, $K = \pi r^2$

   Substitute 6 for r.

   $$K = \pi(6)^2 = 36\pi \quad \textit{Ans.}\ \text{(D)}$$

(4) Find the area of a circle whose center is at the origin if its circumference passes through the point $(2, 4)$.

(A) 6π (B) 8π (C) 12π (D) 20π (E) 24π

Solution:

In right triangle OAB, $r^2 = x^2 + y^2$
$= 2^2 + 4^2$
$= 4 + 16 = 20$

Area of circle $O = \pi r^2$
$= 20\pi$ *Ans.* (D)

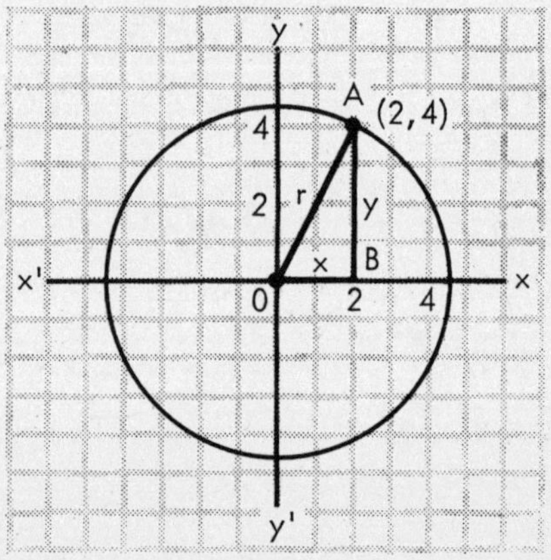

Matching Test

After completing this test, match your answers with the numbers in the box.

Find the area of the circle, if

1. its radius is 10.

2. its circumference is 10π.

3. its center is at the origin and it passes through the point $(3, 5)$.

4. its center is at the origin and it passes through the point $(4, 5)$.

5. its center is at the origin and it passes through the point $(3, 3\sqrt{3})$.

a. 25π	*b.* 41π	*c.* 100π	*d.* 36π	*e.* 34π

3.2 AREA OF A RING

Model Problem

Find the area of the circular ring formed by two concentric circles, one with a radius of 7 and the other with a radius of 10.

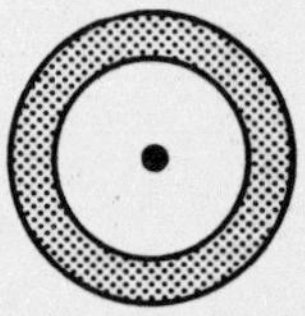

(A) 9π (B) $12\frac{3}{4}\pi$ (C) 35π (D) 51π (E) 149π

Solution:

To find the area of the circular ring, first find the areas of the concentric circles as follows:

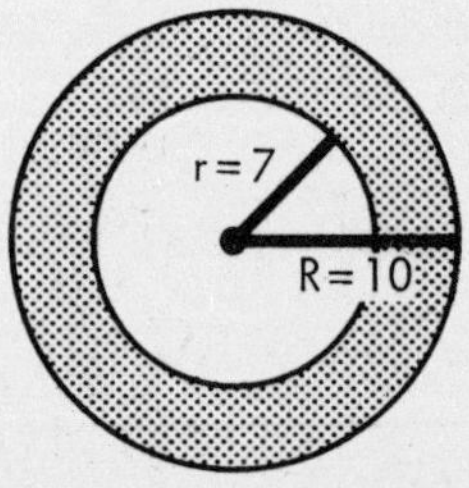

1. Area of larger circle $= \pi R^2$
$= \pi(10^2) = 100\pi$

2. Area of smaller circle $= \pi r^2$
$= \pi(7^2) = 49\pi$

The area of the circular ring is found by subtracting the area of the smaller circle from that of the larger circle.

Hence, the area of the ring $= 100\pi - 49\pi$

$= 51\pi$ *Ans.* (D)

> *Note.* The area of the circular ring may also be found as follows:
>
> $$\begin{aligned}\text{Area of circular ring} &= \pi(R^2 - r^2)\\ &= \pi(10^2 - 7^2)\\ &= \pi(100 - 49)\\ &= 51\pi\end{aligned}$$

After completing each test, match your answers with the numbers in the boxes.

Matching Test A

Find the number of square units in the area of the ring, if

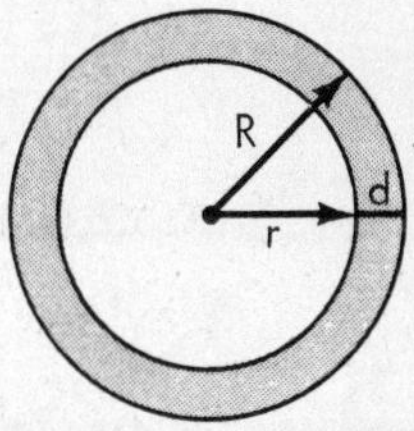

1. the small radius = 6, the large radius = 8.
2. the small radius = 8, the distance between the circles = 2.
3. the area of the small circle $= 36\pi$, the distance between the circles = 4.
4. the circumference of the small circle $= 12\pi$, the large radius = 9.
5. the area of the large circle $= 121\pi$, the small radius = 8.

a. 57π	*b.* 36π	*c.* 45π	*d.* 64π	*e.* 28π

Matching Test B

1. The grooved part of a phonograph record begins 2 inches from the center. If the record has a radius of 6 inches, how many square inches of play does it have?
2. Find the number of square inches in a cross-section of a pipe whose outside diameter is 8 inches and whose thickness is 1 inch.
3. A circular pool having a radius of 20 feet is bounded by a concrete rim that is 1 foot wide. Find the number of square feet in the surface of the rim.
4. A destroyer has been patroling a circular area whose radius is 4 miles. If the destroyer is now required to patrol an area having twice that radius, how many more square miles must it patrol?
5. A windshield wiper has a 9-inch arm. The last 3 inches of the arm contains the rubber wiper blade. How many square inches of the window are covered by the blade if the rotation is $\frac{2}{9}$ of a complete circle?

a. 7π	*b.* 32π	*c.* 48π	*d.* 10π	*e.* 41π

3.3 RELATED AREAS OF CIRCLES AND POLYGONS

Model Problem

Find the length of the diagonal of a square if the area of the square equals twice the area of a circle whose radius is 5.

(A) 5π (B) $5\sqrt{\pi}$ (C) 10π (D) $10\sqrt{2\pi}$ (E) $10\sqrt{\pi}$

Solution:

Let $d =$ the diagonal of the square.

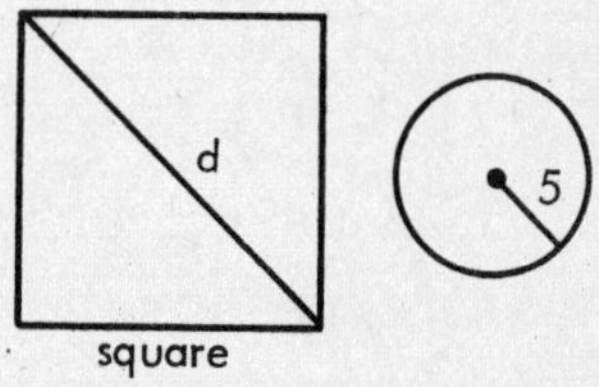

Since the area of the square equals twice the area of the circle,

$$\frac{1}{2}d^2 = 2\pi r^2$$

Hence, $d^2 = 4\pi r^2$

$$= 4\pi \cdot 5^2 = 100\pi$$

Take the square root of each side.

$$d = 10\sqrt{\pi} \quad \textit{Ans.}\ \text{(E)}$$

Matching Test

After completing this test, match your answers with the numbers in the box.

1. The area of a rectangle equals the area of a circle whose radius is 6. If the base of the rectangle is 12, find its altitude.
2. The area of a triangle equals the area of a circle whose radius is 6. If the base of the triangle is 9, find its altitude.
3. The area of a square equals the area of a circle whose radius is 9. Find the side of the square.
4. The area of a square equals one-half the area of a circle whose radius is 6. Find the diagonal of the square.
5. Find the radius of a circle whose area is the sum of the areas of two circles having radii of 6 and 8.

a. 8π	*b.* 3π	*c.* $6\sqrt{\pi}$	*d.* 10	*e.* $9\sqrt{\pi}$

4. AREA OF A SECTOR

4.1 PROBLEMS INVOLVING THE AREA OF A SECTOR

Model Problems

(1) Find the area of a sector of 45° in a circle of radius 4 inches.

(A) π (B) 2π (C) 4π (D) 8π (E) 12π

Solution:

Find the area of the sector as follows:

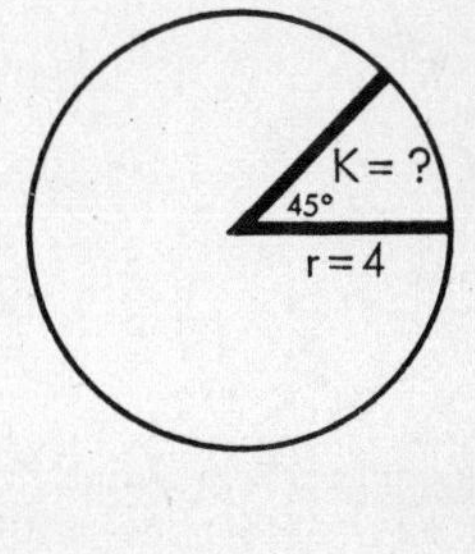

The area of a sector of $n° = \frac{n}{360}(\pi r^2)$.

Substitute 45 for n and 4 for r.

If K = area of sector,

$$K = \frac{45}{360}(\pi 4^2)$$

$$= \frac{1}{8}(16\pi) = 2\pi \quad \textit{Ans. } \text{(B)}$$

(2) The area of circle O is 18 square inches. If there are 40 degrees in central angle AOB, what is the number of square inches in the area of sector AOB?

(A) 2 (B) 2π (C) 3 (D) 4 (E) 4π

Solution:

Find the area of the sector as follows:

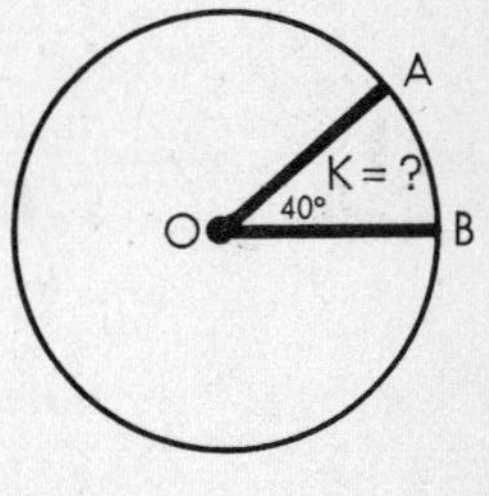

The area of a sector equals $\frac{n}{360}$ of the area of the circle.

Hence, the area of the sector $= \frac{40}{360}$, or $\frac{1}{9}$, of the area of the circle.

Area of sector $= \frac{1}{9}$(area of circle)

$$= \frac{1}{9}(18) = 2 \quad \textit{Ans. } \text{(A)}$$

(3) The radius of a circle is 8. Find the number of degrees in the central angle of a sector having an area of 16π.

(A) 45 (B) 60 (C) 90 (D) 120 (E) 180

Solution:

Find n by applying the following principle:

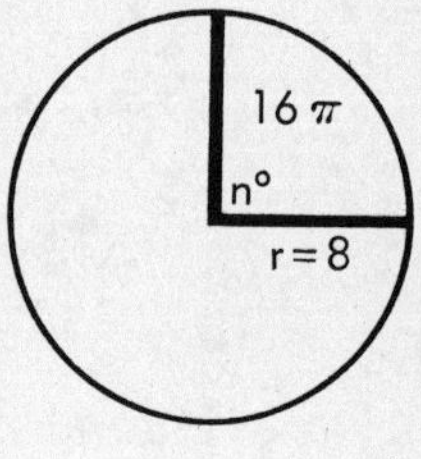

$$\frac{\text{area of sector of } n°}{\text{area of circle}} = \frac{n}{360}$$

The area of the circle $= \pi r^2$. If $r = 8$, the area $= 64\pi$.

Substitute 16π for the area of the sector and 64π for the area of the circle.

$$\frac{16\pi}{64\pi} = \frac{n}{360}$$

Reduce the fraction $\frac{16\pi}{64\pi}$ to $\frac{1}{4}$.

$$\frac{1}{4} = \frac{n}{360}$$

$$4n = 360, \quad n = 90 \quad \textit{Ans.}\ \text{(C)}$$

(4) The angle of a sector of a circle is 90° and the area of the sector is 16π. Find the radius of the circle.

(A) 2 (B) $2\sqrt{2}$ (C) 4 (D) $4\sqrt{2}$ (E) 8

Solution:

Find r as follows:

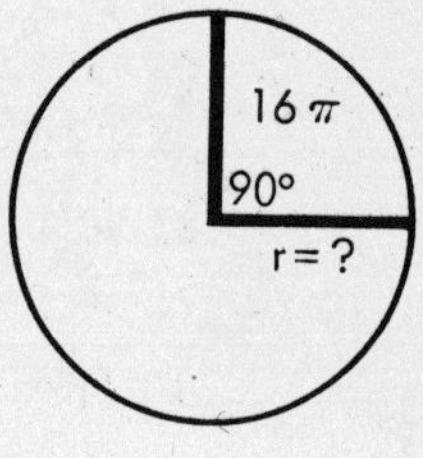

The area of a sector of $n°$ is to the area of the circle as $n°$ is to 360°.

Hence, $\frac{\text{area of } 90° \text{ sector}}{\text{area of circle}} = \frac{90°}{360°}$

Substitute 16π for the area of the 90° sector, and πr^2 for the area of the circle. Reduce $\frac{90°}{360°}$ to $\frac{1}{4}$.

$$\frac{16\pi}{\pi r^2} = \frac{1}{4}$$

Cancel the common factor, π, and cross-multiply to obtain equal products.

$$r^2 = 64, \quad r = 8 \quad \textit{Ans.}\ \text{(E)}$$

After completing each test, match your answers with the answers in the boxes.

Matching Test A

Refer to the figure at the right to find the missing values indicated by the question marks in the following table.

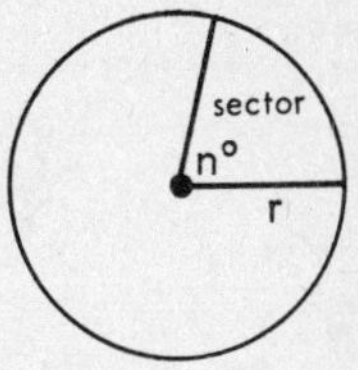

	Central angle, n°	*Radius r*	*Area of the circle*	*Area of the sector*
1.	60°		48π	?
2.	?		48π	6π
3.	40°	?		4π
4.	90°	6		?
5.	?	6		12π

a. 120°
b. 45°
c. 9π
d. 6
e. 8π

Matching Test B

1. Over how many square yards may a horse roam if it is tethered to a 10-yard rope tied to a post at the corner of two fences meeting at a 72° angle?
2. Over how many square feet may a goat graze if it is tethered to a 12-foot rope tied to a post at the intersection of two walls at right angles to each other?
3. How many square rods are lit by a searchlight mounted on a swivel capable of turning through an angle of 120° if the length of the light beam is 6 rods?
4. How many square feet of ground can be watered by a hose that rotates 30° and throws a stream of water 18 feet long?
5. How many square inches of window are dried by a windshield wiper that has a full rubber arm of 10 inches and rotates through an angle of 108°?

a. 30π *b.* 12π *c.* 27π *d.* 36π *e.* 20π

4.2 ADDING TO FIND THE TOTAL AREA OF A FIGURE COMPOSED OF RECTANGLES AND SECTORS

Model Problem

A semicircle surmounts a rectangle whose length is $2a$ and whose width is a, as shown in the diagram. A formula for finding the area of the whole figure is

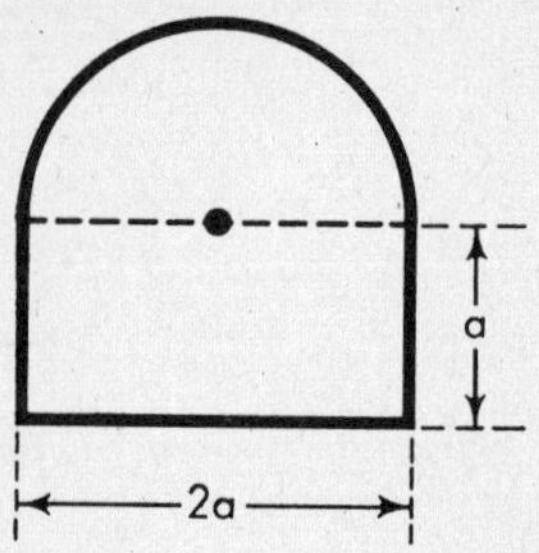

(A) $2a^2 + \frac{\pi a^2}{2}$ (B) $2\pi a^2$ (C) $3\pi a^2$ (D) $2a^2 + \pi a^2$

(E) $a^2 + \frac{\pi a^2}{2}$

Solution:

The area of the entire figure can be found by adding the area of the semicircle to the area of the rectangle. Find the separate areas of each figure as follows:

1. The area of the semicircle is one-half the area of a circle whose diameter equals $2a$. If the diameter is $2a$, then the radius $= \frac{1}{2}(2a) = a$.

   The area of a circle equals π multiplied by the square of the radius. Hence, the area of the circle $= \pi a^2$.

   The area of the semicircle $= \frac{1}{2}$(area of circle) $= \frac{1}{2}\pi a^2$.

2. The length and width of the rectangle are $2a$ and a, respectively. Since the area of a rectangle equals the product of the length and width, the area of the rectangle in the figure $= (2a)(a) = 2a^2$.

3. The sum of the areas of the semicircle and the rectangle $= \frac{1}{2}\pi a^2 + 2a^2$. *Ans.* (A)

Matching Test

The five diagrams below contain arcs of a small circle and arcs of a large circle. The radius of the small circle is 4; the radius of the large circle is 8. Determine the entire area of each figure. After completing this test, match your answers with the values in the box.

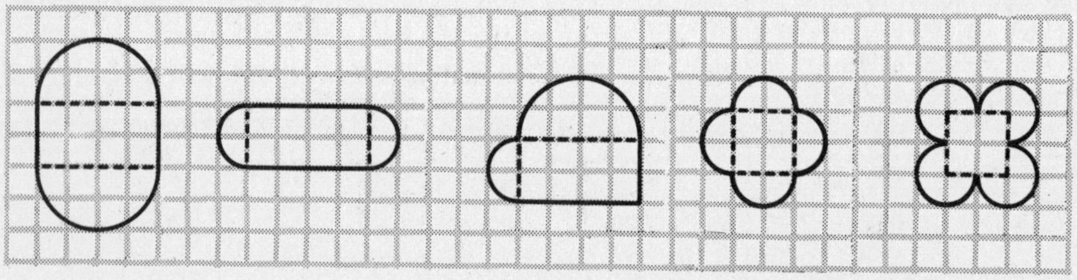

1. **2.** **3.** **4.** **5.**

a. $128 + 64\pi$	*b*. $128 + 40\pi$	*c*. $64 + 32\pi$	*d*. $64 + 48\pi$	*e*. $128 + 16\pi$

4.3 SUBTRACTING TO FIND THE REMAINING AREA IN A FIGURE COMPOSED OF RECTANGLES AND SECTORS

Model Problem

Semicircles are cut from opposite ends of a rectangular piece of cardboard with dimensions as shown in the diagram. The area represented by the shaded portion of the figure is

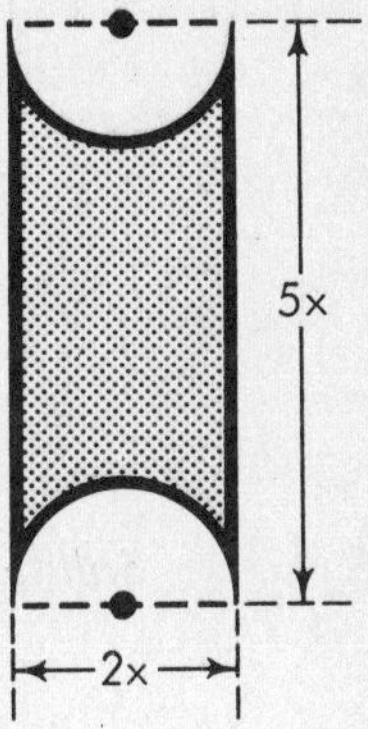

(A) $x^2(10 - 2\pi)$ (B) $x^2(10 - \pi)$ (C) $2x(7 - \pi)$ (D) $2x(7 - 2\pi)$

(E) $10x^2 - 4\pi x^2$

Solution:

The area of the shaded portion of the figure equals the area of the rectangular piece of cardboard minus the area of the two semicircles that are cut out from the opposite ends.

1. The area of a rectangle equals the product of its length and width. Since the length is $5x$ and the width is $2x$, the area of the rectangle $= (5x)(2x) = 10x^2$.

2. The area of the two semicircles equals the area of a single circle whose diameter is $2x$. Hence the radius of this circle is $\frac{1}{2}(2x)$, or x.

   Using r for the radius, the formula for the area of a circle $= \pi r^2$.

   Hence, if x is the radius, the area of the circle $= \pi x^2$.

3. The area of the shaded portion = area of rectangle − area of semicircles

   $= 10x^2 - \pi x^2 = x^2(10 - \pi)$ *Ans.* (B)

Matching Test

The five diagrams below contain arcs of a small circle and arcs of a large circle. The radius of the small circle is 4; the radius of the large circle is 8. In each figure, find the area of the shaded part. After completing this test, match your answers with the values in the box.

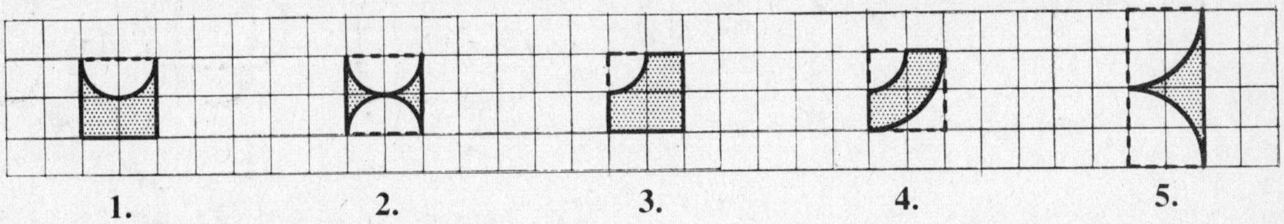

a. $128 - 32\pi$	*b.* $64 - 4\pi$	*c.* $64 - 8\pi$	*d.* $64 - 16\pi$	*e.* 12π

4.4 SUBTRACTING TO FIND THE AREA BETWEEN A CIRCLE AND AN INSCRIBED SQUARE

Model Problem

A square is inscribed in circle O whose radius is 4, as shown in the diagram. The area of the shaded portion is

(A) $16\pi - 16$ (B) $16\pi - 32$ (C) $8\pi - 16$ (D) $8\pi - 32$ (E) $4\pi - 8$

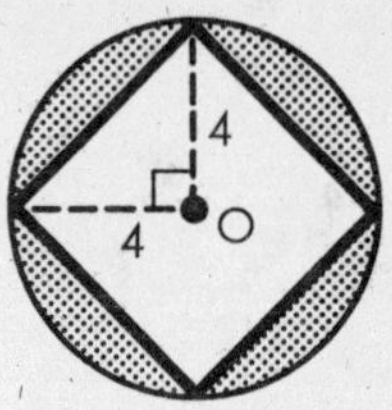

Solution:

To choose the correct formula, first find the areas of the circle and the inscribed square as follows:

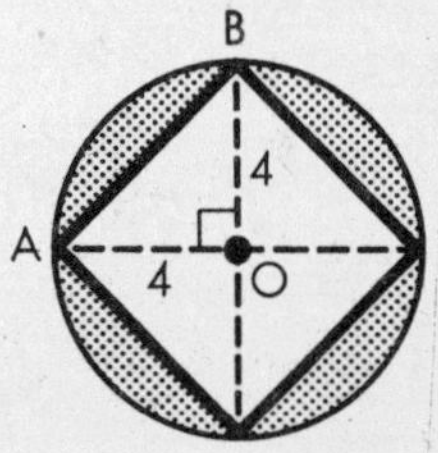

1. The area of a circle equals the product of π and the square of the radius.

   Hence, the area of the circle $= \pi(4^2) = 16\pi$.

2. The area of the square is four times the area of right triangle AOB shown in the figure. Note the four equal triangles formed by extending the radius. The area of each of these four triangles is one-half the product of the base and the altitude. Since the base and altitude are each 4, the area of each triangle is $\frac{1}{2}(4)(4) = 8$.

   Hence, the area of the square is $4(8) = 32$.

3. Area of the shaded part = area of the circle − area of the square

   $= 16\pi - 32$ *Ans.* (B)

> *Note.* The area of the square may be found by using the formula $K = \frac{1}{2}d^2$. Here, $d = 8$. Hence, $K = \frac{1}{2}(64) = 32$.

Matching Test

After completing this test, match your answers with the values in the box.

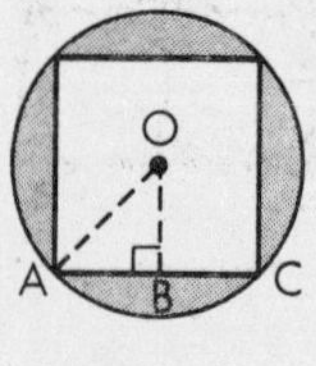

Fig. 1

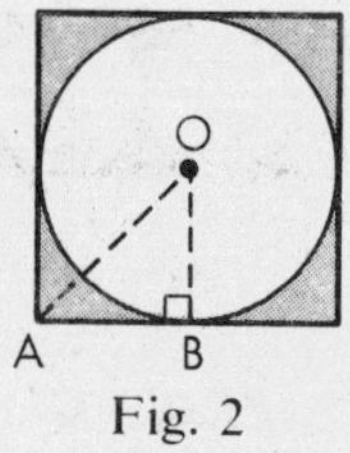

Fig. 2

Find the area of the shaded part between the square and the circle if

1. (Fig. 1) $OA = 6$. **2.** (Fig. 1) $OB = 6$. **3.** (Fig. 1) $AC = 6$.

4. (Fig. 2) $OA = 6$. **5.** (Fig. 2) $OB = 6$.

a. $36\pi - 72$ *b.* $18\pi - 36$ *c.* $144 - 36\pi$ *d.* $72 - 18\pi$ *e.* $72\pi - 144$

4.5 SUBTRACTING TO FIND THE REMAINING AREA INSIDE A RECTANGLE CONTAINING TANGENT CIRCLES

Model Problem

Two circles are tangent to each other and to the sides of a rectangle as shown. The area of the shaded portion is

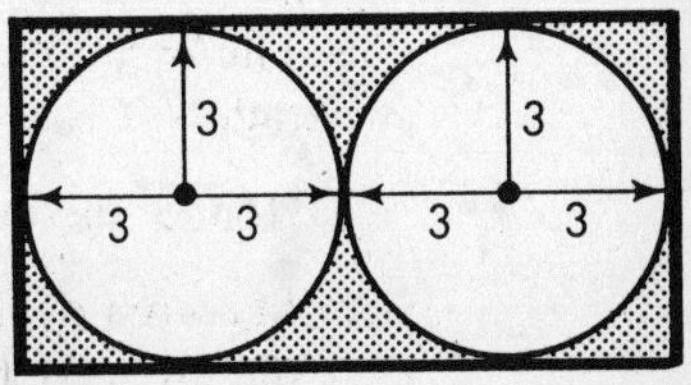

(A) $36 - 6\pi$ (B) $36 - 12\pi$ (C) $72 - 6\pi$
(D) $72 - 18\pi$ (E) $72 - 12\pi$

Solution:

The area of the shaded portion is the area of the rectangle decreased by the area of the two circles. Hence, find the area of the shaded portion as follows:

1. The length of the rectangle is 12 and its width is 6. Since the area of a rectangle is the product of its length and width, the area of the rectangle is $12(6)$, or 72.
2. The area of each circle is πr^2, where r is the radius. Since the radius is 3, the area of each circle is 9π. Hence, the area of both circles is $2(9\pi) = 18\pi$.
3. The area of the shaded portion is the area of the rectangle decreased by the area of both circles. Hence, the area of the shaded portion is $72 - 18\pi$. *Ans.* (D)

Matching Test

In the five figures below, the radius of each tangent circle is 4. In each figure, find the area of the shaded part. After completing this test, match your answers with the values in the box.

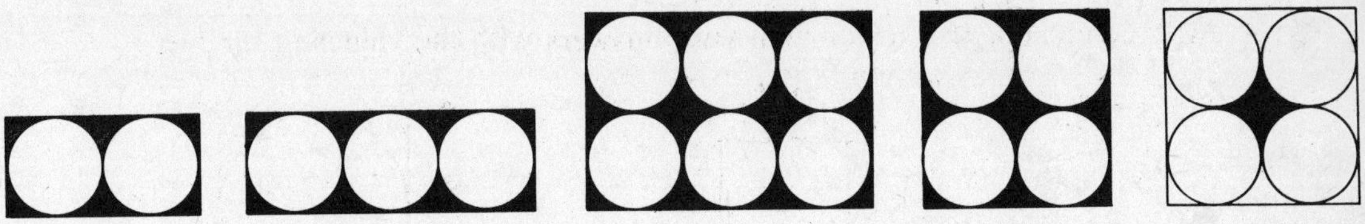

1. **2.** **3.** **4.** **5.**

a. $256 - 64\pi$ *b.* $128 - 32\pi$ *c.* $384 - 96\pi$ *d.* $192 - 48\pi$ *e.* $64 - 16\pi$

4.6 FINDING RATIOS OF AREAS FORMED BY CIRCLES WITHIN CIRCLES

Model Problem

In the figure at the right, the radius of the large circle is R and the radius of each small circle is r. Write, in terms of R and r, a formula that can be used to find the area of the shaded portion.

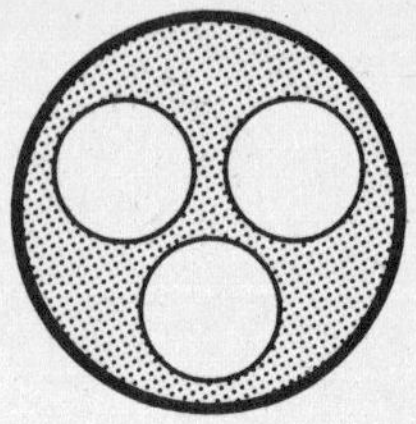

(A) $3\pi r^2 - \pi R^2$ (B) $\pi R^2 - 6\pi r^2$ (C) $\frac{\pi r^2}{7}$ (D) $\pi r^2 - 3\pi R^2$

(E) $\pi R^2 - 3\pi r^2$

Solution:

To write the required formula, first find the areas of the circles in the figure by applying the following rule:

The area of a circle equals the product of π and the square of the radius.

Hence, the area of the large circle $= \pi R^2$.

Since the area of each small circle $= \pi r^2$, the area of the three small circles $= 3\pi r^2$.

Area of the shaded part = area of the large circle − area of the three small circles

$= \pi R^2 - 3\pi r^2$ *Ans.* (E)

Matching Test

In each figure below, find the ratio of the area of the shaded part to that of the largest circle. After completing this test, match your answers with the ratios in the box.

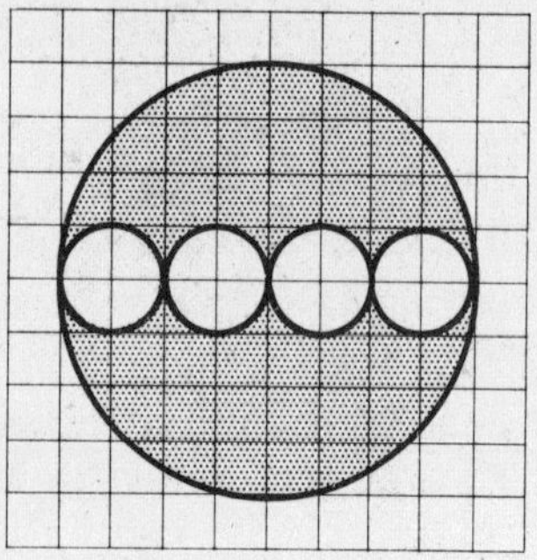

1.

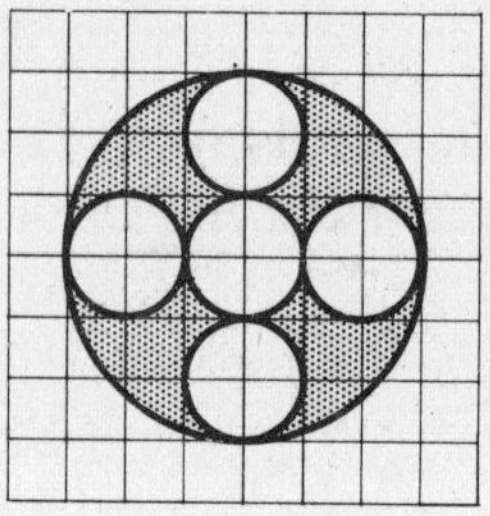

2.

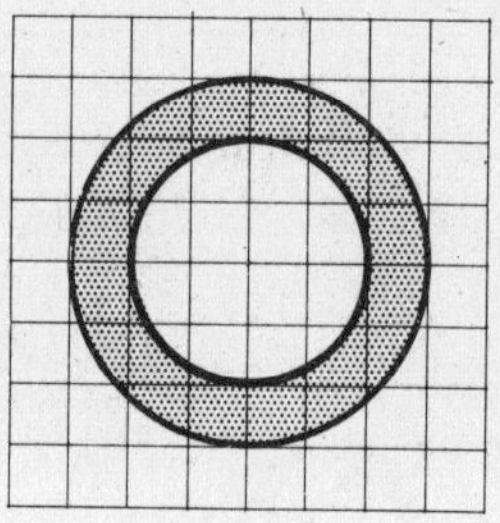

3.

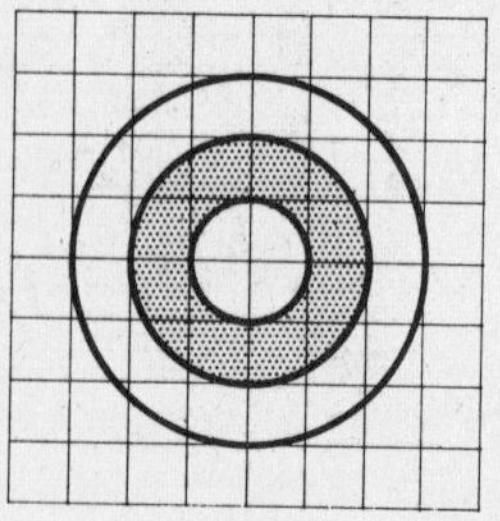

4.

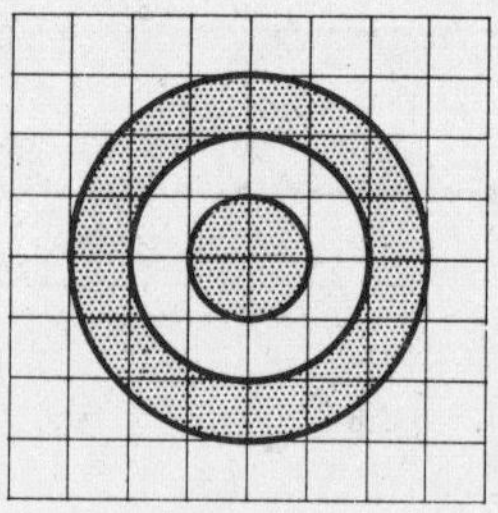

5.

a. $\frac{1}{3}$ *b.* $\frac{3}{4}$ *c.* $\frac{4}{9}$ *d.* $\frac{5}{9}$ *e.* $\frac{2}{3}$

5. VOLUME OF A CYLINDER

Model Problem

A cylinder is enlarged by increasing its radius from 2 to 3 and its height from 3 to 5. How many more cubic units of volume does it now have?

(A) 9π (B) 22π (C) 33π (D) 66π (E) 132π

Solution:

The volume of a cylinder is the product of the area of the base and the height. If the radius of the base is r and the height of the cylinder is h, then the volume, $V = \pi r^2 h$.

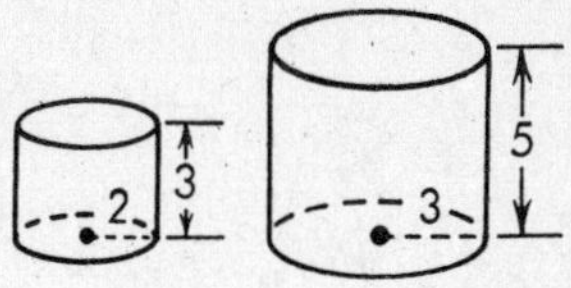

If $r = 2$ and $h = 3$, then the old volume was $\pi \cdot 2^2 \cdot 3$, or 12π.

If $r = 3$ and $h = 5$, then the new volume is $\pi \cdot 3^2 \cdot 5$, or 45π.

Hence, the increase in volume $= 45\pi - 12\pi = 33\pi$. *Ans.* (C)

Matching Test

After completing this test, match your answers with the numbers in the box.

1. How many more cubic inches of volume does a cylindrical jar have if its radius is increased from 3 inches to 4 inches and its height from 5 inches to 6 inches?
2. How many more cubic feet of volume does a cylindrical tank have if its diameter is increased from 4 feet to 6 feet and its height from 5 feet to 6 feet?
3. How many cubic feet of water does a cylindrical pool contain if its radius is 7 feet and the height of the water in the pool is 6 inches? $\left(\text{Use } \pi = \frac{22}{7}.\right)$
4. If 1 gallon is equivalent to 231 cubic inches, how many gallons fill a pail that is 7 inches in radius and 9 inches high? $\left(\text{Use } \pi = \frac{22}{7}.\right)$
5. Water is poured into a cylinder at the rate of 8 cubic inches every minute. How many minutes will it take to fill the cylinder if its radius is 4 inches and its height is 7 inches? $\left(\text{Use } \pi = \frac{22}{7}.\right)$

a. 34π	*b.* 44	*c.* 51π	*d.* 6	*e.* 77

6. PRACTICE PROBLEMS

1. Rectangle $ABCD$ represents a strip of cardboard 16 inches long that is rolled up to form a cylindrical surface in which edges AB and CD meet in a single line. Find the number of inches in the diameter of the base of the cylinder.

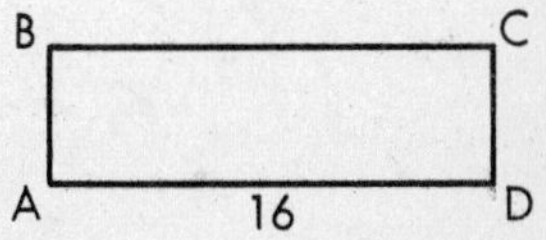

(A) $\frac{8}{\pi}$ (B) $\frac{16}{\pi}$ (C) 8π (D) 16π (E) 32π **[1.1]**

2. A pipe of inner diameter 6 inches fits snugly into another pipe of inner diameter 9 inches. Find the number of inches in the thickness of the smaller pipe.

(A) .75 (B) 1.5 (C) 3 (D) 5 (E) 7.5 **[1.2]**

3. A train wheel of diameter 3 feet makes 28,000 revolutions per hour. At what speed in miles per hour is the train traveling? $\left(\text{Use } \pi = \frac{22}{7}.\right)$

(A) 25 (B) 40 (C) 50 (D) 75 (E) 100 **[1.3]**

4. Find the length of an arc cut off by the side of a regular decagon inscribed in a circle of circumference 36π.

(A) 3π (B) 3.6π (C) 12 (D) 6π (E) 7.2π **[2.1]**

5. In the diagram, $ABCD$ is a square of side 6. Each of the arcs is a semicircle. Find the total length of the arcs and the full lines in the figure. (Disregard the dotted lines.)

(A) $12 + 9\pi$ (B) $10 + 9\pi$ (C) $12 + 4\frac{1}{2}\pi$ (D) $10 + 4\frac{1}{2}\pi$ (E) 9π

[2.2]

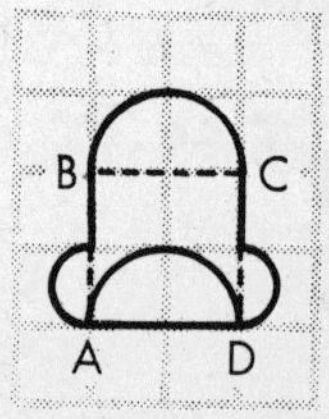

6. A nautical mile is the length of a minute of arc along the circumference of a great circle of the earth. If a ship sails a distance of $2\frac{1}{3}$ degrees of arc along the equator, how many nautical miles has it traveled?

(A) 70 (B) 140 (C) 150 (D) 220 (E) $233\frac{1}{3}$ **[2.3]**

7. Find the area of a circle having its center at the origin and passing through the point $(4\sqrt{2}, 4\sqrt{2})$.

(A) 8π (B) 16π (C) 64 (D) 64π (E) $64\sqrt{2}\pi$ **[3.1]**

8. Find the number of square units in the area of the ring shown at the right, if the circumference of the larger circle is 18π and the distance between the circles is 2.

(A) 17π (B) 32π (C) 64π (D) 68π (E) 128π **[3.2]**

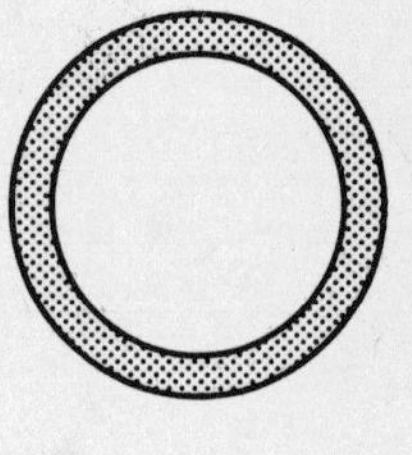

9. A cylindrical wooden block in a lathe is cut from a 12-inch diameter to a 10-inch diameter. How many square inches in each circular cross-section of the original block have been cut away?

(A) 11π (B) 12π (C) 22π (D) 24π (E) 44π **[3.2]**

10. Find the radius of a circle whose area is the sum of the areas of three circles having radii of 2, 3, and 6.

(A) 7 (B) 11 (C) 14 (D) 7π (E) 11π **[3.3]**

11. A sector of a circle has a central angle of 120°. If the area of the sector is 12π, find the radius of the circle.

(A) 3 (B) 6 (C) 9 (D) 12 (E) 18 **[4.1]**

12. How many square feet are there in the area of a children's play yard that has the form of a sector having a radius of 8 feet and an angle of 135°?

(A) 6π (B) 12π (C) 15π (D) $21\frac{1}{3}\pi$ (E) 24π **[4.1]**

13. If the radius of each arc is 4, determine the area of the entire figure.

(A) $128 + 16\pi$ (B) $128 + 32\pi$ (C) $128 + 48\pi$ (D) $64 + 48\pi$ (E) 48π **[4.2]**

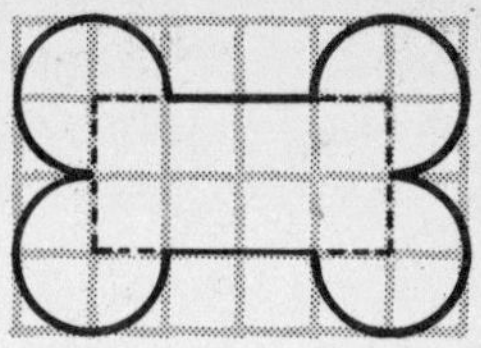

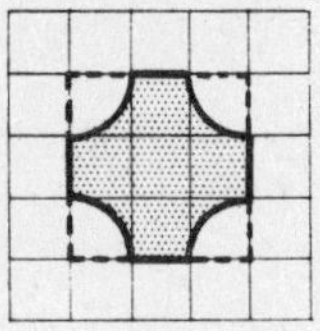

14. If the radius of each arc is 4, find the area of the shaded part. (Each of the arcs is a quarter-circle.)

(A) 128π (B) $72 - 8\pi$ (C) $72 - 16\pi$ (D) $144 - 16\pi$ (E) $144 - 8\pi$ **[4.3]**

15. If the side of the square is 6, find the area of the shaded part between the square and the inscribed circle.

(A) $36 - 6\pi$ (B) $24 - 9\pi$ (C) 27π (D) 30π (E) $36 - 9\pi$ **[4.4]**

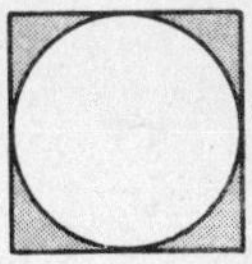

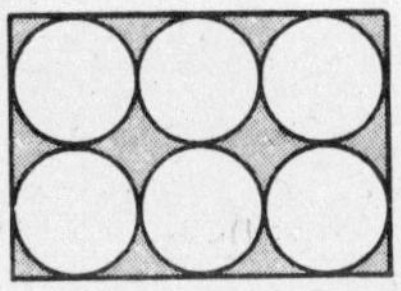

16. The radius of each circle is 3. If the circles are tangent to each other, as shown, find the area of the shaded portion.

(A) 162π (B) $216 - 54\pi$ (C) $24 - 6\pi$ (D) $96 - 24\pi$ (E) $216 - 216\pi$ **[4.5]**

17. Find the ratio of the area of the shaded part to that of the entire circle. (The inner arcs are semicircles.)

(A) $\frac{1}{3}$ (B) $\frac{1}{4}$ (C) $\frac{2}{3}$ (D) $\frac{3}{8}$ (E) $\frac{1}{2}$ **[4.6]**

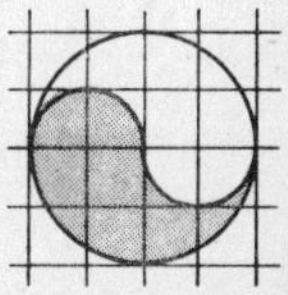

18. A cylinder filled with sand has a radius of 7 inches and a height of 8 inches. The entire contents of the cylinder is poured into a rectangular sand box whose base is 22 inches by 14 inches. By how many inches is the height of the sand in the box increased? $\left(\text{Use } \pi = \frac{22}{7}.\right)$

(A) 16 (B) 8 (C) 4 (D) 2 (E) $1\frac{1}{7}$ **[5]**

X. REVIEWING GRAPHS

A ***number line***, or ***number scale***, is a line on which distances from a point are numbered in equal units, positively in one direction and negatively in the other. The point from which the distances are numbered is the ***origin***, or zero point.

Number Line

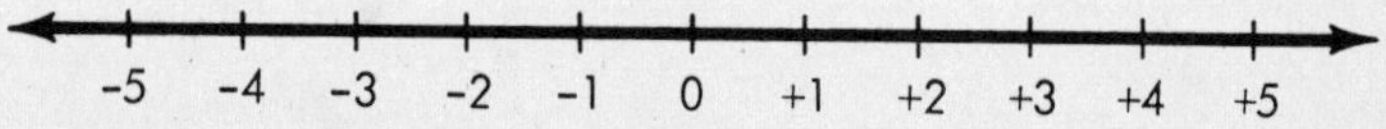

Fig. 1

Two-Dimensional Graph

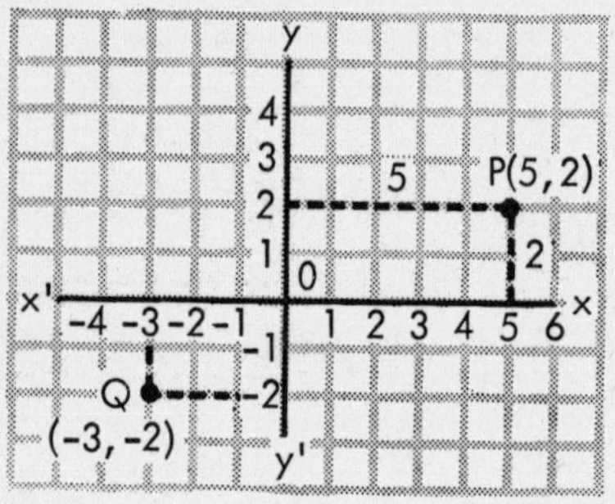

Fig. 2

A ***graph*** is formed by combining two number lines at right angles to each other so that their zero points coincide. The horizontal number line is the x-axis and the vertical number line is the y-axis.

A point is located on a graph by means of its two coordinates. The ***coordinates*** of a point are its distances from the axes. The x-coordinate, or ***abscissa***, of a point is its distance from the y-axis; the y-coordinate, or ***ordinate***, of a point is its distance from the x-axis. In writing the coordinates of a point, the x-coordinate is placed before the y-coordinate. For example, $P(5, 2)$ means that the x-coordinate of point P is 5 and the y-coordinate of P is 2. In Fig. 2, note that 5 and 2 are the distances of point P from the two axes. Note also the position of point $Q(-3, -2)$, whose x-coordinate is -3 and whose y-coordinate is -2.

The two axes divide the graph into four ***quadrants***, numbered I, II, III, and IV (Fig. 3). Note that the quadrants are numbered in a counterclockwise direction. The signs under each quadrant number indicate the signs of the coordinates of every point in that quadrant. For example, in quadrant II, $(-, +)$ indicates that any point in this quadrant has a negative x-coordinate and a positive y-coordinate. Note that $R(-1, 2)$ is such a point.

Quadrants

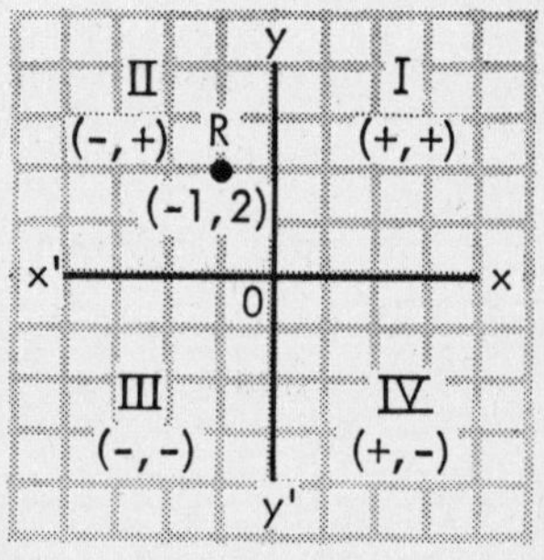

Fig. 3

X-Intercept and Y-Intercept

The ***x-intercept*** of a line is the value of x at the point where the line crosses, or intersects, the x-axis. The ***y-intercept*** of a line is the value of y at the point where the line intersects the y-axis.

For example, in Fig. 4 the x-intercept of the line $x + y = 3$ is the x-value of the point Q (3, 0), namely, 3. Also, the y-intercept of the line $y = x + 1$ is the y-value of the point $S(0, 1)$, namely, 1.

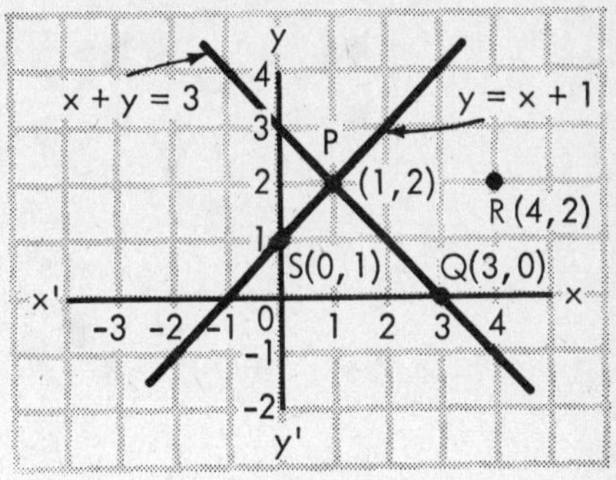

Fig. 4

To find the intercepts of a line from the equation of a line, solve the equation for x when $y = 0$ and for y when $x = 0$. For example, to find the x-intercept of the line $x + y = 3$, let $y = 0$ and solve the equation for x as follows:

$$x + y = 3$$
$$x + 0 = 3$$
$$x = 3$$

Graphs of Linear Equations

The graph of a first-degree equation in two unknowns is a straight line. Hence, first-degree equations like $y = x + 1$ and $x + y = 3$ are called ***linear equations.*** (See Fig. 4.)

The graph of a first-degree equation in one unknown is any one of the following:

(1) the x-axis, whose equation is $y = 0$.

(2) the y-axis, whose equation is $x = 0$.

(3) a line parallel to the x-axis having the equation $y = k$, where k is a constant.

(4) a line parallel to the y-axis having the equation $x = k$, where k is a constant.

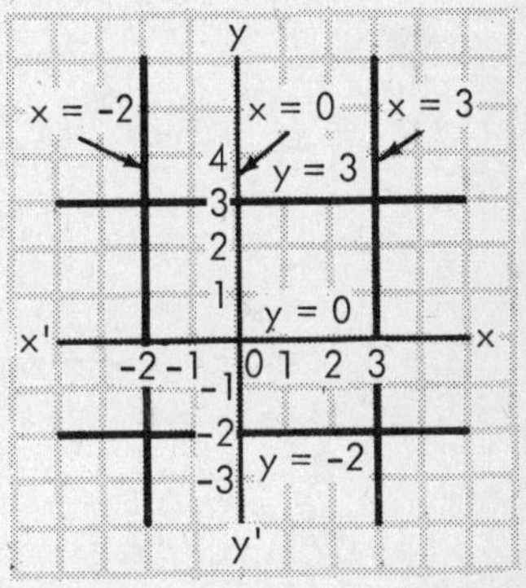

Fig. 5

In Fig. 5, note that the graphs of $y = 3$ and $y = -2$ are parallel to the x-axis, while the graphs of $x = 3$ and $x = -2$ are parallel to the y-axis.

1. POINTS ON A LINE

If a point is on the graph of an equation, its coordinates satisfy the equation. For example, in Fig. 4, the coordinates of $P(1, 2)$ satisfy the equations $x + y = 3$ and $y = x + 1$.

If a point is not on the graph of an equation, its coordinates do not satisfy the equation. For example, in Fig. 4, the coordinates of $R(4, 2)$ do not satisfy the equation $x + y = 3$ nor the equation $y = x + 1$. Thus, R is not on the graph of either of these equations.

1.1 LINE JOINING TWO GIVEN POINTS

Model Problem

On a sheet of graph paper, locate the points (8, 3) and (−4, −6) and connect them with a straight line. This line will intersect the x-axis at the point

(A) (0, 4) (B) (−3, 0) (C) (4, 0) (D) (3, 0) (E) (5, 0)

Solution:

On the graph, note that the line joining point (8, 3) and point (−4, −6) intersects the x-axis at point $P(4, 0)$, that is, at the point where $x = 4$ and $y = 0$.

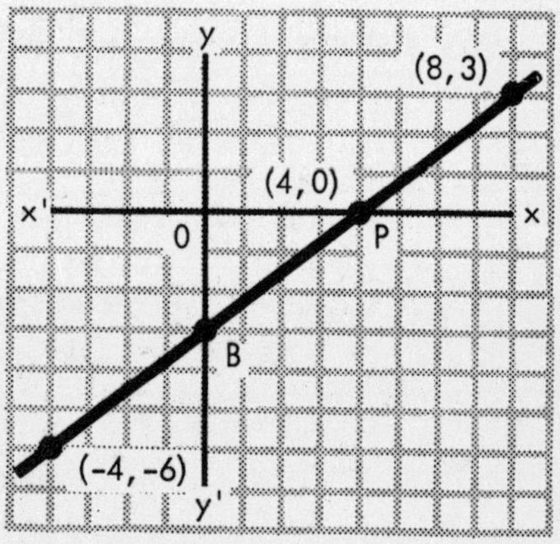

Ans. (C)

Matching Test

After completing this test, match your answers with the pairs of coordinates in the box.

1. Find the coordinates of point B, where the line in the preceding model problem intersects the y-axis.
2. A line joins (8, 3) and (8, −3). Find the coordinates of the point where the line intersects the x-axis.
3. A line joins (8, 3) and (−8, 3). Find the coordinates of the point where the line intersects the y-axis.
4. A line joins (−4, −6) and (−4, 6). Find the coordinates of the point where the line intersects the x-axis.
5. A line joins (−4, −6) and (4, −6). Find the coordinates of the point where the line intersects the y-axis.

a. (8, 0)	*b.* (0, −3)	*c.* (0, 3)	*d.* (−4, 0)	*e.* (0, −6)

1.2 TESTING POINTS ON A LINE AND FINDING POINTS OF INTERSECTION

Model Problem

The following number pairs are the coordinates of points on the same straight line: (0, 2), (−1, 0), (3, 8). An equation of the line is:

(A) $x + y = 11$ (B) $y = 2x + 2$ (C) $x = y - 1$ (D) $y = x + 2$ (E) $y = 3x - 2$

Solution:

Make the correct choice by applying the following principle:

If a point lies on the graph of an equation, then the coordinates of the point satisfy the equation.

The coordinates of each of the three given points can be shown to satisfy the equation given in choice (B), $y = 2x + 2$. Show this by substituting the first coordinate for x and the second coordinate for y.

Point (0, 2)	Point (−1, 0)	Point (3, 8)
$y \stackrel{?}{=} 2x + 2$	$y \stackrel{?}{=} 2x + 2$	$y \stackrel{?}{=} 2x + 2$
$2 \stackrel{?}{=} 2(0) + 2$	$0 \stackrel{?}{=} 2(-1) + 2$	$8 \stackrel{?}{=} 2(3) + 2$
$2 \stackrel{?}{=} 0 + 2$	$0 \stackrel{?}{=} -2 + 2$	$8 \stackrel{?}{=} 6 + 2$
$2 = 2$	$0 = 0$	$8 = 8$

Ans. (B)

Wrong Choices:

(A) Only the coordinates of the point (3, 8) satisfy the equation $x + y = 11$.
(C) Only the coordinates of the point (−1, 0) satisfy the equation $x = y - 1$.
(D) Only the coordinates of the point (0, 2) satisfy the equation $y = x + 2$.
(E) None of the given points has coordinates that satisfy the equation $y = 3x - 2$.

Matching Test

After completing this test, match your answers with the pairs of coordinates in the box.

1. What are the coordinates of the y-intercept of $y = 2x + 2$?

2. What are the coordinates of the x-intercept of $y = 2x + 2$?

3. What are the coordinates of the point of intersection of $y = 2x + 2$ and $x + y = 11$?

4. What are the coordinates of the point of intersection of $y = 2x + 2$ and $y = x - 1$?

5. What are the coordinates of the point of intersection of $y = 2x + 2$ and $x = y - 1$?

a. (−1, 0) *b.* (−3, −4) *c.* (3, 8) *d.* (−1, 0) *e.* (0, 2)

1.3 FINDING COORDINATES OF POINTS ON A GIVEN LINE

Model Problem

The coordinates of one of the points which will lie on the graph of the equation $3x - 2y = 0$ are

(A) $\left(0, 1\frac{1}{2}\right)$ (B) $\left(\frac{2}{3}, 0\right)$ (C) $(2, 3)$ (D) $(4, -6)$ (E) $\left(1, \frac{2}{3}\right)$

Solution:

Determine the point that lies on the graph of the equation $3x - 2y = 0$ by applying the following principle:

A point lies on the graph of an equation if the coordinates of the point satisfy the equation.

The coordinates of the point $(2, 3)$ are $x = 2$ and $y = 3$. These values are found to satisfy the given equation by substituting 2 for x and 3 for y.

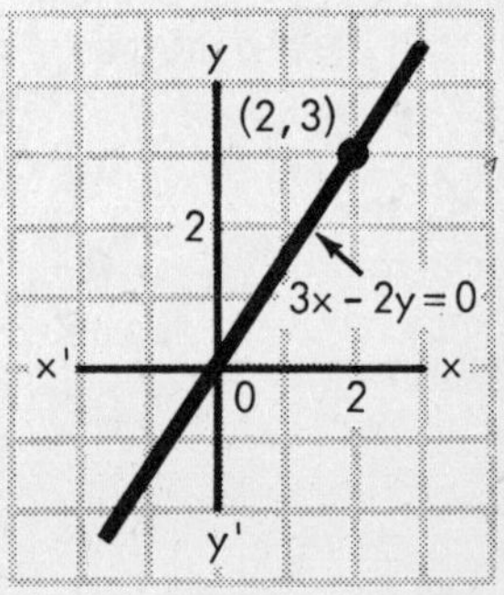

$$3x - 2y = 0$$
$$3(2) - 2(3) \stackrel{?}{=} 0$$
$$6 - 6 \stackrel{?}{=} 0$$
$$0 = 0$$

Ans. (C)

> *Note.* The graph of the line $3x - 2y = 0$ passes through the origin. If the graph is drawn, it is seen that the point $(2, 3)$ is on the graph. However, none of the other four points lies on the graph.

Matching Test

In each of the following, one of the coordinates of a point on the line $3x - 2y = 0$ is given. Find the value of the other coordinate, k. After completing this test, match your answers with the numbers in the box.

1. $(6, k)$ **2.** $(k, 6)$ **3.** $(12, k)$ **4.** $(k, 12)$ **5.** $(-6, k)$

a. −9	*b.* 4	*c.* 8	*d.* 18	*e.* 9

2. STRAIGHT-LINE GRAPHS, CIRCLE GRAPHS, AND BAR GRAPHS

Straight-line graphs are useful in solving certain kinds of motion and work problems. Circle graphs and bar graphs have important uses in the presentation of statistical data. Sets 2.1 and 2.2 show some of the ways in which these types of graphs are used.

2.1 USING GRAPHS IN MOTION PROBLEMS

Model Problem

The accompanying graph shows that Mr. *A* started from Oldtown at 12 noon and arrived at Newtown at 1 P.M.; Mr. *B* started from Oldtown at 12 noon and had traveled 12 miles by 3 P.M.; Mr. *C* started from Newtown at 12 noon and arrived at Oldtown at 3 P.M.

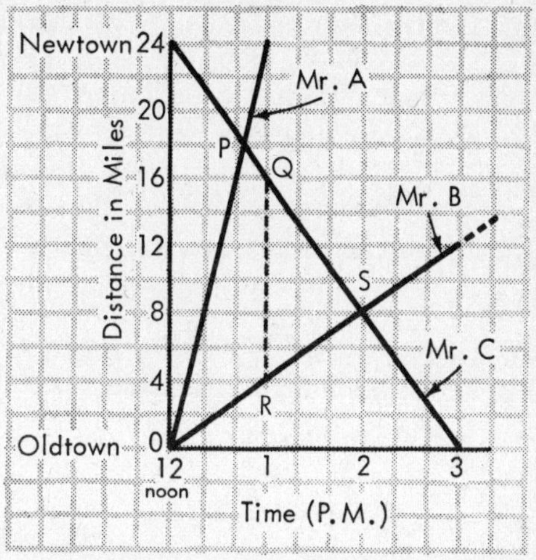

Which of the following statements is false?

(A) Mr. *C*'s speed was 8 mph.

(B) Mr. *A* and Mr. *C* met 6 miles from Newtown.

(C) If Mr. *B* maintains his speed, he will arrive in Newtown at 6 P.M.

(D) At 1 P.M., Mr. *B* and Mr. *C* were 12 miles apart.

(E) Mr. *B* and Mr. *C* met 8 miles from Newtown.

Solution:

(A) Note that Mr. *C* covered 24 miles in 3 hours. Hence, his speed was 8 mph. On the graph, note that, at 1 P.M., Mr. *C* was at *Q*, 8 miles from Newtown. (A) *true*

(B) Mr. *A* and Mr. *C* met at point *P*, 6 miles from Newtown. (B) *true*

(C) Mr. *B* traveled 12 miles in 3 hours; hence, if he maintains his speed, he will travel the remaining 12 miles in another 3 hours and arrive in Newtown at 6 P.M. (C) *true*

(D) On the graph, note the length of line *QR* between the graphs of Mr. *C* and Mr. *B*. Since *QR* represents a distance of 12 miles, Mr. *B* and Mr. *C* at 1 P.M. were 12 miles apart. (D) *true*

(E) Mr. *B* and Mr. *C* met at point *S*, 8 miles from Oldtown. (E) *false*

Ans. (E)

Matching Test

After completing this test, match your answers with the numbers in the box.

The following questions refer to the preceding model problem.

1. What is the speed of Mr. *A* in miles per hour?

2. What is the speed of Mr. *B* in miles per hour?

3. How many miles from Oldtown were Mr. *B* and Mr. *C* when they met?

4. At 2:20 P.M., how far was Mr. *C* from Oldtown, to the nearest mile?

5. At 12:20 P.M., how far apart were Mr. *A* and Mr. *C*, to the nearest mile?

a. 4
b. 8
c. 5
d. 13
e. 24

Model Problem

The accompanying graph pictures the relationships described as follows:

"At noon, Bob started from his home in Kayville to ride to camp at Big Lake on his bicycle, traveling at a speed of 8 miles per hour. At 2:00 P.M., he stopped to rest for 30 minutes and then continued on his way at one-half his original speed. At 4:30 P.M., Bob's father left Kayville in the family car, following the same route and traveling at a constant speed. He overtook Bob after 45 minutes, and together they rode on in the car to Big Lake."

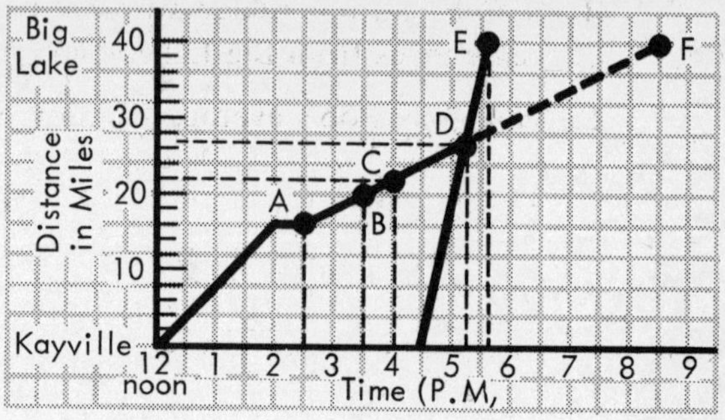

Refer to the graph to determine which of the following statements is false:

(A) Bob traveled 4 miles between 2:30 P.M. and 3:30 P.M.

(B) At 4:00 P.M., Bob was 22 miles from Kayville.

(C) Bob and his father rode 13 miles together in the car.

(D) At approximately 5:35 P.M., Bob and his father arrived at Big Lake.

(E) If Bob had not ridden with his father for part of the way but instead had completed the journey to Big Lake on his bicycle, it would have taken him 9 hours for the entire trip.

Solution:

(A) Point A indicates that at 2:30 P.M., Bob was 16 miles from Kayville. Point B indicates that at 3:30 P.M., Bob was 20 miles from Kayville.
Hence, between 2:30 P.M. and 3:30 P.M., Bob traveled 4 miles. (A) *true*

(B) Point C indicates that at 4 P.M. Bob was 22 miles from Kayville. (B) *true*

(C) Point D, the point of intersection of the two graphs, indicates that Bob and his father met at 5:15 P.M., 27 miles from Kayville. Since they traveled to Big Lake, which is 40 miles from Kayville, the distance traveled together from Kayville to Big Lake was 13 miles. (C) *true*

(D) Point E indicates that Bob and his father arrived at Big Lake at about 5:35 P.M. (D) *true*

(E) Point F indicates that Bob would have arrived at Big Lake at 8:30 P.M. Since he began his trip at 12 noon, the entire trip would have taken him $8\frac{1}{2}$ hours, not 9 hours. (E) *false*

Ans. (E)

Matching Test

After completing this test, match your answers with the numbers in the box.

The following questions refer to the preceding model problem.

1. How many miles did Bob travel between 12 and 4:30 P.M.?

2. At 3 P.M., how many miles was Bob from Big Lake?

3. At what speed did Bob's father travel in miles per hour?

4. If Bob had not ridden with his father but had completed the journey to Big Lake, how many miles from Big Lake would he have been at 6:30 P.M.?

5. What was Bob's average speed in miles per hour from noon to 4 P.M.?

a. 8	*b*. 24	*c*. $5\frac{1}{2}$	*d*. 22	*e*. 36

2.2 CIRCLE GRAPHS AND BAR GRAPHS

Model Problem

The circle and bar graphs at the right represent the distribution of 3600 books in a certain school library. Which of the following statements is false?

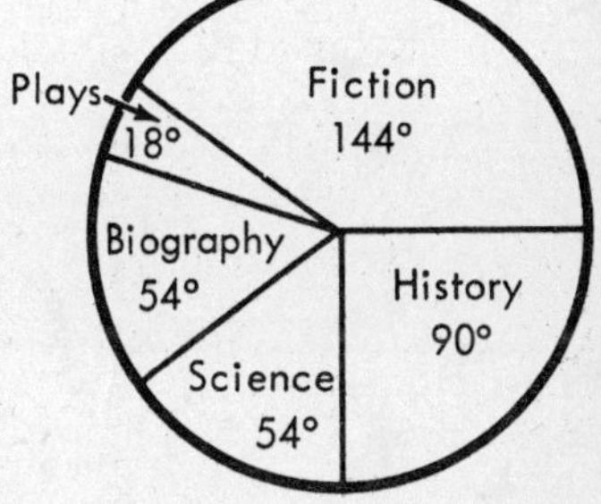

(A) The number of plays is $\frac{1}{20}$ of the total number of books.

(B) The number of biographies is three times the number of plays.

(C) The ratio of the number of fiction books to the number of plays and biographies combined is 2 to 1.

(D) There are 540 more fiction books than history books.

(E) The number of fiction and history books combined is 980 more than the number of the remaining books.

BAR GRAPH SHOWING DISTRIBUTION OF BOOKS IN SCHOOL LIBRARY

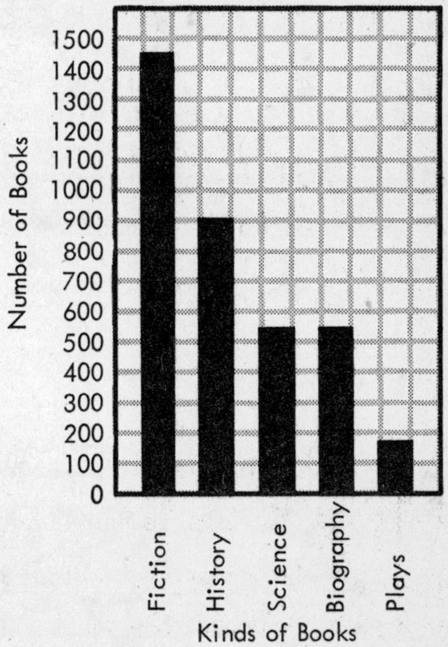

Solution:

(A) The circle graph shows that the play sector has a central angle of 18°. Since the number of degrees in a complete rotation is 360, the sector of 18° is $\frac{18}{360}$, or $\frac{1}{20}$, of the complete circle. Hence, the number of plays is $\frac{1}{20}$ of the total number of books. (A) *true*

(B) The circle graph shows that the biography sector has a central angle of 54°. Since this angle is three times the 18° central angle of the play sector, the number of biographies is three times the number of plays. (B) *true*

(C) If the play and biography sectors are combined, the resulting sector has a central angle of 72°. Since the central angle of the fiction sector is 144°, the ratio of the central angles is $\frac{144}{72}$, or 2 to 1. Hence, the number of fiction books is twice the number of plays and biographies combined. (C) *true*

(D) The bar graph indicates that the number of fiction books is 1440 and the number of history books is 900. Hence, there are 540 more fiction books than history books. (D) *true*

(E) The combined number of fiction and history books is 1440 + 900, or 2340. Hence, the remaining books number 3600 − 2340, or 1260. Since 2340 is 1080 more than 1260, statement (E) is false. (E) *false*

Ans. (E)

Matching Test

After completing this test, match your answers with the fractions or ratios in the box.

The following questions refer to the preceding model problem.

1. What is the ratio of the number of plays to the number of fiction books?
2. What is the ratio of the number of plays to the number of history books?
3. What is the ratio of the number of science books to the number of history books?
4. What part of the total number of books is the number of fiction books?
5. What part of the total number of books is the number of non-history books?

a. $\frac{3}{4}$
b. $\frac{1}{5}$
c. $\frac{2}{5}$
d. $\frac{3}{5}$
e. $\frac{1}{8}$

3. TABLES OF VALUES

A table of values displays two related sets of numbers. In the table, each number in one set is paired with a corresponding number in the other set, and these number pairs are arranged in an ordered sequence.

3.1 FINDING A MISSING NUMBER IN A TABLE OF VALUES

Model Problem

Find the missing number in the following table:

(A) 56 (B) 80 (C) 108 (D) 162 (E) 486

y	2	4	6	8	12
x	2	6	18	54	?

Solution:

Note that as each value of y is increased by 2, each value of x is tripled. For example, when y equals 6 + 2, or 8, x equals 18 × 3, or 54. Similarly, when y equals 8 + 2, or 10, x equals 54 × 3, or 162.

Hence, when y equals 12, x equals 162 × 3, or 486. *Ans.* (E)

Matching Test

In each of the following tables, find the missing value represented by the question mark. After completing this test, match your answers with the numbers in the box.

Hint. In **1**, as y increases by 2, x is doubled.
In **2**, as y increases by 1, x is halved.
In **3**, as r increases by 1, t is the next larger square.
In **4**, as p increases by 1, q is the next smaller cube.
In **5**, as f increases by 1, g is being multiplied by successively larger integers.

1.

y	2	4	6	8	?
x	1	2	4	8	32

2.

y	1	2	3	4	5
x	24	12	6	3	?

3.

r	2	3	4	5	7
t	9	16	25	36	?

4.

p	1	2	3	4	5
q	125	64	27	8	?

5.

f	3	4	5	6	7
g	1	2	6	24	?

a. 12 *b.* $1\frac{1}{2}$ *c.* 1 *d.* 120 *e.* 64

3.2 DERIVING A LINEAR EQUATION BY INSPECTING THE VALUES IN A TABLE

Model Problem

Write an equation expressing the relation between x and y as shown in the table:

x	0	1	2	3
y	2	3	4	5

(A) $x + y = 2$ (B) $y = 3x$ (C) $y = 2x$ (D) $y = x + 2$ (E) $x - y = 2$

Solution:

The relation between each x and each corresponding y can be seen by inspecting each pair of values. Note that each y is 2 more than the corresponding x; that is, the value of 2 for y is 2 more than the value of 0 for x, the value of 3 for y is 2 more than the value of 1 for x, etc.

Since each y is 2 more than the corresponding x, then $y = x + 2$. *Ans.* (D)

Matching Test

For each table, find an equation that may be used to express the relation between x and y. After completing this test, match your answers with the equations in the box.

1.

y	3	4	10
x	0	1	7

2.

y	2	0	−4
x	4	2	−2

3.

x	1	3	6
y	5	15	30

4.

y	3	6	12	24
x	$\frac{1}{2}$	1	2	4

5.

y	2	5	7	10
x	8	20	28	40

a. $y = x - 2$, or $x = y + 2$ *b.* $y = \frac{1}{4}x$, or $x = 4y$ *c.* $y = x + 3$, or $x = y - 3$

d. $y = 6x$, or $x = \frac{1}{6}y$ *e.* $y = 5x$, or $x = \frac{1}{5}y$

3.3 DERIVING A LINEAR EQUATION IN THE FORM $y = mx + b$

Model Problem

The table at the right is a set of ordered pairs representing points on a straight line. An equation of the line is

x	−2	0	2	3
y	−1	3	7	9

(A) $y = 3x + 1$ (B) $y = x + 3$ (C) $x = 3y$ (D) $y = 2x + 3$
(E) $y = 3x + 3$

Solution:

The correct equation is the one that is satisfied by all of the four pairs of ordered values. Since the equation of a straight line may be written in the form $y = mx + b$, make the correct choice by applying the following principle:

To derive a linear equation in the form $y = mx + b$, the value of m can be obtained from a table by using the ratio of the y-difference to the corresponding x-difference; that is,

$$m = \frac{y\text{-difference}}{x\text{-difference}} = \frac{\text{difference of two } y\text{-values}}{\text{corresponding difference of two } x\text{-values}}$$

(Think of $\boldsymbol{m}$ as the ***m***ultiplier of x in the equation $y = \boldsymbol{m}x + b$.)

Note in the table that the ratio of any y-difference to the corresponding x-difference is 2 : 1; that is, $m = 2$.

y-differences: →4 4 2

y	−1	3	7	9
x	−2	0	2	3

x-differences: →2 2 1

$$m = \frac{y\text{-difference}}{x\text{-difference}} = \frac{4}{2} = \frac{2}{1} = 2$$

To obtain the value of b, note that $y = 3$ when $x = 0$. In the equation $y = mx + b$, when $x = 0$, the values of y and b are the same. Hence, $b = 3$.

Since $m = 2$ and $b = 3$, then $y = mx + b$ becomes $y = 2x + 3$. *Ans.* (D)

Matching Test

Assume that each table below is a set of ordered pairs representing points that lie on a straight line. For each table, find the equation of the line. After completing this test, match your answers with the equations in the box.

1.

y	-1	1	5	9
x	1	2	4	6

2.

y	2	4	5	8
x	2	6	8	14

3.

y	2	4	9	10
x	5	3	-2	-3

4.

y	-2	13	23	38
x	0	3	5	8

5.

y	2	5	8	14
x	0	1	2	4

a. $y = -x + 7$, or $x + y = 7$ *b.* $y = 5x - 2$ *c.* $y = \frac{1}{2}x + 1$, or $2y - x = 2$

d. $y = 3x + 2$ *e.* $y = 2x - 3$

4. PRACTICE PROBLEMS

1. A line joins the points $(8, 3)$ and $(-4, -1)$. Find the coordinates of the point where the line intersects the x-axis.

(A) $(-1, 0)$ (B) $(0, -1)$ (C) $(-2, 0)$ (D) $(0, -2)$ (E) $\left(-\frac{1}{2}, 0\right)$ **[1.1]**

2. What are the coordinates of the point on the graph of $y = 2x + 2$ that has equal coordinates?

(A) $(1, 1)$ (B) $(-2, -2)$ (C) $(2, 2)$ (D) $(-1, -1)$ (E) $\left(\frac{1}{2}, \frac{1}{2}\right)$ **[1.2]**

3. If $(k, -6)$ is a point on the line that is the graph of $3x - 2y = 0$, find k.

(A) -9 (B) 9 (C) -4 (D) 4 (E) $\frac{2}{3}$ **[1.3]**

4. Towns A and B are 20 miles apart. The accompanying graph shows that Jack started from B at noon and arrived at A at 1 P.M., while Peter started from A at noon and arrived at B at 12:30 P.M. According to the graph, at what time did they meet, and how far were they from A, to the nearest mile, when they met?

(A) 12:20 P.M. 13 mi. (B) 12:30 P.M. 13 mi. (C) 12:30 P.M. 14 mi.

(D) 12:20 P.M. 14 mi. (E) 12:20 P.M. 15 mi.

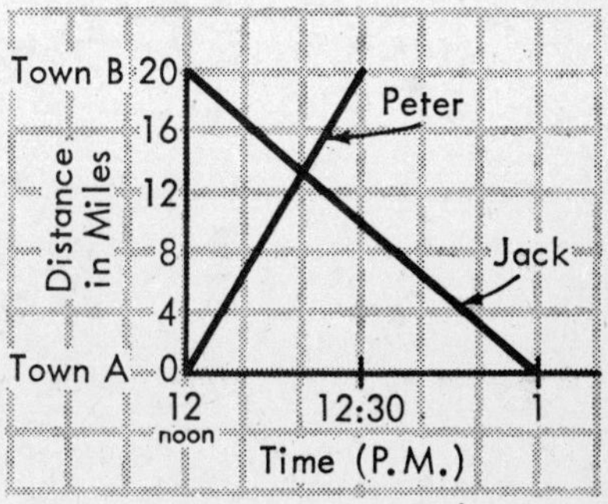

[1.4]

5. Towns A and B are 40 miles apart. The accompanying graph shows that Tom started from A at 12 noon and arrived at B at 4 P.M. In order to meet Tom at B at 4 P.M., Fred started from A at 2 P.M., as shown. During his trip, at what rate in miles per hour was Fred gaining on Tom?

(A) 8 (B) 9 (C) 10 (D) 12 (E) 15

[2.1]

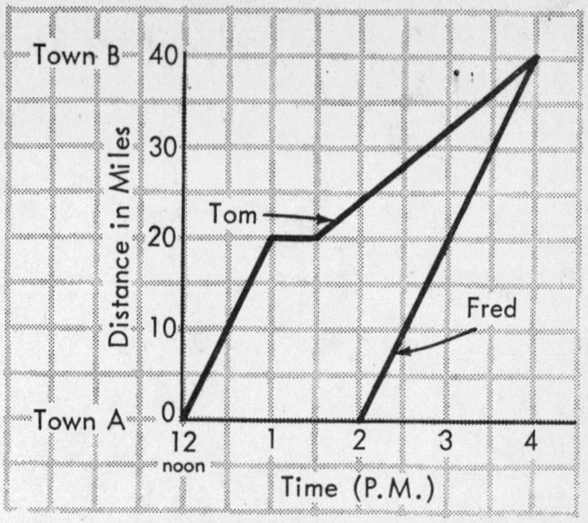

6. The accompanying bar graph shows the number of history books, fiction books, and plays available in a school library. Find the ratio of the number of history books to the number of fiction books and plays combined.

(A) 18:15 (B) 18:13 (C) 9:4 (D) 9:7 (E) 8:7

[2.2]

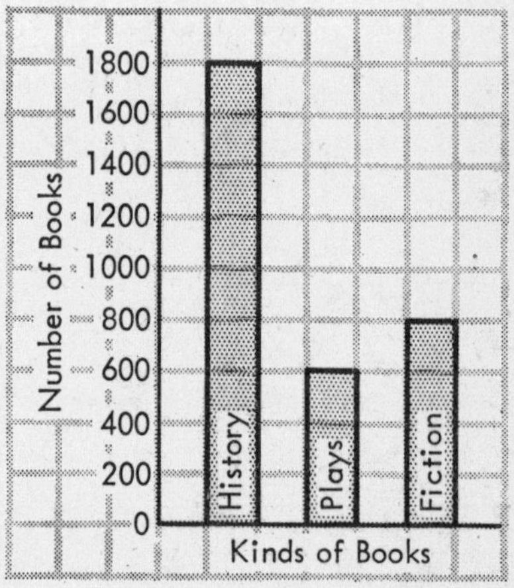

7. Find the missing number in the table:

(A) −.32 (B) −3.2 (C) .16 (D) 2.56 (E) .256 **[3.1]**

y	3	4	5	6	7
x	10	4	1.6	.64	?

8. Find an equation that may be used to express the relation between x and y as shown in the table:

(A) $x = y^2$ (B) $y = x^2$ (C) $x = \sqrt{y}$ (D) $x = 9y$

(E) $x = 16y - 63$ **[3.2]**

y	1	4	7	9
x	1	16	49	81

9. The set of number pairs at the right represents points that lie on a straight line. Find the equation of the line.

(A) $x + 2y = 6$ (B) $x = y - 10$ (C) $x + 2y = -6$

(D) $2x + y = -6$ (E) $y = x + 18$ **[3.3]**

y	−4	−5	−7	4
x	2	4	8	−14

XI. MODERN MATHEMATICS

1. INTEGERS

The set of integers consists of (1) positive integers, such as +1, +2, +3, . . . , (2) negative integers, such as −1, −2, −3, . . . , and (3) zero (0).

The rules for adding or multiplying odd and even integers can be derived from the following tables:

Addition Table

+	odd	even
odd	even	odd
even	odd	even

Multiplication Table

×	odd	even
odd	odd	even
even	even	even

Note, for example, that *the product of any two odd integers is an odd integer.* Hence, if k is an odd integer, any power of k, such as k^2, k^3, or k^{25}, is an odd integer.

Also note that *the sum of any two odd integers is an even integer.* Hence, if k is an odd integer, numbers like $k^2 + 11$ and $k^3 + 99$ are even integers.

Consecutive integers differ by 1. However, consecutive odd integers and consecutive even integers differ by 2. Hence, if k is the middle integer in a set of five consecutive odd integers, these five integers may be represented by $k - 4$, $k - 2$, k, $k + 2$, and $k + 4$.

1.1 DETERMINING ODD AND EVEN INTEGERS

Model Problem

If k represents an odd integer, which expression represents an even integer?

(A) $k + 2$ (B) $3k + 1$ (C) $2k - 1$ (D) $3k - 2$ (E) $k^2 + 2$

Solution:

The correct choice is (B), namely, $3k + 1$. To show this, apply the following properties of odd and even integers:

1. The product of two odd integers is an odd integer. It is given that k is an odd integer; since 3 is also an odd integer, the product $3k$ must be an odd integer.
2. The sum of two odd integers is an even integer.

   Since $3k + 1$ is the sum of $3k$, an odd integer, and 1, another odd integer, $3k + 1$ must be an even integer. *Ans.* (B)

> *Note.* The correct choice may be determined by selecting any odd integer as a value for k in each expression. For example, if $k = 7$, then $3k + 1 = 22$, an even integer. Hence, (B) is the correct choice.

Wrong Choices:

(A) $k + 2$ is the sum of k, an odd integer, and 2, an even integer. The sum of an odd integer and an even integer must be an odd integer.

(C) The term $2k$ contains 2 as a factor. Since an even integer is an integer that contains 2 as a factor, $2k$ is even. Subtracting 1 from an even integer results in an odd integer.

(D) The term $3k$ is an odd integer. (See rule 1.) If 2 is subtracted from an odd integer, the result must be an odd integer.

(E) The term k^2 is odd, since the square of an odd integer is an odd integer. Hence, adding 2 to k^2 results in an odd integer.

Matching Test

In the five groups of expressions listed below, k represents an odd integer. In each group, which of the three expressions represents an odd integer? After completing this test, match your answers with the expressions in the box.

1. $2k + 1$, $3k + 1$, $5k + 1$

2. $2(k + 1)$, $3(k + 2)$, $4(k + 3)$

3. $k^2 + 1$, $k^3 + 2$, $k^4 + 3$

4. $2k^2$, $3k^3$, $4k^4$

5. $k(k + 1)$, $(k + 1)(k + 2)$, $(k + 2)(k + 4)$

a. $k^3 + 2$
b. $3k^3$
c. $2k + 1$
d. $(k + 2)(k + 4)$
e. $3(k + 2)$

1.2 REPRESENTING ODD AND EVEN INTEGERS

Model Problem

Any odd integer can be expressed as $2x + 1$ where x is any integer. Express in terms of x the next larger odd integer.

(A) $2x + 2$ (B) $x + 3$ (C) $3x + 1$ (D) $4x + 1$ (E) $2x + 3$

Solution:

Among consecutive odd integers, any odd integer is two more than the odd integer before it. For example, in the set 19, 21, 23, note that 21 is two more than 19, while 23 is two more than 21.

Hence, if $2x + 1 =$ an odd integer, then the next larger odd integer $= (2x + 1) + 2 = 2x + 3$. *Ans.* (E)

Matching Test

After completing this test, match your answers with the sets of expressions in the box.

If x is any integer and $2x + 1$ represents an odd integer,

1. represent the next two larger odd integers.

2. represent the next two larger even integers.

3. represent the next two smaller odd integers.

4. represent the next two smaller even integers.

5. represent the square of any odd integer.

a. $(2x + 1)^2$, or $4x^2 + 4x + 1$
b. $2x + 2$, $2x + 4$
c. $2x$, $2x - 2$
d. $2x + 3$, $2x + 5$
e. $2x - 1$, $2x - 3$

1.3 UNEQUAL POSITIVE INTEGERS AS FACTORS OR ADDENDS

Model Problem

The product of three unequal positive integers is 84. If the integers are each larger than 1, what is the maximum value of the largest integer?

(A) 6 (B) 7 (C) 12 (D) 14 (E) 21

Solution:

Separate 84 into its prime factors 2, 2, 3, and 7.

Since there are four prime factors, a set of three factors may be obtained by selecting two of the factors and then using the product of the remaining two factors as the third factor.

Hence, the possible sets of 3 unequal factors of 84 are the following:

(1) 2, 3, 14 found by multiplying 2 and 7.

(2) 2, 6, 7 found by multiplying 2 and 3.

(3) 3, 4, 7 found by multiplying 2 and 2.

(The set 2, 2, 21 must be ruled out since it contains 2 equal factors.)

The maximum value of the largest integer is found in set (1) 2, 3, 14. The maximum value is 14. *Ans.* (D)

Matching Test

After completing this test, match your answers with the numbers in the box.

1. If the product of three unequal positive integers is 60, and the integers are each larger than 1, what is the *minimum* value of the largest integer?

2. If the product of three unequal positive integers is 60, and the integers are each larger than 1, what is the maximum value of the largest integer?

3. If the sum of three unequal positive integers is 30, what is the maximum value of the largest integer?

4. If the sum of three unequal positive integers is 30, what is the minimum value of the largest integer?

5. If the average of three unequal positive integers is 8, what is the maximum value of the smallest integer?

a. 10	*b.* 5	*c.* 11	*d.* 27	*e.* 7

1.4 REPRESENTING POWERS OF INTEGERS

Model Problems

(1) If n represents an integer, the square of the next larger integer is

(A) $n^2 + 1$ (B) $n^2 + 4$ (C) $n^2 + 2n + 1$ (D) $n^2 + 4n + 4$ (E) $n^2 + 2n + 2$

Solution:

If n represents an integer, then $n + 1$ represents the next larger integer. Hence, $(n + 1)^2$ is the square of the next larger integer.

As shown below, $(n + 1)^2$ or $(n + 1)(n + 1) = n^2 + 2n + 1$.

$$\begin{array}{r} n + 1 \\ n + 1 \\ \hline n^2 + n \\ + n + 1 \\ \hline n^2 + 2n + 1 \end{array}$$

Ans. (C)

(2) The last digit of the fourth power of 273 ends in

(A) 0 (B) 1 (C) 4 (D) 6 (E) 9

Solution:

The squares of the ten digits from 0 to 9 end in 0, 1, 4, 5, 6, or 9, as shown in the following table:

$0^2 = \mathbf{0}$	$1^2 = \mathbf{1}$	$2^2 = \mathbf{4}$	$3^2 = \mathbf{9}$	$4^2 = 1\mathbf{6}$	$5^2 = 25$
	$9^2 = 8\mathbf{1}$	$8^2 = 6\mathbf{4}$	$7^2 = 4\mathbf{9}$	$6^2 = 3\mathbf{6}$	

The last digit of the square of any integer is determined by squaring the last digit of the integer. Hence, if 273 is squared, the resulting last digit is 9, since this last digit must be obtained by squaring 3, the last digit of 273.

To obtain the fourth power of 273, square the square of 273; that is, $(273^2)^2 = 273^4$. If this is done, the last digit of the result is 1, since this last digit must be obtained by squaring 9, the last digit of 273^2. *Ans.* (B)

After completing these tests, match your answers with the answers in the boxes.

Matching Test A

If n represents an integer, find

1. the square of the next smaller integer.

2. the square of the double of n.

3. the double of the square of n.

4. the square of an integer that is three more than n.

5. the square of an integer that is three times n.

a. $(n + 3)^2$, or $n^2 + 6n + 9$

b. $(n - 1)^2$, or $n^2 - 2n + 1$

c. $2n^2$

d. $(2n)^2$, or $4n^2$

e. $(3n)^2$, or $9n^2$

Matching Test B

1. If the last digit of the square of an integer is 6, what may be the last digit of the integer?

2. If the last digit of the fourth power of an integer is 1, what may be the last digit of the integer besides 1 and 9?

3. How many digits may the square of a three-digit integer have?

4. The square of an integer has 7 digits. How many digits does the integer have?

5. What is the largest positive integer whose cube is less than 8000?

a. 3, 7	*b.* 5, 6	*c.* 19	*d.* 4, 6	*e.* 4

2. INEQUALITIES

Representing Inequalities on a Number Line

On a number line, a point to the right of another point represents the greater number. For example, in Fig. 1, point R, which represents 5, is to the right of point Q, which represents 2. The relation between two unequal numbers such as 5 and 2 may be expressed by using inequality symbols as follows:

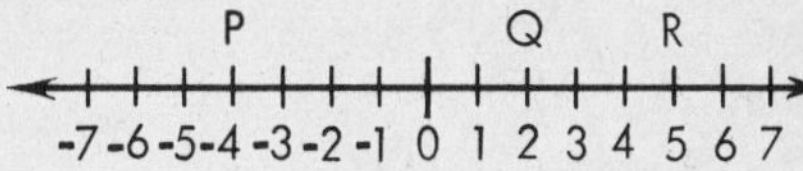

Fig. 1

$5 \neq 2$ means that 5 *is unequal to* 2.

$5 > 2$ means that 5 *is greater than* 2.

$2 < 5$ means that 2 *is less than* 5.

If two inequalities have the same inequality symbol, they are said to be of the same sense, as in the case of $5 > 2$ and $7 > 1$. Two inequalities having different symbols are said to be of opposite sense, as in the case of $2 < 5$ and $7 > 1$.

Number-Line Graphs of Inequalities

On the number line in Fig. 1, note that any number to the left of P has a value that is less than -4. This set, or collection, of numbers may be represented by x, and the inequality $x < -4$ may be used to show that any number in the set is less than -4.

The graph of the inequality $x < -4$ is shown in Fig. 2. Note that the position of P is represented by a hollow dot to indicate that -4 is not a member of the set.

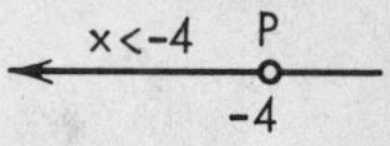

Fig. 2

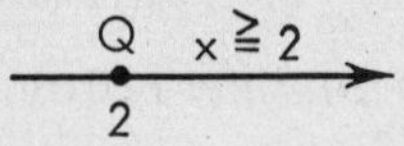

Fig. 3

The graph of the inequality $x \geqq 2$ (that is, x is *greater than or equal to* 2) is shown in Fig. 3. Note that the position of Q is represented by a full, or blackened, dot to indicate that 2 is a member of the set.

The inequality $-4 < x < 2$ indicates that x is any number between -4 and 2; that is, -4 is less than x, and also x is less than 2. The graph of this inequality is shown in Fig. 4. Note the hollow dots at P and Q to show that -4 and 2 are not members of the set.

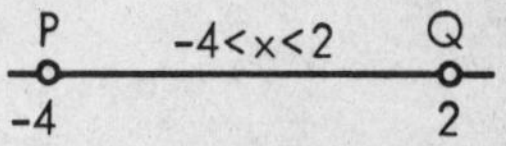

Fig. 4

Operations on Inequalities

1. Do not change the sense of an inequality if the same number is added to or subtracted from each side. For example:

$$\begin{array}{rcl} x+5 & > & 15 \\ -5 & = & -5 \\ \hline x & > & 10 \end{array} \qquad \begin{array}{rcl} x-5 & < & 15 \\ +5 & = & +5 \\ \hline x & < & 20 \end{array}$$

2. Do not change the sense of an inequality if each side is multiplied by or divided by the same positive number. For example:

$$5x > 15 \qquad \frac{x}{5} < 15$$

$$\frac{5x}{5} > \frac{15}{5} \qquad 5\left(\frac{x}{5}\right) < 5(15)$$

$$x > 3 \qquad x < 75$$

3. Reverse the sense of an inequality if each side is multiplied by or divided by the same negative number. For example:

$$-5x > 15 \qquad -\frac{x}{5} < 15$$

$$\frac{-5x}{-5} < \frac{15}{-5} \qquad -5\left(-\frac{x}{5}\right) > -5(15)$$

$$x < -3 \qquad x > -75$$

Two-Dimensional Inequality Graphs

Any line drawn in a plane divides the plane into two sections called ***half-planes.*** Thus, in Fig. 5, the graph of $y = 3$ is a line that divides the entire plane into the half-planes represented by the two shaded areas. The inequality represented by the half-plane above the graph of $y = 3$ is $y > 3$ because the y-value of any point in this half-plane is greater than 3. The half-plane below the graph of $y = 3$ represents the inequality $y < 3$.

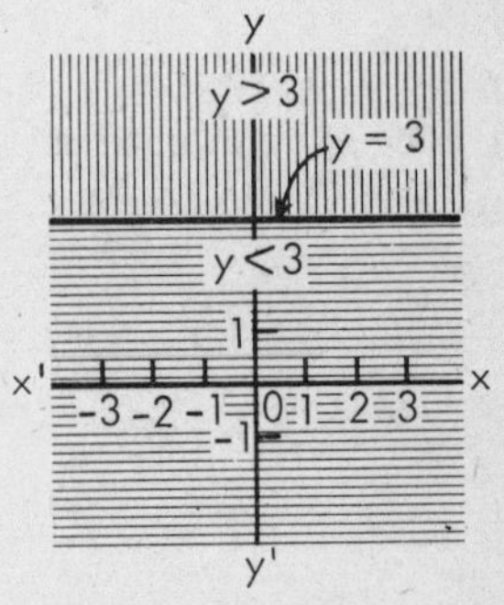

Fig. 5

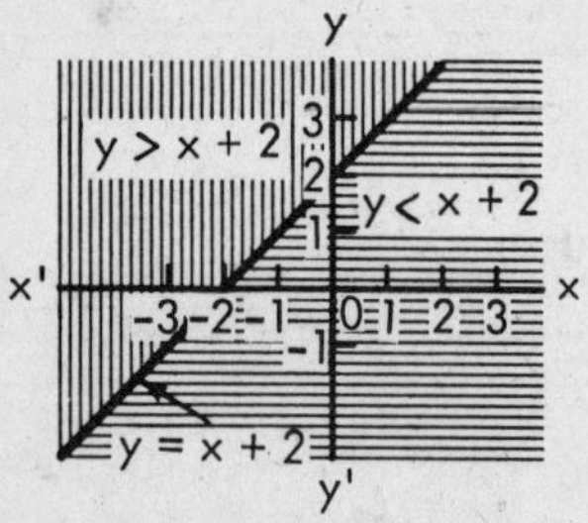

Fig. 6

In like fashion, the graph of $y = x + 2$ divides the plane into the half-planes that represent the inequalities $y > x + 2$ and $y < x + 2$. (See Fig. 6.) Test any point in either half-plane to see if the coordinates of the point satisfy the equation of the half-plane.

2.1 SOLVING INEQUALITIES

Model Problems

(1) Find all values of x for which $4x - 3 > x + 5$.

(A) $x > \frac{8}{3}$ (B) $x > \frac{2}{3}$ (C) $x > \frac{3}{8}$ (D) $x < \frac{8}{3}$ (E) $x < \frac{3}{8}$

Solution:

Solve the inequality $4x - 3 > x + 5$ for x as follows:

Add 3 to each side. (Addition does not change the sense of an inequality.)

$$4x > x + 8$$

Subtract x from each side. (Subtraction does not change the sense of an inequality.)

$$3x > 8$$

Divide each side by 3. (Division by a positive number does not change the sense of an inequality.)

$$x > \frac{8}{3}$$

Ans. (A)

(2) The entire set of values of x that satisfies the inequality $5 - 3x < 2$ also satisfies the inequality.

(A) $x > \frac{7}{3}$ (B) $x < -\frac{7}{3}$ (C) $x > 1$ (D) $x < 1$ (E) $x > -1$

Solution:

Solve the inequality $5 - 3x < 2$ for x as follows:

Subtract 5 from each side. (Subtraction does not change the sense of an inequality.)

$$-3x < -3$$

Divide each side by -3. (Division by a negative number reverses the sense of an inequality.)

$$x > 1$$

Ans. (C)

(3) Solve for x the inequality $4 - 3x \leqq 12 + x$.

(A) $x \geqq 8$ (B) $x \geqq -8$ (C) $x \leqq -2$ (D) $x > -2$ (E) $x \geqq -2$

Solution:

The given statement is a combination of the equality $4 - 3x = 12 + x$ and the inequality $4 - 3x < 12 + x$. The operations to be performed on these can be done at the same time as follows:

Add $-12 + 3x$ to each side.

$$\begin{array}{rcr} 4 - 3x & \leqq & 12 + x \\ -12 + 3x & = & -12 + 3x \\ \hline -8 & \leqq & +4x \end{array}$$

(Addition does not change the sense of an inequality.)

Divide each side by 4.

$$\frac{-8}{4} \leqq \frac{4x}{4}$$

(Division by a positive number does not change the sense of an inequality.)

$$-2 \leqq x, \text{ or } x \geqq -2$$

Ans. (E)

Matching Test

Find all the values of x that satisfy each given inequality. After completing this test, match your answers with the inequalities in the box.

1. $5 + 3x > 2$ **2.** $2 - 3x > 5$ **3.** $4x + 3 < x + 9$ **4.** $x - 3 < 4x + 9$ **5.** $2x - 4 \geqq 2$

a. $x \geqq 3$	*b.* $x < 2$	*c.* $x < -1$	*d.* $x > -1$	*e.* $x > -4$

2.2 GRAPHS OF INEQUALITIES

Model Problem

The graphs of $y = x$ and $y = 1$, and part of the y-axis, divide the plane at the right into five sections labeled A to E. (Note, for example, that sections B and E are shaded, and that section A, which is bounded by the lines $y = 1$ and $y = x$, contains segments of the two axes.) Name the section that represents the graph of the solution set of the following system of inequalities:

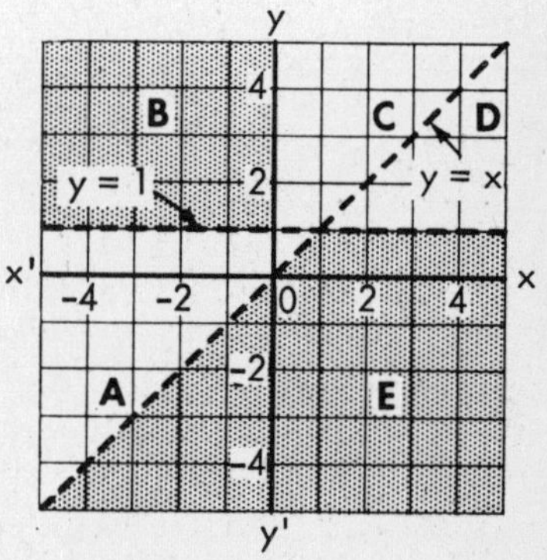

$$y > x$$
$$y - 1 < 0$$

(A) A (B) B (C) C (D) D (E) E

Solution:

1. The graph of the inequality $y > x$ is the half-plane above the line $y = x$. Any point in the half-plane above the line $y = x$ has a y-value that is greater than its corresponding x-value.

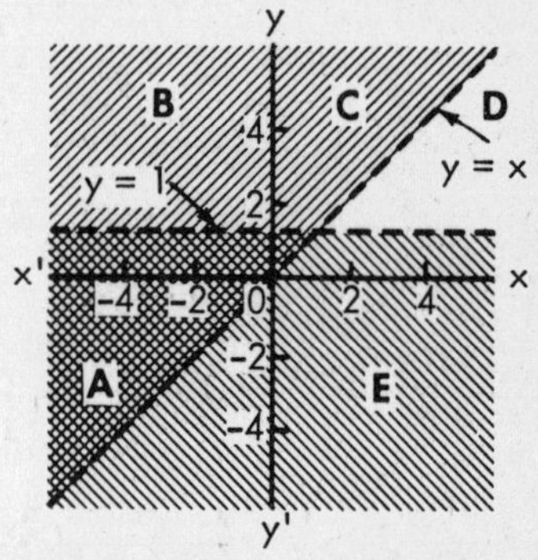

2. Change the inequality $y - 1 < 0$ to the form $y < 1$ by adding 1 to each side of the inequality. The graph of the inequality $y < 1$ is the half-plane below the line $y = 1$. Any point in the half-plane below the line $y = 1$ has a y-value that is less than 1.

3. The intersection of the two half-planes obtained in steps 1 and 2 is the graph of the solution set for the system of inequalities $y > x$ and $y - 1 < 0$.

   The crosshatched section of the graph in the second figure is the required solution set. *Ans.* (A)

Matching Test A

The first and fourth quadrants in the figure at the right are divided into sections labeled A to F by the coordinate axes and the graphs of $y = x$ and $x = 4$. Match each set of inequalities below with the section representing the graph of its solution set in quadrants I and IV.

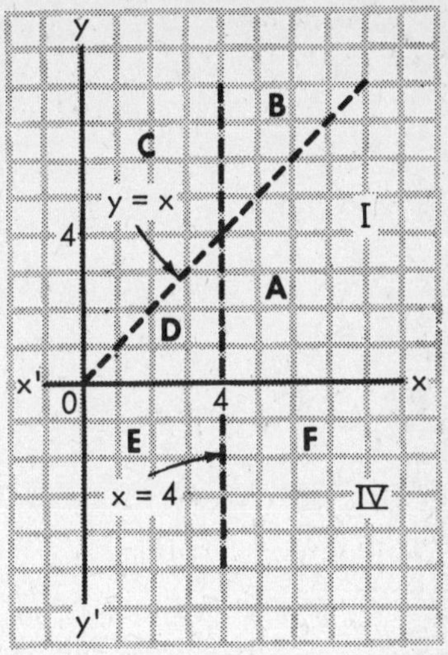

1. $y < x,\ x < 4,\ y < 0$ **2.** $y < x,\ x > 4,\ y < 0$

3. $y > x,\ x > 4,\ y > 0$ **4.** $y < x,\ x > 4,\ y > 0$

5. $y < x,\ x < 4,\ y > 0$

Matching Test B

The first and second quadrants in the figure at the right are divided into sections labeled A to F by the coordinate axes and the graphs of $x + y = 4$ and $y = x + 4$. Match each set of inequalities below with the section representing the graph of its solution set in quadrants I and II.

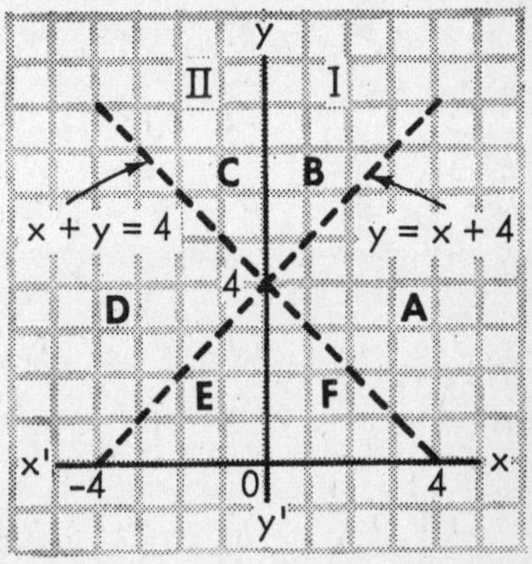

1. $x + y > 4,\ y > x + 4,\ x > 0$

2. $x + y > 4,\ y > x + 4,\ x < 0$

3. $x + y < 4,\ y > x + 4,\ x < 0$

4. $x + y < 4,\ y < x + 4,\ x > 0$

5. $x + y < 4,\ y < x + 4,\ x < 0$

3. OTHER TOPICS OF MODERN MATHEMATICS

3.1 FUNDAMENTAL LAWS OF MATHEMATICS

Associative Law of Addition. Changing the order in which addends are added does not change their sum.

For example, $5 + 6 + 7$ may be added as $(5 + 6) + 7$ or as $5 + (6 + 7)$, since

$$(5 + 6) + 7 = 11 + 7 = 18 \text{ and } 5 + (6 + 7) = 5 + 13 = 18$$

In general, $a + b + c = (a + b) + c = a + (b + c)$.

Associative Law of Multiplication. Changing the order in which factors are multiplied does not change their product.

For example, $5 \cdot 6 \cdot 7$ may be multiplied as $(5 \cdot 6) \cdot 7$ or as $5 \cdot (6 \cdot 7)$ since

$$(5 \cdot 6) \cdot 7 = 30 \cdot 7 = 210 \text{ and } 5 \cdot (6 \cdot 7) = 5 \cdot 42 = 210$$

In general, $a \cdot b \cdot c = (a \cdot b) \cdot c = a \cdot (b \cdot c)$, or $abc = (ab)c = a(bc)$.

Commutative Law of Addition. Interchanging addends does not change their sum.

For example, $5 + 6 = 6 + 5$.

In general, $a + b = b + a$.

Commutative Law of Multiplication. Interchanging factors does not change their product.

For example, $5 \cdot 6 = 6 \cdot 5$.

In general, $a \cdot b = b \cdot a$, or $ab = ba$.

Distributive Law. To multiply a polynomial by a monomial, multiply each term of the polynomial by the monomial.

For example, $5(6 + 7) = 5 \cdot 6 + 5 \cdot 7$

$$= 30 + 35 = 65$$

In general, $a(b + c) = ab + ac$.

> *Note.* By means of the distributive law, the monomial is "distributed" to each term of the polynomial to be multiplied.

Model Problems

(1) To the right of each numbered expression below is the name of a law of mathematics that was applied to obtain that expression from the one above it. The names of the laws are labeled from (A) to (E). One of the laws is incorrectly paired with the expression on its left; that is, it cannot be used to derive the expression from the preceding one. Indicate by its letter the law that is incorrectly paired.

$a(b + c)$	
1. $ab + ac$	(A) Distributive Law
2. $ac + ab$	(B) Associative Law of Multiplication
3. $ca + ba$	(C) Commutative Law of Multiplication
$\frac{1}{4} + 5 + \frac{3}{4}$	
4. $\left(5 + \frac{1}{4}\right) + \frac{3}{4}$	(D) Commutative Law of Addition
5. $5 + \left(\frac{1}{4} + \frac{3}{4}\right)$	(E) Associative Law of Addition

Solution:

1. When the polynomial $(b + c)$ is multiplied by a, the product, $ab + ac$, is the result of "distributing" the multiplication over the two terms in the polynomial. Hence, expression 1 results from the application of the Distributive Law.
2. The expression $ac + ab$ is obtained from the preceding expression by interchanging addends, that is, the terms of $ab + ac$. Hence, expression 2 is the result of applying the Commutative Law of Addition.
3. The expression $ca + ba$ is obtained from the preceding expression by interchanging factors. If the factors of each term in $ac + ab$ are interchanged, ac becomes ca and ab becomes ba. Hence, expression 3 results from the application of the Commutative Law of Multiplication.

4. The expression $5 + \frac{1}{4} + \frac{3}{4}$ is obtained by interchanging the first two terms of $\frac{1}{4} + 5 + \frac{3}{4}$. Hence, expression 4 results from the application of the Commutative Law of Addition.

5. The expression $5 + \left(\frac{1}{4} + \frac{3}{4}\right)$ is obtained by changing the order in which the terms of $\left(5 + \frac{1}{4}\right) + \frac{3}{4}$ are to be added. Hence, expression 5 results from the application of the Associative Law of Addition.

Ans. (B)

(2) One of the following is *not always* true. Select the one that is *not always* true.

(A) $x + y = y + x$ (B) $x - y = y - x$ (C) $xy = yx$ (D) $x(y + z) = xy + xz$

(E) $(x + y) + z = x + (y + z)$

Solution:

Choice (A) is an application of the Commutative Law of Addition. Hence, it is always true.

Choice (B) is not always true, since $(x - y)$ is the negative of $(y - x)$. For example, $5 - 2 = 3$ but $2 - 5 = -3$. Choice (B) is true only if $x = y$.

Choice (C) is an application of the Commutative Law of Multiplication. Hence, it is always true.

Choice (D) is an application of the Distributive Law. Hence, it is always true.

Choice (E) is an application of the Associative Law of Addition. Hence, it is always true.

Ans. (B)

Matching Test

Match each law with the equality in the box to which it applies.

1. Associative Law of Addition
2. Associative Law of Multiplication
3. Commutative Law of Addition
4. Commutative Law of Multiplication
5. Distributive Law

a. $\frac{1}{2} \times 25 + \frac{1}{2} \times 5 = \frac{1}{2}(25 + 5)$

b. $25 \times \frac{1}{2} \times 4 = 25 \times 4 \times \frac{1}{2}$

c. $\left(\frac{1}{2} \times 25\right) \times 4 = \frac{1}{2} \times (25 \times 4)$

d. $\left(4 + 3\frac{1}{2}\right) + 6\frac{1}{2} = 4 + \left(3\frac{1}{2} + 6\frac{1}{2}\right)$

e. $3\frac{1}{2} + 4 + 6\frac{1}{2} = 4 + 3\frac{1}{2} + 6\frac{1}{2}$

3.2 SETS AND SUBSETS

Sets

The concept of the set is of fundamental importance in every branch of mathematics.

A ***set*** is a well-defined collection of things. Each thing in a set is an ***element***, or ***member***, of the set. For example, the elements of the set of vowels in English are a, e, i, o, and u. If V is used to represent the set of vowels, and e is a member of set V, we write $e \in V$. The symbol $\in$ means "is a member of." Since f is *not* a member of set V, we write $f \notin V$. A slanted line (/) drawn through a symbol negates the symbol.

A set may be defined by listing, or ***tabulating***, all the members. For example, to define the set of vowels by listing its members, we write $V = \{a, e, i, o, u\}$. Note that the elements are separated by commas and enclosed in braces.

A set may also be defined by stating a property or rule to determine whether a particular thing is a member of the set. For example, a letter such as x, may be used to denote any member of the set. Thus, to define the set of even integers, we write $E = \{x | x \text{ is an even integer}\}$, which is read "$E$ is the set of numbers x such that x is an even integer." Note the use of the vertical bar (|) to represent "such that."

A set like the set of vowels is a ***finite set***, since the elements of the set can be counted; that is, a finite set consists of a specific number of elements.

A set like the set of even integers is an ***infinite set***, since it has an unlimited number of elements. The set of even integers can be written as $\{2, 4, 6, \ldots\}$. The three dots, . . ., indicate that the set continues without limit.

A ***null set*** is a set that has no members. Such a set is empty, or void, and is denoted by the symbol $\varnothing$. For example, the set of people 20 feet tall is a null set.

A ***solution set of an equation*** is the set whose elements are the roots of the equation. For example, if $A = \{1, 2\}$, then A is the solution set of the equation $(x-1)(x-2) = 0$, or $x^2 - 3x + 2 = 0$.

A ***solution set of an inequality*** is the set whose elements satisfy the inequality. For example, if $A = \{x | x > 7\}$, then A is the solution set of the inequality $x + 3 > 10$.

Subsets

If every member of set A is a member of set B, then A is a ***subset*** of B. To state that A is a subset of B, we write $A \subset B$. To state that P is not a subset of B, we write $P \not\subset B$. A null set is considered to be a subset of every set. Also, a set is considered to be a subset of itself. Hence, a subset S of a set U may contain all, some, or none of the elements of U.

A ***proper subset*** of a set U does not contain all of the elements of U. Hence, if $U = \{\text{red, white, blue}\}$, there are 8 subsets of U, and all but one of these are proper subsets. Of the following 8 subsets, sets B to H are proper subsets of U:

1. $A = \{\text{red, white, blue}\}$	5. $E = \{\text{red}\}$
2. $B = \{\text{red, white}\}$	6. $F = \{\text{white}\}$
3. $C = \{\text{red, blue}\}$	7. $G = \{\text{blue}\}$
4. $D = \{\text{white, blue}\}$	8. $H = \varnothing$

A ***universal set*** , or a ***universe***, is a set that contains all the elements that are used to form subsets. Thus, $U = \{\text{red, white, blue}\}$ is the universe that was used to form the 8 subsets shown.

Equal and Equivalent Sets

Equal sets are sets that have the same members. For example, if $A = \{1, 2\}$, $B = \{2, 1\}$, and $C = \{x | (x - 1)(x - 2) = 0\}$, then $A = B = C$.

Equivalent sets are sets in which every element of one set can be paired with one and only one element of the other set. For example, the set of integers from 1 to 7 and the set of the days of the week are equivalent sets. All finite sets having the same number of members are equivalent to each other.

Union and Intersection of Sets

The ***union*** of sets A and B is the set of all the elements that belong to either set or to both sets. The set that is the union of sets A and B is denoted by "$A \cup B$," read "A union B." In the diagram at the right, which is called a Venn diagram, the union of sets A and B is shown by shading both sets.

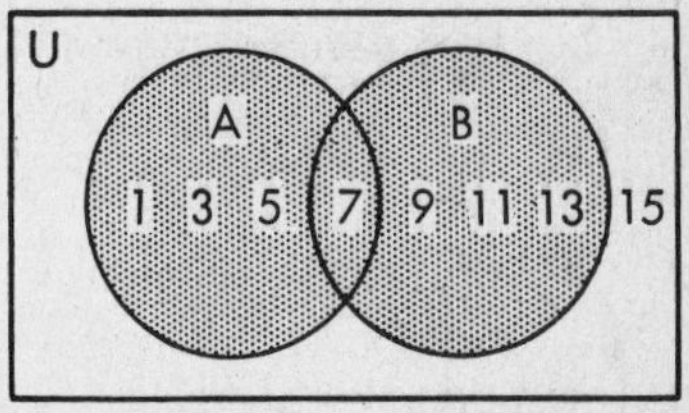

$A \cup B$ is the shaded part.

$$\text{If } A = \{1, 3, 5, 7\}$$
$$\text{and } B = \{7, 9, 11, 13\},$$
$$\text{then } A \cup B = \{1, 3, 5, 7, 9, 11, 13\}.$$

The ***intersection*** of sets A and B is the set of those elements that are common to both A and B. The set that is the intersection of sets A and B is denoted by "$A \cap B$," read "A intersection B." In the Venn diagram at the left, the intersection of sets A and B is shown by shading the area that is common to both sets.

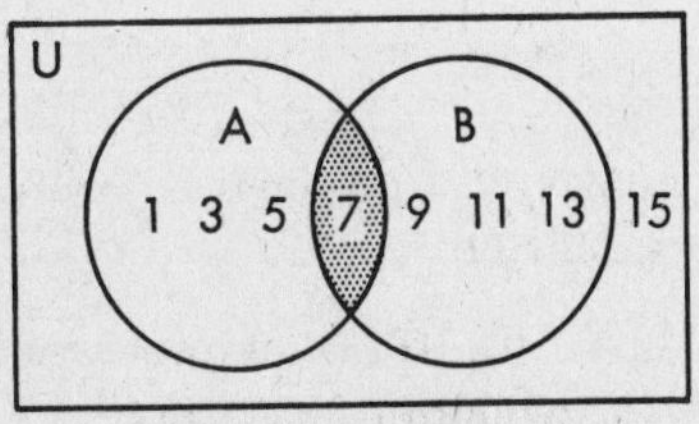

$A \cap B$ is the shaded part.

$$\text{If } A = \{1, 3, 5, 7\}$$
$$\text{and } B = \{7, 9, 11, 13\}$$
$$\text{then } A \cap B = \{7\}.$$

The ***complement*** of a set A is the set of elements in the universe U that do not belong to A. The set that is the complement of A is denoted by A', read "complement of A." In the Venn diagram at the right, the complement of A is shown by shading the part of U that is not A.

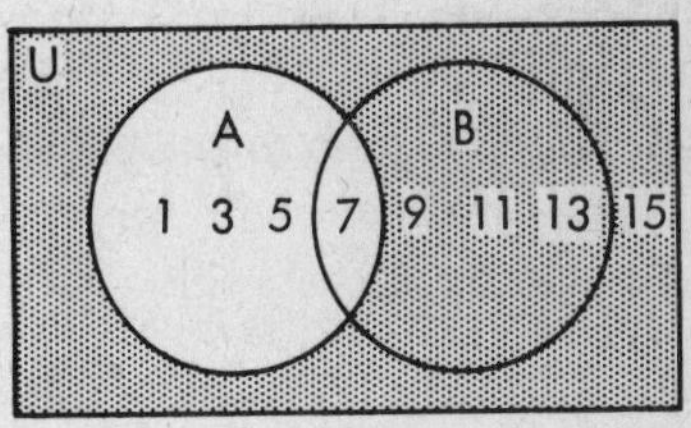

A' is the shaded part.

$$\text{If } U = \{1, 3, 5, 7, 9, 11, 13, 15\}$$
$$\text{and } A = \{1, 3, 5, 7\},$$
$$\text{then } A' = \{9, 11, 13, 15\}.$$

Note that 9, 11, and 13 are the elements of B that are in A'.

The ***product set*** of sets A and B consists of all the ordered pairs (a, b) where a is a member of A and b is a member of B. The product set of A and B is denoted by "$A \times B$," read "A cross B." For example, if $A = \{1, 2\}$ and $B = \{x, y, z\}$, then:

$$A \times B = \{(1, x), (1, y), (1, z), (2, x), (2, y), (2, z)\}$$

Model Problems

(1) Given: $A = \{2, 4, 5, 7\}$.

Which of the following statements is false?

(A) There are 6 different subsets of A that can be formed to contain exactly two elements (or members) of A.

(B) There are 2 elements in the subset of A that contains all the prime numbers of A.

(C) There are no elements in the subset of A that contains all the multiples of 3 in A.

(D) There are 2 elements in the subset of A that contains all the even integers in A.

(E) A set that is equivalent to set A is $\{a, b, c, d\}$.

Solution:

(A) The 6 different subsets of A containing two elements of A are $\{2, 4\}$, $\{2, 5\}$, $\{2, 7\}$, $\{4, 5\}$, $\{4, 7\}$, and $\{5, 7\}$. (A) *true*

(B) Of the elements of A, those that are prime numbers are 2, 5, and 7, since each of these is divisible only by 1 and itself. (The element 4 is not a prime number, since it is divisible by 2.) Hence, the subset of A that contains the prime numbers of A is $\{2, 5, 7\}$, which contains 3 elements. (B) *false*

(C) Since no element of A is divisible by 3, there are no elements in the subset of A that contains the multiples of 3 in A. (C) *true*

(D) The even integers in A are 2 and 4. Hence, 2 and 4 are the elements in the subset of A that contains the even integers in A. (D) *true*

(E) $B = \{a, b, c, d\}$ is a set equivalent to set A because each element of set B can be paired with one and only one element of the other set. Note that a and 2, b and 4, c and 5, d and 7 are such pairs of elements. (E) *true*

Ans. (B)

(2) Given: $A = \{-2, 3\}$, $B = \{1, 2, 3\}$, $C = \{-1, 1, 2\}$

Which of the following statements is false?

(A) $A \cap C = \varnothing$.

(B) $B \cup C = \{-1, 1, 2, 3\}$.

(C) $A \times B = \{(-2, 1), (-2, 2), (-2, 3), (3, 1), (3, 2)\}$.

(D) An equation that has A for its solution set is $x^2 - x - 6 = 0$.

(E) Among the given sets, only set B has elements that are all non-negative.

Solution:

(A) The set $A \cap C$ is the set that contains the elements that are in A and also in C. Since there are no elements common to A and C, the set $A \cap C$ is an empty set. Hence, $A \cap C = \varnothing$. ($\varnothing$ is the symbol for an empty set.) (A) *true*

(B) The set $B \cup C$ is the set that contains the elements that are either in B or in C. Since the elements in either B or C are -1, 1, 2, and 3, $B \cup C = \{-1, 1, 2, 3\}$. (B) *true*

(C) The set $A \times B$ is the set of ordered pairs formed by taking the first element from A and the second from B.

-2 from A may be paired with 1, 2, or 3 from B to give $(-2, 1)$, $(-2, 2)$, or $(-2, 3)$. Also, 3 from A may be paired with 1, 2, or 3 from B to give $(3, 1)$, $(3, 2)$, or $(3, 3)$.

Hence, $A \times B = \{(-2, 1), (-2, 2), (-2, 3), (3, 1), (3, 2), (3, 3)\}$.
Note that this result differs from statement (C), which does not contain the pair $(3, 3)$. (C) *false*

(D) If A is the solution set of an equation, then the elements of A are the roots of that equation.
The elements of A are -2 and 3. If the given equation, $x^2 - x - 6 = 0$, is written in the factored form $(x+2)(x-3) = 0$, it can be seen that its roots are -2 and 3. Hence, A is the solution set of the equation. (D) *true*

(E) The elements of B are the positive integers 1, 2, and 3.
Positive integers and zero are non-negative numbers. Note that A has the negative integer -2, and C has the negative integer -1. (E) *true*

Ans. (C)

After completing these tests, match your answers with the sets in the boxes.

Matching Test A

Given: $A = \{2, 3, 7, 9\}$, $B = \{2, 4, 5\}$, $C = \{-1, 4\}$

1. Find $A \cap B$.

2. Find $B \cup C$.

3. Find $C \times B$.

4. Find the set whose elements are all the prime numbers in A and B.

5. Find the set whose elements are all the odd numbers in A and B.

a. $\{-1, 2, 4, 5\}$
b. $\{2, 3, 5, 7\}$
c. $\{(-1, 2), (-1, 4), (-1, 5), (4, 2), (4, 4), (4, 5)\}$
d. $\{3, 5, 7, 9\}$
e. $\{2\}$

Matching Test B

The elements of the following sets are integers in the interval *beginning with 7 and ending with 23*. Find each set.

1. the set of primes

2. the set of multiples of 7 or 11

3. the set of integers divisible by 2 and 3

4. the set of integers the sum of whose digits is divisible by 9

5. the set of integers that have a remainder of 1 when divided by 4

a. $\{7, 11, 14, 21, 22\}$
b. $\{7, 11, 13, 17, 19, 23\}$
c. $\{9, 18\}$
d. $\{12, 18\}$
e. $\{9, 13, 17, 21\}$

3.3 DIVISION BY ZERO

It is impossible to divide by zero. This can be shown as follows:

Let n be the result when a number, such as 7, is divided by 0; that is, let $n = \frac{7}{0}$. If each side of this equation is multiplied by 0, then $0 \cdot n = 7$. Since 0 multiplied by any number is 0, we arrive at the absurd conclusion that $0 = 7$.

Hence, a fraction like $\frac{4}{0}$ is meaningless. Similarly, $\frac{4}{n}$ is meaningless if $n = 0$.

Model Problem

For what value of x is the fraction $\frac{3x-9}{3x+9}$ meaningless?

(A) 6 (B) $\frac{1}{3}$ (C) $-\frac{1}{3}$ (D) -3 (E) 3

Solution:

Since division by zero is impossible, $\frac{3x-9}{3x+9}$ is meaningless when the denominator, $3x+9$, is equal to 0.

If $3x+9=0$,

$3x=-9$ and $x=-3$. *Ans.* (D)

Matching Test

Determine for what value of x each given fraction is meaningless. After completing this test, match your answer with the values in the box.

1. $\frac{x+3}{x-3}$ **2.** $\frac{3x}{x+3}$ **3.** $\frac{x-3}{3x}$ **4.** $\frac{x+3y}{x-3y}$ **5.** $\frac{3xy}{x+3y}$

a. $x=-3$	*b.* $x=0$	*c.* $x=3$	*d.* $x=3y$	*e.* $x=-3y$

3.4 NUMBER BASES

The number 10 is the base of the number system we commonly use, since ordinary numerals are formed from powers of ten. For example, the numeral 333 represents a combination of 3 hundreds, 3 tens, and 3 ones:

$$\begin{aligned} 333 &= 3 \cdot 100 + 3 \cdot 10 + 3 \\ &= 3 \cdot 10^2 + 3 \cdot 10 + 3 \end{aligned}$$

A numeral may have a base other than 10. If a base other than 10 is used, the base is indicated as a subscript. For example, 333 to the base 5 is written 333_5. (This is read "three-three-three to the base 5," to avoid confusion with ordinary numerals. Similarly, 251_8 is read "two-five-one to the base 8.") The value of 333_5 is found as follows:

$$\begin{aligned} 333_5 &= 3 \cdot \mathbf{5^2} + 3 \cdot \mathbf{5} + 3 \\ &= 3 \cdot 25 + 3 \cdot 5 + 3 \\ &= 75 + 15 + 3 \\ &= 93 \end{aligned}$$

Hence, $333_5 = 93_{10}$, or simply 93, since the base 10 need not be indicated.

In general, $333_x = 3x^2 + 3x + 3$.

Model Problem

Find the sum of 23_5 and 21_3 to the base 10.

(A) 20 (B) 22 (C) 24 (D) 28 (E) 30

Solution:

Since 23_5 is the value of 23 to the base 5, while 21_3 is the value of 21 to the base 3, obtain the values of 23_5 and 21_3 as follows:

$23_5 = 2$ fives $+ 3$. Hence, to the base 10, the value of $23_5 = 13$.

$21_3 = 2$ threes $+ 1$. Hence, to the base 10, the value of $21_3 = 7$.

The sum of 23_5 and $21_3 = 13 + 7$, or 20. *Ans.* (A)

Matching Test

Perform the indicated operations and write each answer to the base 10. After completing this test, match your answers with the numbers in the box.

1. $23_6 + 23_4$ **2.** $32_6 + 32_4$ **3.** $34_7 - 20_5$ **4.** $(12_3)(21_4)$ **5.** $\frac{61_8}{21_3}$

a. 15	*b.* 7	*c.* 34	*d.* 26	*e.* 45

3.5 CLOCK ARITHMETIC

A clock like the one shown in the following model problem is a circular number line divided into equal parts. Motion along the number line may be clockwise or counterclockwise; in either case, the set of numbers can be repeated endlessly. The two directions, clockwise and counterclockwise, may be represented respectively by the signs + and −. Thus, +5 means a clockwise movement of 5 units, while −5 means a counterclockwise movement of 5 units.

Addition on a clock scale can be illustrated as follows: To add +2 and −7, begin at "2" on the scale and move 7 units counterclockwise to "3." Thus, $2 - 7 = 3$.

This type of arithmetic, in which computations can be represented by motions along a clock scale, is also known as ***modular arithmetic.***

Model Problem

On the clock scale at the right, find the sum of $(5 + 7)$ and $(4 - 6)$.

(A) 8 (B) 6 (C) 4 (D) 2 (E) 1

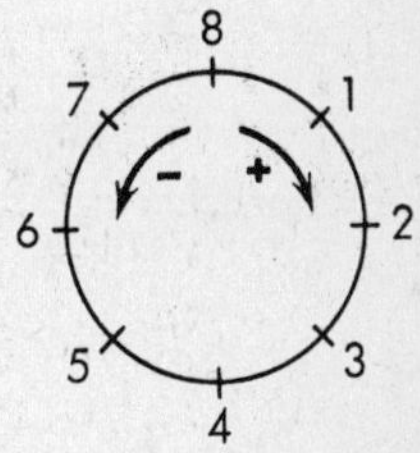

Solution:

The clock scale shown is a circular number line divided into 8 equal units. On this scale, +7 indicates a movement of 7 units in a clockwise direction, and −6 indicates a movement of 6 units in a counterclockwise direction.

Hence, $5 + 7$ means that, beginning at 5, a movement of 7 units clockwise is made. The result is the termination of the movement at 4. Therefore, $5 + 7 = 4$.

Also, $4 - 6$ means that, beginning at 4, a movement of 6 units counterclockwise is made. The result is the termination of the movement at 6. Therefore, $4 - 6 = 6$.

The sum of $(5 + 7)$ and $(4 - 6)$ is the sum of the results, 4 and 6.

The sum of 4 and 6 is the termination of a movement that begins at 4 and moves 6 units clockwise. The result is 2. *Ans.* (D)

Matching Test

In this test, use the clock scale shown in the preceding model problem.

Perform the indicated operations, noting that "+" indicates a clockwise motion and "−" indicates a counterclockwise motion. After completing this test, match your answers with the numbers in the box.

1. $6+2+7$ **2.** $6-2-7$ **3.** $6+2(7)$ **4.** $6-2(7)$ **5.** $(7+6)+(5-7)$

a. 4	*b*. 3	*c*. 5	*d*. 7	*e*. 8

3.6 GRAPHS INVOLVING ABSOLUTE VALUES

The numbers +12 and −12 are each 12 units distant from the origin on a number line. This distance, 12, is the ***absolute value*** of either number, +12 or −12.

The absolute value of a number is either positive or equal to 0. For example, the absolute value of −2 is 2, that of $(-5+1)$ is 4. The absolute value of a number is indicated by placing the number between two vertical bars. For example, $|-2| = 2$, $|-5+1| = 4$.

Model Problem

The broken line *PQR* is the graph of

(A) $y = |1-x|$ (B) $y = |x| + 1$ (C) $y = |x+1|$

(D) $x = |y+1|$ (E) $y = |x-1|$

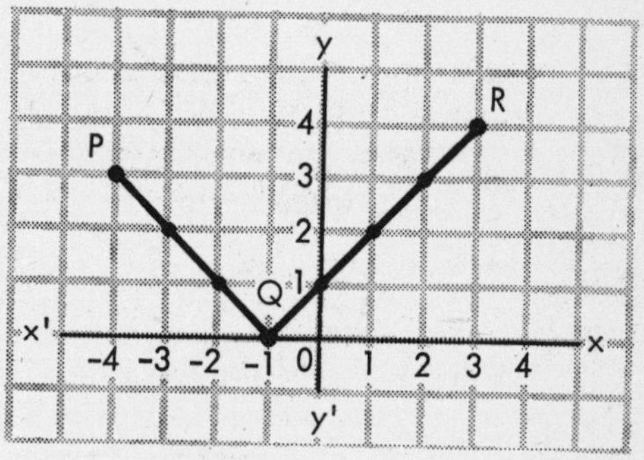

Solution:

Note that the points on the broken line *PQR* have the following coordinates:

$(-4, 3)$, $(-3, 2)$, $(-2, 1)$, $(-1, 0)$,

$(0, 1)$, $(1, 2)$, $(2, 3)$, $(3, 4)$

We see two relationships between the corresponding *x*- and *y*-values:

1. The *y*-values are either positive or zero.
2. Every *y*-value equals the absolute value of 1 more than the corresponding *x*-value. For example, in $(-3, 2)$, note that $2 = |-3+1| = |-2|$.

The only equation among the choices that expresses the two relationships is $y = |x+1|$, in choice (C). Read this as "*y* is the absolute value of $x+1$." Keep in mind that the absolute value of a number must be either positive or zero.

Choice (C) can be tested by plotting the graph of $y = |x + 1|$. The following tables show how x may be given convenient values, and how the corresponding values of y may then be found:

x	$\|x + 1\| = y$
3	$\|3 + 1\| = 4$
2	$\|2 + 1\| = 3$
1	$\|1 + 1\| = 2$
0	$\|0 + 1\| = 1$

x	$\|x + 1\| \; = \; y$
-1	$\|-1 + 1\| = \|0\| \;\; = 0$
-2	$\|-2 + 1\| = \|-1\| = 1$
-3	$\|-3 + 1\| = \|-2\| = 2$
-4	$\|-4 + 1\| = \|-3\| = 3$

Ans. (C)

Matching Test

Determine the equation of each graph below. After completing this test, match your answers with the equations in the box.

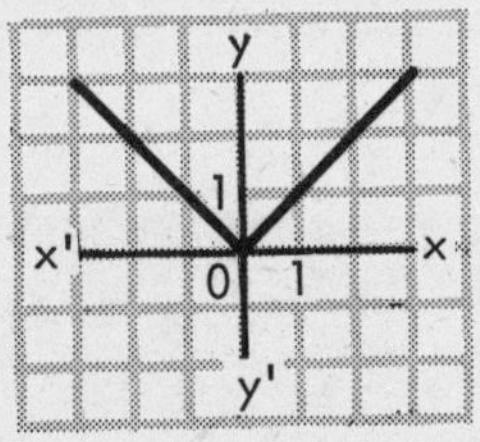

1.

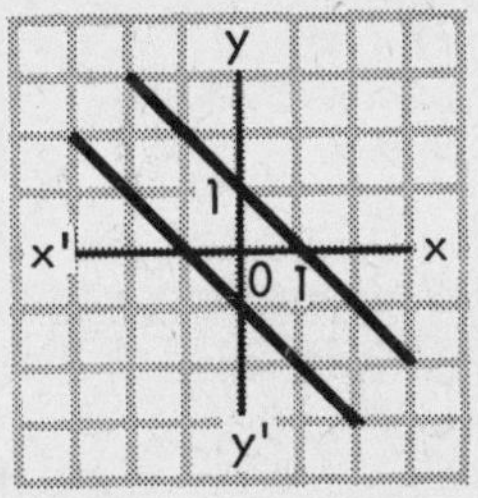

2.

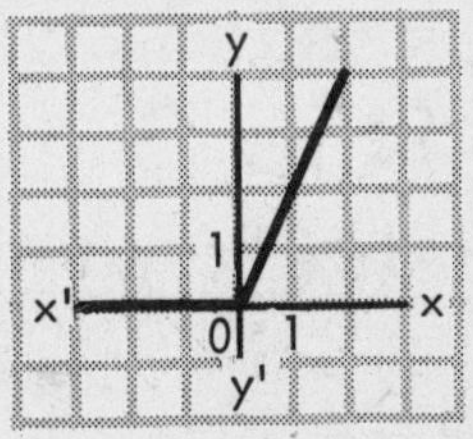

3.

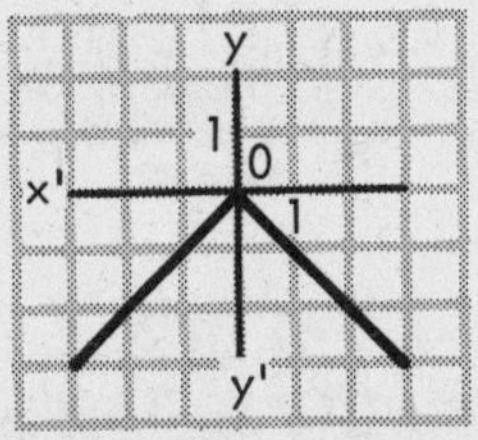

4.

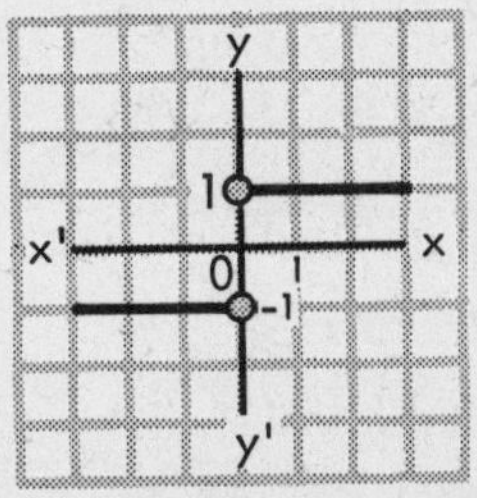

5.

a. $y = \|x\|$	*b*. $y = \|x\| + x$	**c*. $\|x + y\| = 1$	*d*. $y = \dfrac{x}{\|x\|}$	*e*. $y = -\|x\|$

*(*Hint*. In c, $x + y = 1$ or $x + y = -1$.)

3.7 UNFAMILIAR MATHEMATICAL SYMBOLS

Aptitude tests in mathematics often contain problems that introduce special symbols. The symbols are briefly defined and the student is asked to solve problems in which they are used. Such problems are designed to test the student's ability to understand and manipulate new mathematical symbols.

Model Problem

The symbols $\tau(n)$ and $\sigma(n)$, used below, contain the Greek letters *tau* (τ) and *sigma* (σ). They are read, respectively, "tau of n" and "sigma of n."

$\tau(n)$ is defined as the number of positive integral divisors of n, while $\sigma(n)$ is defined as the sum of the positive integral divisors of n. For example, the positive integral divisors of 8 are 1, 2, 4, and 8; hence, $\tau(n) = 4$ and $\sigma(n) = 1 + 2 + 4 + 8 = 15$.

Find the value of $\tau(10) + \sigma(10)$.

(A) 6 (B) 8 (C) 11 (D) 22 (E) 36

Solution:

The positive integral divisors of 10 are 1, 2, 5, and 10.

Since there are 4 positive integral divisors of 10, $\tau(10) = 4$.

Since the sum of the positive integral divisors is $1 + 2 + 5 + 10$, or 18, $\sigma(10) = 18$.

Hence, $\tau(10) + \sigma(10) = 4 + 18 = 22$. *Ans.* (D)

Matching Test

After completing this test, match your answers with the numbers in the box.

1. (See the model problem for the definitions of $\tau(n)$ and $\sigma(n)$.)

Find the value of $\tau(12) + \sigma(12)$.

2. x^{-y} is defined as $\frac{1}{x^y}$, where x and y are positive integers. For example, $3^{-2} = \frac{1}{3^2} = \frac{1}{9}$.

Find the value of $2^3 + 16(2^{-3})$.

3. $[n]$ is defined as the greatest of the integers that is less than or equal to n. For example, $\left[2\frac{1}{4}\right] = 2$ and $\left[-2\frac{1}{4}\right] = -3$.

Find the value of $\left[3\frac{1}{2}\right] + \left[-1\frac{1}{2}\right]$.

4. $\overset{+}{x}$ is defined as $x + 1$, and $\bar{x}$ is defined as $x - 1$. For example, $\overset{+}{7} = 8$ and $\bar{7} = 6$.

Find the value of $(\overset{+}{5})(\bar{6}) - (\overset{+}{6})(\bar{5})$.

5. $n!$ is read "factorial n," and is defined as $n(n-1)(n-2)\cdots 3 \cdot 2 \cdot 1$. For example, $5! = 5 \cdot 4 \cdot 3 \cdot 2 \cdot 1 = 120$.

Find the value of $4! - 3!$.

a. 1	*b.* 10	*c.* 2	*d.* 18	*e.* 34

3.8 IDENTITIES

There is an important difference between two kinds of equalities like $3x = 9$ and $3x = 2x + x$. In the case of $3x = 9$, x can have only one value, 3. In the case of $3x = 2x + x$, x can have any value. Equalities like $3x = 2x + x$ and $x + y = y + x$ that are unconditionally true are called ***identities***.

Two equalities may appear to be merely alternate forms of each other; that is, it may seem that they are satisfied by the same values of the unknown. But this may not be so. For example, the identity $5(x - 2) = 5x - 10$, which is true for all values of x, must be distinguished from the equality $5 = \frac{5x - 10}{x - 2}$, which is true for all values of x except 2. In the second equality, x cannot assume a value that would make the denominator, $x - 2$, equal to 0, since division by zero is impossible.

Model Problems

(1) If $6x + 2y = ax + by$ is an identity, what is the value of ab?

(A) 40 (B) 3 (C) 4 (D) 8 (E) 12

Solution:

The equality $6x + 2y = ax + by$ will become an identity if values substituted for a and b make the two sides identical. This will be true only if $a = 6$ and $b = 2$. In that case, $6x + 2y$ will equal $6x + 2y$.

If $a = 6$ and $b = 2$, then $ab = 6(2) = 12$. *Ans.* (E)

(2) $\frac{3b - 6}{b - 2} = 3$ is true

(A) for all values of b
(B) for all values of b except $b = 2$
(C) for all values of b except $b = 0$
(D) for all positive values of b
(E) for all negative values of b

Solution:

To make the correct choice, multiply each side of the equality by $b - 2$.

$$\frac{3b - 6}{b - 2} = 3$$

$$3b - 6 = 3(b - 2)$$

Eliminate parentheses.

$$3b - 6 = 3b - 6$$

Since the two sides are identical, the resulting equality is true for all values of b. Hence, the original equality must be true for all values of b except a value that would make the denominator 0. Such a value is 2.

Ans. (B)

Matching Test

After completing this test, match your answers with the numbers in the box.

1. If $6x + 5y = ax + by$ is an identity, what is the value of ab?

2. If $5\frac{1}{2}x - 2y = ax + by$ is an identity, what is the value of $2a + b^2$?

3. If $6y + 2x = ax + by$ is an identity, what is the value of $\frac{b}{a}$?

4. $\frac{3b + 12}{b + 4} = 3$ is true for all values of b except . . . .

5. $\frac{b^2 - 16}{b - 4} = b + 4$ is true for all values of b except . . . .

a. 3	*b*. −4	*c*. 15	*d*. 30	*e*. 4

3.9 CHOICES AND COMBINATIONS

In a problem involving choices and combinations, it is necessary to find the number of ways in which a sequence of acts can be performed when each act may be performed in more than one way. To solve problems of this kind, apply the following fundamental principle:

If an act can be performed in r ways, and, after it is performed in one of these ways, a second act can be performed in s ways, and then a third act can be performed in t ways, and so on, the acts can be performed successively in rst . . . ways.

Thus, as illustrated in the first Model Problem that follows, if we can choose either of 2 hats, then choose any one of 3 skirts, and then choose any one of 4 sweaters, we can obtain any one of $2 \cdot 3 \cdot 4$, or 24, outfits consisting of a hat, a skirt, and a sweater.

Model Problems

(1) How many different outfits consisting of a hat, a skirt, and a sweater can be chosen from 2 hats, 3 skirts, and 4 sweaters?

(A) 5 (B) 9 (C) 12 (D) 24 (E) 48

Solution:

Each of the 2 hats may be combined with each of the 3 skirts to obtain $2 \cdot 3$, or 6, different combinations of a hat and a skirt.

Each of the 6 different combinations of a hat and a skirt may be combined with each of the 4 sweaters to obtain $6 \cdot 4$, or 24, different outfits consisting of a hat, a skirt, and a sweater.

Note that the answer, 24, is the product of 2, 3, and 4. *Ans.* (D)

(2) Four contestants in a tennis match are to be paired successively so that each contestant will play against each of the other three. In how many ways can the contestants be paired?

(A) 4 (B) 6 (C) 7 (D) 12 (E) 24

Solution:

Each of the 4 contestants must play each of the other 3.

If A, B, C, and D represent the contestants, the possible pairings are as follows:

(1) AB, (2) AC, (3) AD, (4) BC, (5) BD, (6) CD. *Ans.* (B)

Alternate Method

If the 4 letters A, B, C, and D are arranged so that each letter is paired with each of the other 3, the number of possible arrangements is $4 \cdot 3$, or 12. This can be shown as follows:

AB	AC	AD	BC	BD	CD
BA	CA	DA	CB	DB	DC

Note that each pair in the second row has the same letters as the pair above it in the first row.

However, if the letters represent tennis players, one of the rows may be eliminated, since it does not matter in which order any two players are paired. Hence, to find in how many ways the contestants can be paired, divide 12 by 2 to obtain 6.

Matching Test

After completing this test, match your answers with the numbers in the box.

1. How many different outfits consisting of a skirt, a sweater, and a pair of shoes can be obtained from 3 skirts, 3 sweaters, and 4 pairs of shoes?
2. How many different outfits consisting of a sport shirt, a sport jacket, and a pair of slacks can be obtained from 2 sport shirts, 3 sport jackets and 2 pairs of slacks?
3. In a tennis match, each of 6 contestants is to play each of the five other contestants. In how many ways can the contestants be paired?
4. In a baseball league, each team played each of the other teams. If 28 games were played, how many teams are there in the league?
5. How many different lines can be drawn between 5 points on the circumference of a circle?

a. 15   *b.* 36   *c.* 12   *d.* 8   *e.* 10

4. PRACTICE PROBLEMS

1. If k represents an odd integer, which of the following expressions represents another odd integer?
   (A) $k+1$ (B) $2k+2$ (C) $3k+3$ (D) k^2+4 (E) k^3+3 **[1.1]**

2. If $2x+1$ is used to represent any odd integer, write expressions to represent, respectively, twice and three times an odd integer.
   (A) $4x+2$, $6x+3$ (B) $2x+3$, $2x+5$ (C) $4x+1$, $6x+1$ (D) $2x+2$, $2x+3$
   (E) $(2x+1)^2$, $(2x+1)^3$ **[1.2]**

3. If the average of three unequal positive even integers is 8, what is the maximum value of the largest integer?
   (A) 12 (B) 14 (C) 16 (D) 18 (E) 21 **[1.3]**

4. If n is an integer, write an expression that can be used to represent three times the square of an integer that is 3 more than n.
(A) $3n^2+3$ (B) $3(n+3)^2$ (C) $3n^2+9$ (D) $9n^2+9$ (E) $9(n+3)^2$ **[1.4]**

5. What is the largest positive integer whose square is less than 14,400?
(A) 37 (B) 119 (C) 121 (D) 1199 (E) 1430 **[1.4]**

6. Find all the values of x that satisfy the inequality $-2x+2 \geqq 4$.
(A) $x<-1$ (B) $x \leqq -1$ (C) $x \geqq -1$ (D) $x \geqq -3$ (E) $x \leqq -3$ **[2.1]**

7. In the graph plane at the right, the coordinate axes and the graphs of $x=2$ and $y=x$ divide the first and fourth quadrants into sections labeled A to F. One of these sections contains all points in quadrants I and IV whose coordinates satisfy the system of inequalities

$$y > x$$
$$x < 2$$
$$y > 0$$

Which section contains these points?
(A) A (B) B (C) C (D) D (E) E (F) F **[2.2]**

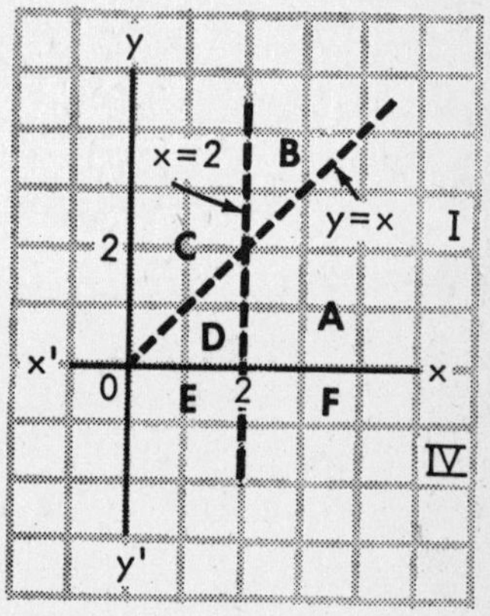

8. In the graph plane at the right, the coordinate axes and the graphs of $x+y=2$ and $y=x+2$ divide the first and second quadrants into sections labeled A to F. One of these sections contains all points in quadrants I and II whose coordinates satisfy the system of inequalities

$$x+y<2$$
$$y<x+2$$
$$x>0$$

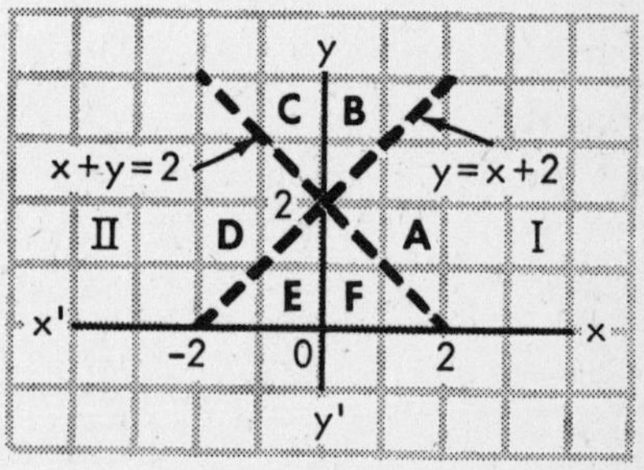

Which section contains these points?
(A) A (B) B (C) C (D) D (E) E (F) F **[2.2]**

9. The equality $\left(7+2\frac{1}{2}\right)+7\frac{1}{2}=7+\left(2\frac{1}{2}+7\frac{1}{2}\right)$ is an illustration of which law?
(A) Associative Law of Addition
(B) Associative Law of Multiplication
(C) Commutative Law of Addition
(D) Commutative Law of Multiplication
(E) Distributive Law **[3.1]**

10. Given: $B=\{2,4,5\}$ and $C=\{-1,4,9\}$.
Find the set whose elements are all the perfect squares in B and C.
(A) $\{-1,4,9\}$ (B) $\{4,4,9\}$ (C) $\{4,9\}$ (D) $\{4\}$ (E) 4 and 9 **[3.2]**

11. In the interval 7 through 30, find the set of integers that are either perfect squares or perfect cubes.
(A) $\{9,16,25\}$ (B) $\{8,27\}$ (C) $\{9,25,27\}$ (D) $\{8,9,16,25,27\}$
(E) $\{9,16,25,27\}$ **[3.2]**

12. The fraction $\frac{x-3y}{3xy}$ is meaningless when

(A) $x = 3y$ (B) $x = \frac{y}{3}$ (C) $y = \frac{x}{3}$ (D) $xy = \frac{1}{3}$ (E) $x = 0$ or $y = 0$ [3.3]

13. Expressing the answer to the base 10, find the positive square root of 41_6.
(A) 5 (B) 6 (C) 7 (D) 8 (E) 9 [3.4]

14. On the clock scale at the right, the plus sign indicates a clockwise motion and the minus sign indicates a counterclockwise motion. Using the clock scale, find the value of $(7+8)-(5-7)$.
(A) 1 (B) 3 (C) 5 (D) 9 (E) 17 [3.5]

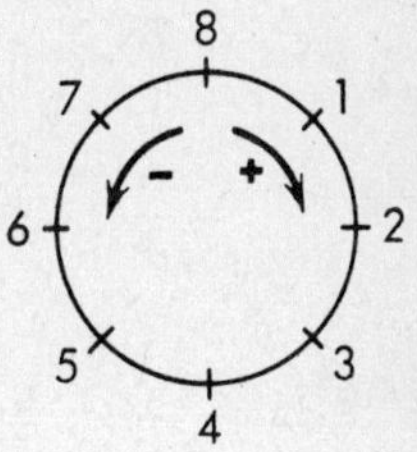

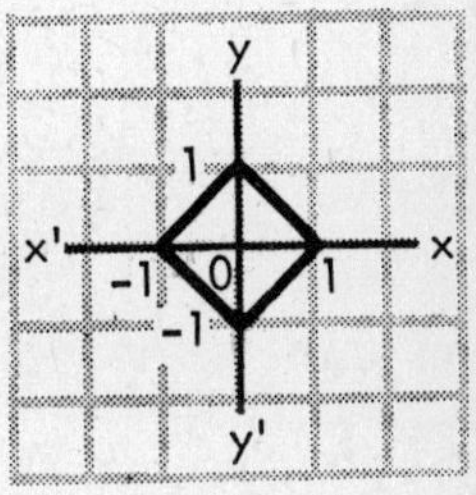

15. The equation of the graph at the right is

(A) $|x|+|y|=1$ (B) $|x+y|=1$ (C) $|x|+y=1$

(D) $x+|y|=1$ (E) $|xy|=1$ [3.6]

16. The repeating decimal .636363 . . . may be written as $.\overline{63}$, and .333 . . . may be written as $.\overline{3}$. Find the value of $.\overline{65}+.\overline{2}$.

(A) $.\overline{652}$ (B) .85 (C) $.\overline{85}$ (D) $.\overline{87}$ (E) .87 [3.7]

17. The equation $\frac{2b^2-32}{b^2-16}=2$ is true for all values of b except

(A) 4 and -4 (B) 4 (C) -4 (D) 2 and -2 (E) 2 [3.8]

18. How many different triangles may be constructed by joining four points, no three of which are in the same straight line?
(A) 2 (B) 3 (C) 4 (D) 5 (E) none [3.9]

XII. DATA-SUFFICIENCY PROBLEMS

A data-sufficiency problem is one in which the student is asked to determine whether a certain problem can be solved or a certain question answered by using given information. He is *not* required to find the answer itself.

In this chapter, each problem is followed by two statements, labeled (1) and (2). The student must determine how useful these two statements are in solving the problem. There are five possible answers:

(A) Only statement (1) is needed to solve the problem; statement (2) alone is not sufficient.

(B) Only statement (2) is needed to solve the problem; statement (1) alone is not sufficient.

(C) Both (1) and (2) are needed to solve the problem; either statement alone is not sufficient.

(D) Either (1) or (2) can be used to solve the problem.

(E) Both (1) and (2) together are not sufficient to solve the problem.

Keep in mind that the problem need not be solved. What must be done is to decide how useful statements (1) and (2) are in obtaining a solution.

The following five solved problems are presented to help the student to acquire an understanding of the nature of these data-sufficiency problems. Tests *A* to *F* are intended to provide practice in handling arithmetic, algebraic, and geometric data-sufficiency problems.

Model Problems

Each of the five problems below consists of a question followed by two statements, numbered (1) and (2), containing information that may be needed to solve the problem. In each case, determine how useful the statements are and select one of the following alternatives as your answer:

(A) Only statement (1) is needed to solve the problem; statement (2) alone is not sufficient.

(B) Only statement (2) is needed to solve the problem; statement (1) alone is not sufficient.

(C) Both (1) and (2) are needed to solve the problem; either statement alone is not sufficient.

(D) Either (1) or (2) can be used to solve the problem.

(E) Both (1) and (2) together are not sufficient to solve the problem.

1. What day of the week is March 1 in a certain year?
(1) That year, Feb. 1 was a Monday.
(2) The year before was a leap year.

Solution:

Since the year before was a leap year, that certain year cannot be a leap year. Hence, the month of February in that year has 28 days. Since March 1 is 28 days after Feb. 1, March 1 must fall on the same day of the week; in this case, Monday.

Both (1) and (2) are needed to solve the problem. *Ans.* (C)

2. What day of the week is March 1 in a certain year?

(1) That year, Feb. 1 was a Monday.

(2) The year before was not a leap year.

Solution:

Since the year before was not a leap year, that certain year may or may not be a leap year. Hence, in that year, February may have 28 or 29 days. The exact day of the week for March 1 cannot be determined from the given information; that is, both (1) and (2) together are not sufficient to solve the problem. *Ans.* (E)

3. What day of the week is March 1 in a certain year?

(1) That year, March 15 was a Monday.

(2) The year before was a leap year.

Solution:

Since March 1 is exactly two weeks before March 15 in *any* year, March 1 must fall on the same day of the week as March 15; in this case, Monday. Statement (2) is of no value in this problem.

Only statement (1) is needed to solve the problem. *Ans.* (A)

4. What day of the week was March 1, 1965?

(1) Feb. 1, 1965 was a Monday.

(2) April 1, 1965 was a Thursday.

Solution:

Since 1965 is not divisible by 4, the year 1965 was not a leap year. Hence, March 1, 1965 was exactly 28 days after Feb. 1, 1965 and 31 days before April 1, 1965. Therefore, the exact day of the week for March 1, 1965 can be found by using either statement (1) or statement (2). *Ans.* (D)

5. What day of the week is March 1 of a certain year?

(1) Feb. 1 of that year is a Monday.

(2) March 8 of that year is a Tuesday.

Solution:

Since March 1 is exactly seven days before a Tuesday, March 1 must be a Tuesday. Hence, only statement (2) is needed to solve the problem. Statement (1) alone does not suffice because the year in question may or may not be a leap year. *Ans.* (B)

Tests

In tests *A* to *F*, each problem consists of a question followed by two statements, numbered (1) and (2), containing information that may be needed to solve the problem. In each case, determine how useful the statements are and select one of the following alternatives:

(A) Only statement (1) is needed to solve the problem; statement (2) alone is not sufficient.

(B) Only statement (2) is needed to solve the problem; statement (1) alone is not sufficient.

(C) Both (1) and (2) are needed to solve the problem; either statement alone is not sufficient.

(D) Either (1) or (2) can be used to solve the problem.

(E) Both (1) and (2) together are not sufficient to solve the problem.

Test A

1. How many students are there in the class?
(1) 27, or 75% of them, are boys.
(2) 12 students take both mathematics and science.

2. How many students are there in the class?
(1) 12 students take both mathematics and science.
(2) The remaining 60% of the class take only science.

3. How many students are there in the school?
(1) One-third of the students are freshmen.
(2) There are as many sophomores as there are juniors, and as many juniors as seniors.

4. What is the ratio of the number of girls to the number of boys in the school?
(1) There are twice as many girls as boys on the honor roll.
(2) The number of boys is 75% of the number of girls.

5. In a class, 36 students take either mathematics or science, or both. What is the ratio of the number who take only mathematics to the number who take only science?
(1) 12 students take both mathematics and science and 18 take only science.
(2) Twice as many students take both mathematics and science as those who take only mathematics. Three times the number who take only mathematics take only science.

Test B

1. Find x.
(1) $y = x + 2$
(2) $4 = 2y - 2x$

2. Find x.
(1) $y = 2 + x$
(2) $y = 2 - x$

3. Find the ratio of x to y.
(1) $2x = 3y$
(2) $5x = 3(x + y)$

4. Find $x^2 + y^2$.
(1) $(x + y)^2 = 2(xy + 10)$
(2) $x + y = y + x$

5. Find xy.
(1) $(x + y)^2 - x^2 = y^2 + 2xy$
(2) $\dfrac{x}{x + 2} = \dfrac{5}{y + 5}$

Test C

In each of the following problems, the price of a roll is 5¢ and the price of a doughnut is 10¢. What is the cost of the rolls and doughnuts?

1. (1) The number of rolls is twice the number of doughnuts.
(2) The cost of the rolls is equal to the cost of the doughnuts.

2. (1) Twice as many rolls as doughnuts were purchased. The rolls cost 50¢.
(2) The doughnuts were 5 fewer than the rolls. The doughnuts cost 50¢.

3. (1) Twice as many rolls as doughnuts were purchased.
(2) There were 4 more rolls than doughnuts.

4. (1) The number of rolls and the number of doughnuts were equal but the doughnuts cost 25¢ more.
(2) The doughnuts cost twice as much as the rolls.

5. (1) The number of rolls was twice the number of doughnuts.
(2) The cost of the rolls equalled the cost of the doughnuts and there were 4 more rolls than doughnuts.

Test D

In quadrilateral $ABCD$, $AB = BC$ and $AD = DC$. How many degrees does angle D contain?

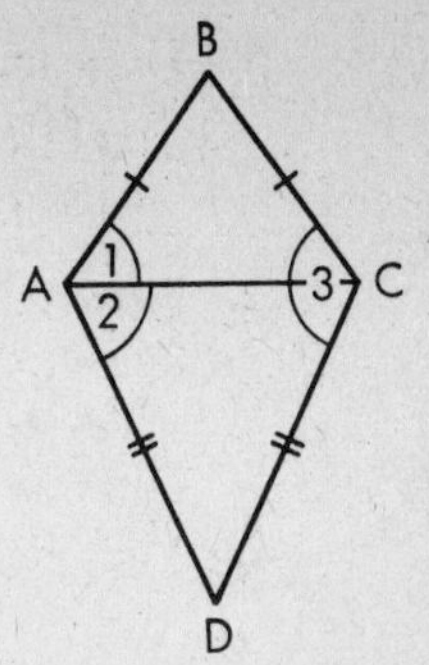

1. (1) $\angle 3 = 120°$.
   (2) $\angle 1 = 55°$.

2. (1) $\angle B = 70°$.
   (2) $\angle B$ is 15° more than $\angle 1$.

3. (1) $\angle 2 = 65°$.
   (2) $\angle 1 = 55°$.

4. (1) The sum of $\angle D$ and $\angle 2$ is 115°.
   (2) $\angle D$ is 15° less than $\angle 2$.

5. (1) $\angle 3 = 120°$.
   (2) The ratio of $\angle D$ to $\angle 2$ is 10:13.

Test E

1. Is the number divisible by 25?
   (1) The last two digits are 25.
   (2) The number is a perfect square.

2. Is the number divisible by 25?
   (1) The last digit is 5.
   (2) The number is a perfect square.

3. Is the number divisible by 9?
   (1) The number ends in the two digits 99.
   (2) The sum of the digits is divisible by 3.

4. Is the number divisible by 6?
   (1) The number is divisible by 3.
   (2) The number is divisible by 12.

5. Is the number divisible by 4?
   (1) The number ends in 24.
   (2) The number is even and a perfect square.

Test F

In each of the following problems, is the number an odd number or is it an even number?

1. (1) The square of the number is an even number.
   (2) The triple of the number is an even number.

2. (1) The square of the number is an odd number.
   (2) The double of the number is an even number.

3. (1) The square of the double of the number is an even number.
   (2) The double of the square of the number is an even number.

4. (1) The number is the smallest of five consecutive integers.
   (2) The largest of the integers is an even number.

5. (1) If 10 is added to the double of the number, the result is an even number.
   (2) If 10 is added to the triple of the number, the result is an odd number.

PRACTICE PROBLEMS

Each problem contains two statements, numbered (1) and (2), supplying information that may be needed to solve the problem. In each case, determine how useful the statements are and select one of the following alternatives as your answer:

(A) Only statement (1) is needed to solve the problem; statement (2) alone is not sufficient.
(B) Only statement (2) is needed to solve the problem; statement (1) alone is not sufficient.
(C) Both (1) and (2) are needed to solve the problem; either statement alone is not sufficient.
(D) Either (1) or (2) can be used to solve the problem.
(E) Both (1) and (2) together are not sufficient to solve the problem.

1. In a certain leap year, what day of the week was April 1?
(1) The 70th day later was a Monday.
(2) The 50th day earlier was a Sunday.

2. How many students are there in a class?
(1) One-third of the class are girls.
(2) The boys in the class outnumber the girls by 12.

3. How many games did a team win during the baseball season?
(1) The team won 20 more games that season than it had won during the previous season.
(2) Of the 100 games played during the previous season, the team had lost 42.

4. Find x.
(1) $x - y = -(y - x)$
(2) $\dfrac{3x(2x - 3y)}{3y - 2x} = -36$

5. A man paid \$1.20 for some rolls and doughnuts. How many rolls did he buy?
(1) The price of a roll was 8¢ and the price of a doughnut was 10¢.
(2) The cost of the rolls was twice the cost of the doughnuts.

6. In quadrilateral $ABCD$, $AB = BC$ and $AD = DC$. Find the number of degrees in angle B.
(1) Triangle ADC is an equilateral triangle and angle 3 is twice as large as angle 2.
(2) Angle D is 30° more than angle 1, and angles 1 and 3 are complementary.

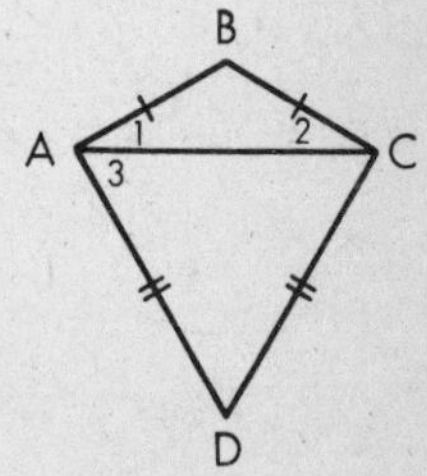

7. In quadrilateral $ABCD$, $AB = AD$ and $CD = BC$. Find the number of degrees in angle C.
(1) Angle B is a right angle and angle $A = 120°$.
(2) Angles B and D are supplementary and angle 2 is twice as large as angle 1.

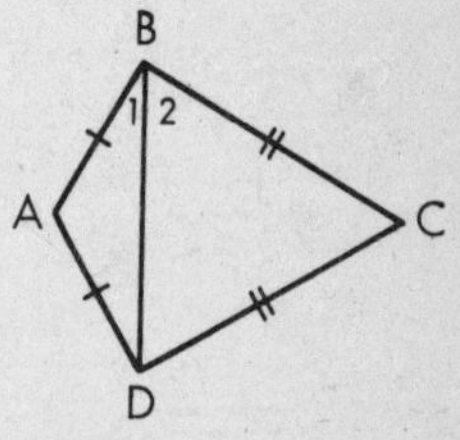

8. Is the triangle an obtuse triangle?
(1) The smallest angle of the triangle is 85° less than the largest angle.
(2) The sum of the largest angle and the smallest angle is 95°.

9. Is 1 the last digit of a certain number?
(1) The square of the number ends in the digit 1.
(2) If the number is squared and then doubled, the last digit of the resulting number is 2.

10. Is the number an odd number or an even number?
(1) The fifth power of the number is an odd number.
(2) The square of twice the number is an even number.

XIII. QUANTITATIVE COMPARISON PROBLEMS

A quantitative comparison problem is one in which the student is asked to compare two quantities to determine whether they are equal or which of them is the greater. The following four solved problems are presented to help the student acquire an understanding of the nature of such problems. Tests *A* to *P* provide practice in handling arithmetic, algebraic, and geometric quantitative comparison problems.

Model Problems

In each problem, compare the two given quantities, choosing answer

(A) if the quantity in Column *A* is the greater;

(B) if the quantity in Column *B* is the greater;

(C) if the quantities are equal;

(D) if the relation cannot be determined from the information given.

	Column A	*Column B*
1.	Given: $5x = 0$	
	1	x
2.	$\frac{x^2y}{y^2x}$	$\frac{xy^2}{yx^2}$
3.	Given: $\frac{1}{x}$ is a negative integer	
	x	$-\frac{1}{x}$
4.	Given: $y = x - 1$	
	the average of y, $y + 1$, and $y + 2$	the average of $x - 1$, x, and $x + 1$

Note. When a quantitative comparison problem contains letters such as x and y, the letters stand for real numbers. Keep in mind that real numbers may be positive, negative, or zero.

Solutions:

1. Since $5x = 0$, then $x = 0$. Hence, 1 is greater than x. *Ans.* (A)
2. By simplifying, we obtain $\frac{x}{y}$ in Column *A* and $\frac{y}{x}$ in Column *B*. Until we know what values x and y represent, the two expressions cannot be compared. *Ans.* (D)
3. Since $\frac{1}{x}$ is negative, x is negative and $-\frac{1}{x}$ is positive. Hence, $-\frac{1}{x}$ is greater than x. *Ans.* (B)
4. The averages are $y + 1$ and x, respectively. Since $y = x - 1$, then $y + 1 = x$. Hence, the averages are equal. *Ans.* (C)

Tests

In Tests *A* to *P*, each problem consists of two quantities, one in Column *A* and the other in Column *B*. In each case, choose one of the following answers:

(A) if the quantity in Column *A* is the greater;

(B) if the quantity in Column *B* is the greater;

(C) if the quantities are equal;

(D) if the relation cannot be determined from the information given.

Note. All four choices occur in the set of answers to each test; that is, in every set of four problems, each choice must be used once.

Test A

	Column A	Column B
1.	$4 \times 3\frac{1}{2} \times \frac{2}{3} \times 6$	$\frac{2}{3} \times 3\frac{1}{2} \times 4 \times 5$
2.	$\frac{1}{3} \times \frac{1}{5} \times \frac{2}{7}$	$\frac{2}{3} \times \frac{1}{5} \times \frac{1}{7}$
3.	75% of 16	7.5% of 165
4.	Given: $y = x^3$	
	y	x

Test B

1.	$\sqrt{16 + 9}$	$\sqrt{16} + \sqrt{9}$
2.	$\sqrt{169 - 144}$	$\sqrt{169} - \sqrt{144}$
3.	Given: $a = 2, b = 3, c = 6$	
	$\sqrt{a^2 + b^2 + c^2}$	$b + c - a$
4.	Given: a and b are positive	
	$\sqrt{a + b}$	$\sqrt{ab}$

Test C

1.	Given: $x + y = 5, 2x + y = 4$	
	x	y
2.	Given: $x^2 - 4x + 4 = 0, 4y^2 - 4y + 1 = 0$	
	x	$4y$
3.	Given: $x + y = z, z = 2y - 1$	
	y	x
4.	Given: $5x - 5y = 50, z = y^2$	
	x	z

Test D

	Column A	*Column B*
1.	The average of 19, 21, 23	the average of 19, 20, 21, 22, 23
2.	The average of $x - 2, x, x + 2$	the average of $y, y + 2, y + 4$
3.	The average of 22%, 25%, 28%	the average of 20%, 24%, 28%, 32%
4.	The average of $\frac{1}{4}, \frac{1}{2}, \frac{3}{4}$	the average of $\frac{1}{4}, \frac{1}{2}, \frac{2}{3}$

Test E

In each of the following, a, b, and c are consecutive integers.

1.	b	the average of a and c
2.	$b + c$	ab
3.	c	$a + 3$
4.	The average of b and c	$a + 1$

Test F

1. Given: $x \neq 0$

$\frac{x + 5}{x}$ $\qquad$ $\frac{5}{x}$

2. Given: $y > 5$

$\frac{y}{y + 5}$ $\qquad$ $\frac{y}{y - 5}$

3. Given: $x \neq 0$

$\frac{x^3 + x^2}{x}$ $\qquad$ $\frac{x^3 + x^2}{x^2}$

4. Given: $x \neq 0$

$\frac{x - y}{x}$ $\qquad$ $1 - \frac{y}{x}$

Test G

1. Given: $a > 0, b > 0$

ab $\qquad$ $\frac{a}{b}$

2. Given: $a > 0, b < 0$

$a + b$ $\qquad$ $a - b$

3. Given: $a > 0, a > b$

a^2 $\qquad$ ab

4. Given: $a = b$

$a^2 + b^2$ $\qquad$ $2ab$

Test H

	Column A	Column B
1.	Given: $5x > 0$	
	1	x
2.	Given: $2x - 3 > 7$	
	5	x
3.	Given: $10 - 2x < 6$	
	x	2
4.	Given: x is a positive integer and $2x > x^2$	
	x	1

Test I

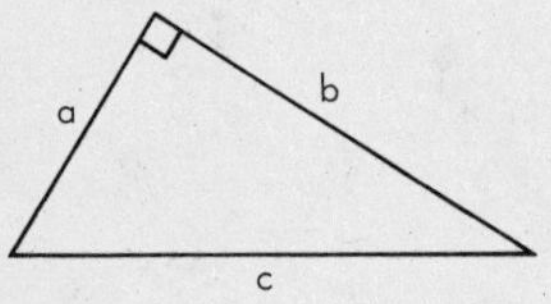

	Column A	Column B
1.	$a + b$	c
2.	$a - b$	c
3.	$b\sqrt{2}$	c
4.	$c^2 - b^2$	a^2

Test J

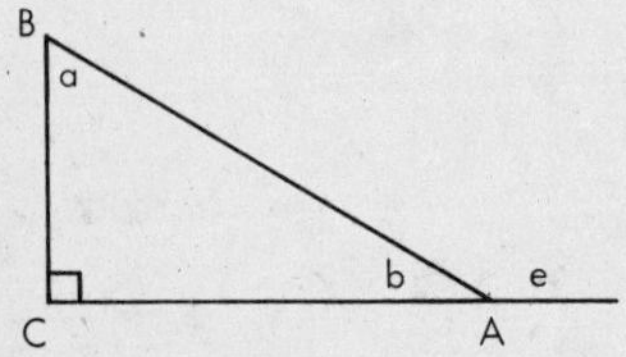

	Column A		Column B
1.	$e - a$		90
2.	$e - b$		90
3.		Given: $AC > BC$	
	a		45
4.		Given: $AB = 2BC$	
	b		35

Test K

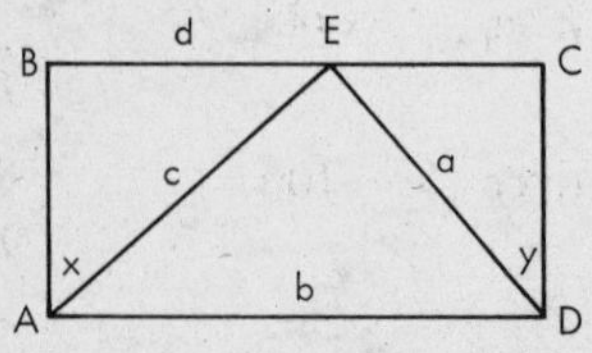

ABCD is a rectangle.

1.		Given: $c < a$	
	x		y
2.		Given: $x = y$	
	d		$\frac{1}{2}b$
3.		Given: $x < y$	
	b		c
4.		Given: $x > y$	
	c		a

Test L

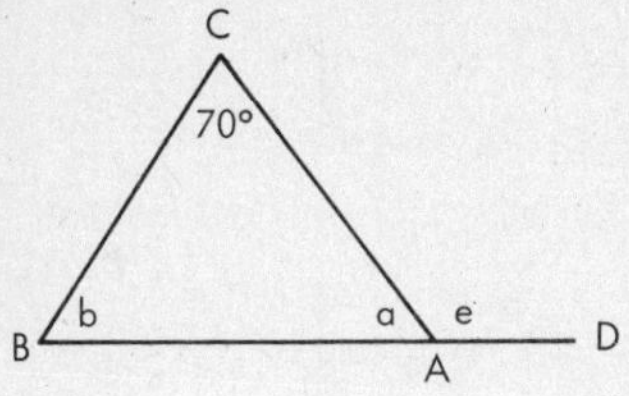

	Column A		Column B
1.		Given: $AB = AC$	
	e		140
2.		Given: $BC = AC$	
	e		140
3.		Given: $AB < AC$	
	e		140
4.		Given: $BC < AC$	
	e		140

Test M

ABCD is a parallelogram.

1.		Given: $x > 90$	
	A		90
2.		Given: $x = 90$	
	B		90
3.		Given: $x < 90$	
	90		B
4.	B		x

Test N

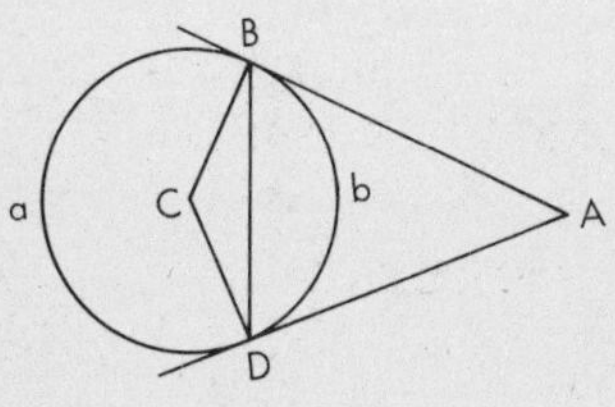

AB and *AD* are tangents to circle *C*.

1.		Given: $A = 60$	
	AB		BD
2.		Given: $b = 100$	
	A		90
3.	A		C
4.		Given: $b < 90$	
	A		90

Test O

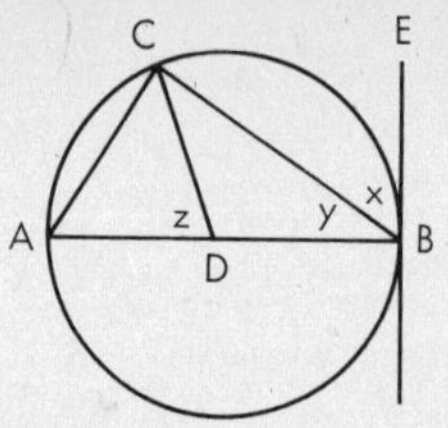

EB is tangent to circle D. Diameter $AB = 20$.

	Column A		Column B
1.	x		A
2.	A		89
3.	AC	Given: $y > 30$	10
4.	z	Given: $x > 45$	90

Test P

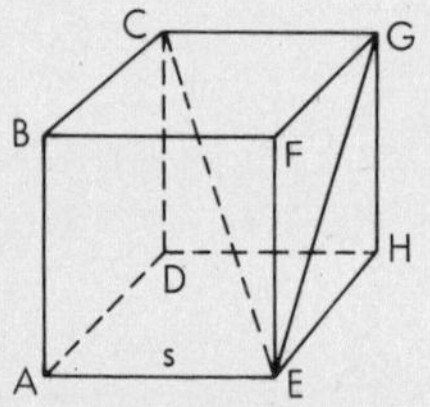

The figure is a cube having side s.

	Column A	Column B
1.	CE	$s\sqrt{2}$
2.	GE	$s\sqrt{2}$
3.	The perimeter of square $ABCD$	the perimeter of triangle ECG
4.	The number of cubic feet in the volume of the cube	the number of square feet in the total area of the cube

PRACTICE PROBLEMS

In each problem, a quantity in Column *A* is to be compared with a quantity in Column *B*. Choose answer

(A) if the quantity in Column *A* is the greater;
(B) if the quantity in Column *B* is the greater;
(C) if the quantities are equal;
(D) if the relation cannot be determined from the information given.

	Column A		*Column B*
1.	$5 \times 87\frac{1}{2} \times 12$		$87\frac{1}{2} \times 11\frac{1}{4} \times 5$
2.		Given: $x > 0$ $y = x^2 + x$	
	x		y
3.		Given: $y = 2x$	
	the average of y, $2y$, and $3y$		the average of x, $4x$, and $7x$
4.		Given: $3x + 4 > 8$	
	2		x
5.		Given: $2x - y = 9$ $3x + y = 6$	
	x		$-y$
6.	$\sqrt{64} + \sqrt{25} - \sqrt{16}$		$\sqrt{64 + 25 - 16}$

7.

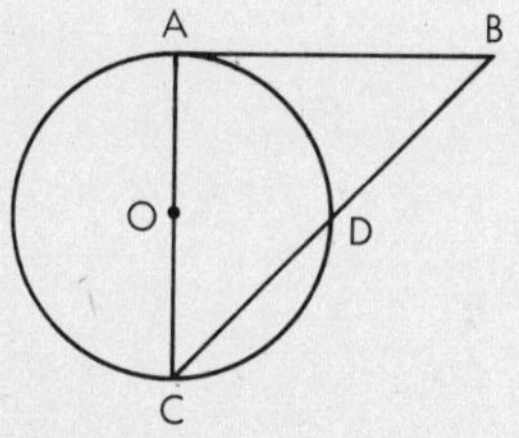

$\overline{AB}$ is tangent to circle O.
$AB = AC$

Column A	*Column B*
arc AD	$90°$

	Column A		*Column B*

8.

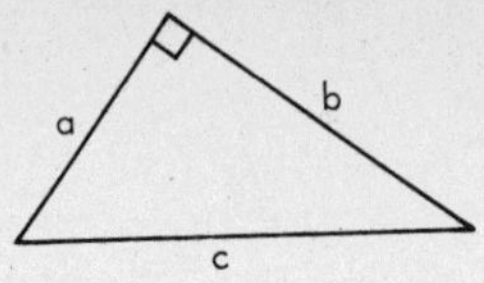

$\sqrt{c^2 - a^2}$ | a

9.

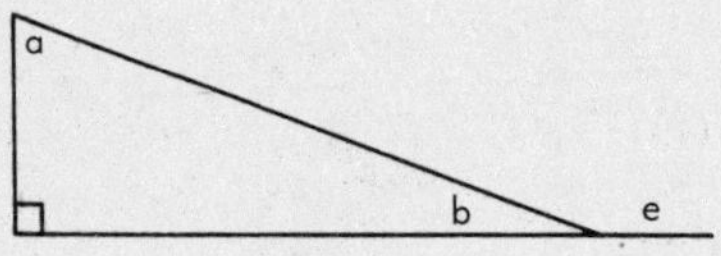

$a = 4b$

$160°$ | e

10.

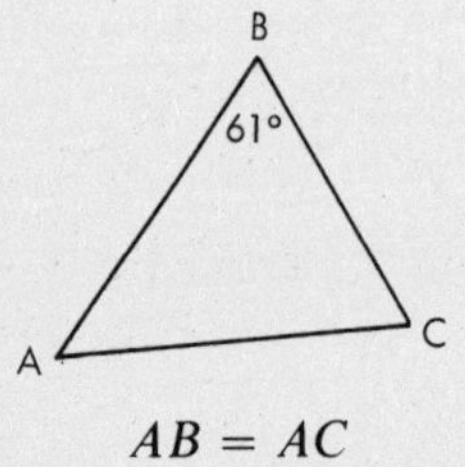

$AB = AC$

AB | BC

APTITUDE TESTS

I

1. An estimate of the value of $\frac{591 \times 5017}{992 \times 305}$ is

(A) 5 (B) 8 (C) 1 (D) 10 (E) 100

2. Combine: $3\frac{2}{3} - 2\frac{1}{6} - \frac{1}{4}$

(A) $\frac{1}{4}$ (B) $1\frac{1}{4}$ (C) $1\frac{7}{12}$ (D) $1\frac{3}{4}$ (E) $5\frac{1}{4}$

3. Simplify: $.06 \div \frac{6}{.5}$

(A) .005 (B) .05 (C) .5 (D) 5 (E) .72

4. How many thousandths are there in 3.4%?

(A) 3.4 (B) 34 (C) 340 (D) .34 (E) 30.4

5. The library charges for a borrowed book are 7¢ for the first day and 4¢ for each additional day. If a borrower paid $1.19 for the loan of a book, for how many days was the book borrowed?

(A) 11 (B) 28 (C) 29 (D) 30 (E) 32

6. Simplify: $3p + 2p(p - 4) + p^2$

(A) $3p^2 + 5p$ (B) $3p^2 - 5p$ (C) $-7p^2 + 3p$ (D) $6p^2 - 20p$ (E) $2p^2 - 4p$

7. Find the sum of $4x^2 - 3$ and $3x^2 - 4x - 4$.

(A) $7x^2 - 4x - 7$ (B) $7x^2 + 4x - 7$ (C) $7x^4 - 4x - 7$ (D) $7x^2 + 4x + 7$
(E) $7x^2 - 7x - 4$

8. Factor: $25st^4 - 50s^2t^3 + 75s^3t^2$

(A) $25st(t^3 - 2st^2 + 3s^2t)$ (B) $5st(5t^3 - 10st^2 + 15s^2t^2)$ (C) $25st^2(t^2 - 2st + 3s^2)$
(D) $5st^2(5t^2 - 10st + 15st^2)$ (E) $25t^2(st^2 - 2s^2t + 3s^3)$

9. Solve for x: $3(2 - x) = 12$

(A) 2 (B) −6 (C) 6 (D) −2 (E) −7

10. Solve for y: $\frac{2y}{5} = 20$

(A) 50 (B) 35 (C) 8 (D) 98 (E) 5

11. The sum of two numbers is 15. If s represents the smaller number, write in simplest form the expression for the larger number decreased by 7.

(A) $s - 8$ (B) $8 - s$ (C) $s + 8$ (D) $22 - s$ (E) $s - 22$

12. If n represents the first of a set of four consecutive even integers, write an expression that may be used to represent the sum of the set.

(A) $4n + 20$ (B) $4n - 12$ (C) $4n$ (D) $4n + 12$ (E) $13n$

13. Mr. Jones was 35 years old 10 years ago. How old was he y years ago?

(A) $35 - y$ (B) $y - 25$ (C) $25 - y$ (D) $y - 45$ (E) $45 - y$

14. Find the average of $\frac{1}{2}$, $\frac{1}{3}$, and $\frac{1}{6}$.

(A) $\frac{1}{6}$ (B) 1 (C) $\frac{1}{3}$ (D) $\frac{1}{2}$ (E) $\frac{2}{3}$

15. The length of a tennis court is 8 feet shorter than three times its width, w. Express in terms of w the number of feet in the semiperimeter of the court.

(A) $3w - 8$ (B) $3w - 16$ (C) $4w - 8$ (D) $4w - 16$ (E) $8w - 16$

16. Express in lowest terms the ratio 40 seconds:$1\frac{1}{2}$ minutes:$\frac{1}{6}$ hour.

(A) 4:9:6 (B) 4:9:60 (C) 40:90:600 (D) $\frac{2}{3}:1\frac{1}{2}:10$ (E) 60:9:4

17. In triangle ABC, what is the ratio of the shaded part to the unshaded part?

(A) $\frac{1}{4}$ (B) $\frac{1}{3}$ (C) $\frac{2}{3}$ (D) $\frac{3}{8}$ (E) $\frac{1}{8}$

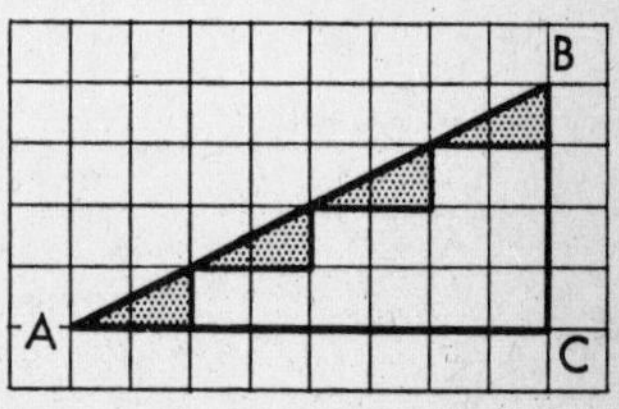

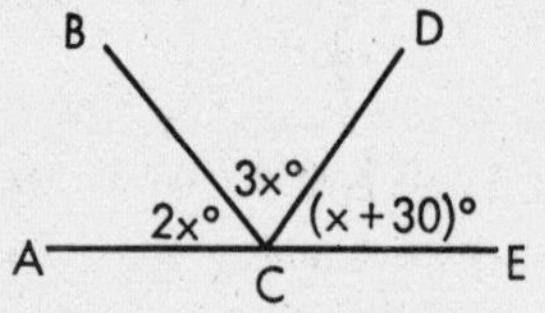

18. ACE is a straight line. Find the number of degrees in angle BCE.

(A) 25 (B) 55 (C) 105 (D) 125 (E) 130

19. The sum of two angles of a triangle is 160°. One of these angles is three times the other. Find the number of degrees in the smallest angle of the triangle.

(A) 15 (B) 20 (C) 25 (D) 35 (E) 40

20. In the figure at the right, $\angle A = \angle A'$, $\angle B = \angle B'$, and $\angle C = \angle C'$. Find a.

(A) 6 (B) $6\frac{3}{4}$ (C) $7\frac{1}{2}$ (D) $10\frac{2}{3}$ (E) 12

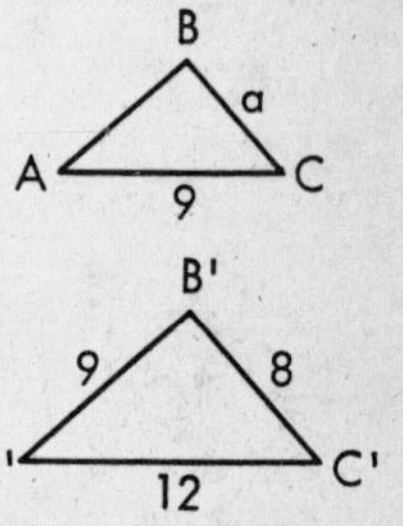

21. If the ratio of two corresponding diagonals of two similar pentagons is 4:9, then the ratio of their perimeters is

(A) 2:3 (B) 4:9 (C) 9:4 (D) 16:81 (E) 3:2

22. A square is inscribed in a circle of radius 6. The side and the area of the square are

(A) $3\sqrt{2}$, 18 (B) $6\sqrt{2}$, 36 (C) $6\sqrt{2}$, 72 (D) $3\sqrt{2}$, 36 (E) $3\sqrt{2}$, 72

23. To cover a floor, 405 tiles are used. If each tile is a square 4 inches on a side, then the floor area in square feet is

(A) 45 (B) 54 (C) 60 (D) 540 (E) 6480

24. Find the circumference of a circle whose area is 49π.

(A) 7π (B) 14π (C) 7 (D) 14 (E) 42

25. Find the number of inches in the thickness of a pipe whose outer circumference is 10π inches and whose inner diameter is 7.2 inches.

(A) 1.4 (B) 2.8 (C) 3.6 (D) 6.4 (E) 12.8

26. A line joins the points (8, 3) and (−8, −1). The coordinates of the point where the line intersects the y-axis are
(A) (0, 1) (B) (1, 0) (C) (0, −1) (D) (0, 2) (E) (2, 0)

27. If k is an odd integer, then an expression that represents an even integer is
(A) $3k$ (B) $k^3 - 3$ (C) $3k + 4$ (D) k^3 (E) $k^3 - 4$

28. If an odd integer is expressed as $2x + 1$, where x is any integer, then the square of the next larger even integer is
(A) $(2x + 2)^2$ (B) $(2x + 3)^2$ (C) $4x + 2$ (D) $(4x + 1)^2$ (E) $(4x + 2)^2$

29. *Quantitative Comparison Problem:* Compare the quantity in Column A with the quantity in Column B. Choose answer

(A) if the quantity in Column A is the greater;
(B) if the quantity in Column B is the greater;
(C) if the quantities are equal;
(D) if the relation cannot be determined from the information given.

Column A	*Column B*
65% of 15.8	15.8% of 65

30. A bookseller buys 15 books, selecting them from a list of books offered at prices ranging from \$2 to \$2.65 per book. He plans to sell them in his shop as part of a collection of books marked for sale at prices ranging from \$3.75 to \$5.25. Find the maximum and the minimum profit that he can make if he sells all of the 15 books.
(A) \$39, \$26.25 (B) \$48.75, \$39 (C) \$48.75, \$16.50 (D) \$39, \$16.50
(E) \$16.50, \$48.75

II

1. $\frac{3}{4}$ of 8 is what percent of $\frac{1}{4}$ of 32?

(A) 7.5 (B) 25 (C) $66\frac{2}{3}$ (D) 75 (E) $133\frac{1}{3}$

2. What is the social security tax over a 12-week period on a salary of \$120 a week if the tax is 2.5% of the salary?
(A) \$3.60 (B) \$30 (C) \$36 (D) \$300 (E) \$360

3. Add twice 2 ft. 3 in. and one-half 1 yd. 2 ft.
(A) 2 yd. 1 ft. (B) 2 yd. 9 in. (C) 2 yd. 1 ft. 9 in. (D) 2 yd. (E) 2 yd. 2 ft.

4. A club has 500 members. What is the smallest number of members that can be chosen at random from this club to form a group that is sure to contain at least two members whose birthdays fall on the same day of the year?
(A) 365 (B) 366 (C) 367 (D) 368 (E) 370

5. The expression $2m^3$ is equivalent to
(A) $mmm + mmm$ (B) $(2m)(2m)(2m)$ (C) $2(m + m + m)$ (D) $mm + mm + mm$
(E) $(mmm)(mmm)$

6. Evaluate $2a^2b^2$ when $a = 5$ and $b = -2$.
(A) −200 (B) 400 (C) −400 (D) 200 (E) 46

7. Simplify: $3(x + 2) - (x - 4y) + (x + y)$
(A) $3x + 11y$ (B) $3x - 3y + 6$ (C) $x + 5y + 6$ (D) $3x + 5y + 6$ (E) $3x + 11y + 6$

8. Find the sum of $a - 2b + 3c$, $-4a - 5b + 5c$, and $3a - b - 2c$.
(A) $-7b + 6c$ (B) $-a - 8b + 6c$ (C) $-8b + 6c$ (D) $-a - 7b + 6c$ (E) $-6b + 10c$

9. Solve for x: $0.08x = 0.448$
(A) 56 (B) .56 (C) 5.06 (D) 5.6 (E) 6

10. If 80% of x is 24, find the value of x.
(A) 3 (B) 30 (C) 19.8 (D) .3 (E) 20

11. Solve for x: $y = mx - b$
(A) $\frac{y}{m - b}$ (B) $\frac{y + b}{m}$ (C) $\frac{y}{m} + b$ (D) $\frac{y}{m} - b$ (E) $\frac{by}{m}$

12. What is the value per pound of a mixture of 40 pounds of tea worth 60¢ a pound and 20 pounds of tea worth 90¢ a pound?
(A) 64¢ (B) 65¢ (C) 68¢ (D) 70¢ (E) 75¢

13. Mr. White deposited a sum of money in a bank at 4% annual interest. If this sum increased to $546 at the end of the year, how many dollars had he originally deposited?
(A) $525 (B) $515 (C) $520 (D) $502.50 (E) $510

14. If a jet plane travels 1.95 miles in 10 seconds, find its speed in miles per hour.
(A) 580 (B) 660 (C) 690 (D) 702 (E) 1170

15. A man paid $62 for 8 shirts and 2 hats. Each shirt cost the same, and the hats were also alike in price. If half as many shirts and twice as many hats had been bought at the same unit prices, the total cost would have been $76. What was the price per shirt?
(A) $3.25 (B) $4.00 (C) $3.50 (D) $4.25 (E) $4.50

16. Let t and u represent, respectively, the tens and units digit of a two-digit number, and write the equations that may be used to solve the following problem:
The difference between the units digit and the tens digit of a number is 6. Twice the number increased by 16 is 50. Find the number.
(A) $t + u = 6$, $2(10t + u) + 16 = 50$
(B) $u - t = 6$, $2(10t + u) + 16 = 50$
(C) $t - u = 6$, $2(10t + u) + 16 = 50$
(D) $u - t = 6$, $(10t + u) + 16 = 50$
(E) $u - t = 6$, $2(t + u) + 16 = 50$

17. Henry works half as fast as John, and John requires 10 days to do a certain job. If Henry spent 3 days on that job, what part of the job did he complete?
(A) $\frac{3}{20}$ (B) $\frac{3}{5}$ (C) $\frac{1}{2}$ (D) $\frac{3}{10}$ (E) $\frac{1}{4}$

18. A secretary works on a job from 10:30 A.M. to 12:45 P.M. If the entire job takes 3 hours, what is the ratio of the work remaining to be done to the work completed?
(A) $\frac{3}{1}$ (B) $\frac{2}{3}$ (C) $\frac{3}{4}$ (D) $\frac{1}{3}$ (E) $\frac{1}{4}$

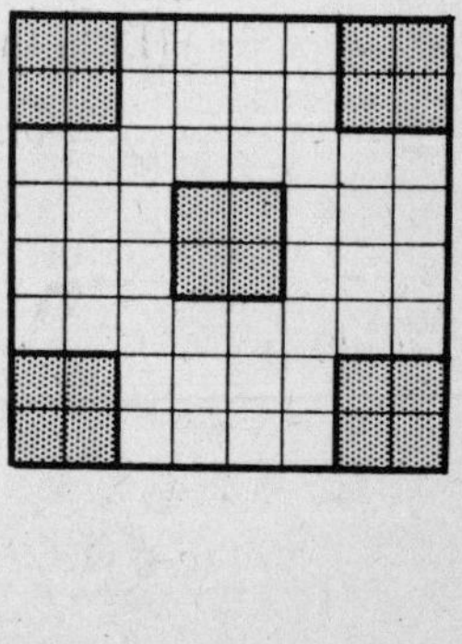

19. The ratio of the shaded part of the square to the entire square is
(A) 1:3 (B) 5:16 (C) 5:11 (D) 11:16 (E) 1:4

20. The difference between one-half the supplement and one-half the complement of an angle is

(A) 10 (B) 15 (C) 30 (D) 45 (E) 75

21. If the ratio of two corresponding medians in two similar triangles is 2:3, and an altitude of the smaller triangle is 10, then the corresponding altitude of the larger triangle is

(A) $6\frac{2}{3}$ (B) 12 (C) 15 (D) 16 (E) 20

22. An upright pole 5 feet high casts a shadow 6 feet long at the same time that a nearby building 45 feet high casts a shadow whose length in feet is

(A) $37\frac{1}{2}$ (B) 48 (C) 51 (D) 54 (E) 60

23. The perimeter of the entire figure, which consists of 7 squares, is 48. Its area is

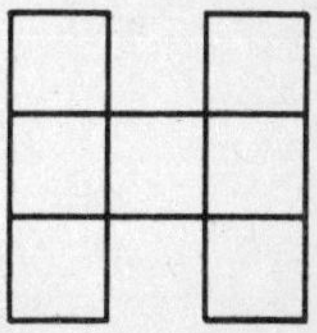

(A) 9 (B) 48 (C) 63 (D) 81 (E) 112

24. A pool 9 feet by 17 feet is bordered by a walk that is $1\frac{1}{2}$ feet wide. The number of square feet in the total area of the pool and the walk is

(A) 153 (B) 180 (C) 204 (D) 221 (E) 240

25. If the diameter of a wheel is 21 inches, then the number of revolutions made by the wheel in rolling a distance of 1 mile is

(A) 240 (B) 480 (C) 600 (D) 720 (E) 960

26. Find the length of an arc of 66° in a circle whose circumference is 180.

(A) 22 (B) 33 (C) 33π (D) 66 (E) 66π

27. The following number pairs are the coordinates of points on the same straight line: $(0,-1)$, $(1,2)$, and $(-1,-4)$. An equation of the line is

(A) $x+y=3$ (B) $y=3x-1$ (C) $x=y+3$ (D) $y=x-1$ (E) $x=3y+3$

28. The product of three unequal positive integers is 90. If the integers are each larger than 1, then the maximum value of the largest integer is

(A) 5 (B) 6 (C) 9 (D) 10 (E) 15

29. If n represents an odd integer, then the square of the next larger odd integer is

(A) $4n^2$ (B) $2(n+2)^2$ (C) $(n+1)^2$ (D) $(n+2)^2$ (E) $(n+3)^2$

30. *Quantitative Comparison Problem:* Compare the quantity in Column *A* with the quantity in Column *B*. Choose answer

(A) if the quantity in Column *A* is the greater;
(B) if the quantity in Column *B* is the greater;
(C) if the quantities are equal;
(D) if the relation cannot be determined from the information given.

Column A		*Column B*
	Given: $a > b$ $ax = b$	
1		x

III

1. How much is saved in buying an article priced at \$100 less 50%, rather than an article priced at \$100 less 30% and 20%?

   (A) \$0 (B) \$5 (C) \$6 (D) \$750 (E) \$10

2. How many 15-gallon cans have the same capacity as eighteen 10-gallon cans and twelve 5-gallon cans?

   (A) 12 (B) 16 (C) 20 (D) 22 (E) 24

3. Multiply and divide as indicated: $\frac{7}{9} \times .54 \div \frac{2}{3}$

   (A) .28 (B) 2.8 (C) .63 (D) 6.3 (E) 63

4. What is $\frac{2}{3}\%$ of 2400?

   (A) 1.6 (B) 16 (C) 36 (D) 360 (E) 1600

5. Evaluate $3a^2 - 2a + 1$ when $a = -3$.

   (A) 34 (B) −20 (C) 22 (D) 88 (E) 43

6. Subtract $x^2 + x + 1$ from $3x - 5x^2$.

   (A) $-x^2 + 2x + 1$ (B) $x^2 - 2x + 1$ (C) $-6x^2 + 2x - 1$ (D) $6x^2 - 2x + 1$
   (E) $2x^2 - 6x - 1$

7. Factor: $9b^2 - 25a^2$

   (A) $(3b + 5a)(3b - 5a)$ (B) $(3b - 5a)(3b - 5a)$ (C) $(9b + 25a)(b - a)$
   (D) $(9b + 5a)(b - 5a)$ (E) $(9b - 5a)(b - 5a)$

8. Combine into one fraction: $\frac{1}{a} + \frac{b}{a} + 1$

   (A) $\frac{a+b}{a}$ (B) $\frac{a+b+1}{a}$ (C) $\frac{ab+1}{a}$ (D) $\frac{a+b+1}{ab}$ (E) $\frac{2a+b}{a}$

9. Solve for x: $ax - b = c + b$

   (A) $\frac{c}{a}$ (B) $\frac{c-2b}{a}$ (C) $\frac{c+b}{a-b}$ (D) $\frac{2b+c}{a}$ (E) $\frac{2bc}{a}$

10. Solve for m: $\frac{3-m}{2-m} = \frac{5}{3}$

    (A) $-\frac{3}{2}$ (B) $-\frac{1}{2}$ (C) $\frac{3}{2}$ (D) $\frac{1}{2}$ (E) $\frac{1}{8}$

11. If $m = \frac{12}{3-x}$, find the value of m when $x = -3$.

    (A) 2 (B) 3 (C) 4 (D) $1\frac{1}{2}$ (E) $\frac{4}{3}$

12. Let x represent the denominator of the fraction sought in the following problem, and write an equation that may be used to solve the problem:

    The numerator of a fraction is three times as large as the denominator. If the numerator is left unchanged and the denominator is decreased by 2, the value of the resulting fraction is 5. Find the original fraction.

    (A) $\frac{3x-2}{x} = 5$ (B) $\frac{3x}{x-2} = 5$ (C) $\frac{x}{3x-2} = 5$ (D) $\frac{x+2}{3x-2} = 5$ (E) $\frac{3x+2}{x} = 5$

13. A fruit grower had 65 trees, of which there were twice as many orange trees as grapefruit trees, while the number of tangerine trees was 10 less than twice the number of grapefruit trees. How many tangerine trees were there?
(A) 12 (B) 15 (C) 18 (D) 20 (E) 30

14. The product of two consecutive integers is 182. If n represents the smaller integer, then an equation that may be used to find the integers is
(A) $n(n+2) = 182$ (B) $n(n+1) = 182$ (C) $2n+2 = 182$ (D) $2n+1 = 182$
(E) $n(n-1) = 182$

15. Find the average of the numbers represented by $x - 8$, $3x - 2$, and $2x + 4$.
(A) $2x - 2$ (B) $x - 1$ (C) $2x - 4$ (D) $2x + 4$ (E) $x - 2$

16. Find the perimeter of a rectangle whose width is represented by w and whose length is represented by $5w + 6$.
(A) $w(5w + 6)$ (B) $6w + 6$ (C) $6w + 12$ (D) $12w + 6$ (E) $12w + 12$

17. A collection of 21 coins consists of nickels and dimes. What is its value in cents if there are six times as many dimes as nickels?
(A) 195 (B) 120 (C) 180 (D) 135 (E) 155

18. Cashews, almonds, and walnuts are mixed in the ratio 2:3:5. What part of the mixture are the almonds?
(A) $\frac{3}{10}$ (B) $\frac{3}{5}$ (C) $\frac{3}{7}$ (D) $\frac{1}{10}$ (E) $\frac{1}{3}$

19. The perimeter of a triangle is 37 inches. If one side is 7 inches and the other two sides are in the ratio 3:7, find the number of inches in the longest side of the triangle.
(A) 3 (B) 9 (C) 12 (D) 20 (E) 21

20. A scale of 1:72 is used in a scale model of a plane. How many feet is the length of the plane if the length of the model is 12.5 inches?
(A) 25 (B) 37.5 (C) 50 (D) 62.5 (E) 75

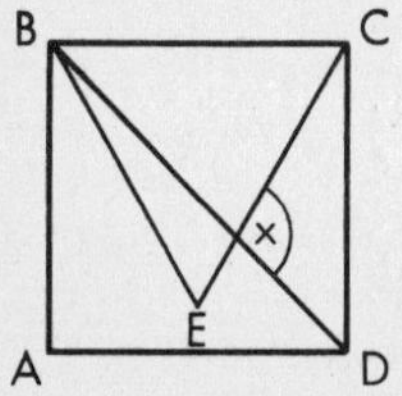

21. Find the number of degrees in x if, as shown in the diagram, equilateral triangle BEC is inside square $ABCD$.
(A) 60 (B) 75 (C) 85 (D) 90 (E) 105

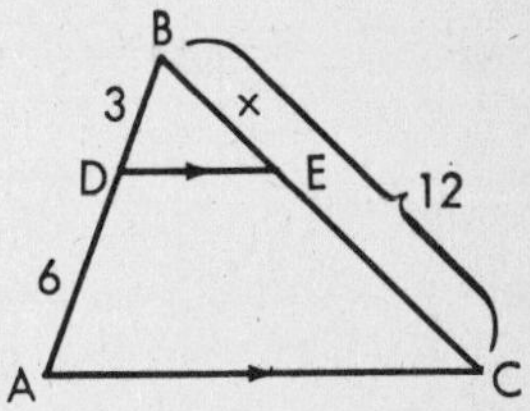

22. DE is parallel to AC. Find x.
(A) 4 (B) $3\frac{1}{2}$ (C) $4\frac{1}{2}$ (D) 6 (E) 8

23. The area of a square is 12 more than twice the area of a rectangle. If the side of the square is a and the dimensions of the rectangle are b and 12, express the relation between a and b.
(A) $a^2 = 12b + 12$ (B) $2a = 24b + 12$ (C) $a^2 = 24b - 12$ (D) $a^2 = 24b + 12$
(E) $a^2 = 12b - 12$

24. Represent the area of a square whose side is $3x + 2$.
(A) $6x + 4$ (B) $12x + 8$ (C) $6x^2 + 12x + 4$ (D) $9x^2 + 6x + 4$ (E) $9x^2 + 12x + 4$

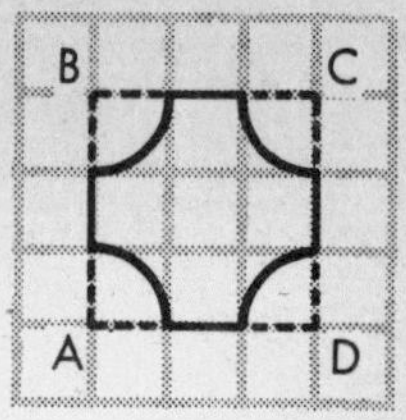

25. In the figure at the right, $ABCD$ is a square of side 9. Find the total length of the arcs and the full lines in the figure.
(A) 6π (B) $6\pi + 6$ (C) $6\pi + 12$ (D) $3\pi + 6$ (E) $12\pi + 12$

26. Through how many degrees does the minute hand move in 24 minutes?
(A) 12 (B) 36 (C) 72 (D) 144 (E) 288

27. If the point $(k, 2k)$ lies on the graph of the equation $2x + y = 8$, then k is
(A) 1 (B) $\frac{8}{5}$ (C) 2 (D) 4 (E) 8

28. The square of an integer has 9 digits. How many digits does the integer have?
(A) 3 (B) 4 (C) 5 (D) 6 (E) 7

29. Find all the values of x for which $8 - 3x \leqq 4 + x$.
(A) $x > 1$ (B) $x \geqq 1$ (C) $x \leqq 1$ (D) $x \geqq 2$ (E) $x \geqq 3$

30. *Quantitative Comparison Problem:* Compare the quantity in Column *A* with the quantity in Column *B*. Choose answer

(A) if the quantity in Column *A* is the greater;
(B) if the quantity in Column *B* is the greater;
(C) if the quantities are equal;
(D) if the relation cannot be determined from the information given.

Column A		*Column B*
	Given: $x + 2y = z$ $z = 3y - 5$	
y		$x + 3$

IV

1. An 8-ounce can of orange juice costs 20¢, and a 20-ounce can costs 45¢. How much cheaper is it to buy 15 pints of juice in cans of the more economical size? (1 liquid pint = 16 ounces.)
(A) \$.54 (B) \$.60 (C) \$.75 (D) \$5.40 (E) \$6.00

2. 4.2 is $83\frac{1}{3}$% of
(A) 3.5 (B) 5.04 (C) 35 (D) 50.4 (E) 3.675

3. A dealer mixed 20 pounds of tea costing 75¢ a pound with 16 pounds of tea costing \$1.15 a pound. If the entire mixture was sold at \$1.25 per pound, what was his profit?
(A) \$1.16 (B) \$10.80 (C) \$11.60 (D) \$12.50 (E) \$18.00

4. To the nearest tenth of a mile, how many nautical miles are there in 20 statute miles? (1 nautical mile = 1.15 statute miles.)
(A) 16.9 (B) 17.0 (C) 17.3 (D) 17.4 (E) 23.0

5. A bushel of tomatoes may range in weight from 40 to 45 pounds, while the weight of a bushel of potatoes may range from 45 to 55 pounds. Find the maximum and minimum ratios of the weight of a bushel of tomatoes to that of a bushel of potatoes.
(A) $\frac{9}{11}, \frac{8}{9}$ (B) $\frac{1}{1}, \frac{8}{11}$ (C) $\frac{8}{11}, \frac{1}{1}$ (D) $\frac{8}{9}, \frac{9}{11}$ (E) $\frac{1}{1}, \frac{8}{9}$

6. The product of $-2a^3$ and $-6a^4$ is
(A) $-8a^7$ (B) $12a^{12}$ (C) $-8a^{12}$ (D) $12a^7$ (E) $-12a^7$

7. Evaluate $5x - 2(x - 5)$ when $x = 8$.
(A) 96 (B) 19 (C) 34 (D) 90 (E) 114

8. Subtract $x - y - 2$ from $2y - 4x - 5$.
(A) $x - 3y - 3$ (B) $-5x + 3y - 3$ (C) $5x - 3y + 3$ (D) $-3x + y - 7$ (E) $y - 3x - 3$

9. Factor: $x^2 - 11x - 12$
(A) $(x + 12)(x - 1)$ (B) $(x - 4)(x + 3)$ (C) $(x - 12)(x + 1)$ (D) $(x - 6)(x - 2)$
(E) $(x + 4)(x - 3)$

10. Solve for y: $9(y - 5) = 2y + 11$

(A) -8 (B) $3\frac{1}{4}$ (C) $5\frac{1}{2}$ (D) 7 (E) 8

11. Solve for d: $\frac{3}{2}d + 4 = 16$

(A) 18 (B) 12 (C) 16 (D) 8 (E) 30

12. The sum of \$4000 is invested, part at 5% and the rest at 6%. How much is invested at 6% if the total annual interest is \$225?
(A) \$1500 (B) \$2000 (C) \$2500 (D) \$2250 (E) \$2400

13. Find the average speed in miles per hour required to fly 420 miles in 1 hour and 20 minutes.
(A) 310 (B) 315 (C) 325 (D) 330 (E) 350

14. A fraction is equal to $\frac{1}{2}$. If its numerator is increased by 1 and its denominator is increased by 5, the value of the resulting fraction is $\frac{4}{9}$. Find the original fraction.

(A) $\frac{9}{18}$ (B) $\frac{11}{22}$ (C) $\frac{6}{12}$ (D) $\frac{7}{13}$ (E) $\frac{15}{31}$

15. The tens digit of a two-digit number is 1 more than the units digit. If the digits are reversed, the new number is five times the sum of the digits. Find the original number.
(A) 87 (B) 45 (C) 56 (D) 54 (E) 65

16. Find the number of hours required by two typists working together to complete a report, if one can do the report alone in 8 hours and the other requires 12 hours.

(A) 4 (B) $4\frac{2}{5}$ (C) $4\frac{4}{5}$ (D) $5\frac{4}{5}$ (E) $6\frac{1}{5}$

17. A team won five-eighths of all the games it played. The ratio of the games lost to the games won is

(A) $\frac{3}{13}$ (B) $\frac{3}{5}$ (C) $\frac{3}{8}$ (D) $\frac{8}{3}$ (E) $\frac{5}{3}$

18. Two chemicals are mixed to form 20 quarts of a solution. If the volumes of the chemicals in the solution are in the ratio 1:4, find the number of quarts in the greater volume.
(A) 4 (B) 5 (C) 12 (D) 15 (E) 16

19. If the angles of a triangle are in the ratio 3:4:7, the triangle is
(A) an obtuse triangle (B) a right triangle (C) an acute triangle that has no equal angles (D) an acute triangle that has two equal angles (E) an equiangular triangle

20. AB is parallel to CD. Find x.
(A) 28 (B) 40 (C) 64 (D) 84 (E) 96

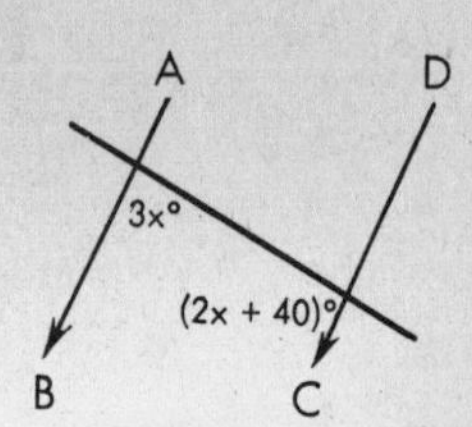

21. DE is parallel to AC. Find x.
(A) 10 (B) 9 (C) 4 (D) 6 (E) 8

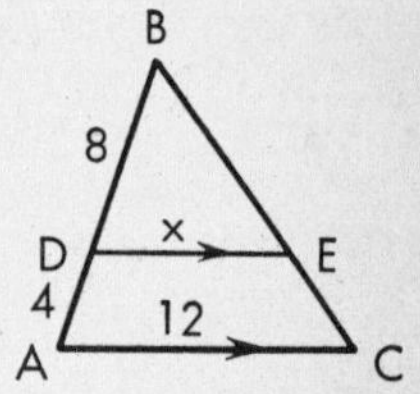

22. The legs of a right triangle are 3 and $3\sqrt{3}$. The hypotenuse is
(A) $\frac{6}{\sqrt{3}}$ (B) 6 (C) $6\sqrt{3}$ (D) 9 (E) 12

23. Two sides of a triangle are 12 and 13 and include an angle of 30°. Find the area of the triangle.
(A) $32\frac{1}{2}$ (B) 36 (C) 39 (D) $39\sqrt{3}$ (E) 78

24. The hypotenuse of a right triangle is 15 and one of the legs is 12. The area of the triangle is
(A) 48 (B) 54 (C) 72 (D) 78 (E) 108

25. Find the area of a circle that has its center at the origin and passes through the point (2, 3).
(A) 13π (B) $2\pi\sqrt{13}$ (C) $\frac{13}{4}\pi$ (D) 8π (E) 4π

26. The area of the smaller circle of a ring is 16π, and the circumference of the larger circle is 16π. The number of square units in the area of the ring is
(A) 24π (B) 28π (C) 48π (D) 56π (E) 96π

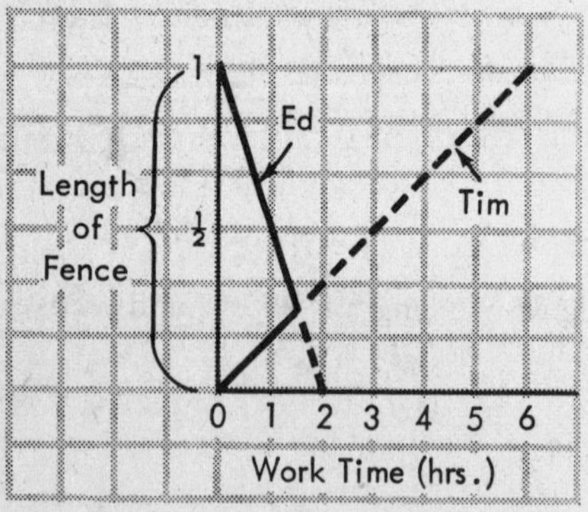

27. Working alone, Ed can paint a fence of a certain size in 2 hours, and Tim can paint it in 6 hours. The graph on the left shows that Ed and Tim painted the fence together, Ed starting at one end of the fence and Tim starting at the other end. How long did it take them to paint the whole fence, and what part of the work was done by Ed?
(A) $1\frac{1}{2}$ hrs., $\frac{3}{4}$ (B) $1\frac{1}{2}$ hrs., $\frac{1}{4}$ (C) $1\frac{3}{4}$ hrs., $\frac{3}{4}$
(D) $1\frac{3}{4}$ hrs., $\frac{1}{4}$ (E) $1\frac{1}{4}$ hrs., $\frac{3}{4}$

28. The graphs of $x = 2$ and $x + y = 4$ are shown in the first and fourth quadrants. Designate by its letter the section of the plane that contains all points in quadrants I and IV whose coordinates satisfy the following set of inequalities:

$$x > 2$$
$$x + y < 4$$
$$y > 0$$

(A) A (B) B (C) C (D) D (E) E (F) F

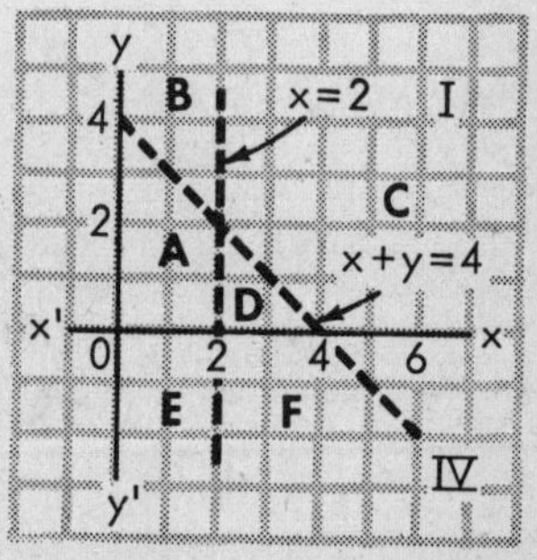

29. The equality $\frac{1}{7} \times 11 \times 28 = \frac{1}{7} \times 28 \times 11$ is an illustration of the

(A) associative law of addition (B) associative law of multiplication
(C) commutative law of addition (D) commutative law of multiplication
(E) distributive law

30. *Quantitative Comparison Problem:* Compare the quantity in Column *A* with the quantity in Column *B*. Choose answer

(A) if the quantity in Column *A* is the greater;
(B) if the quantity in Column *B* is the greater;
(C) if the quantities are equal;
(D) if the relation cannot be determined from the information given.

Column A	*Column B*
Given: *x*, *y*, and *z* are consecutive integers	
$x + z$	$2y$

V

1. Arrange the following fractions in order of size beginning with the smallest: $\frac{1}{2}, \frac{5}{9}, \frac{7}{12}, \frac{5}{8}$

(A) $\frac{1}{2}, \frac{5}{9}, \frac{5}{8}, \frac{7}{12}$ (B) $\frac{1}{2}, \frac{7}{12}, \frac{5}{9}, \frac{5}{8}$ (C) $\frac{1}{2}, \frac{5}{8}, \frac{5}{9}, \frac{7}{12}$ (D) $\frac{1}{2}, \frac{5}{8}, \frac{7}{12}, \frac{5}{9}$ (E) $\frac{1}{2}, \frac{5}{9}, \frac{7}{12}, \frac{5}{8}$

2. A man worked 60 hours one week. What did he earn that week if his pay was \$2.40 an hour for the first 40 hours and 50% more for each hour thereafter?
(A) \$216 (B) \$144 (C) \$168 (D) \$180 (E) \$192

3. Of the numbers $\frac{3}{5}, \frac{16}{25}, \frac{5}{8}$, .627, and $62\frac{2}{3}\%$, the largest is

(A) $\frac{3}{5}$ (B) $\frac{16}{25}$ (C) $\frac{5}{8}$ (D) .627 (E) $62\frac{2}{3}\%$

4. A salesman earns a commission of 6% on all sales between \$300 and \$500, and 10% on all sales over \$500. If his sales were \$1440, how much did he earn?
(A) \$39.40 (B) \$91.20 (C) \$106 (D) \$124 (E) \$10.60

5. The expression $(3x^3)^3$ is equivalent to
(A) $9x^9$ (B) $27x^6$ (C) $9x^{27}$ (D) $9x^6$ (E) $27x^9$

6. Write an expression for the area of a rectangle whose length is $3x + 2$ and whose width is $5x - 1$.
(A) $15x^2 - 7x - 2$ (B) $8x^2 + 7x - 2$ (C) $15x^2 - 2$ (D) $15x^2 + 7x - 2$ (E) $16x + 2$

7. Factor: $x^2 - 16x + 64$
(A) $(x - 32)^2$ (B) $(x - 8)^2$ (C) $(x - 16)(x - 4)$ (D) $(x - 32)(x - 2)$
(E) $(x - 64)(x - 1)$

8. Multiply: $\frac{3b}{a} \cdot \frac{c}{9b} \cdot \frac{3a}{4c}$

(A) $\frac{1}{4}$ (B) $\frac{a}{4}$ (C) $\frac{a}{4c}$ (D) $\frac{b}{4}$ (E) $\frac{c}{8}$

9. Solve for x: $4x + 2.7 = 0.6 - 2x$
(A) .35 (B) −.35 (C) −1.05 (D) 1.05 (E) 3.5

10. A pair of skis was sold for \$24, which was 75% of the original price. What was the original price?

(A) \$18 (B) \$30 (C) \$36 (D) \$32 (E) \$27.50

11. If $\frac{a+b}{b}=\frac{9}{5}$, find the value of $\frac{a}{b}$.

(A) $\frac{5}{4}$ (B) $\frac{4}{5}$ (C) $\frac{14}{5}$ (D) $\frac{5}{14}$ (E) $-\frac{4}{5}$

12. The difference between two numbers is 12. If the smaller number is increased by 8 times the larger, the result is 6. Find the numbers.

(A) 10, 22 (B) 2, −10 (C) 14, 2 (D) 4, −8 (E) 6, −6

13. Find three consecutive odd integers such that the sum of the first two is 7 more than the last.

(A) 9, 10, 11 (B) −9, −7, −5 (C) 7, 9, 11 (D) 9, 11, 13 (E) 5, 7, 9

14. The average of two consecutive integers is $\frac{21}{2}$. Find the larger of these two integers.

(A) $11\frac{1}{2}$ (B) 10 (C) 11 (D) 12 (E) 21

15. A wholesale merchant wishes to blend two kinds of tea, one priced at 80¢ a pound and the other at 90¢ a pound. How many pounds of each should he use to make 250 pounds of a mixture to sell at 84¢ a pound?

(A) 130 lb. @ 80¢ / 120 lb. @ 90¢ (B) 130 lb. @ 90¢ / 120 lb. @ 80¢ (C) 150 lb. @ 80¢ / 100 lb. @ 90¢
(D) 150 lb. @ 90¢ / 100 lb. @ 80¢ (E) 144 lb. @ 80¢ / 106 lb. @ 90¢

16. A sum of \$8000 is invested in two parts, the first part at 4% and the second at 5%. The annual interest on the first part is \$50 more than that on the second. How much was invested at 5%?

(A) \$2850 (B) \$2900 (C) \$3200 (D) \$5000 (E) \$3000

17. A legacy of \$1435 is to be divided among three heirs in the ratio 1:2:4. Find the number of dollars in the largest share.

(A) 205 (B) 615 (C) 715 (D) 820 (E) 1025

18. In the graph on the right, each x-unit equals 1 and each y-unit equals 5. Find the area of the triangle.

(A) 56 (B) 70 (C) 80 (D) 140 (E) 350

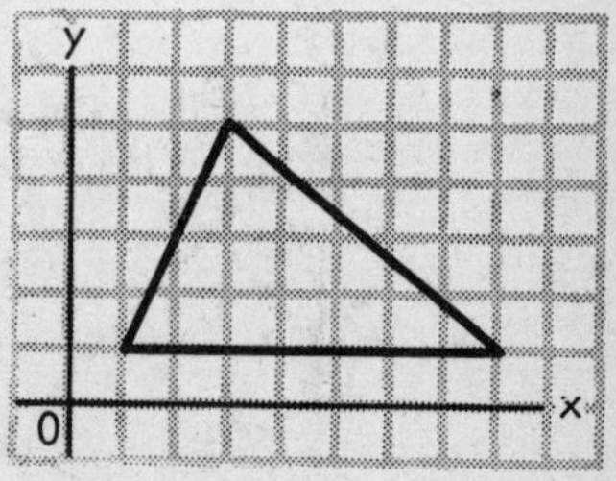

19. Find x: $\frac{x+3}{2x}=\frac{4}{5}$

(A) −5 (B) 3 (C) $3\frac{1}{2}$ (D) 5 (E) 6

20. In triangle ABC, $AB = BC$. Express x in terms of a.

(A) $180-2a$ (B) $\frac{a-180}{2}$ (C) $2(a-180)$ (D) $\frac{a+180}{2}$
(E) $2a-90$

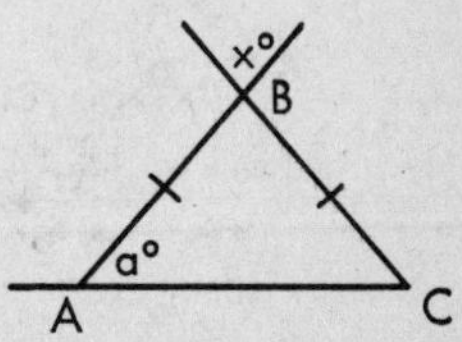

21. If the hypotenuse of a right triangle is $5\sqrt{5}$ and one leg is 5, then the other leg is
(A) 5 (B) 7.5 (C) 10 (D) $5\sqrt{6}$ (E) $10\sqrt{5}$

22. Three numbers that may represent the lengths of the sides of a right triangle are
(A) 6, 7, 9 (B) 6, 8, 10 (C) 6, 8, 9 (D) 7, 9, 10 (E) 7, 8, 10

23. If the perimeter of an equilateral triangle is 30, than its area is
(A) 25 (B) $25\sqrt{2}$ (C) $25\sqrt{3}$ (D) $50\sqrt{3}$ (E) $225\sqrt{3}$

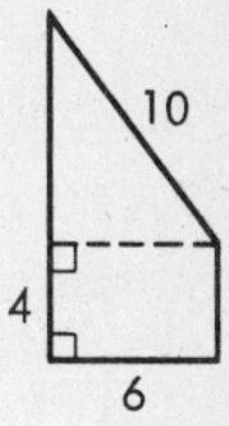

24. The area of the entire figure is
(A) 24 (B) 36 (C) 48 (D) 60 (E) 64

25. A destroyer patrols a circular area whose radius is 3 miles. If the destroyer must increase the area of its patrol duty to one whose radius is twice as great, how many more square miles must it patrol?
(A) 9π (B) 12π (C) 18π (D) 27π (E) 36π

26. The area of a rectangle equals the area of a circle whose radius is 8. If the base of the rectangle is 16, then its altitude is
(A) π (B) 2π (C) 4π (D) $\frac{4}{\pi}$ (E) 16π

27. Towns A and B are 40 miles apart. The accompanying graph shows that Joe started from A at noon, reached town B at 4 P.M., and started back at 4:30 P.M. Harry started from B at 3 P.M. and arrived at A at 5 P.M. At what time did Harry meet Joe, and how far were they from town B at that time?

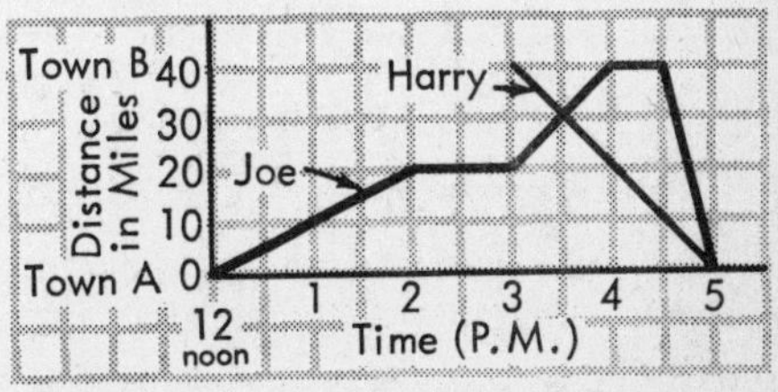

(A) 3:20 P.M., 30 mi. (B) 3:20 P.M., 10 mi. (C) 4 P.M., 30 mi. (D) 3:30 P.M., 30 mi. (E) 3:30 P.M., 10 mi.

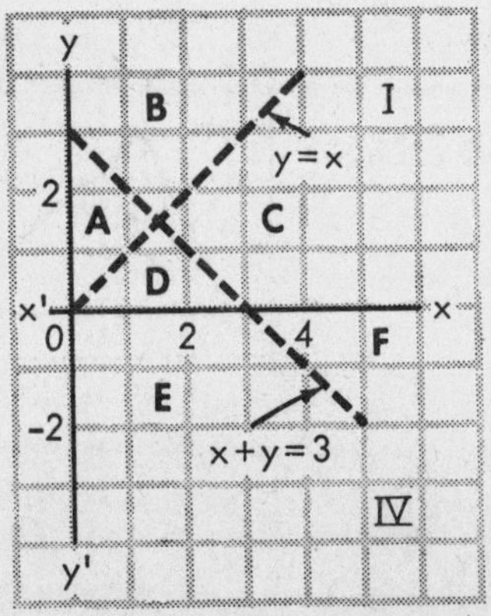

28. The graphs of $y = x$ and $x + y = 3$ are shown in the first and fourth quadrants. Designate by its letter the section of the plane that contains all points in quadrants I and IV whose coordinates satisfy the following set of inequalities:

$$x + y < 3$$
$$y < x$$
$$y < 0$$

(A) A (B) B (C) C (D) D (E) E (F) F

29. Given: $A = \{3, 5, 7, 9, 10\}$ and $B = \{1, 7, 9\}$. Find $A \cup B$.
(A) {7, 9} (B) {1, 3, 5, 7, 9} (C) {9, 10} (D) {1, 9} (E) {1, 3, 5, 7, 9, 10}

30. *Quantitative Comparison Problem:* Compare the quantity in Column *A* with the quantity in Column *B*. Choose answer

(A) if the quantity in Column *A* is the greater;
(B) if the quantity in Column *B* is the greater;
(C) if the quantities are equal;
(D) if the relation cannot be determined from the information given.

Column A		*Column B*
	Given: $x \neq 0$	
$\dfrac{x^3 + x^2}{x}$		$\dfrac{x^3 - x^2}{x}$

VI

1. Of a group of 1200 people, $\frac{1}{6}$ are men and $\frac{1}{4}$ of the women are not married. What fraction of the entire group are married women?

(A) $\frac{3}{8}$ (B) $\frac{5}{24}$ (C) $\frac{1}{2}$ (D) $\frac{5}{8}$ (E) $\frac{3}{4}$

2. At a school game, receipts amounted to \$425. If 400 tickets were sold to students at 75¢ each, how many tickets were sold to adults at \$1.25 each?

(A) 100 (B) 125 (C) 425 (D) 500 (E) 525

3. If a watch is set at 2 P.M. and loses $1\frac{1}{2}$ minutes every 2 hours, what will the watch read at 2 P.M. the following day?

(A) 1:42 (B) 1:51 (C) 1:82 (D) 2:09 (E) 2:18

4. A school contains 20 evening classes with enrollments that range from 25 to 35 students. On certain evenings, the absentee rate is 20%. What are the maximum and minimum numbers of students that attend classes on such evenings?

(A) 560, 500 (B) 700, 400 (C) 400, 360 (D) 560, 400 (E) 700, 560

5. Represent the number of days in y weeks and x hours.

(A) $\frac{y}{7} + 24x$ (B) $7y + \frac{x}{24}$ (C) $7y + 24x$ (D) $7x + \frac{y}{24}$ (E) $\frac{x}{7} + 24y$

6. The expression $(2 - 5x)^2$ is equal to

(A) $4 - 20x + 25x^2$ (B) $4 + 25x^2$ (C) $4 - 25x^2$ (D) $4 - 10x + 25x^2$
(E) $4 + 20x - 25x^2$

7. Find the value of $(a + c)(a + b)$ when $a = -2$, $b = -3$, and $c = 1$.

(A) 5 (B) -12 (C) -5 (D) -1 (E) -7

8. Divide: $\dfrac{a - b}{a + b} \div \dfrac{b - a}{b + a}$

(A) $\dfrac{b - a}{b + a}$ (B) $\dfrac{b + a}{b - a}$ (C) -1 (D) $\dfrac{a + b}{a - b}$ (E) $\dfrac{a - b}{a + b}$

9. If $\frac{x}{2} = 3y$, find the value of $\frac{x}{y}$.

(A) $\frac{3}{2}$ (B) $\frac{2}{3}$ (C) $\frac{1}{6}$ (D) 6 (E) $-\frac{3}{2}$

10. Solve for x: $\frac{a}{b} = \frac{x}{c}$

(A) $\frac{ab}{c}$ (B) $\frac{ac}{b}$ (C) $\frac{bc}{a}$ (D) $\frac{a}{bc}$ (E) $\frac{b}{ac}$

11. A traveler made two journeys. On the second journey, he traveled twice as fast and took $\frac{3}{2}$ as much time to reach his destination as he had done on the first journey. How many times greater was the distance traveled on the second journey than that covered on the first?

(A) $1\frac{1}{3}$ (B) 2 (C) $2\frac{1}{2}$ (D) $3\frac{1}{2}$ (E) 3

12. If the average of three numbers is $2x + 1$, and two of the numbers are x and $x + 4$, what is the third number?

(A) $2x - 2$ (B) $2x + 2$ (C) $4x - 2$ (D) $4x + 2$ (E) $4x - 1$

13. If $\frac{1}{4}s - 1\frac{1}{2}$ represents the side of a square, then an expression for the perimeter of the square is

(A) $s - 6$ (B) $\frac{1}{2}s - 3$ (C) $2s - 12$ (D) $2s - 6$ (E) $s - 3$

14. Mr. Mayer has $3200 invested at 6%. How much money must he invest at 3% so that his total annual income will be 4% of his two investments?

(A) $6000 (B) $6400 (C) $6500 (D) $6750 (E) $7000

15. On a trip, Mr. Henry traveled for 4 hours at 65 mph. Returning over the same route, he took 20 minutes longer. How many miles per hour was his average speed returning?

(A) 45 (B) 50 (C) 54 (D) 55 (E) 60

16. The numerator of a certain fraction is 6 less than the denominator. If the numerator is tripled and the denominator is increased by 7, the value of the resulting fraction is $\frac{3}{2}$. Find the original fraction.

(A) $\frac{12}{18}$ (B) $\frac{26}{39}$ (C) $\frac{15}{21}$ (D) $\frac{14}{20}$ (E) $\frac{13}{19}$

17. A recipe requires 3 parts of sugar to 10 parts of flour. If 15 ounces of flour are to be used, how many ounces of sugar are needed?

(A) 3 (B) $3\frac{1}{3}$ (C) 4 (D) $4\frac{1}{2}$ (E) 5

18. Solve for r: $9 : r + 1 = 12 : 8$

(A) 4 (B) 5 (C) 6 (D) 7 (E) 7.5

19. When a ray of light is directed at a mirror, the original direction of the ray and the new direction in which it is reflected make equal angles with the surface of the mirror.

In the diagram, a beam of light directed at mirror A is reflected successively from mirrors A and B. The light beam strikes the surface of A at an angle of $x°$. Find x in terms of a.

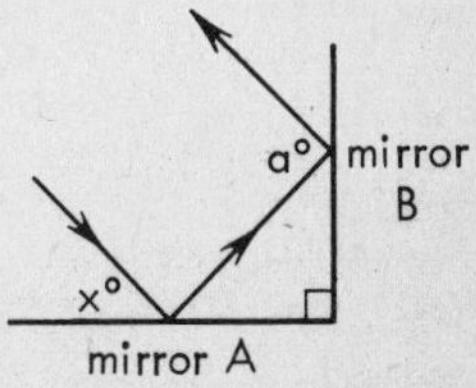

(A) $90 - \frac{a}{2}$ (B) $\frac{90 - a}{2}$ (C) $90 - a$ (D) $\frac{a}{2}$ (E) $90 + \frac{a}{2}$

20. The legs of a right triangle are in the ratio 3:4. If the longer leg is 24, then the hypotenuse is

(A) 27 (B) 30 (C) 32 (D) 35 (E) 36

21. A leg and the hypotenuse of a right triangle are in the ratio 5:13. If the hypotenuse is 91, then the other leg is

(A) 70 (B) 56 (C) 35 (D) 12 (E) 84

22. The perimeter of a rhombus is 52 and its longer diagonal is 24. The area of the rhombus is

(A) 60 (B) 78 (C) 120 (D) 130 (E) 240

23. The coordinates of the vertices of parallelogram $ABCD$ are $A(0, 2)$, $B(2, 5)$, $C(10, 5)$, and $D(8, 2)$. The area of the parallelogram is

(A) 12 (B) 24 (C) 36 (D) 40 (E) 48

24. A square and a triangle have equal areas. If the side of the square and the base of the triangle are each 8, then the altitude of the triangle is

(A) 8 (B) 12 (C) 16 (D) 24 (E) 32

25. The central angle of a sector is 40° and the area of the sector is 16π. The area of the circle is

(A) 36π (B) 64π (C) 72π (D) 144π (E) 288π

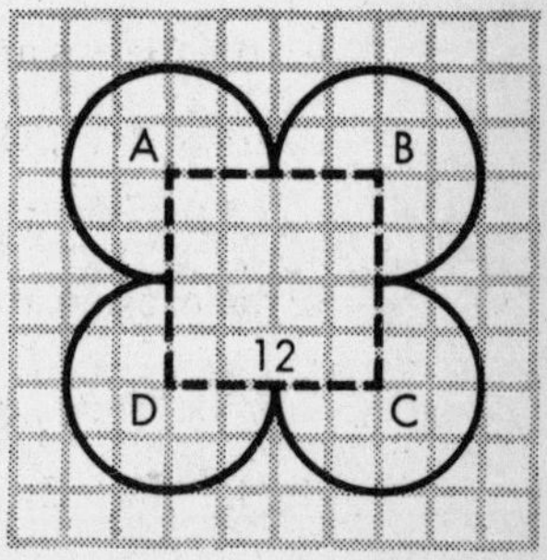

26. The side of square $ABCD$ is 12. The area of the entire figure is

(A) 108π (B) $108\pi + 144$ (C) $27\pi + 36$ (D) $27\pi + 144$
(E) 252π

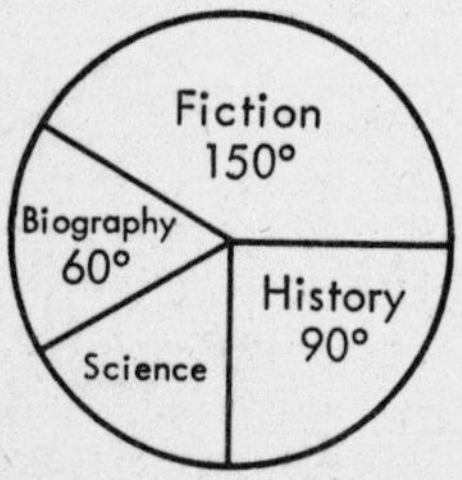

27. The accompanying circle graph shows the distribution of books in a school library. If the library contains 350 science books, what is the total number of books in the library?

(A) 1400 (B) 2000 (C) 2100 (D) 2400 (E) 4200

28. In the interval beginning with 11 and ending with 57, find the set of integers divisible by 3 and 5 or by 3 and 7.

(A) {15, 30, 45} (B) {21, 42} (C) {21, 35, 42, 45}
(D) {15, 21, 35, 42, 45} (E) {15, 21, 30, 42, 45}

29. The fraction $\dfrac{2x-12}{2x+12}$ is meaningless when x equals

(A) $\frac{1}{6}$ (B) -6 (C) 6 (D) -10 (E) 10

30. *Data-Sufficiency Problem:* In quadrilateral $ABCD$, is diagonal BD perpendicular to diagonal AC?

(1) $AB = AD$, and $\triangle BCD$ is an equilateral triangle.
(2) AC bisects $\angle A$ and $\angle C$.

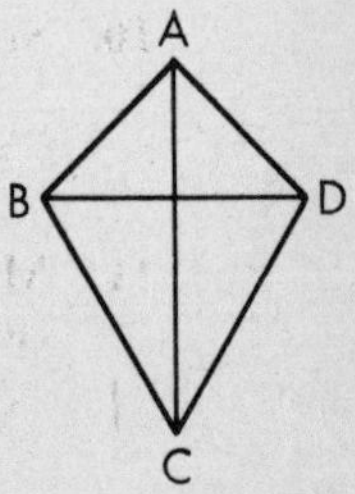

To find the answer to the preceding question:

(A) Only statement (1) can be used.
(B) Only statement (2) can be used.
(C) Both statements are needed.
(D) Either statement can be used.
(E) Both statements together would be insufficient.

VII

1. A red light at one corner goes on every 3 minutes, while another red light at the next corner goes on every 4 minutes. How many minutes are the intervals between the times that they go on simultaneously?

(A) $\frac{3}{4}$ (B) $1\frac{1}{3}$ (C) 7 (D) 12 (E) 24

2. Two classes of equal size have a total enrollment of 72. On a certain day, $\frac{3}{4}$ of one class and $\frac{5}{6}$ of the other are present. How many students are absent in both classes?

(A) 15 (B) 27 (C) 33 (D) 39 (E) 57

3. How many pairs of stockings were rejected in a shipment of 8500 pairs if .6% of them were returned to the manufacturer because of defects?

(A) 5 (B) 50 (C) 51 (D) 510 (E) 5100

4. A family's annual income is $10,000. If 35% of this income is spent for food and $27\frac{1}{2}$% for rent, how many dollars remain for other purposes?

(A) 37.5 (B) 375 (C) 400 (D) 3750 (E) 6250

5. Represent the number of dimes equivalent in value to h half-dollars, n nickels, and c cents.

(A) $5h+\frac{n}{2}+\frac{c}{10}$ (B) $\frac{h}{5}+2n+10c$ (C) $50h+5n+c$ (D) $10h+n+\frac{c}{5}$
(E) $5h+2n+10c$

6. If $x^2+y^2=36$ and $xy=6$, find $(x+y)^2$.

(A) 48 (B) 42 (C) 24 (D) 36 (E) 30

7. Simplify: $\dfrac{\frac{1}{2}+\frac{1}{4}}{\frac{1}{3}-\frac{1}{5}}$

(A) $\frac{1}{10}$ (B) $\frac{45}{8}$ (C) $\frac{45}{16}$ (D) $\frac{15}{8}$ (E) $\frac{15}{16}$

8. If 3 is a root of the equation $x^2+5x+k=0$, then the value of k is

(A) -18 (B) 16 (C) -16 (D) 24 (E) -24

9. Solve for x: $4(1-x)+3(x-2)=10$

(A) 12 (B) -12 (C) 8 (D) -8 (E) $\frac{-12}{7}$

10. Solve for n: $\frac{n}{3}-\frac{n}{5}=4$

(A) 15 (B) -15 (C) 16 (D) -30 (E) 30

11. Mr. Dodge drove for 5 hours at an average speed of 35 mph and for 7 hours at an average speed of 47 mph. Find his average speed for the entire trip.

(A) 39 mph (B) 42 mph (C) 40 mph (D) 44 mph (E) 45 mph

12. If $\frac{3s}{2}-5$ represents the side of a regular octagon, find an expression for the perimeter.

(A) $9s-5$ (B) $9s-30$ (C) $9s-40$ (D) $12s-5$ (E) $12s-40$

13. How many quarters must be added to 24 nickels so that the average value of each coin in the collection is 10¢?

(A) 9 (B) 10 (C) 12 (D) 16 (E) 8

14. How many feet per second are traveled at 30 mph?

(A) 22 (B) 33 (C) 44 (D) 66 (E) 88

15. How many minutes does it take to fly 700 miles at 600 mph?

(A) 35 (B) 56 (C) 63 (D) 140 (E) 70

16. In the proportion $\frac{a}{b} = \frac{c+3}{2d}$, find a if $b = 6$, $c = 5$, and $d = 8$.

(A) 2 (B) 2.5 (C) 3 (D) 4 (E) 12

17. Using a scale of $\frac{1}{4}$ inch to 10 yards, how many inches long should a line on a blueprint be if the line is to represent a distance of 300 feet?

(A) $2\frac{1}{2}$ (B) $7\frac{1}{2}$ (C) 15 (D) 30 (E) 40

18. If n books cost d dollars, write an expression that represents the cost in dollars of 30 books.

(A) $\frac{30d}{n}$ (B) $\frac{30n}{d}$ (C) $\frac{n}{30d}$ (D) $\frac{d}{30n}$ (E) $30nd$

19. AB is parallel to CD. Find x.

(A) 25 (B) 50 (C) 55 (D) 75 (E) 125

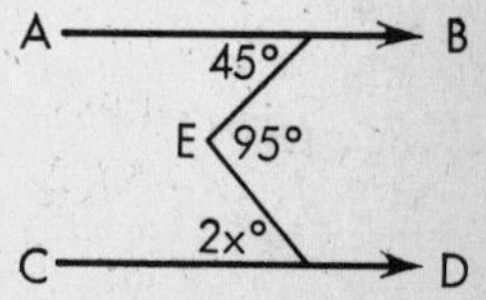

20. $\widehat{AB}$ is one-fifth of the circumference. Find x.

(A) 18 (B) 36 (C) 54 (D) 72 (E) 108

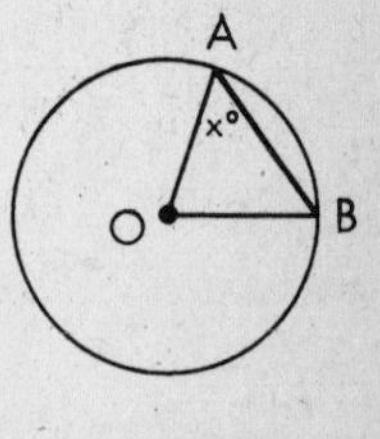

21. Two boats leave the same dock at the same time, one sailing due south at the rate of 6 mph, the other due west at the rate of 8 mph. At the end of 4 hours, the number of miles between them will be

(A) 56 (B) 48 (C) 40 (D) 32 (E) 24

22. The sides of a rectangular field are 25 feet and 60 feet. If a man walking at the rate of 4 feet per second takes the diagonal short-cut across the field instead of walking along the two sides, the number of seconds he will save is

(A) 3 (B) 3.5 (C) 4 (D) 4.5 (E) 5

23. The vertices of the shaded triangle are midpoints of the sides of the rectangle. Find the ratio of the area of the shaded triangle to the area of the rectangle.

(A) $\frac{1}{8}$ (B) $\frac{3}{8}$ (C) $\frac{1}{3}$ (D) $\frac{1}{4}$ (E) $\frac{1}{2}$

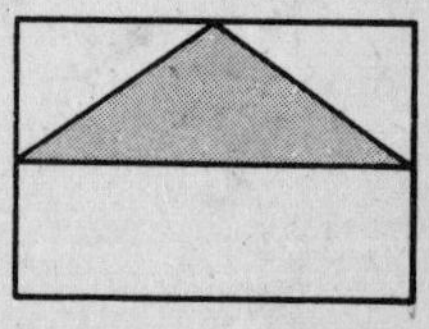

24. The bases of a trapezoid are 8 and 12. If the area of the trapezoid is 50, then the length of the altitude is

(A) $2\frac{1}{2}$ (B) 4 (C) 5 (D) 8 (E) 10

25. How many square yards of terrain can be swept by a light beam 20 yards long, if the beam is thrown by a searchlight mounted on a swivel that can be turned through an angle of 90°?

(A) 20π (B) 80π (C) 100π (D) 200π (E) 400π

26. Two semicircles are cut from rectangle *ABCD*, whose dimensions are 8 and 10. The area of the remaining part (shaded) is

(A) $80 - 8\pi$ (B) $80 - 16\pi$ (C) $80 - 64\pi$ (D) $40 - 16\pi$
(E) 64π

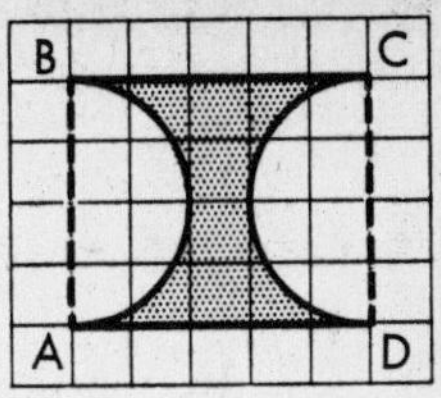

27. Find the missing value in the table of values:

(A) 20 (B) 24 (C) 32 (D) 64 (E) 128

y	1	4	7	10	16
x	2	4	8	16	?

28. Find the sum of 13_4, 20_5, and 21_6 to the base 10.

(A) 28 (B) 30 (C) 33 (D) 45 (E) 60

29. On the clock scale at the right, "+" indicates a clockwise motion and "−" indicates a counterclockwise motion. On this scale, find the sum of $(3 + 2)$ and $(2 - 3)$.

(A) 2 (B) 3 (C) 4 (D) 5 (E) 6

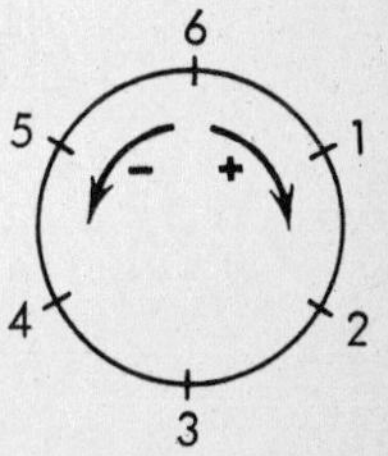

30. *Data-Sufficiency Problem:* What is the number of degrees in the smallest angle of the triangle?

(1) The angles of the triangle are in the ratio of 1:3:5.

(2) The triangle is an obtuse triangle, and the smallest angle is 80° less than the largest angle.

To find the answer to the preceding question:

(A) Only statement (1) can be used.
(B) Only statement (2) can be used.
(C) Both statements are needed.
(D) Either statement can be used.
(E) Both statements together would be insufficient.

ANSWER KEY

PREVIEW

1	*Page 1:*	**1.** *c*	**2.** *e*	**3.** *b*	**4.** *d*	**5.** *a*
2	*Page 2:*	**1.** *b*	**2.** *d*	**3.** *e*	**4.** *c*	**5.** *a*
3	*Page 3:*	**1.** *d*	**2.** *e*	**3.** *b*	**4.** *c*	**5.** *a*
4	*Page 4:*	**1.** *d*	**2.** *b*	**3.** *a*	**4.** *e*	**5.** *c*
5	*Page 7:*	**1.** *e*	**2.** *b*	**3.** *d*	**4.** *a*	**5.** *c*
6	*Page 8:*	**1.** *b*	**2.** *e*	**3.** *c*	**4.** *d*	**5.** *a*
7	*Page 9:*	**1.** *d*	**2.** *a*	**3.** *b*	**4.** *e*	**5.** *c*
8	*Page 9:*	**1.** *b*	**2.** *e*	**3.** *a*	**4.** *c*	**5.** *d*
9	*Page 10:*	**1.** *c*	**2.** *b*	**3.** *d*	**4.** *a*	**5.** *e*
10	*Page 11:*	**1.** E	**2.** C	**3.** A	**4.** B	**5.** D
11	*Page 12:*	**1.** C	**2.** A	**3.** B	**4.** D	

CHAPTER I

1.1	*Page 14:*	**1.** *b*	**2.** *d*	**3.** *e*	**4.** *a*	**5.** *c*
1.2	*Page 15:*	**1.** *d*	**2.** *b*	**3.** *c*	**4.** *e*	**5.** *a*
1.3	*Page 16:*	**1.** *e*	**2.** *a*	**3.** *b*	**4.** *c*	**5.** *d*
1.4	*Page 16:*	**1.** *e*	**2.** *c*	**3.** *b*	**4.** *a*	**5.** *d*
1.5	*Page 17:*	**1.** *d*	**2.** *b*	**3.** *e*	**4.** *c*	**5.** *a*
2.1	*Page 21:*	**1.** *b*	**2.** *d*	**3.** *e*	**4.** *c*	**5.** *a*
2.2	*Page 21:*	**1.** *c*	**2.** *a*	**3.** *e*	**4.** *d*	**5.** *b*
2.3	*Page 22:*	**1.** *e*	**2.** *c*	**3.** *b*	**4.** *a*	**5.** *d*
2.4	*Page 23:*	**1.** *d*	**2.** *b*	**3.** *a*	**4.** *c*	**5.** *e*
2.5	*Page 24:*	**1.** *c*	**2.** *e*	**3.** *a*	**4.** *b*	**5.** *d*
2.6	*Page 25:*	**1.** *c*	**2.** *a*	**3.** *d*	**4.** *e*	**5.** *b*
2.7	*Page 26:*	**1.** C	**2.** B	**3.** C	**4.** D	**5.** D
3.1	*Page 28:*	**1.** *d*	**2.** *e*	**3.** *c*	**4.** *b*	**5.** *a*
3.2	*Page 29:*	**1.** *b*	**2.** *a*	**3.** *e*	**4.** *c*	**5.** *d*
3.3	*Page 29:*	**1.** *d*	**2.** *b*	**3.** *c*	**4.** *e*	**5.** *a*
3.4	*Page 30:*	**1.** *b*	**2.** *d*	**3.** *e*	**4.** *a*	**5.** *c*
3.5	*Page 31:*	**1.** *b*	**2.** *d*	**3.** *c*	**4.** *e*	**5.** *a*
3.6	*Page 31:*	**1.** *c*	**2.** *e*	**3.** *a*	**4.** *b*	**5.** *d*
3.7	*Page 32:*	**1.** *e*	**2.** *a*	**3.** *d*	**4.** *c*	**5.** *b*
4.1	*Page 36:*	**1.** *d*	**2.** *b*	**3.** *e*	**4.** *c*	**5.** *a*
4.2	*Page 35:*	**1.** *c*	**2.** *d*	**3.** *b*	**4.** *a*	**5.** *e*
4.3	*Page 36:*	**1.** *b*	**2.** *a*	**3.** *d*	**4.** *e*	**5.** *c*
4.4	*Page 36:*	**1.** *c*	**2.** *e*	**3.** *a*	**4.** *b*	**5.** *d*
4.5	*Page 37:*	**1.** *c*	**2.** *e*	**3.** *a*	**4.** *b*	**5.** *d*
4.6	*Page 38:*					
	Test A:	**1.** *c*	**2.** *a*	**3.** *d*	**4.** *e*	**5.** *b*
	Test B:	**1.** *d*	**2.** *b*	**3.** *c*	**4.** *e*	**5.** *a*
4.7	*Page 39:*	**1.** *d*	**2.** *b*	**3.** *c*	**4.** *e*	**5.** *a*
4.8	*Page 40:*					
	Test A:	**1.** *e*	**2.** *b*	**3.** *d*	**4.** *c*	**5.** *a*
	Test B:	**1.** *c*	**2.** *a*	**3.** *e*	**4.** *b*	**5.** *d*
4.9	*Page 42:*					
	Test A:	**1.** *d*	**2.** *e*	**3.** *b*	**4.** *a*	**5.** *c*
	Test B:	**1.** *e*	**2.** *c*	**3.** *a*	**4.** *b*	**5.** *d*
4.10	*Page 44:*	**1.** *d*	**2.** *b*	**3.** *c*	**4.** *e*	**5.** *a*
4.11	*Page 45:*	**1.** *d*	**2.** *c*	**3.** *a*	**4.** *b*	**5.** *e*
5.1	*Page 45:*	**1.** *c*	**2.** *d*	**3.** *b*	**4.** *a*	**5.** *e*
5.2	*Page 46:*	**1.** *d*	**2.** *b*	**3.** *e*	**4.** *a*	**5.** *c*
5.3	*Page 47:*	**1.** *d*	**2.** *c*	**3.** *a*	**4.** *b*	**5.** *e*
5.4	*Page 48:*	**1.** *d*	**2.** *e*	**3.** *b*	**4.** *a*	**5.** *c*
6.1	*Page 51:*	**1.** *b*	**2.** *d*	**3.** *e*	**4.** *a*	**5.** *c*
6.2	*Page 51:*	**1.** *d*	**2.** *c*	**3.** *a*	**4.** *b*	**5.** *e*
6.3	*Page 52:*	**1.** *d*	**2.** *e*	**3.** *b*	**4.** *c*	**5.** *a*
6.4	*Page 53:*	**1.** *e*	**2.** *b*	**3.** *d*	**4.** *c*	**5.** *a*
6.5	*Page 53:*	**1.** *c*	**2.** *e*	**3.** *a*	**4.** *b*	**5.** *d*
7.1	*Page 54:*	**1.** *e*	**2.** *a*	**3.** *c*	**4.** *d*	**5.** *b*
7.2	*Page 55:*	**1.** *c*	**2.** *d*	**3.** *e*	**4.** *a*	**5.** *b*
7.3	*Page 56:*	**1.** *d*	**2.** *b*	**3.** *e*	**4.** *c*	**5.** *a*
7.4	*Page 57:*	**1.** *c*	**2.** *e*	**3.** *b*	**4.** *a*	**5.** *d*
8	*Page 58:*	**1.** C	**2.** A	**3.** D	**4.** A	**5.** D
		6. B	**7.** B	**8.** B	**9.** B	**10.** D
		11. B	**12.** E	**13.** C	**14.** C	**15.** D
		16. D	**17.** D	**18.** C	**19.** D	**20.** C
		21. C	**22.** B	**23.** D	**24.** B	**25.** B
		26. D	**27.** B	**28.** E	**29.** D	**30.** C
		31. C	**32.** B	**33.** D	**34.** B	**35.** A
		36. C	**37.** D	**38.** D	**39.** B	**40.** C
		41. C	**42.** E	**43.** C	**44.** C	**45.** B
		46. B				

CHAPTER II

1.1	*Page 66:*	**1.** *b*	**2.** *e*	**3.** *c*	**4.** *a*	**5.** *d*
1.2	*Page 66:*	**1.** *a*	**2.** *d*	**3.** *c*	**4.** *e*	**5.** *b*
1.3	*Page 67:*	**1.** *d*	**2.** *a*	**3.** *e*	**4.** *b*	**5.** *c*
1.4	*Page 68:*	**1.** *c*	**2.** *a*	**3.** *e*	**4.** *d*	**5.** *b*
1.5	*Page 68:*	**1.** *b*	**2.** *d*	**3.** *c*	**4.** *a*	**5.** *e*
1.6	*Page 69:*					
	Test A:	**1.** *d*	**2.** *b*	**3.** *a*	**4.** *e*	**5.** *c*
	Test B:	**1.** *e*	**2.** *b*	**3.** *d*	**4.** *c*	**5.** *a*
1.7	*Page 69:*	**1.** *d*	**2.** *a*	**3.** *e*	**4.** *b*	**5.** *c*
1.8	*Page 70:*	**1.** *b*	**2.** *e*	**3.** *d*	**4.** *a*	**5.** *c*
2.1	*Page 71:*	**1.** *d*	**2.** *b*	**3.** *e*	**4.** *a*	**5.** *c*
2.2	*Page 72:*	**1.** *d*	**2.** *b*	**3.** *c*	**4.** *a*	**5.** *e*
2.3	*Page 72:*	**1.** *c*	**2.** *d*	**3.** *e*	**4.** *a*	**5.** *b*
2.4	*Page 73:*	**1.** *c*	**2.** *a*	**3.** *d*	**4.** *b*	**5.** *e*
2.5	*Page 74:*					
	Test A:	**1.** *e*	**2.** *b*	**3.** *d*	**4.** *c*	**5.** *a*
	Test B:	**1.** *e*	**2.** *b*	**3.** *d*	**4.** *c*	**5.** *a*
3.1	*Page 75:*	**1.** *d*	**2.** *a*	**3.** *c*	**4.** *b*	**5.** *e*
3.2	*Page 76:*	**1.** *d*	**2.** *e*	**3.** *c*	**4.** *a*	**5.** *b*
4.1	*Page 77:*	**1.** *c*	**2.** *a*	**3.** *e*	**4.** *b*	**5.** *d*
4.2	*Page 77:*	**1.** *d*	**2.** *c*	**3.** *a*	**4.** *e*	**5.** *b*
4.3	*Page 78:*	**1.** *c*	**2.** *b*	**3.** *d*	**4.** *a*	**5.** *e*
4.4	*Page 78:*	**1.** *d*	**2.** *c*	**3.** *a*	**4.** *e*	**5.** *b*
4.5	*Page 79:*	**1.** *a*	**2.** *c*	**3.** *e*	**4.** *d*	**5.** *b*
4.6	*Page 80:*					
	Test A:	**1.** *d*	**2.** *e*	**3.** *b*	**4.** *c*	**5.** *a*
	Test B:	**1.** *c*	**2.** *a*	**3.** *d*	**4.** *e*	**5.** *b*
4.7	*Page 81:*	**1.** *d*	**2.** *a*	**3.** *c*	**4.** *b*	**5.** *e*
4.8	*Page 82:*	**1.** *e*	**2.** *b*	**3.** *c*	**4.** *d*	**5.** *a*
4.9	*Page 82:*	**1.** *e*	**2.** *d*	**3.** *b*	**4.** *a*	**5.** *c*
4.10	*Page 83:*	**1.** *b*	**2.** *c*	**3.** *d*	**4.** *e*	**5.** *a*
4.11	*Page 84:*	**1.** *d*	**2.** *c*	**3.** *a*	**4.** *b*	**5.** *e*
4.12	*Page 85:*	**1.** *e*	**2.** *c*	**3.** *b*	**4.** *d*	**5.** *a*
4.13	*Page 85:*	**1.** *b*	**2.** *e*	**3.** *c*	**4.** *d*	**5.** *a*
5.1	*Page 86:*	**1.** *e*	**2.** *b*	**3.** *a*	**4.** *c*	**5.** *d*
5.2	*Page 87:*	**1.** *c*	**2.** *a*	**3.** *d*	**4.** *e*	**5.** *b*
5.3	*Page 88:*	**1.** *e*	**2.** *a*	**3.** *b*	**4.** *c*	**5.** *d*
5.4	*Page 88:*	**1.** *c*	**2.** *e*	**3.** *a*	**4.** *b*	**5.** *d*
5.5	*Page 89:*	**1.** *e*	**2.** *b*	**3.** *d*	**4.** *a*	**5.** *c*
6.1	*Page 91:*					
	Test A:	**1.** *c*	**2.** *e*	**3.** *a*	**4.** *b*	**5.** *d*
	Test B:	**1.** *e*	**2.** *b*	**3.** *d*	**4.** *c*	**5.** *a*
6.2	*Page 92:*	**1.** *d*	**2.** *a*	**3.** *c*	**4.** *e*	**5.** *b*
6.3	*Page 92:*	**1.** *b*	**2.** *d*	**3.** *a*	**4.** *c*	**5.** *e*
6.4	*Page 94:*					
	Test A:	**1.** *c*	**2.** *e*	**3.** *a*	**4.** *b*	**5.** *d*
	Test B:	**1.** *b*	**2.** *d*	**3.** *c*	**4.** *a*	**5.** *e*
	Test C:	**1.** *b*	**2.** *d*	**3.** *e*	**4.** *c*	**5.** *a*
	Test D:	**1.** *d*	**2.** *c*	**3.** *a*	**4.** *b*	**5.** *e*
7	*Page 95:*	**1.** A	**2.** B	**3.** D	**4.** C	**5.** E
		6. B	**7.** A	**8.** D	**9.** C	**10.** A
		11. C	**12.** D	**13.** B	**14.** D	**15.** A
		16. E	**17.** B	**18.** B	**19.** D	**20.** A
		21. C	**22.** D	**23.** B	**24.** B	**25.** D
		26. D	**27.** B	**28.** A	**29.** A	**30.** B
		31. D	**32.** B	**33.** D	**34.** A	**35.** C
		36. A	**37.** E	**38.** B	**39.** A	**40.** C
		41. E	**42.** B			

CHAPTER III

1.1	*Page 100:*	**1.** *c*	**2.** *a*	**3.** *b*	**4.** *e*	**5.** *d*
1.2	*Page 101:*	**1.** *c*	**2.** *a*	**3.** *d*	**4.** *e*	**5.** *b*
1.3	*Page 101:*					
	Test A:	**1.** *e*	**2.** *a*	**3.** *c*	**4.** *d*	**5.** *b*
	Test B:	**1.** *b*	**2.** *e*	**3.** *d*	**4.** *a*	**5.** *c*
1.4	*Page 102:*	**1.** *c*	**2.** *e*	**3.** *a*	**4.** *b*	**5.** *d*
2.1	*Page 103:*					
	Test A:	**1.** *d*	**2.** *b*	**3.** *e*	**4.** *c*	**5.** *a*
	Test B:	**1.** *d*	**2.** *a*	**3.** *b*	**4.** *e*	**5.** *c*
2.2	*Page 104:*	**1.** *d*	**2.** *b*	**3.** *c*	**4.** *a*	**5.** *e*
3.1	*Page 106:*					
	Test A:	**1.** *b*	**2.** *d*	**3.** *e*	**4.** *c*	**5.** *a*
	Test B:	**1.** *c*	**2.** *a*	**3.** *e*	**4.** *d*	**5.** *b*
3.2	*Page 107:*					
	Test A:	**1.** *e*	**2.** *d*	**3.** *a*	**4.** *b*	**5.** *c*
	Test B:	**1.** *e*	**2.** *d*	**3.** *a*	**4.** *b*	**5.** *c*
3.3	*Page 107:*	**1.** *d*	**2.** *a*	**3.** *b*	**4.** *e*	**5.** *c*
4.1	*Page 108:*					
	Test A:	**1.** *a*	**2.** *d*	**3.** *b*	**4.** *c*	**5.** *e*
	Test B:	**1.** *b*	**2.** *a*	**3.** *e*	**4.** *c*	**5.** *d*
4.2	*Page 109:*	**1.** *b*	**2.** *e*	**3.** *d*	**4.** *c*	**5.** *a*
5.1	*Page 110:*	**1.** *c*	**2.** *b*	**3.** *e*	**4.** *a*	**5.** *d*
5.2	*Page 111:*	**1.** *e*	**2.** *d*	**3.** *b*	**4.** *a*	**5.** *c*
6.1	*Page 113:*					
	Test A:	**1.** *e*	**2.** *b*	**3.** *d*	**4.** *c*	**5.** *a*
	Test B:	**1.** *b*	**2.** *e*	**3.** *d*	**4.** *a*	**5.** *c*
	Test C:	**1.** *d*	**2.** *a*	**3.** *e*	**4.** *c*	**5.** *b*
	Test D:	**1.** *d*	**2.** *b*	**3.** *e*	**4.** *c*	**5.** *a*
6.2	*Page 115:*	**1.** *e*	**2.** *d*	**3.** *a*	**4.** *c*	**5.** *b*
6.3	*Page 115:*	**1.** *c*	**2.** *a*	**3.** *e*	**4.** *b*	**5.** *d*
7.1	*Page 116:*	**1.** *d*	**2.** *e*	**3.** *b*	**4.** *a*	**5.** *c*
7.2	*Page 117:*	**1.** *c*	**2.** *d*	**3.** *b*	**4.** *a*	**5.** *e*
8	*Page 118:*	**1.** *b*	**2.** *c*	**3.** *e*	**4.** *d*	**5.** *a*
8.1	*Page 118:*	**1.** *e*	**2.** *d*	**3.** *c*	**4.** *a*	**5.** *b*
8.2	*Page 119:*	**1.** *b*	**2.** *d*	**3.** *e*	**4.** *c*	**5.** *a*
8.3	*Page 120:*	**1.** *b*	**2.** *c*	**3.** *e*	**4.** *a*	**5.** *d*
9	*Page 120:*	**1.** A	**2.** C	**3.** D	**4.** C	**5.** C
		6. A	**7.** A	**8.** C	**9.** C	**10.** D
		11. B	**12.** A	**13.** C	**14.** C	**15.** A
		16. A	**17.** C	**18.** D	**19.** A	**20.** A
		21. A	**22.** D	**23.** A	**24.** B	**25.** A
		26. C	**27.** A	**28.** B	**29.** B	**30.** C
		31. B				

CHAPTER IV

1.1	*Page 124:*	**1.** *b*	**2.** *a*	**3.** *c*	**4.** *d*	**5.** *e*
1.2	*Page 126:*	**1.** *c*	**2.** *b*	**3.** *d*	**4.** *a*	**5.** *e*
1.3	*Page 126:*	**1.** *b*	**2.** *a*	**3.** *c*	**4.** *d*	**5.** *e*
2.1	*Page 127:*	**1.** *e*	**2.** *c*	**3.** *b*	**4.** *a*	**5.** *d*
2.2	*Page 128:*	**1.** *c*	**2.** *a*	**3.** *d*	**4.** *e*	**5.** *b*
2.3	*Page 129:*	**1.** *d*	**2.** *e*	**3.** *a*	**4.** *b*	**5.** *c*
3.1	*Page 130:*	**1.** *e*	**2.** *d*	**3.** *b*	**4.** *c*	**5.** *a*
4.1	*Page 131:*	**1.** *b*	**2.** *d*	**3.** *c*	**4.** *a*	**5.** *e*
4.2	*Page 132:*	**1.** *e*	**2.** *d*	**3.** *b*	**4.** *c*	**5.** *a*
4.3	*Page 133:*	**1.** *d*	**2.** *b*	**3.** *e*	**4.** *c*	**5.** *a*
4.4	*Page 133:*	**1.** *c*	**2.** *e*	**3.** *a*	**4.** *b*	**5.** *d*
4.5	*Page 134:*	**1.** *b*	**2.** *a*	**3.** *e*	**4.** *d*	**5.** *c*
5.1	*Page 135:*	**1.** *a*	**2.** *c*	**3.** *e*	**4.** *d*	**5.** *b*
5.2	*Page 136:*	**1.** *a*	**2.** *c*	**3.** *e*	**4.** *d*	**5.** *b*
5.3	*Page 136:*	**1.** *e*	**2.** *b*	**3.** *a*	**4.** *d*	**5.** *c*
5.4	*Page 137:*					
	Test A:	**1.** *c*	**2.** *a*	**3.** *d*	**4.** *e*	**5.** *b*
	Test B:	**1.** *c*	**2.** *a*	**3.** *d*	**4.** *e*	**5.** *b*
5.5	*Page 138:*	**1.** *b*	**2.** *d*	**3.** *c*	**4.** *a*	**5.** *e*
5.6	*Page 139:*	**1.** *d*	**2.** *c*	**3.** *b*	**4.** *e*	**5.** *a*
5.7	*Page 140:*	**1.** *d*	**2.** *c*	**3.** *a*	**4.** *b*	**5.** *e*
5.8	*Page 142:*	**1.** *d*	**2.** *e*	**3.** *b*	**4.** *a*	**5.** *c*
5.9	*Page 143:*					
	Test A:	**1.** *d*	**2.** *c*	**3.** *a*	**4.** *b*	**5.** *e*
	Test B:	**1.** *e*	**2.** *a*	**3.** *d*	**4.** *c*	**5.** *b*
6.1	*Page 145:*	**1.** *e*	**2.** *d*	**3.** *b*	**4.** *a*	**5.** *c*
6.2	*Page 146:*	**1.** *c*	**2.** *a*	**3.** *d*	**4.** *e*	**5.** *b*
6.3	*Page 148:*	**1.** *c*	**2.** *e*	**3.** *b*	**4.** *a*	**5.** *d*
6.4	*Page 150:*	**1.** *e*	**2.** *a*	**3.** *c*	**4.** *b*	**5.** *d*
6.5	*Page 151:*	**1.** *d*	**2.** *c*	**3.** *b*	**4.** *a*	**5.** *e*
6.6	*Page 152:*	**1.** *c*	**2.** *d*	**3.** *b*	**4.** *a*	**5.** *e*
7.1	*Page 153:*	**1.** *b*	**2.** *a*	**3.** *d*	**4.** *e*	**5.** *c*
7.2	*Page 154:*	**1.** *b*	**2.** *c*	**3.** *d*	**4.** *e*	**5.** *a*
7.3	*Page 155:*	**1.** *e*	**2.** *a*	**3.** *c*	**4.** *d*	**5.** *b*
7.4	*Page 156:*	**1.** *e*	**2.** *a*	**3.** *b*	**4.** *c*	**5.** *d*
8.1	*Page 158:*	**1.** *d*	**2.** *b*	**3.** *e*	**4.** *a*	**5.** *c*
8.2	*Page 159:*					
	Test A:	**1.** *a*	**2.** *e*	**3.** *b*	**4.** *c*	**5.** *d*
	Test B:	**1.** *e*	**2.** *a*	**3.** *d*	**4.** *b*	**5.** *c*
8.3	*Page 160:*	**1.** *d*	**2.** *c*	**3.** *a*	**4.** *b*	**5.** *e*
8.4	*Page 161:*	**1.** *d*	**2.** *b*	**3.** *e*	**4.** *c*	**5.** *a*
8.5	*Page 162:*	**1.** *e*	**2.** *a*	**3.** *b*	**4.** *c*	**5.** *d*
8.6	*Page 162:*	**1.** *e*	**2.** *a*	**3.** *d*	**4.** *c*	**5.** *b*
8.7	*Page 163:*	**1.** *d*	**2.** *e*	**3.** *b*	**4.** *a*	**5.** *c*
8.8	*Page 164:*	**1.** *d*	**2.** *c*	**3.** *a*	**4.** *b*	**5.** *e*
8.9	*Page 165:*	**1.** *d*	**2.** *e*	**3.** *c*	**4.** *a*	**5.** *b*
8.10	*Page 166:*	**1.** *e*	**2.** *b*	**3.** *a*	**4.** *c*	**5.** *d*
8.11	*Page 167:*	**1.** *e*	**2.** *d*	**3.** *b*	**4.** *a*	**5.** *c*
8.12	*Page 169:*	**1.** *d*	**2.** *c*	**3.** *a*	**4.** *b*	**5.** *e*
8.13	*Page 171:*	**1.** *d*	**2.** *c*	**3.** *a*	**4.** *b*	**5.** *e*
8.14	*Page 172:*	**1.** *b*	**2.** *e*	**3.** *a*	**4.** *d*	**5.** *c*
9.1	*Page 173:*	**1.** *d*	**2.** *c*	**3.** *b*	**4.** *e*	**5.** *a*
10.1	*Page 174:*					
	Test A:	**1.** *a*	**2.** *b*	**3.** *d*	**4.** *e*	**5.** *c*
	Test B:	**1.** *d*	**2.** *e*	**3.** *b*	**4.** *a*	**5.** *c*
11.1	*Page 176:*	**1.** *b*	**2.** *a*	**3.** *d*	**4.** *e*	**5.** *c*
11.2	*Page 177:*	**1.** *e*	**2.** *d*	**3.** *b*	**4.** *a*	**5.** *c*
12.1	*Page 179:*	**1.** *d*	**2.** *b*	**3.** *e*	**4.** *a*	**5.** *c*
12.2	*Page 180:*	**1.** *c*	**2.** *b*	**3.** *a*	**4.** *e*	**5.** *d*
13	*Page 181:*	**1.** A	**2.** C	**3.** A	**4.** C	**5.** A
		6. C	**7.** A	**8.** B	**9.** D	**10.** B
		11. D	**12.** A	**13.** E	**14.** D	**15.** A
		16. B	**17.** D	**18.** C	**19.** E	**20.** B
		21. D	**22.** D	**23.** A	**24.** C	**25.** E
		26. B	**27.** D	**28.** A	**29.** C	**30.** E
		31. B	**32.** D	**33.** A	**34.** C	**35.** A
		36. E	**37.** A	**38.** D	**39.** A	**40.** C
		41. E	**42.** B	**43.** D	**44.** A	**45.** C
		46. E	**47.** B	**48.** D	**49.** A	**50.** C
		51. C	**52.** E	**53.** B	**54.** A	**55.** A
		56. D				

CHAPTER V

1.1	*Page 188:*					
	Test A:	**1.** *c*	**2.** *d*	**3.** *b*	**4.** *a*	**5.** *e*
	Test B:	**1.** *e*	**2.** *c*	**3.** *a*	**4.** *d*	**5.** *b*
	Test C:	**1.** *c*	**2.** *a*	**3.** *e*	**4.** *d*	**5.** *b*
	Test D:	**1.** *c*	**2.** *a*	**3.** *d*	**4.** *e*	**5.** *b*
1.2	*Page 190:*					
	Test A:	**1.** *d*	**2.** *e*	**3.** *b*	**4.** *a*	**5.** *c*
	Test B:	**1.** *e*	**2.** *d*	**3.** *a*	**4.** *c*	**5.** *b*
1.3	*Page 193:*					
	Test A:	**1.** *c*	**2.** *a*	**3.** *b*	**4.** *e*	**5.** *d*
	Test B:	**1.** *d*	**2.** *e*	**3.** *b*	**4.** *a*	**5.** *c*
	Test C:	**1.** *e*	**2.** *d*	**3.** *a*	**4.** *b*	**5.** *c*
	Test D:	**1.** *d*	**2.** *a*	**3.** *b*	**4.** *e*	**5.** *c*
1.4	*Page 195:*	**1.** *e*	**2.** *a*	**3.** *b*	**4.** *c*	**5.** *d*
1.5	*Page 196:*	**1.** *e*	**2.** *d*	**3.** *a*	**4.** *b*	**5.** *c*
2.1	*Page 198:*	**1.** *d*	**2.** *b*	**3.** *c*	**4.** *e*	**5.** *a*
2.2	*Page 198:*	**1.** *e*	**2.** *d*	**3.** *b*	**4.** *a*	**5.** *c*
2.3	*Page 199:*	**1.** *c*	**2.** *a*	**3.** *e*	**4.** *d*	**5.** *b*

2.4	Page 200:	1. e	2. b	3. d	4. c	5. a
2.5	Page 202:					
	Test A:	1. b	2. d	3. c	4. e	5. a
	Test B:	1. e	2. a	3. c	4. b	5. d
	Test C:	1. b	2. c	3. d	4. e	5. a
2.6	Page 204:	1. e	2. a	3. b	4. c	5. d
3.1	Page 208:	1. c	2. a	3. e	4. b	5. d
3.2	Page 209:	1. e	2. b	3. a	4. c	5. d
3.3	Page 212:	1. d	2. c	3. a	4. e	5. b
3.4	Page 213:	1. c	2. a	3. d	4. e	5. b
3.5	Page 214:	1. e	2. c	3. d	4. a	5. b
4	Page 215:	1. A	2. C	3. A	4. B	5. D
		6. A	7. D	8. B	9. E	10. E
		11. B	12. C	13. D	14. B	15. E
		16. B	17. E	18. B	19. D	20. B
		21. D	22. B	23. D	24. A	25. D

CHAPTER VI

1.1	Page 230:					
	Test A:	1. c	2. a	3. d	4. e	5. b
	Test B:	1. b	2. e	3. d	4. c	5. a
1.2	Page 234:					
	Test A:	1. c	2. e	3. d	4. b	5. a
	Test B:	1. d	2. e	3. b	4. c	5. a
	Test C:	1. d	2. c	3. e	4. a	5. b
	Test D:	1. c	2. a	3. d	4. e	5. b
	Test E:	1. d	2. e	3. b	4. a	5. c
1.3	Page 237:					
	Test A:	1. d	2. c	3. a	4. b	5. e
	Test B:	1. c	2. d	3. b	4. a	5. e
1.4	Page 240:					
	Test A:	1. b	2. e	3. d	4. c	5. a
	Test B:	1. c	2. a	3. d	4. b	5. e
2.1	Page 241:	1. c	2. e	3. a	4. d	5. b
2.2	Page 242:	1. b	2. d	3. e	4. a	5. c
2.3	Page 243:	1. d	2. e	3. a	4. b	5. c
3	Page 244:	1. B	2. D	3. A	4. B	5. B
		6. E	7. D	8. A	9. B	10. D
		11. C	12. A	13. E	14. B	

CHAPTER VII

1.1	Page 249:	1. e	2. d	3. b	4. a	5. c
1.2	Page 251:					
	Test A:	1. c	2. a	3. b	4. e	5. d
	Test B:	1. e	2. a	3. c	4. d	5. b
1.3	Page 252:	1. d	2. e	3. b	4. a	5. c
2.1	Page 253:	1. d	2. b	3. e	4. c	5. a
2.2	Page 254:	1. a	2. c	3. d	4. e	5. b
3.1	Page 257:					
	Test A:	1. e	2. a	3. d	4. b	5. c
	Test B:	1. d	2. b	3. e	4. c	5. a
3.3	Page 259:					
	Test A:	1. a	2. c	3. e	4. b	5. d
	Test B:	1. b	2. e	3. a	4. d	5. c
3.4	Page 261:					
	Test A:	1. c	2. a	3. d	4. e	5. b
	Test B:	1. e	2. d	3. b	4. a	5. c
3.5	Page 264:	1. e	2. d	3. a	4. b	5. c
3.6	Page 268:					
	Test A:	1. e	2. b	3. d	4. a	5. c
	Test B:	1. e	2. d	3. b	4. c	5. a
4	Page 268:	1. A	2. B	3. D	4. D	5. C
		6. A	7. B	8. C	9. A	10. C
		11. C	12. B	13. C	14. D	15. E
		16. B	17. C	18. D		

CHAPTER VIII

1.1	Page 276:	1. d	2. e	3. b	4. a	5. c
1.2	Page 277:	1. c	2. e	3. a	4. b	5. d
1.3	Page 277:	1. d	2. c	3. a	4. b	5. e
1.4	Page 278:	1. c	2. e	3. a	4. b	5. d
1.5	Page 279:	1. c	2. d	3. a	4. b	5. e
2.1	Page 281:	1. c	2. e	3. a	4. b	5. d
2.2	Page 282:	1. d	2. e	3. a	4. c	5. b
2.3	Page 284:	1. c	2. a	3. e	4. d	5. b
2.4	Page 285:	1. c	2. b	3. d	4. a	5. e
2.5	Page 286:	1. e	2. a	3. d	4. c	5. b
2.6	Page 288:	1. e	2. b	3. c	4. a	5. d
2.7	Page 289:	1. b	2. d	3. c	4. a	5. e
2.8	Page 290:	1. c	2. a	3. e	4. b	5. d
3.1	Page 292:	1. e	2. b	3. d	4. c	5. a
3.2	Page 296:	1. a	2. e	3. d	4. b	5. c
3.3	Page 297:	1. e	2. a	3. c	4. b	5. d
4	Page 299:	1. c	2. a	3. e	4. b	5. d
5.1	Page 301:	1. e	2. d	3. b	4. a	5. c
5.2	Page 303:	1. e	2. d	3. a	4. b	5. c
5.3	Page 305:	1. e	2. d	3. a	4. b	5. c
6	Page 306:	1. b	2. d	3. e	4. a	5. c
7	Page 307:	1. B	2. B	3. A	4. C	5. B
		6. C	7. C	8. B	9. C	10. B
		11. C	12. B	13. A	14. A	15. C
		16. C	17. D	18. B	19. C	20. C
		21. B	22. C			

CHAPTER IX

1.1	Page 313:	1. c	2. e	3. b	4. d	5. a
1.2	Page 314:	1. d	2. a	3. b	4. c	5. e
1.3	Page 315:	1. c	2. a	3. e	4. b	5. d
2.1	Page 316:	1. b	2. a	3. e	4. c	5. d
2.2	Page 317:	1. b	2. d	3. a	4. c	5. e
2.3	Page 318:	1. c	2. d	3. a	4. e	5. b
3.1	Page 320:	1. c	2. a	3. e	4. b	5. d
3.2	Page 321:					
	Test A:	1. e	2. b	3. d	4. c	5. a
	Test B:	1. b	2. a	3. e	4. c	5. d
3.3	Page 322:	1. b	2. a	3. e	4. c	5. d
4.1	Page 325:					
	Test A:	1. e	2. b	3. d	4. c	5. a
	Test B:	1. e	2. d	3. b	4. c	5. a
4.2	Page 326:	1. a	2. e	3. b	4. c	5. d
4.3	Page 327:	1. c	2. d	3. b	4. e	5. a
4.4	Page 328:	1. a	2. e	3. b	4. d	5. c
4.5	Page 329:	1. b	2. d	3. c	4. a	5. e
4.6	Page 330:	1. b	2. c	3. d	4. a	5. e
5	Page 331:	1. c	2. a	3. e	4. d	5. b
6	Page 331:	1. B	2. B	3. C	4. B	5. A
		6. B	7. D	8. B	9. A	10. A
		11. B	12. E	13. C	14. D	15. E
		16. B	17. E	18. C		

CHAPTER X

1.1	Page 336:	1. b	2. a	3. c	4. d	5. e
1.2	Page 337:	1. e	2. d	3. c	4. b	5. a
1.3	Page 338:	1. e	2. b	3. d	4. c	5. a
2.1	Page 339:	1. e	2. a	3. b	4. c	5. d
	Page 340:	1. b	2. d	3. e	4. a	5. c
2.2	Page 342:	1. e	2. b	3. d	4. c	5. a
3.1	Page 343:	1. a	2. b	3. e	4. c	5. d
3.2	Page 344:	1. c	2. a	3. e	4. d	5. b
3.3	Page 345:	1. e	2. c	3. a	4. b	5. d
4	Page 345:	1. A	2. B	3. C	4. A	5. D
		6. D	7. E	8. A	9. C	

CHAPTER XI

1.1	*Page 348:*	**1.** *c*	**2.** *e*	**3.** *a*	**4.** *b*	**5.** *d*
1.2	*Page 348:*	**1.** *d*	**2.** *b*	**3.** *e*	**4.** *c*	**5.** *a*
1.3	*Page 349:*	**1.** *b*	**2.** *a*	**3.** *d*	**4.** *c*	**5.** *e*
1.4	*Page 350:*					
	Test A:	**1.** *b*	**2.** *d*	**3.** *c*	**4.** *a*	**5.** *e*
	Test B:	**1.** *d*	**2.** *a*	**3.** *b*	**4.** *e*	**5.** *c*
2.1	*Page 354:*	**1.** *d*	**2.** *c*	**3.** *b*	**4.** *e*	**5.** *a*
2.2	*Page 355:*					
	Test A:	**1.** *E*	**2.** *F*	**3.** *B*	**4.** *A*	**5.** *D*
	Test B:	**1.** *B*	**2.** *C*	**3.** *D*	**4.** *F*	**5.** *E*
3.1	*Page 357:*	**1.** *d*	**2.** *c*	**3.** *e*	**4.** *b*	**5.** *a*
3.2	*Page 361:*					
	Test A:	**1.** *e*	**2.** *a*	**3.** *c*	**4.** *b*	**5.** *d*
	Test B:	**1.** *b*	**2.** *a*	**3.** *d*	**4.** *c*	**5.** *e*
3.3	*Page 362:*	**1.** *c*	**2.** *a*	**3.** *b*	**4.** *d*	**5.** *e*
3.4	*Page 363:*	**1.** *d*	**2.** *c*	**3.** *a*	**4.** *e*	**5.** *b*
3.5	*Page 364:*	**1.** *d*	**2.** *c*	**3.** *a*	**4.** *e*	**5.** *b*
3.6	*Page 365:*	**1.** *a*	**2.** *c*	**3.** *b*	**4.** *e*	**5.** *d*
3.7	*Page 366:*	**1.** *e*	**2.** *b*	**3.** *a*	**4.** *c*	**5.** *d*
3.8	*Page 367:*	**1.** *d*	**2.** *c*	**3.** *a*	**4.** *b*	**5.** *e*
3.9	*Page 369:*	**1.** *b*	**2.** *c*	**3.** *a*	**4.** *d*	**5.** *e*
4	*Page 369:*	**1.** D	**2.** A	**3.** D	**4.** B	**5.** B
		6. B	**7.** C	**8.** F	**9.** A	**10.** C
		11. D	**12.** E	**13.** A	**14.** A	**15.** A
		16. D	**17.** A	**18.** C		

CHAPTER XII

Tests, *page 373*

Test A:	**1.** A	**2.** C	**3.** E	**4.** B	**5.** D
Test B:	**1.** E	**2.** C	**3.** D	**4.** A	**5.** B
Test C:	**1.** E	**2.** D	**3.** C	**4.** A	**5.** B
Test D:	**1.** C	**2.** E	**3.** A	**4.** D	**5.** B
Test E:	**1.** A	**2.** C	**3.** E	**4.** B	**5.** D
Test F:	**1.** D	**2.** A	**3.** E	**4.** C	**5.** B

Practice Problems

Page 375:	**1.** D	**2.** C	**3.** C	**4.** B	**5.** C
	6. D	**7.** D	**8.** C	**9.** E	**10.** A

CHAPTER XIII

Tests, *page 377:*

Test A:	**1.** A	**2.** C	**3.** B	**4.** D
Test B:	**1.** B	**2.** A	**3.** C	**4.** D
Test C:	**1.** B	**2.** C	**3.** A	**4.** D
Test D:	**1.** C	**2.** D	**3.** B	**4.** A
Test E:	**1.** C	**2.** D	**3.** B	**4.** A
Test F:	**1.** A	**2.** B	**3.** D	**4.** C
Test G:	**1.** D	**2.** B	**3.** A	**4.** C
Test H:	**1.** D	**2.** B	**3.** A	**4.** C
Test I:	**1.** A	**2.** B	**3.** D	**4.** C
Test J:	**1.** C	**2.** D	**3.** A	**4.** B
Test K:	**1.** B	**2.** C	**3.** D	**4.** A
Test L:	**1.** C	**2.** B	**3.** A	**4.** D
Test M:	**1.** A	**2.** C	**3.** B	**4.** D
Test N:	**1.** C	**2.** B	**3.** D	**4.** A
Test O:	**1.** C	**2.** D	**3.** A	**4.** B
Test P:	**1.** A	**2.** C	**3.** B	**4.** D

Practice Problems

Page 384:	**1.** A	**2.** B	**3.** C	**4.** D	**5.** C
	6. A	**7.** C	**8.** D	**9.** B	**10.** A

APTITUDE TESTS

I. *Page 385:*	**1.** D	**2.** B	**3.** A	**4.** B	**5.** C
	6. B	**7.** A	**8.** C	**9.** D	**10.** A
	11. B	**12.** D	**13.** E	**14.** C	**15.** C
	16. B	**17.** B	**18.** E	**19.** B	**20.** A
	21. B	**22.** C	**23.** A	**24.** B	**25.** A
	26. A	**27.** B	**28.** A	**29.** A	**30.** C
II. *Page 387:*	**1.** D	**2.** C	**3.** A	**4.** C	**5.** A
	6. D	**7.** D	**8.** C	**9.** D	**10.** B
	11. B	**12.** D	**13.** A	**14.** D	**15.** B
	16. B	**17.** A	**18.** D	**19.** B	**20.** D
	21. C	**22.** D	**23.** C	**24.** E	**25.** E
	26. B	**27.** B	**28.** E	**29.** D	**30.** A
III. *Page 390:*	**1.** C	**2.** B	**3.** C	**4.** B	**5.** A
	6. C	**7.** A	**8.** B	**9.** D	**10.** D
	11. A	**12.** B	**13.** D	**14.** B	**15.** A
	16. E	**17.** A	**18.** A	**19.** E	**20.** E
	21. E	**22.** A	**23.** D	**24.** E	**25.** C
	26. D	**27.** C	**28.** C	**29.** B	**30.** A
IV. *Page 392:*	**1.** B	**2.** B	**3.** C	**4.** D	**5.** B
	6. D	**7.** C	**8.** B	**9.** C	**10.** E
	11. D	**12.** C	**13.** B	**14.** B	**15.** D
	16. C	**17.** B	**18.** E	**19.** B	**20.** A
	21. E	**22.** B	**23.** C	**24.** B	**25.** A
	26. C	**27.** A	**28.** D	**29.** D	**30.** C
V. *Page 395:*	**1.** E	**2.** C	**3.** B	**4.** C	**5.** E
	6. D	**7.** B	**8.** A	**9.** B	**10.** D
	11. B	**12.** B	**13.** D	**14.** C	**15.** C
	16. E	**17.** D	**18.** B	**19.** D	**20.** A
	21. C	**22.** B	**23.** C	**24.** C	**25.** D
	26. C	**27.** E	**28.** E	**29.** E	**30.** D
VI. *Page 398:*	**1.** D	**2.** A	**3.** A	**4.** D	**5.** B
	6. A	**7.** A	**8.** C	**9.** D	**10.** B
	11. E	**12.** E	**13.** A	**14.** B	**15.** E
	16. E	**17.** D	**18.** B	**19.** D	**20.** B
	21. E	**22.** C	**23.** B	**24.** C	**25.** D
	26. B	**27.** C	**28.** E	**29.** B	**30.** D
VII. *Page 401:*	**1.** D	**2.** A	**3.** C	**4.** D	**5.** A
	6. A	**7.** B	**8.** E	**9.** B	**10.** E
	11. B	**12.** E	**13.** E	**14.** C	**15.** E
	16. C	**17.** A	**18.** A	**19.** A	**20.** C
	21. C	**22.** E	**23.** D	**24.** C	**25.** C
	26. B	**27.** D	**28.** B	**29.** C	**30.** A

INDEX